# New Chromosomal Syndromes

**CHROMOSOMES IN BIOLOGY AND MEDICINE**

Edited by

JORGE J. YUNIS, M.D.

*New Chromosomal Syndromes,* 1977

*Molecular Structure of Human Chromosomes,* 1977

# New Chromosomal Syndromes

*Edited by*

**JORGE J. YUNIS, M.D.**

Medical Genetics Division
Department of Laboratory Medicine and Pathology
University of Minnesota Medical School
Minneapolis, Minnesota

ACADEMIC PRESS New York San Francisco London 1977

A Subsidiary of Harcourt Brace Jovanovich, Publishers

ACADEMIC PRESS, INC.
111 Fifth Avenue, New York, New York 10003

*United Kingdom Edition published by*
ACADEMIC PRESS, INC. (LONDON) LTD.
24/28 Oval Road, London NW1

**Library of Congress Cataloging in Publication Data**

Main entry under title:

New chromosomal syndromes.

(Chromosomes in biology and medicine)
Includes bibliographies and index.
1. Human chromosomal abnormalities. 2. Cytogenetics.
I. Yunis, Jorge J. II. Series. [DNLM: 1. Chromosome aberrations. 2. Chromosome abnormalities. QS675 N532]
RB155.N49 616'.042 76-43378
ISBN 0-12-775165-3

PRINTED IN THE UNITED STATES OF AMERICA

# Contents

# List of Contributors

Numbers in parentheses indicate the pages on which author's contributions begin.


UTA FRANCKE (245), Department of Pediatrics, School of Medicine, University of California, San Diego, La Jolla, California

ROBERT J. GORLIN (59), Department of Oral Pathology, University of Minnesota School of Dentistry, Minneapolis, Minnesota

FREDERICK HECHT (301), Crippled Children's Division and Department of Pediatrics, Child Development and Rehabilitation Center, University of Oregon Health Sciences Center, Portland, Oregon

KURT HIRSCHHORN (339), Department of Pediatrics, Division of Medical Genetics, Mount Sinai School of Medicine of the City University of New York, New York, New York

LILLIAN Y. F. HSU (339), Department of Pediatrics, Division of Medical Genetics, Mount Sinai School of Medicine of the City University of New York, New York, New York

RAYMOND C. LEWANDOWSKI, JR. (219, 369), Medical Genetics Division, Department of Laboratory Medicine and Pathology, University of Minnesota Medical School, Minneapolis, Minnesota

R. ELLEN MAGENIS (301), Child Development and Rehabilitation Center, Crippled Children's Division, University of Oregon Health Sciences Center, Portland, Oregon

E. NIEBUHR (273), University Institute of Medical Genetics, Copenhagen, Denmark

R. A. PFEIFFER (197), Medizinische Hochschule Lübeck, Institut für Humangenetik, Lübeck, Federal Republic of Germany.

MARIE-ODILE RETHORÉ (119), Institut de Progénèse, Paris, France

OTTO SANCHEZ (1), Medical Genetics Division, Department of Laboratory Medicine and Pathology, University of Minnesota Medical School, Minneapolis, Minnesota

DAVID W. SMITH (55), Department of Pediatrics, University of Washington, School of Medicine, Seattle, Washington

WALTHER VOGEL (185), Institut für Humangenetik und Anthropologie der Universität Freiburg, Federal Republic of Germany.

HERMAN E. WYANDT (301), Division of Medical Genetics, University of Oregon Health Sciences Center, Portland, Oregon

JORGE J. YUNIS (1, 219, 369), Medical Genetics Division, Department of Laboratory Medicine and Pathology, University of Minnesota Medical School, Minneapolis, Minnesota

# Preface

In the last few years, technical advances have taken place that have revolutionized the field of medical cytogenetics. Amniocentesis is now a common and safe procedure, allowing early study of the fetus. Banding techniques have provided easy identification of each chromosome and the means for recognition of chromosome abnormalities ranging from minute defects to complex rearrangements. Thousands of case reports have appeared in the recent literature describing partial trisomies and partial deletions for every chromosome. Of clinical importance is the fact that close to one percent of newborns are now found to have a chromosome defect, and, in many instances, the affected individuals have a parent with a balanced chromosome translocation, allowing possible prevention of recurrence of the disease with counseling and prenatal diagnosis.

This volume brings together, for the first time, all the clinically relevant information related to birth defects and chromosomes that has accumulated during the last six years. Emphasis is placed on the detailed description of approximately thirty new chromosomal disorders for which there are sufficient data to substantiate their clinical delineation. In addition, the reader will find extensive review of the classic chromosome disorders, the new chromosome techniques, and phenotype–chromosome relationships.

For geneticists, human biologists, cytologists, and members of the medical profession involved in the study of birth defects—pediatricians, obstetricians, and pathologists alike—this book should serve as a valuable, authoritative guide.

The secretarial and editorial assistance of Marylyn S. Hoglund is gratefully acknowledged.

Jorge J. Yunis

# 1

# New Chromosome Techniques and Their Medical Applications

OTTO SANCHEZ and JORGE J. YUNIS



## I. INTRODUCTION

In 1956, a major turning point occurred in the field of human cytogenetics when Tjio and Levan reported the correct number of chromosomes in man. In the following years, the introduction of relatively simple techniques for lymphocyte cultures and chromosome preparations (Hungerford *et al.*, 1959; Moorhead *et al.*, 1960) allowed human cytogenetics to become a discipline on its own. Several well-defined autosomal syndromes (Lejeune, 1959; Patau *et al.,* 1960; Edwards *et al.,* 1960; Lejeune *et al.,* 1963) were then described, sex chromosome anomalies (Turner and Klinefelter syndromes) were confirmed by karyotype analysis, and the association between the $Ph^1$ chromosome and chronic myelogenous leukemia was firmly established (Nowell and Hungerford, 1960). Autoradiographic techniques served later to identify the X chromosome as well as the autosomes more frequently involved in autosomal anomalies (Yunis, 1965). By the mid-1960's, however, the original impetus seemed to have been lost and the field reached a plateau of little new information. Most of the difficulties resided in the fact that cytogeneticists were unable to identify reliably most of the human chromosomes and, although many cases bearing abnormal chromosomes or chromosome segments were known, a correlation between clinical signs and chromosomal abnormalities was fraught with difficulties.

In 1968, Caspersson *et al.* reported a distinctive fluorescence pattern in plant chromosomes after quinacrine mustard staining. The importance of these findings was not immediately realized until Caspersson *et al.* (1970a) applied the same technique to human chromosomes and demonstrated a specific banding pattern for each chromosome pair. The pioneer work of Caspersson's group proved to be the beginning of an explosive development in that area. Since that time, numerous other techniques have been described that allow the visualization of differentially stained regions in the chromosomes. The application of these techniques to human chromosomes has dramatically changed the field of human cytogenetics. New chromosomal defects involving almost every chromosome of the human complement have been described (Lewandowski and Yunis, 1975), classical syndromes have been subdivided according to the chromosomal segment involved (Chapter 9), phenotypic mapping of chromosomes now seems possible (Chapter 12), chromosomal polymorphisms have been shown to be of universal occurrence (see Section IV, B), and previously unsuspected chromosomal rearrangements have been demonstrated (Gray *et al.,* 1972; Pasquali, 1973; Koulischer and Lambrotto, 1974).

It is now becoming apparent that there are chromosomal segments that in the trisomic state might not be as harmful as other segments, and that small chromosomal abnormalities may be compatible with normal or slightly affected phenotypes (Friedrich and Nielsen, 1974; Lewandowski *et al.,* 1976; see also Chapter 8).

The impact produced by the banding techniques has not been limited to clinical cytogenetics; these techniques have made important contributions to other scientific areas as well. Human gene mapping, for example, is a rapidly developing field which owes many of its recent advances to the aid given by the banding methods. The study of the molecular organization of chromosomes is another area of research that has been influenced by these advances (Yunis, 1977a; Yunis and Chandler, 1977a). In the following sections, a general description of the banding techniques and chromosomal bands will be given, as well as introductory concepts on chromosome identification and current cytogenetics terminology. Some of the clinical applications are also reviewed.

## II. THE PARIS CONFERENCE

After the first banding techniques were described, a need became evident for international agreement on a nomenclature system to identify the chromosomal bands. In 1971, an international conference was held in Paris for this specific purpose. The recommendations made by this conference (Paris Conference, 1971) have been widely accepted and are amply used in the current cytogenetic literature. Some of the recommendations of the Paris Conference are outlined below, and examples are given to aid in understanding the terminology.

The nomenclature symbols already suggested by the Chicago Conference (1966) were retained with minor modifications and some additional symbols were added (Table I). It was agreed to use the term "Q-bands" to refer to those bands obtained by the quinacrine mustard banding technique. "C-bands" was the name applied to bands shown by the techniques that stain constitutive heterochromatin. "G-bands" was given to those bands obtained with Giemsa stain, with the exception of the "R-bands" which designate those bands resulting from a particular technique and which are, in general, the opposite of the Q- and G-bands.

Banded chromosomes were considered to consist of a continuous series of light and dark bands, so that, by definition, there are no interbands. Thus, a "band" was defined as a part of a chromosome which is clearly distinguishable from its adjacent segments by appearing darker or lighter with the Q-, G-, C- or R-staining methods.

Fluorescent positive bands elicited by the Q-banding technique generally correspond to those darkly stained with the G-banding methods, while the pale fluorescent or negative Q-bands are, in general, the same chromosomal areas that remain lightly stained with the G-banding techniques. The opposite pattern is followed by the R-bands, where those areas that appear positive (fluorescent or dark) with Q- and G-bands are shown as very lightly stained areas and vice-versa. Exceptions to this rule are represented by the heterochromatic regions of chromosomes 1, 9, and 16, which will be discussed below. It should be noted that, in reality, there are no "negative" regions since the whole chromosome is stained; rather, the intensity varies from almost no fluorescence (or very light

**TABLE I**
**Nomenclature Symbols**

| | |
|---|---|
| *Chicago Conference* | |
| A–G | the chromosome groups |
| 1–22 | the autosome numbers |
| X,Y | the sex chromosomes |
| diagonal (/) | separates cell lines in describing mosaicism |
| ? | questionable identification of chromosome or chromosome structure |
| * | chromosome explained in text or footnote |
| ace | acentric |
| cen | centromere |
| dic | dicentric |
| end | endoreduplication |
| h | secondary constriction or negatively staining region |
| i | isochromosome |
| inv | inversion |
| mar | marker chromosome |
| mat | maternal origin |
| p | short arm of chromosome |
| pat | paternal origin |
| q | long arm of chromosome |
| r | ring chromosome |
| s | satellite |
| t | translocation |
| repeated symbols | duplication of chromosome structure |

*continued*

Giemsa staining) to very brilliant fluorescence (or dark Giemsa staining), with other areas showing intermediate degrees of fluorescence or staining (pale, medium, or intense).

The specific banding patterns obtained in each chromosome pair by the use of the Q-banding technique were used to construct an idiogram of the human banded karyotype for the purposes of chromosome identification (Fig. 1). Certain "constant and distinct morphological features that are important aids in identifying a chromosome" were selected as "landmarks." The centromeres, telomeres, and some well-defined bands were included in this definition. The chromosomal arms were divided into "regions"; a region was defined as "any area of a chromosome lying between two adjacent landmarks." A chromosome arm lacking any prominent landmark was considered to consist of only one region. In each chromosome, the centromere served as a reference point for the numbering of regions and bands. The chromosome arms were first divided into regions according to the landmarks selected by the Paris Conference. In each arm, the region closest to the centromere was identified as Number 1, and other regions,

TABLE I (*continued*)

*Paris Conference*

A. Recommended changes in Chicago Conference nomenclature

\+ 1. The + and – signs should be placed *before* the appropriate symbol where they mean
– additional or missing whole chromosomes. They should be placed *after* a symbol where an increase or decrease in length is meant. Increases or decreases in the length of secondary constrictions, or negatively staining regions, should be distinguished from increases or decreases in length owing to other structural alterations by placing the symbol h between the symbol for the arm and the + or – sign (e.g., 16qh+).

2. All symbols for rearrangements are to be placed before the designation of the chromosome(s) involved in the rearrangement, and the rearranged chromosome(s) always should be placed in parentheses, e.g., r(18), i(Xq), dic(Y).

B. Recommended additional nomenclature symbols

| | |
|---|---|
| del | deletion |
| der | derivative chromosome |
| dup | duplication |
| ins | insertion |
| inv ins | inverted insertion |
| rcp | reciprocal translocation[a] |
| rec | recombinant chromosome |
| rob | Robertsonian translocation[a] ("centric fusion") |
| tan | tandem translocation[a] |
| ter | terminal or end ("pter" for end of short arm; "qter" for end of long arm) |
| : | break (no reunion, as in terminal deletion) |
| :: | break and join |
| → | from – to |

[a]Optional, where greater precision is desired than that provided by the use of t as recommended by the Chicago Conference.

if present, numbered consecutively toward the telomeric ends; the bands present within each region were, in turn, numbered following the same rules applied to the regions. A description of the specific banding patterns of each chromosome with emphasis on prominent landmarks and other characteristics useful for chromosome identification can be found in Rowley (1975).

Using the Paris Conference suggestions, any particular band or segment of a chromosome can be easily identified, and only four items are required: chromosome number, an arm symbol (p = short arm; q = long arm), the region and band numbers. No spacing or punctuation is used and the order of the items may not be changed. For example, 6p23 indicates band number 3 of region 2 in the short arm of chromosome 6 (see Fig. 2A). Provisions were also made for the subdivi-

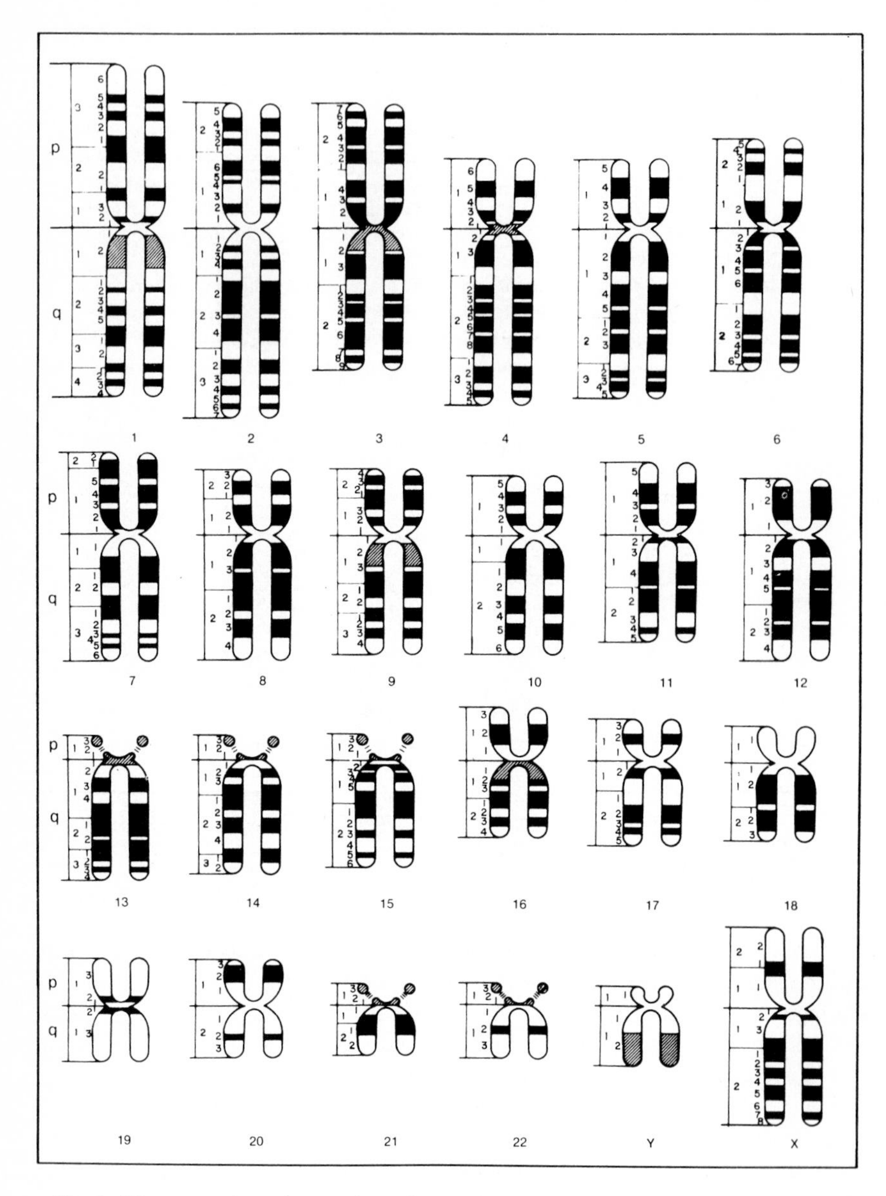


**Fig. 1.** Diagrammatic representation of metaphase chromosome bands as observed with the Q- and G-staining methods; centromere representative of Q-banding method only. Reproduced from the report of the Paris Conference (1971).

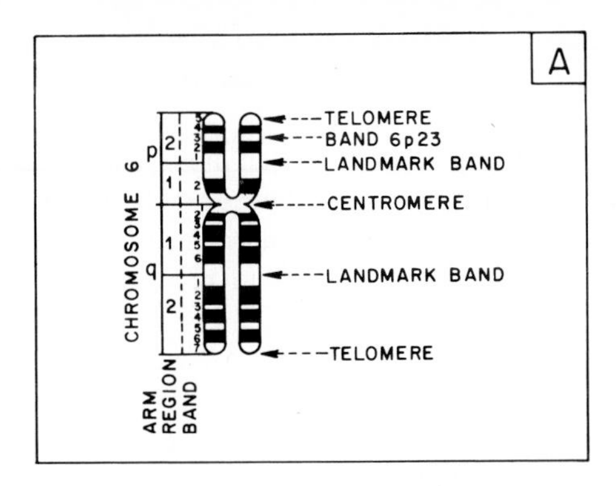


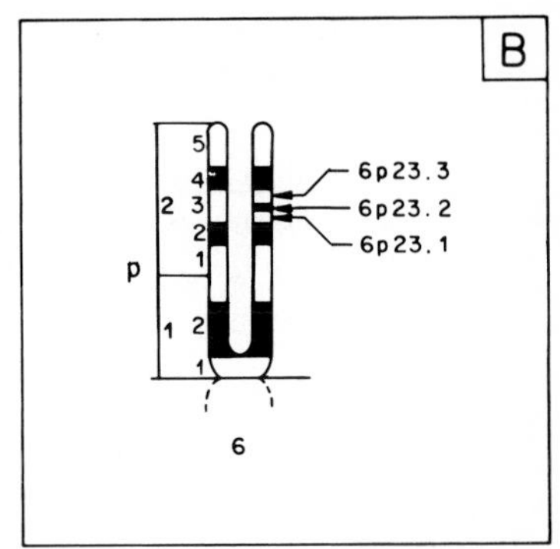


**Fig. 2.** (A) Diagrammatic representation of chromosome 6 in metaphase. (B) Diagrammatic representation of chromosome 6 in an earlier stage of mitosis illustrating the presence of subbands in 6p23.

sion of bands into sub-bands. In such a case, the original band designation would be followed by a decimal point and the sub-bands numbered sequentially, beginning with the subband nearest the centromere. In the previous example, if band 6p23 were subdivided into three subbands, they would be named 6p23.1, 6p23.2, and 6p23.3, with 6p23.1 being proximal and 6p23.3 distal to the centromere (Fig. 2B).

For the designation of structural abnormalities, two systems were recommended: the "short system," which is a simple modification of the Chicago nomenclature (Chicago Conference, 1966), and the "detailed" system, which identifies precisely the type of arrangement and defines each abnormal chromosome by its band composition. The two systems are complementary, although the short system is easier to understand, is most frequently used, and is the most appropriate for clinically oriented workers. In the short system, the number of chromosomes is indicated first, followed by the sex chromosome constitution; next comes the symbol representing the chromosomal abnormality, followed in parentheses by the number(s) of the involved chromosome(s) and, in a second parentheses, the band numbers where the breaks occurred. From this information, the band composition of the abnormal chromosome can be inferred. For example, 46,XY,t(2;6)(q34;p12) indicates a karyotype of 46 chromosomes, male, with a reciprocal translocation involving chromosomes 2 and 6 where the break points have occurred in the long arm of chromosome 2, region 3, band 4, and in the short arm of chromosome 6, region 1, band 2. Notice that in the second parentheses the chromosome numbers have not been repeated. The first band (q34) corresponds to the first chromosome indicated in the preceding parentheses (No. 2) and is separated by a semicolon from the other band (p12) which corresponds to the second chromosome involved (No. 6). In this example, the person carrying this translocation can be inferred to have two abnormal chromosomes, a No. 2 having a segment of chromosome 6 attached to its long arm, and a No. 6 carrying a segment of chromosome 2 attached to its short arm (Fig. 3).

The Paris Conference also provided a way to designate an abnormal karyotype in the offspring of a balanced translocation carrier. For example, if a son received from his father the abnormal chromosome 2 and a normal chromosome 6 (all other paternal and maternal chromosomes assumed to be normal), his

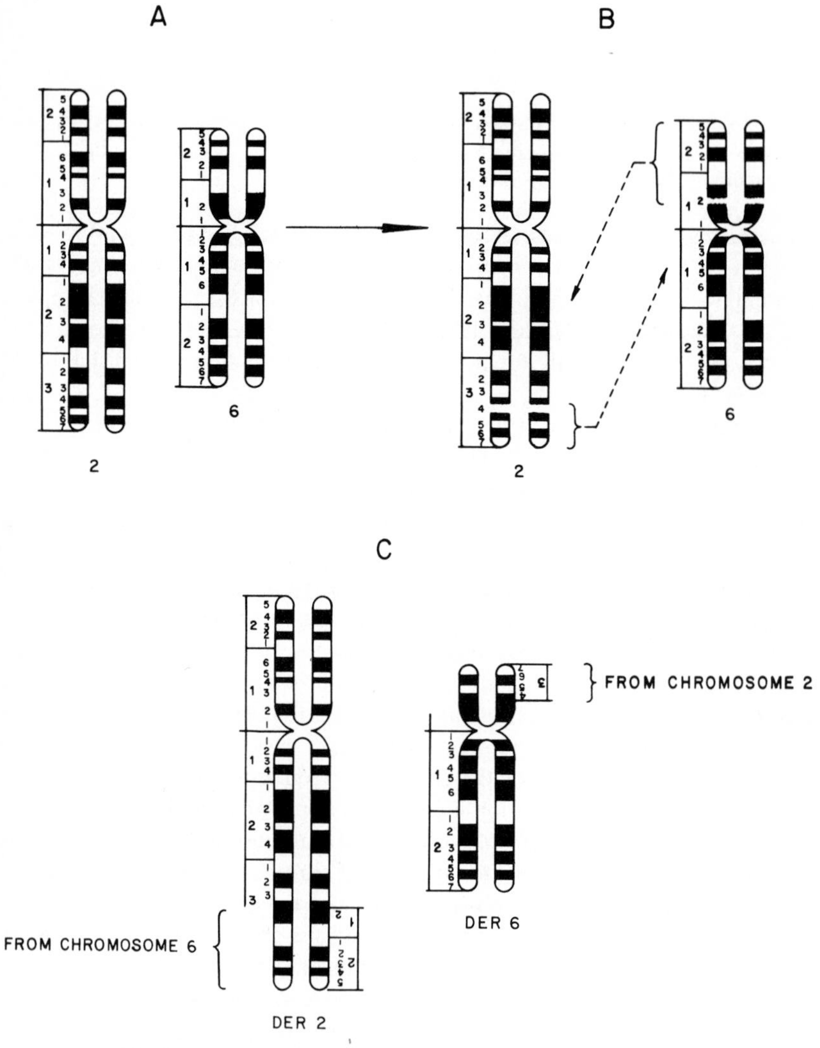


**Fig. 3.** Schematic representation of a balanced translocation 2/6.

karyotype would be written as follows: 46,XY,der(2)t(2;6)(q34;p12)pat, indicating that the boy has 46 chromosomes and an XY sex chromosome complement, and that one of the No. 2 chromosomes is abnormal and derived from a balanced translocation in the father that involved chromosomes 2 and 6 with the indicated break points. As shown in Fig. 4, this nomenclature indicates that the boy has a partial trisomy for a segment of the short arm of chromosome 6 (from band 6p12 to 6pter, attached to the long arm of chromosome 2) and a monosomy for part of the long arm of chromosome 2 (from band 2q34 to 2qter, absent in the derivative chromosome 2, and attached, in the father, to the short arm of derivative chromosome 6).

The recommendations made by the Paris Conference have provided human cytogenetics with a uniform, relatively simple, and generally adequate nomenclature and terminology to describe chromosomal abnormalities. However, the knowledge derived from the application of the banding techniques has uncovered certain problems which cannot be resolved by the use of the Paris Conference nomenclature alone. For example, the use of a simple descriptive term for new partial trisomies has become a matter of disagreement since no clear provisions were made. Prior to the Paris Conference, as well as in the current literature, it became common practice to describe a partial monosomy by writing the chromosome number and arm followed by a minus (−) sign (e.g., 5p− indicating a deletion of the short arm of chromosome 5). To many authors (e.g., Jacobs, 1972; Rethoré *et al.*, 1973a; Niebuhr, 1971; Lewandowski and Yunis, 1975), it seemed logical and appropriate that if a partial monosomy is indicated in the above manner, the partial trisomy syndromes could be indicated by simply adding a plus (+) sign after the corresponding chromosome number and arm (e.g., 9p+ indicating a partial trisomy for the short arm of chromosome 9). Nevertheless, the Paris Conference had suggested that a plus or minus sign following the appropriate symbol indicates *only* an increase or decrease in the length of the chromosomal segment, and that a plus or minus sign preceding the chromosome number should only be used to describe an extra or missing whole chromosome. There is obviously a need for an international conference to resolve this inconsistency of terminology for the use of plus (+) and minus (−) signs. For the present time and in following the Chicago and Paris Conferences, throughout this volume a plus or minus sign preceding the chromosome number indicates a complete trisomy or monosomy, respectively. A partial deletion is described by the use of either a minus sign following the given chromosome arm or by descriptive terminology; for example, del 4p or 4p− designate a partial monosomy of chromosome 4. When describing a partial trisomy, however, no symbol is used and only descriptive terminology is written, such as dup 9p to infer duplication in the short arm of chromosome 9.

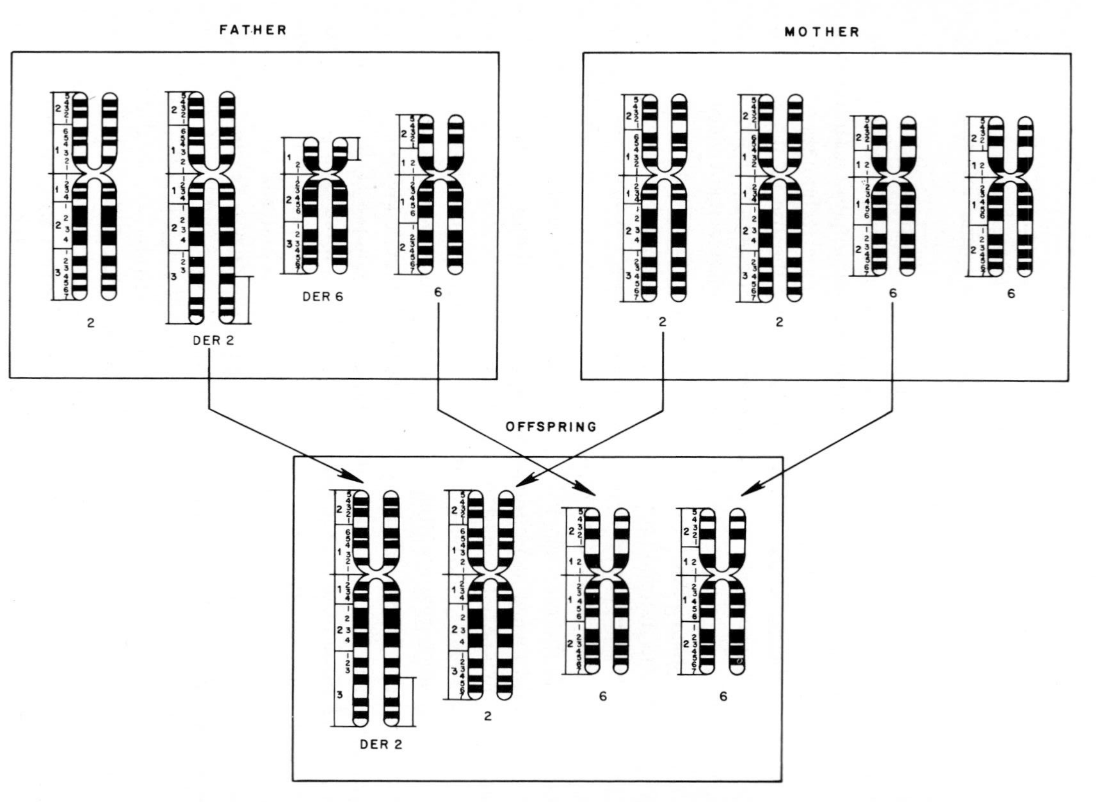


**Fig. 4.** Diagrammatic representation of an unbalanced offspring resulting from a 2/6 balanced translocation carrier.

## III. BANDING TECHNIQUES

Following the Paris Conference, numerous reports describing new banding techniques appeared in the literature. Most of them, however, represent only technical variations of the already established banding methods, and consequently bear the letter identification of C-, Q-, G- or R-band techniques. In the following review, the main classes of chromosomal bands will be described with emphasis on their advantages and/or disadvantages from the clinical point of view. Other banding techniques will be briefly mentioned with the understanding that their clinical application is still limited.

### A. C (Constitutive Heterochromatin)-Bands

Constitutive heterochromatin is a relatively well-defined fraction of chromatin usually located around the centromeres or in blocks in the chromosomal arms. It became a subject of active research due to its peculiar localization, enrichment in repetitive DNA, and function in the eukaryotic genome (Yasmineh and Yunis, 1969; Yunis and Yasmineh, 1971). C-banding techniques selectively stain chromosomal areas known to contain constitutive heterochromatin, primarily of the pericentromeric type. The original C-banding techniques (Yunis *et al.,* 1971; Arrighi and Hsu, 1971) were by-products of basic research in the area of repetitive DNA and were not intended to be used for routine chromosome identification. Nevertheless, these methods have proved useful in the study of chromosomal polymorphism, linkage analysis, population and evolution studies, and in the identification of specific chromosomes in paternity studies (see Section IV, B).

When human chromosomes are subjected to a C-banding technique, there is intensive staining of the pericentromeric areas of all chromosomes and, in particular, there is large and characteristically intensive staining of the secondary constrictions of chromosomes 1, 9, and 16 and of the distal segment of the long arm of the Y chromosome. Less intensive staining can be observed sometimes in the satellites of the acrocentric chromosomes (13-15, 21-22) and in the pericentromeric regions of the remaining chromosomes (Fig. 5). The rest of the chromosomes are lightly stained, although occasionally faint bands may be visible. The size of the areas positively stained by the C-banding techniques may vary from person to person (Fig. 6). The C-bands are usually transmitted unchanged from parents to offspring and constitute one example of polymorphism in man. The variation in size of C-bands may correspond to different amounts of satellite DNA present (Jones, 1977). Although the differences in amount of this kind of DNA may involve thousands of nucleotide pairs, no phenotypic effects are usually observed in the carriers of such polymorphic traits (Pearson *et al.,* 1973).

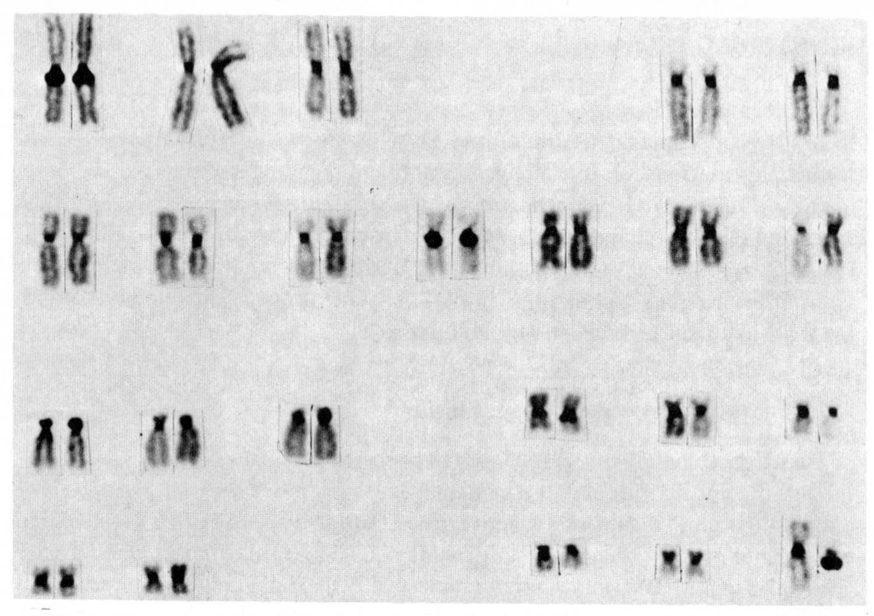

**Fig. 5.** Karyotype of human male showing C-banding pattern. Note the large amount of heterochromatin at the centromeric areas of chromosomes 1, 9, 16 and the distal portion of the Y. (From Chen and Ruddle, 1971.)

## B. Q (Quinacrine)-Bands

The Q-banding technique of Caspersson *et al.* (1968, 1970a) was the first one described. The bands elicited by this technique (Fig. 7) are specific for each species and each chromosome pair (Caspersson and Zech, 1973). Q-bands are similar to G-bands except for the staining of pericentromeric regions and secondary constrictions of chromosomes 1 and 16, which are generally not fluorescent with the Q-bands and are positively stained with the G-bands. A characteristic intensive fluorescence is observed in the distal half of the long arm of the Y chromosome. This intensive fluorescence is also observed in interphase nuclei (Pearson *et al.,* 1970; Mittwoch, 1974) and has provided an easy technique for the visualization of the Y chromosome in interphase nuclei similar to the staining of the Barr body for identification of the inactivated X chromosome. The pericentromeric region of chromosome 3 and the satellites of the acrocentric chromosomes may also show intensive fluorescence which, in interphase nuclei, can be at times confused with the brightly fluorescent spot of the Y chromosome (Uchida and Lin, 1974). As in the case of the C-bands, these intensively stained areas are highly polymorphic and are transmitted as simple Mendelian traits (Bobrow *et al.,* 1971; Genest, 1973; Pearson, 1977) (Fig. 8).

Although the Q-bands were accepted as the reference bands at the Paris Conference and the Q-banding technique remains one of the simplest of the banding methods, it has been largely replaced by the G- and R-banding techniques. These two latter methods do not require any equipment other than that usually present in a regular cytogenetics laboratory; G- and R-bands are permanent and allow a better and finer differentiation of chromosomes, while the fluorescence of small minor bands or sub-bands is difficult to appreciate.

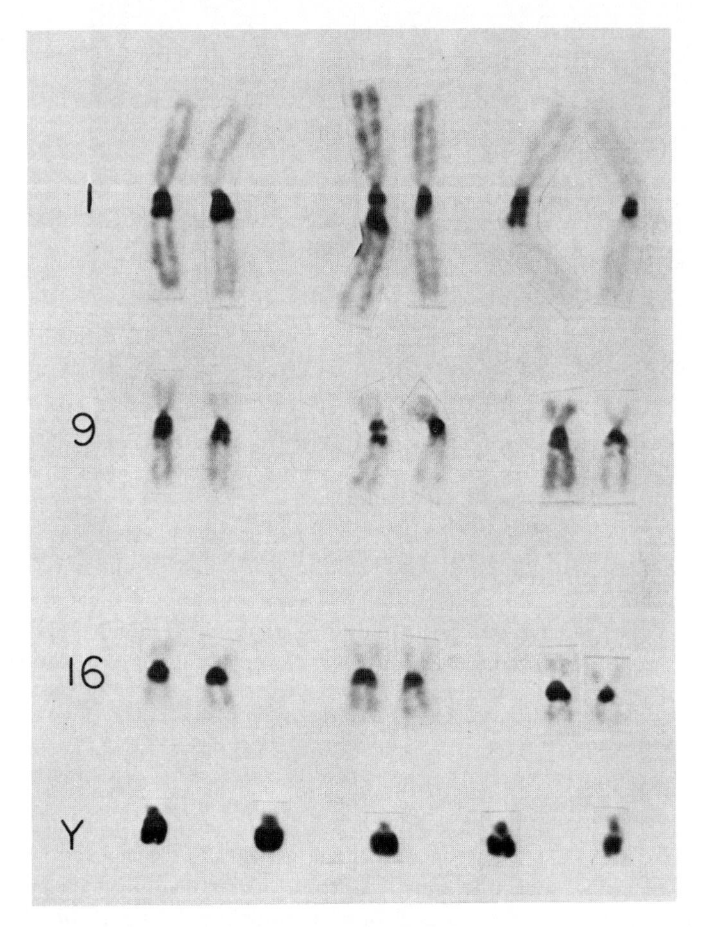


**Fig. 6.** Normal and polymorphic patterns of C-bands in autosomes 1, 9, 16, and in the Y chromosome from human leukocytes. For autosomes 1, 9, and 16, the first pair shows the normal pattern of C-bands, while the other two are polymorphic. The Y chromosomes show a small amount of positive staining on the pericentromeric regions and varying amounts on the distal portion of the long arms. (From Yunis and Yasmineh, 1972.)

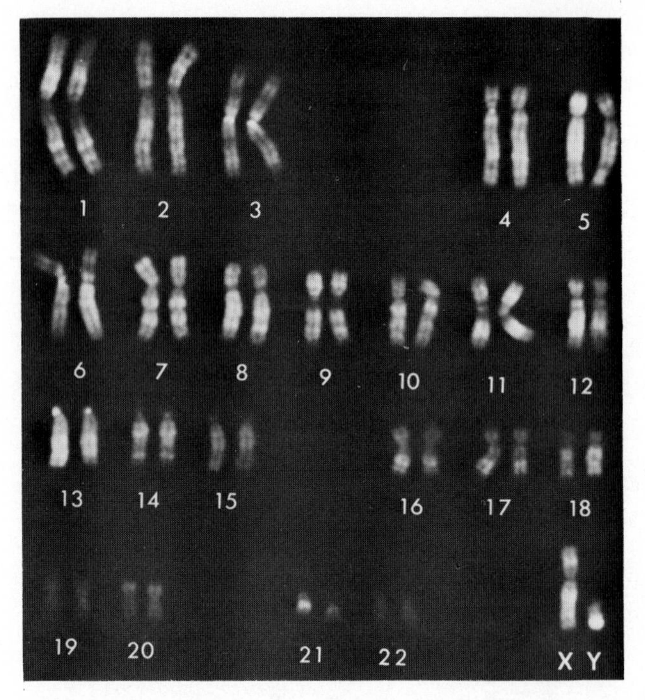


**Fig. 7.** Karyotype of human male showing Q-banding pattern. (From Uchida and Lin, 1974.)

Nevertheless, the Q-banding technique continues to be a useful method for bone marrow studies due to the difficulty in obtaining good G- and R-banding in cells from this tissue. In addition, Q-banding is a method of choice for the ready identification of the Y chromosome and chromosomal polymorphic variants.

### C. G (Giemsa)-Bands

In 1971, Sumner *et al.* (1971b) and Drets and Shaw published techniques based on the denaturation/renaturation techniques of Yunis *et al.* (1971) and Arrighi and Hsu (1971). The banding patterns visualized were very similar to bands found with quinacrine mustard. These techniques produce permanent bands and do not require special optical equipment. The main difference between Q- and G-bands is in the secondary constrictions of chromosomes 1, 9, and 16, which are negatively stained with quinacrine mustard but slightly (9) or intensively (1 and 16) stained under the G-banding techniques (Fig. 9). In the following years, several techniques were described that make use of diverse pretreatment steps to induce the appearance of G-bands. In all cases, the bands obtained are consistently similar to those obtained in the first two original publications. To quote all the available references would take much patience and space, and thus only some of them are mentioned. The pretreatment agents involve enzymes (Seabright, 1971), urea (Shiraishi and Yosida, 1972), alkaline pH (Patil *et al.,* 1971), protein denaturants (Lee *et al.,* 1973), and buffers, bases,

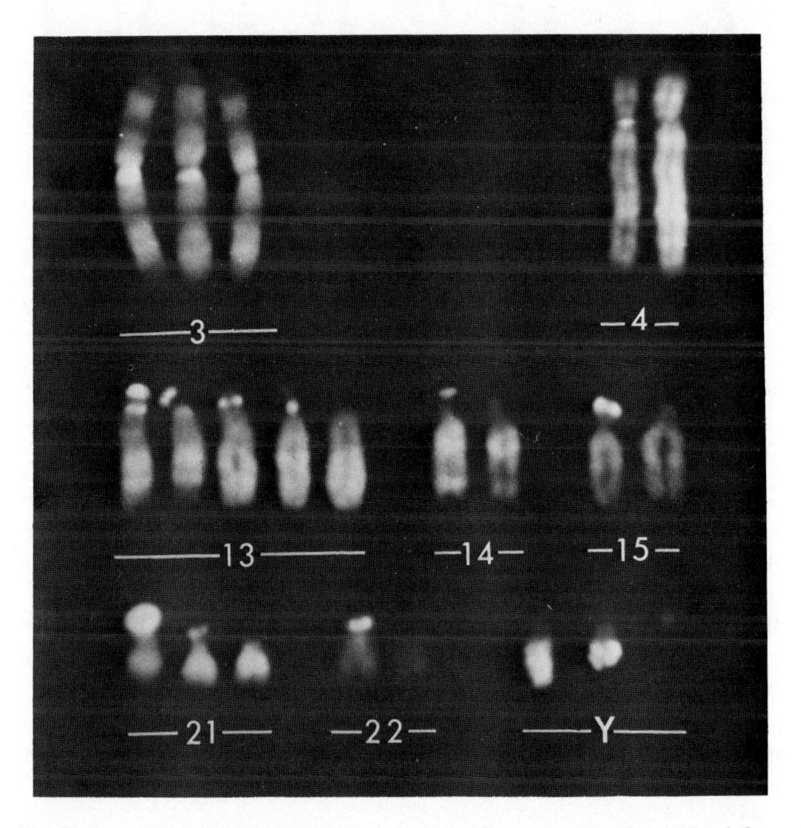


**Fig. 8.** Some normal morphological variants in fluorescent bands obtained from different subjects. The size and intensity of the bright bands are constant for a subject and for those among his relatives who have inherited the same chromosome. Chromosomes 3 and 4 have the variant bands adjacent to the centromere. The acrocentric chromosomes are characterized by satellites of different intensity and size and/or bands near the centromere. Note in particular the unusually large satellite on chromosome No. 21. Three Y chromosomes are shown ranging in size from a large D-like acrocentric Y to a tiny deleted Y that is barely visible with fluorescence. Since these variants are inherited, they are useful as genetic markers. (From Uchida and Lin, 1974.)

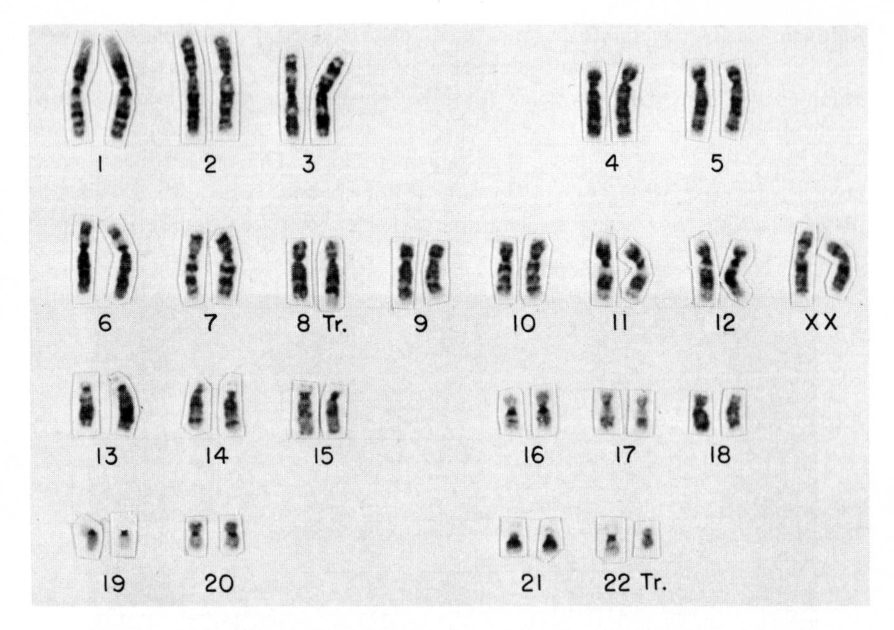


**Fig. 9.** Karyotype of female showing G-banding pattern and a balanced 8/22 translocation. (From Sanchez and Yunis, 1974.)

and salts (Kato and Moriwaki, 1972; Eiberg, 1973). In almost all the described techniques, the pretreatment steps have been claimed to be essential for the development of the chromosomal bands. Perhaps the most popular G-banding technique used is the trypsin method of Seabright (1971). However, with this method, over- or undertrypsinization is frequently observed and the enzymatic treatment appears to destroy some or most of the finer bands. Other G-banding methods are frequently hard to reproduce or are time consuming. In 1973, a simple G-banding technique was described which does not require a pretreatment step, does not alter the fine chromosome structure, and is highly reproducible and consistent (Sanchez *et al.,* 1973; Yunis and Sanchez, 1973, 1975) (Fig. 9).

## D. R (Reverse)- and T (Telomeric)-Bands

Based on minor modifications of the original technique of Yunis *et al.* (1971) for C-bands, Dutrillaux and Lejeune (1971) described a technique that elicited banding patterns on chromosomes with the particular characteristic that are, in general, the reverse (R) of those found by Q- and G-banding techniques. Thus, G- and Q-negatively stained bands appear positively stained and vice versa

(Fig. 10). The exception resides in the secondary constriction of chromosome 9, which is negatively stained with both the Q- and R-banding techniques and slightly stained by the G-banding methods.

Dutrillaux (1973) described a modification of this technique whereby only certain telomeric (T) regions are stained (Fig. 11). The bands observed by these two techniques are known as R- and T-bands, respectively. These two techniques

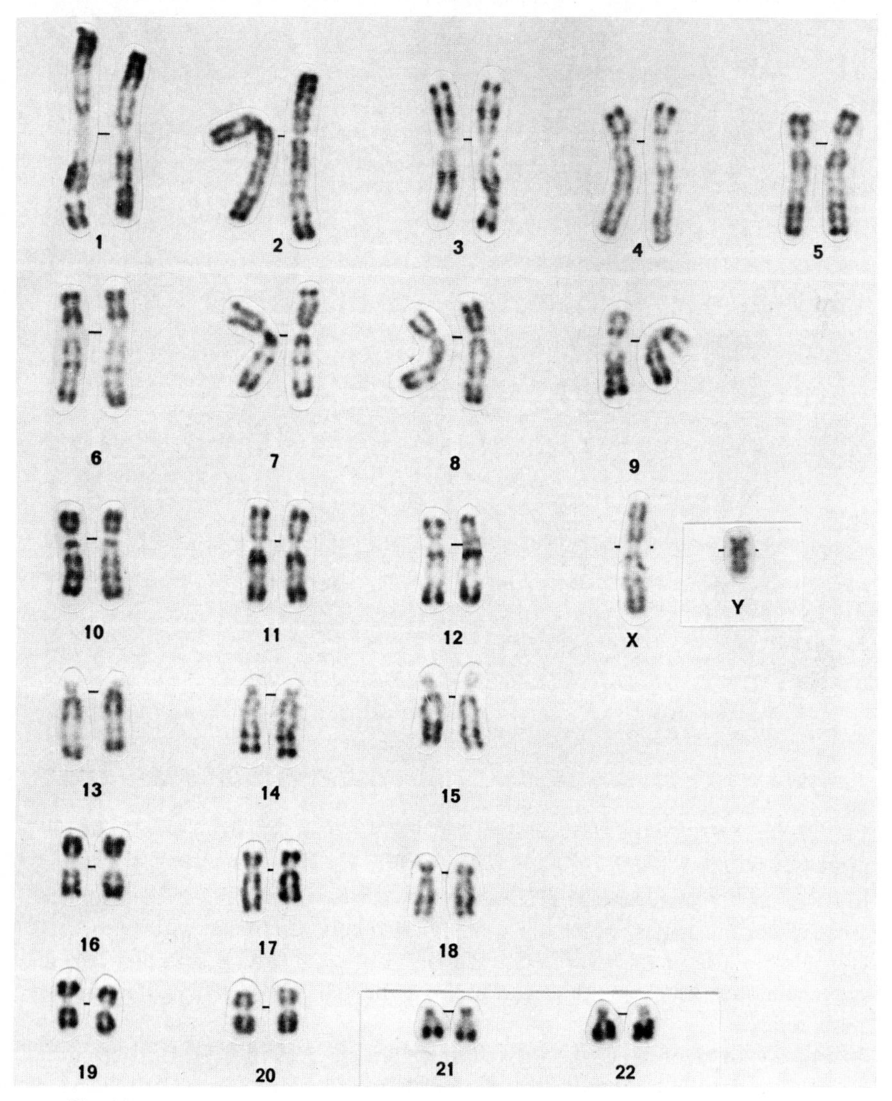


**Fig. 10.** Karyotype of male showing R-banding pattern. (From Dutrillaux, 1977.)

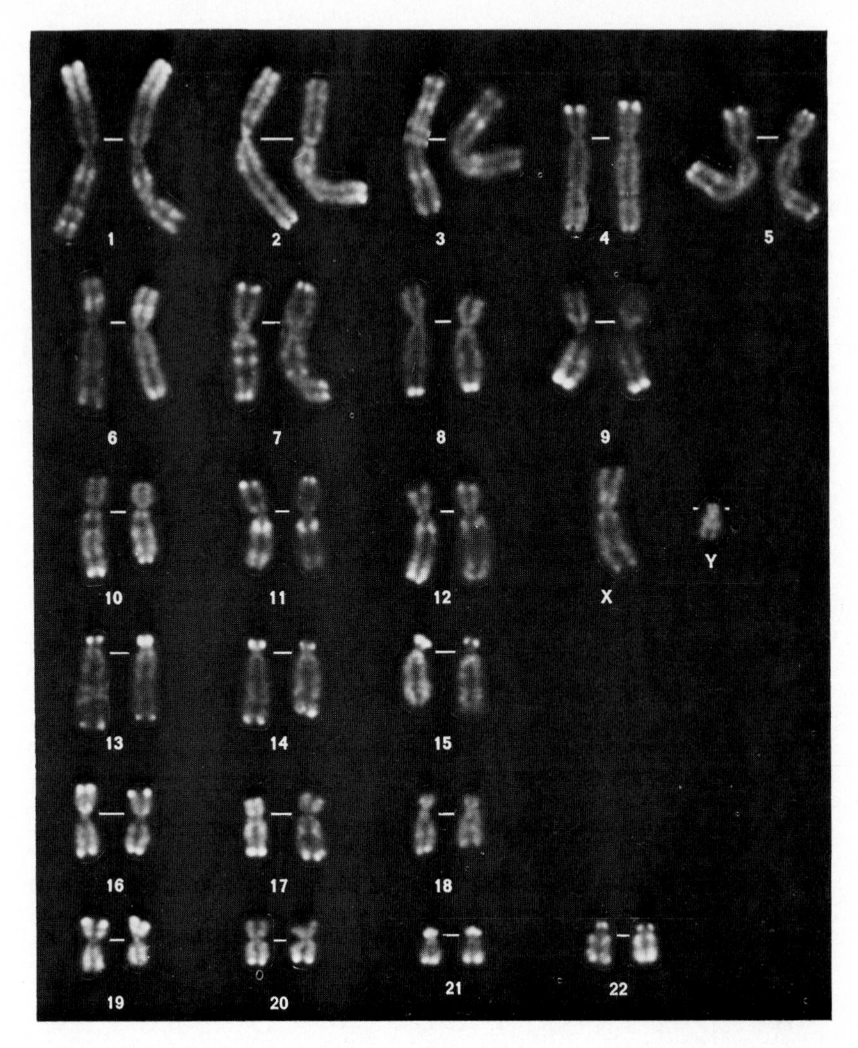


**Fig. 11.** Karyotype of male showing T-banding pattern. (From Dutrillaux, 1977.)

have the particular advantage of showing very clearly the staining of the telomeric areas of chromosomes, which are generally difficult to observe with the Q- and G-bands because of their very light staining. Unfortunately, the R- and T-banding techniques usually require the use of either phase contrast or fluorescence microscopy.

For a period of time, the R-banding technique described by Dutrillaux and Lejeune was the only one available for this type of band. More recently, R-bands have also been obtained with the use of high concentrations of sodium phosphate (Sehested, 1974) and by differential staining with acridine orange (Castoldi *et al.*, 1972; de la Chapelle *et al.*, 1973a; Couturier *et al.*, 1973; Wyandt *et al.*, 1974). As in the case for the G-banding techniques, most of the R-banding methods require pretreatment steps for the demonstration of the bands, with the notable exception of the method described by Couturier *et al.* (1973) where no pretreatment is described. In the case of the T-bands, the technique described by Dutrillaux (1973) remains the only method available.

## E. F (Feulgen)-Bands

The Feulgen stain is an old and well-known specific DNA stain. The application of this dye to mammalian chromosomes has generally resulted in homogeneously stained chromosomes without any internal differentiation. Small modifications of the standard staining technique in humans (Yunis and Sanchez, 1974) evidenced chromosomal bands resembling those elicited by the Q- and G-banding techniques. Similar results were obtained by Rodman (1974). The bands resulting from the use of the Feulgen technique have been named F-bands, since the correspondence between Q-, G-, and F-bands is not exact (Rodman, 1974). While most of the negatively stained landmarks of the Q- and G-banding techniques are also negative with the F-bands, several telomeric segments that are Q- and G-negative stain positively with the F-banding technique. In addition, the distal half of the Y chromosome, which is brightly fluorescent with Q-bands and intensively stained with G-bands, appears negative under the F-banding technique (as previously found with regular Feulgen) (Patau, 1965).

Although the F-bands are interesting when the basic mechanisms for the appearance of the chromosomal bands are considered, F-bands have very little practical application since the Feulgen technique is somewhat difficult to carry out and the resulting bands lack the quality and staining intensity of other banding techniques.

## F. N (Nucleolar Organizer)-Bands

Matsui and Sasaki (1973) developed a technique that apparently induces selective staining of very limited regions of mammalian chromosomes which coincide with those areas recognized as bearing the loci for ribosomal cistrons (nucleolar organizers). These authors referred to these bands as N (nucleolar organizer) bands. Further work led to the identification and/or confirmation of the chromosomal localization of the nucleolar organizer by the use of the N-banding technique in a wide variety of species (marsupials, birds, amphibians, fishes, insects, plants, etc.) (Funaki *et al.*, 1975). Various other techniques have been devised which also claim to stain the same regions specifically, (Eiberg, 1974a; Howell *et al.*, 1975; Goodpasture and Bloom, 1975; Goodpasture *et al.*,

1976). The correspondence between the areas localized by these methods and the loci for ribosomal cistrons has been confirmed by *in situ* hybridization (Evans *et al.,* 1974). The clinical application of these new techniques will be, if any, very limited since no clinical syndrome in man has yet been shown to be related to a deficit in ribosomal RNA. To this end, however, it would be interesting, to study cases of human dwarfism with this technique.

### G. BUdR Techniques

Bromodeoxyuridine (BUdR) is a thymidine analog that can be incorporated into the DNA of replicating cells growing in a medium where such a compound is present. Zakharov *et al.* (1971) reported that, after treatment of cultured human lymphocytes with BUdR, many chromosomes acquire a typical segmentary appearance. This new technique for chromosome identification was set aside, probably because of the burst of reports on relatively simpler and more effective banding techniques which appeared in the same and subsequent years. However, continuous research with BUdR has demonstrated its usefulness for purposes of chromosome identification because: (1) The dye Hoechst 33258 shows a very bright fluorescence when it is bound to poly(dA-dT) nucleotides, but such fluorescence is significantly reduced when bound to poly(dA-dBUdR) (Latt, 1973); (2) Giemsa stains very lightly DNA molecules in which both strands have been BUdR substituted and stains more intensely DNA molecules where no substitution has occurred or where only one strand bears the BUdR instead of thymidine (Ikushima and Wolff, 1974; Perry and Wolff, 1974); (3) BUdR added to the culture medium in the last hours before harvesting is incorporated into late replicating DNA and induces a condensation delay in late replicating DNA segments, allowing the identification of late replicating chromosomes without the need of autoradiography (Zackharov and Egolina, 1972; Baranovskaya *et al.,* 1972; Latt and Gerald, 1973); (4) growing of cells for two cycles in medium containing BUdR and staining with Hoechst 33258 or Giemsa allows the differential staining of sister chromatids (Latt, 1973); and (5) the combined utilization of acridine orange stain and BUdR treatment under variable conditions allows one to obtain R- or Q-bands (Dutrillaux, 1975, 1976). A schematic representation of BUdR substitution on DNA replicating strands and the consequences on staining and condensation delay is depicted in Fig. 12.

The BUdR technique has already found successful application in clinical cytogenetics; identification and delineation of inactive X chromosomes in cases of balanced and unbalanced X-autosomal translocations have been recently achieved (Laurent *et al.,* 1975; Gilgenkrantz *et al.,* 1975). The demonstration of sister chromatid exchanges in cases of Bloom syndrome, Fanconi's anemia, and ataxia-telangiectasia has been made possible by this technique (Chaganti *et al.,*

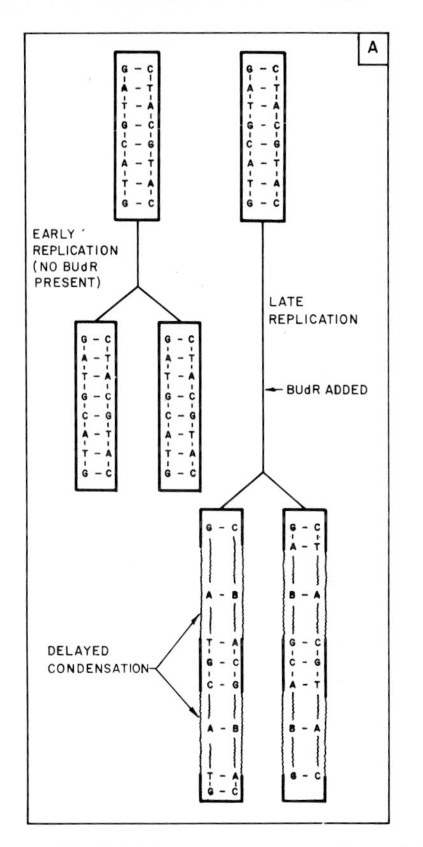


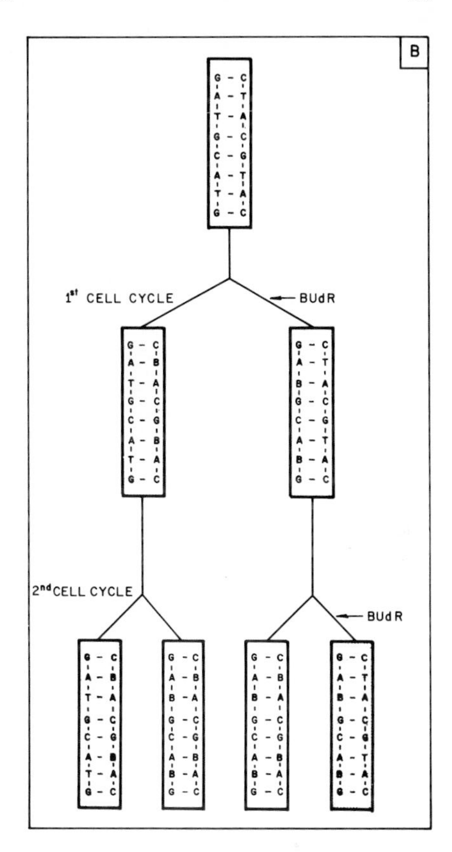


**Fig. 12.** Schematic representation of BUdR substitution on DNA replicating strands. (A) When BUdR is added to a cell at the end of S phase, those areas which are late replicating incorporate BUdR instead of thymidine, leading to delayed condensation and lighter staining of BUdR-rich chromosome segments.

(B) Cells given BUdR for one cell cycle incorporate it into one strand of the DNA leading to elongation of chromosomes. When exposed to BUdR during two cell cycles, chromosomes show one chromatid with single BUdR substitution and the other with double substitution. Although not shown, the chromatid doubly substituted with BUdR stains more lightly than its singly substituted sister chromatid, forming the basis for this technique's use in investigating sister chromatid exchange.

1974; Hayashi and Schmid, 1975a, b; Sperling *et al.*, 1975; Galloway and Evans, 1975) (Fig. 13).

## H. Other Banding Techniques

The description of new staining methods for chromosomes has not been limited to the already mentioned techniques. Others have been reported which

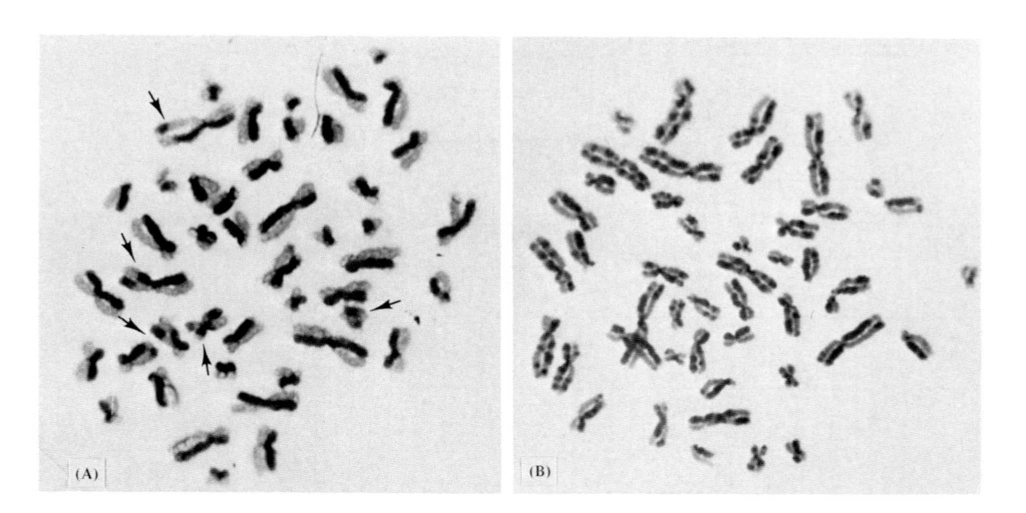


**Fig. 13.** (A) Chromosomes of a normal lymphocyte at the second metaphase after growth in bromodeoxyuridine, fluorodeoxyuridine, and uridine stained first with 33258 Hoechst and a day later with Giemsa. Arrows are at points of exchange between sister chromatids. (From Chaganti *et al.*, 1974). (B) Chromosomes of a Bloom's syndrome lymphocyte stained as in (A), showing many more exchanges between sister chromatids than normal. (From Chaganti *et al.*, 1974.)

claim to be even more restrictive in the chromosomal segments stained. A brief description and comments on some of these techniques follows.

In 1972, Bobrow *et al.* reported a modified Giemsa staining technique (Giemsa 11) by which the secondary constriction of chromosome 9 could be preferentially stained. Similar results were presented by Howell and Denton (1974), who used ammoniacal-silver stain and claimed it to be specifically staining for satellite III DNA-containing regions. Bühler *et al.* (1975) reported a modification of the Giemsa 11 technique by which satelite III DNA is also claimed to be specifically stained and is shown to be present not only in the secondary constriction of chromosome 9 but also in the distal segment of the long arm of the Y chromosome and in the pericentromeric area of chromosome 20. The Giemsa 11 technique has been applied not only to chromosomes but also to interphase nuclei (Pawlowitzki and Pearson, 1972) where the secondary constriction of chromosome 9 can be demonstrated as a condensed body. The Giemsa 11 technique has been used to show that polymorphic variants of this area, including pericentric inversions and large segments, are relatively common. For example, a total of 7.4% of chromosomes 9 in a total population of 282 unrelated individuals were found to possess inversions or variants of the secondary constriction (Madan and Bobrow, 1974). Human spermatozoa studied by the same technique showed 1-2% with two Giemsa 11-positive bodies, suggested that nondisjunction for this chromosome is relatively high (Pearson, 1972; Pawlonitzki and Pearson, 1972).

Geraedts and Pearson (1973) claimed to have specifically stained the secondary constriction of human chromosome 1 by subjecting chromosomal preparations to denaturation in 0.9% NaCl and renaturation in CsCl followed by staining with Leishman stain. The technique was also applied to interphase lymphocytes where two large dots were usually seen and interpreted as representing the secondary constrictions of chromosome 1. When human spermatozoa were stained by this technique, most presented just one stained dot, but 1-2% showed two stained dots, suggesting the presence of two No. 1 chromosomes in a single spermatozoan as a probable result of a nondisjunction event. This value is similar to that reported for spermatozoa containing double Y bodies and double Giemsa 11-stainable bodies (Pearson and Bobrow, 1970; Sumner *et al.*, 1971a; Pearson, 1972; Pawlonitzki and Pearson, 1972). These data seem surprising since they could mean that the proportion of aneuploid spermatozoa would be extremely high (40% as calculated by Pawlonitzki and Pearson, 1972) if other chromosomes are similarly affected. Further work is clearly needed in this area before definitive conclusions can be reached.

Eiberg (1974b) reported a banding technique by which two identical dots were revealed at the location of the centromere in metaphase chromosomes while prometaphase chromosomes showed a narrow long band. The author suggested that this pattern be termed "Cd" (for centromere dots) and also suggested that they may represent organelles associated with the spindle fibers. No clinical use has yet been reported for this technique.

Dallapicola and Ricci (1975) described a method for specific staining of part of the Y chromosome. The segment stained corresponds to the brightly quinacrine fluorescent area of this chromosome and can be demonstrated by this technique even in interphase nuclei. Although such a technique would be useful for the purpose of screening sex chromosomal anomalies without the disadvantages of the quinacrine staining method, it appears to be somewhat difficult to perform and its reliability remains to be proved.

Of the banding techniques available, those eliciting Q-, G-, or R-bands have been more widely used for the study of chromosomal abnormalities, particularly the G- and R-banding techniques. The G-banding methods have received more favor in America, while the R-bands have been more extensively applied in Europe. The selection of a particular banding technique for routine use depends on the user's personal experience and preferences, since both methods are equally suitable for routine studies. However, in special cases, the use of more than one technique separately or in combination on the same slide may be necessary in order to identify precisely the chromosomal abnormality present (Rowley, 1974a; Dutrillaux, 1977).

## I. Banding of Prophase Chromosomes

Of particular importance is the mitotic stage used for chromosome studies. Midmetaphases have been widely used before and after the introduction of the banding techniques. However, at mid-metaphase, only approximately 320 bands per haploid set can be detected (Paris Conference, 1971). When chromosomes are observed in earlier mitotic periods, the major and many minor bands are split into numerous subbands (Schnedl, 1973; Prieur *et al.,* 1973; Bahr *et al.,* 1973; Bahr and Larsen, 1974; Skovby, 1975; Bigger and Savage, 1975). This increased detail makes it easier to establish break points in the study of structural rearrangements and allows a more definitive localization of the chromosomal segments involved. The increased resolution offered by the banding of these earlier mitotic stages of chromosome condensation has become of great help in the study of patients with chromosomal abnormalities. For example, moderately elongated chromosomes have been used to show a maternal complex balanced translocation involving chromosomes 11, 12, and 13, and a total of 7 break points, resulting in partial trisomy for the short arm of chromosome 11 in the child (Sanchez and Yunis, 1974) (Fig. 14).

Until recently, the use of elongated, early mitotic stages had not yet become routine in the study of chromosomal abnormalities. There were two main reasons for this situation. In routine cultures, the yield of mitotic figures is relatively low (about 3 to 5% in most cases). Most chromosomes are found in midmetaphase;

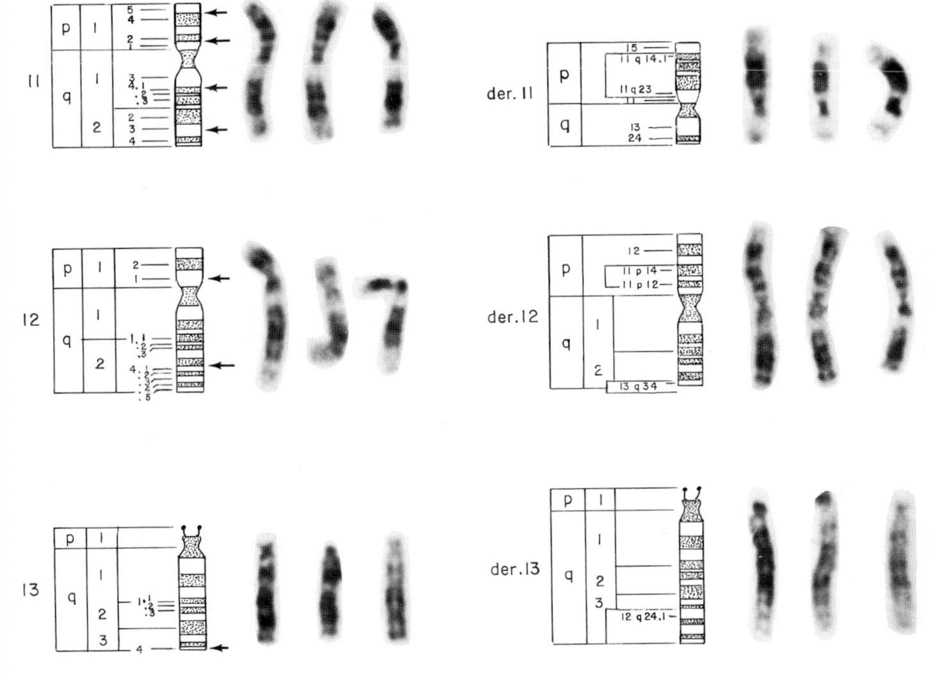


**Fig. 14.** Diagram and representative examples of normal (left) and derivative (right) chromosomes 11, 12, and 13 from early metaphases of a female carrier with a complex balanced translocation. Arrows indicate the break points. (From Sanchez *et al.*, 1974.)

only a few mitoses are obtained in early metaphase and seldom any in earlier stages. Second, the more commonly used G-banding techniques require pretreatment steps that are too harsh for the delicate structures of fine subbands, rendering the observation of minor details virtually impossible in chromosomes stained this way. Work in this laboratory has combined the use of synchronized short-term lymphocyte cultures, which allow an increased number of mitotic cells (12–15%) and a high yield of prometaphases and late prophases (Figs. 15 and 16), with a simple, direct G-banding technique (Yunis, 1976; Yunis and Chandler, 1977b). Such a combination has already allowed the observation of 1256 bands per haploid set of chromosomes and is becoming an important tool for the study of minute chromosome abnormalities and for a more precise definition of break points, inversions, and translocations (Yunis and Chandler, 1977b).

Very recently, a technique which makes use of cell synchronization and actinomycin D has been developed to obtain a larger number of highly elongated and more finely banded prometaphases and late prophases. When administered

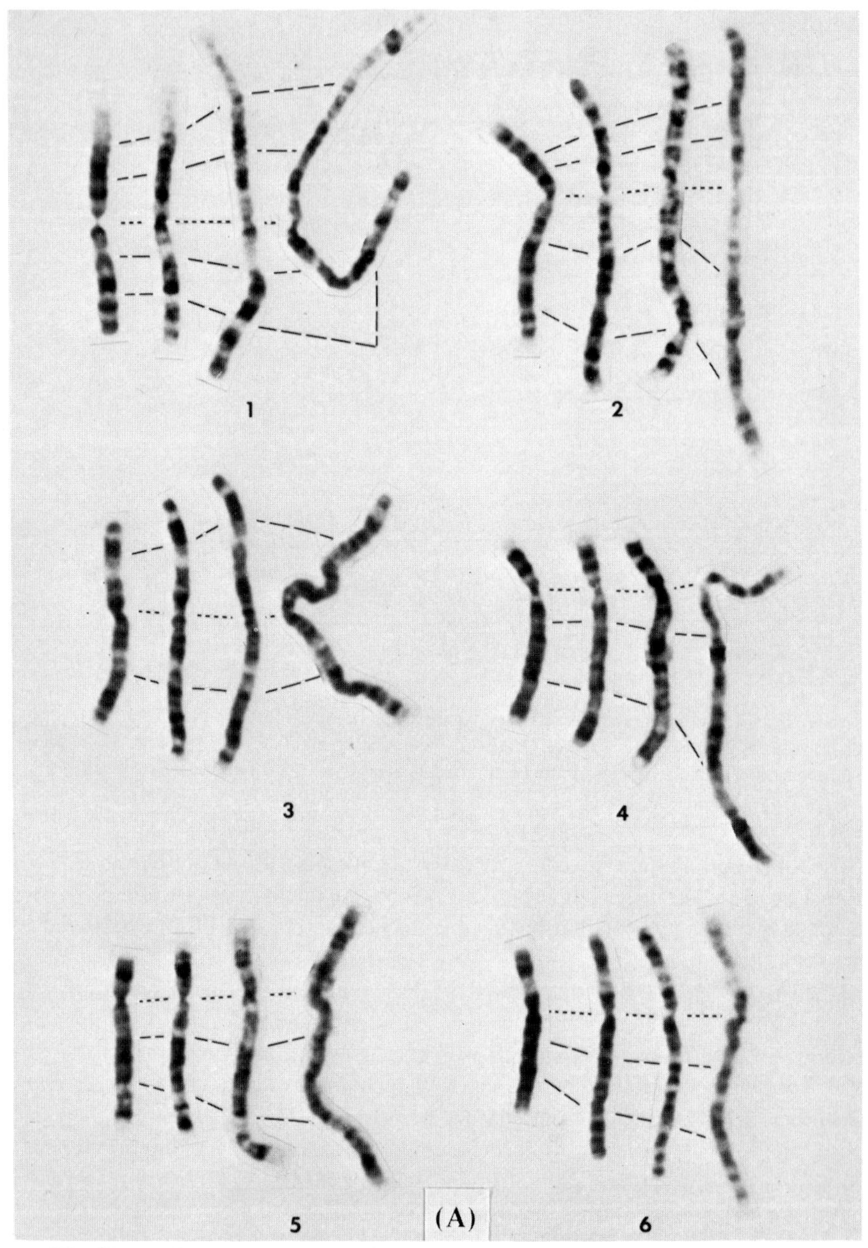


**Fig. 15.** (A–C). Representative G-banded chromosomes at midmetaphase, early metaphase, early prometaphase, and late prophase. Note the progressive coalescence of the multiple fine bands of late prophase into the thicker and fewer dark and light bands of metaphase. (A) Chromosomes 1–6; (B) chromosomes 7–12; (C) chromosomes 13–22, x, and y.

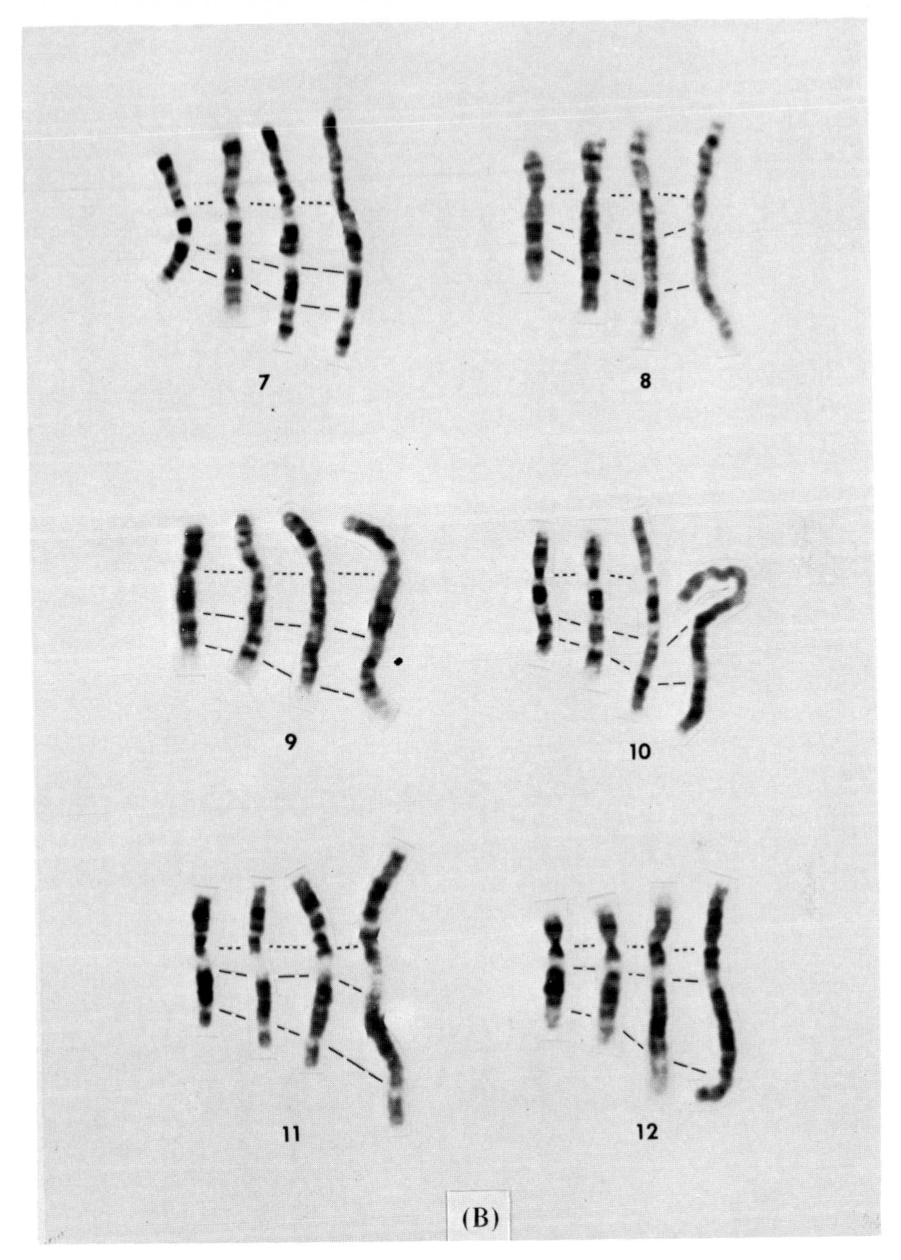


Fig. 15 *Continued*

Figure 15A, B, and C prepared by J. J. Yunis, previously unpublished.

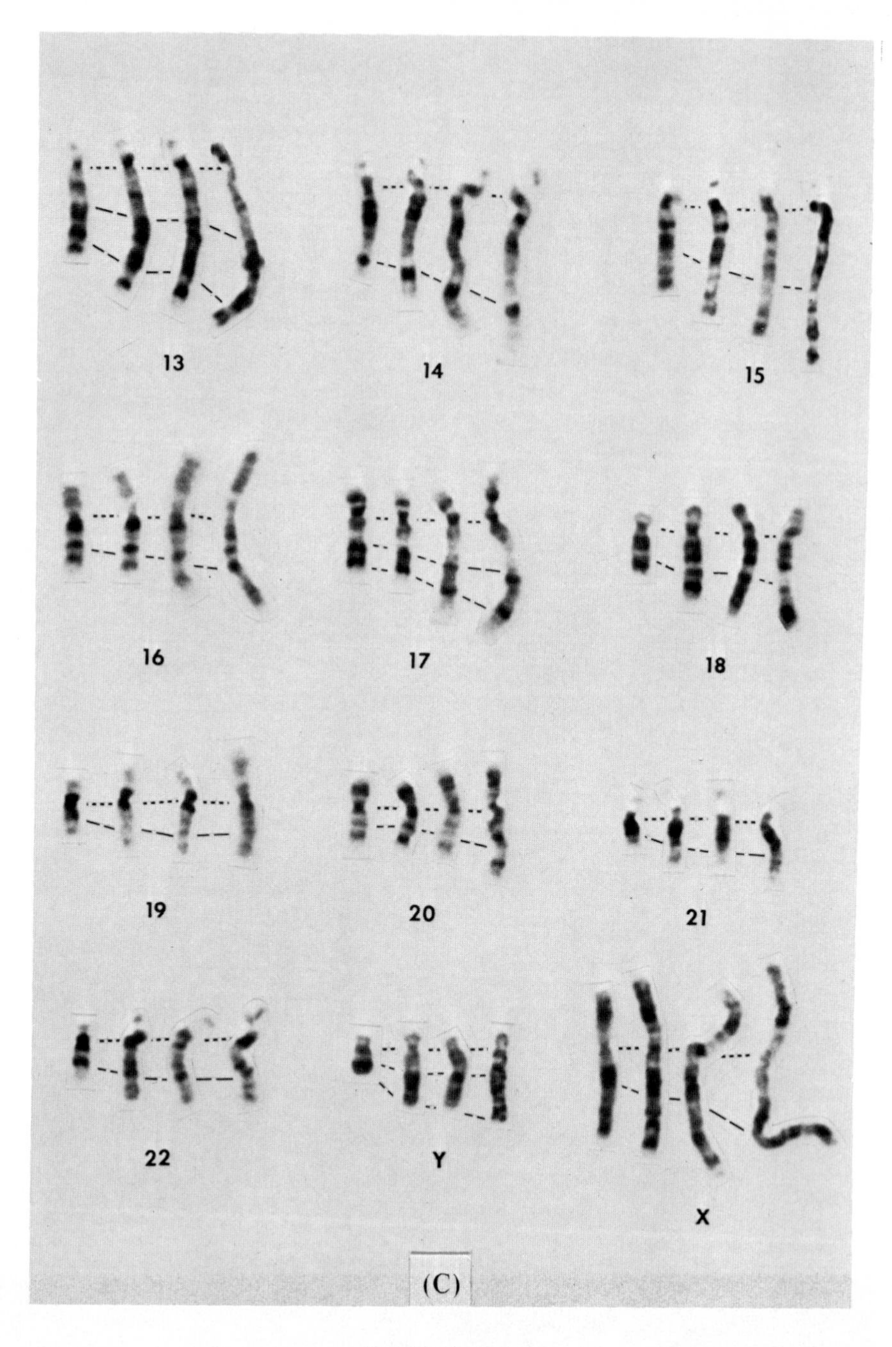


Fig. 15 *Continued*

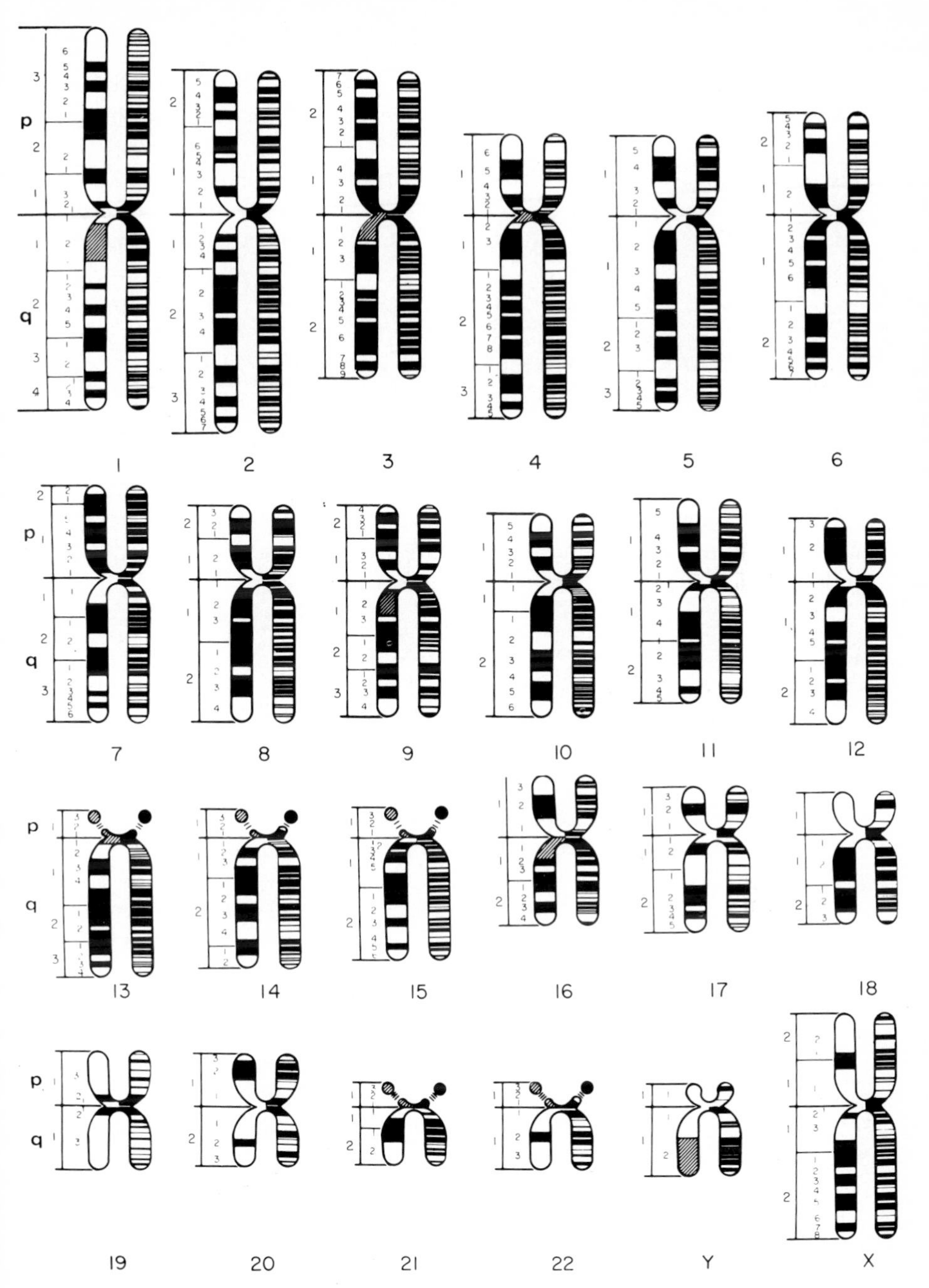


**Fig. 16.** Comparative representation of human chromosomes according to the Paris Conference, 1971 (1972) nomenclature. In each chromosome, the left chromatid represents the banding pattern (320 bands) observed at midmetaphase, and the right chromatid represents the G-banding pattern (1256 subbands) observed in late prophase. (From Yunis, 1976.)

to the cells in $G_2$, actinomycin D complexes with the DNA (Reich and Goldberg, 1964; Muller and Crothers, 1968; Allen *et al.*, 1976) and prevents binding of specific proteins (Kleiman and Huang, 1971) believed to be involved in chromosome condensation (Arrighi and Hsu, 1965; Hsu *et al.*, 1973). Chromosomes exhibit elongation up to 50% and an increase in the number of bands, which coincide with standard G-bands of untreated chromosomes (Yunis and Chandler, 1977b).

## IV. SIGNIFICANCE AND APPLICATION OF BANDING TECHNIQUES

### A. Introduction

The application of the banding techniques in human cytogenetics has produced numerous valuable results which have totally changed the scope of this specialty. The achievements could be summarized as follows:

a. Confirmation of well-established syndromes (e.g., +13, +18, +21, 4p–, 5p–, 13q–, 18p–, 18q–) (see Chapter 3).

b. Subdivision of classical syndromes (e.g., +13 and 13q–) (see Chapter 9).

c. Delineation of new chromosomal syndromes (e.g., +8, duplication 9p and partial duplication 10q) (see Chapters 4–11).

d. Description of patients showing chromosomal anomalies that involve almost every chromosome of the human set (see Table II).

e. A more precise definition of known chromosomal abnormalities (e.g., $Ph^1$ chromosome = 9/22 translocation) (Rowley, 1973c).

f. Demonstration of extensive chromosomal polymorphism in man (see Section IV, B).

g. Association of certain neoplasias with relatively specific chromosomal abnormalities (e.g., lymphomas and 8/14 translocation, retinoblastoma and partial deletion of long arm of chromosome 13) (see Section IV, B).

h. Progress in human gene mapping (Rotterdam Conference, 1974).

i. Evolutionary studies in primates (Dutrillaux, 1977).

j. Advances in the understanding of chromosome structure (Bahr, 1977).

k. Analysis of interphase chromosomes and their localization and associations (e.g., Y chromosome; Wyandt *et al.*, 1971).

l. Demonstration of aneuploidy in human spermatozoa (Sumner *et al.*, 1971a; Pearson, 1972; Pawlowitzki and Pearson, 1972).

m. Demonstration of aneuploidy for most autosomes in early pregnancy wastage (Lauritsen *et al.*, 1972; Kajii *et al.*, 1973; Boué *et al.*, 1975).

n. Identification of "pathogenic" segments of chromosomes (Lewandowski and Yunis, 1975; also see Chapter 12).

o. Advances in the understanding of chromosome-phenotype relationships in man (Chapter 12).

A thorough review of all these aspects is not possible within the limits of this chapter and only three aspects will be briefly touched upon: chromosomal polymorphisms in man; new chromosome abnormalities and birth defects; and specific chromosome defects in cancer.

## B. Chromosomal Polymorphisms

Before the advent of the banding techniques, it was well known that human chromosomes may differ from individual to individual by the presence, absence, or minor variation of certain chromosomal segments that apparently do not reflect abnormal phenotypic manifestations. The differences were limited to satellites and short arms of the acrocentric chromosomes (Ferguson-Smith and Handmaker, 1961), secondary constrictions of chromosomes 1, 16, and a C group chromosome usually identified as No. 9 (Ferguson-Smith *et al.*, 1962), and the long arm of the Y chromosome (Cohen *et al.*, 1966; Court-Brown, 1967). Such variants were believed to occur at a total frequency of about 2% in the general population, although their origin and significance were not clear (Hamerton, 1971).

### *1. Y Chromosome*

The intensive fluorescence of the distal portion of the long arm of the Y chromosome observed with the Q-banding technique (Zech, 1969) provided a most appropriate tool for the study of variants of this chromosome. It was soon learned that the variable size of this chromosome was due to variation in the size of the intensely fluorescent region (Bobrow *et al.*, 1971) (Fig. 8), and normal males may be found having small Y chromosomes and no bright fluorescence in the long arm (Borgaonkar and Hollander, 1971; Meisner and Inhorn, 1972). The frequency of variants for this chromosome have been found to be larger than previously suspected, and values of 9.5%, 0.1%, and 33% have been presented by several authors (Mikelsaar *et al.*, 1973; Soudek *et al.*, 1973; Park and Antley, 1975). The higher value was obtained in an Oriental population and may be due to racial differences.

In the mammalian kingdom, only man and gorilla share a bright fluorescent region in the Y chromosome (Pearson, 1973). This fact suggests that the chromatin of this region is of recent evolutionary acquisition. The absence of phenotypic effects in males lacking the fluorescent segment, the enrichment of this region in highly repetitive DNA (Sanchez and Yunis, 1974; Jones, 1977), and the scarcity of identifiable loci linked to the Y chromosome argue in favor of the nongenic content of this region. However, it has been suggested that the length of the Y chromosome bears some relationship to antisocial behavior, the chromosome being larger in criminals than in randomly selected control groups (Nielsen and Friedrich, 1972; Soudek and Laraya, 1974). Other authors have found no relationship between these parameters (Schwinger and Wild, 1974).

Patil and Lubs (1975) presented data suggesting that the risk for abortions is increased twofold in Caucasian families in which a long Y chromosome was present, although no basic mechanism was suggested. In this regard, it is interesting to note that, prior to the banding techniques, a possible association between long Y chromosomes and the occurrence of Down syndrome was also suggested (Hamerton, 1971). No data is available to confirm or deny this proposition. However, Bott *et al.* (1975), studying polymorphic variants of chromosome 21 in Down syndrome patients, found an increased frequency of Q-band polymorphisms of that chromosome in the families studied. This may indicate a possible association between chromosomal polymorphisms and nondisjunction.

### 2. *Chromosome 9*

The secondary constriction of chromosome 9 is one of the most interesting features of human chromosomes, not only because of its frequent variability, as discussed below, but also because of its heterogeneous response to different banding techniques, probably related to what seems to be a unique composition in terms of repetitive DNA (Jones, 1977).

Three main classes of variants have been described in the secondary constriction of chromosome 9: increase (9qh+) or decrease (9qh–) in size, and pericentric inversions (Fig. 17). Before the development of the banding techniques, only isolated cases of pericentric inversions were reported (Berg *et al.*, 1967; Court-Brown, 1967), and its frequency in the general population was believed to be as low as 0.12% (Lubs and Ruddle, 1970). The presence of large or small secondary constrictions in this chromosome was also known, but their frequency and significance were ill-defined (Hamerton, 1972). Because of the ability to specifically identify chromosome 9, the C- or Giemsa 11-banding techniques have been used to study the frequency and significance of variants and inversions of the

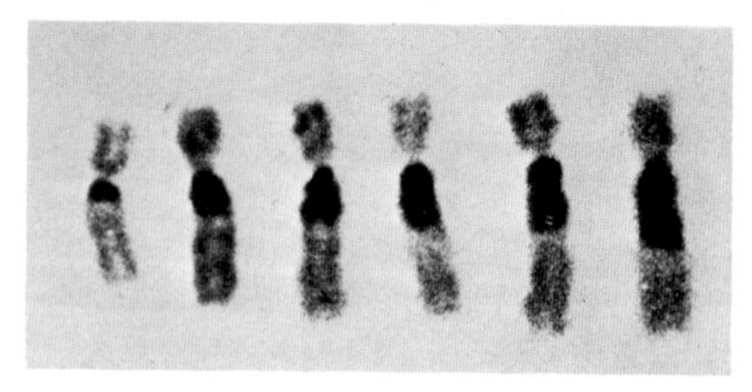

**Fig. 17.** C-band polymorphic variants of chromosome 9. (From Madan and Bobrow, 1974.)

secondary constriction of this chromosome. Several studies have reported between 0.7 and 11.3% of pericentric inversions in selected and random populations (Mutton and Daker, 1973; Craig-Holmes *et al.*, 1973; Madan and Bobrow, 1974; Park and Antley, 1975; McKenzie and Lubs, 1975). Again, the highest figure found is in an Oriental population. Size variations of the secondary constriction of chromosome 9 have been found in a wide range of frequencies in several studies: 0.1% (Nielsen *et al.*, 1974a); 0.13% (Geraedts and Pearson, 1973); 0.28% (Pearson, 1973); 3% (Mikelsaar *et al.*, 1973); 9% (Park and Antley, 1975); 12% (Madan and Bobrow, 1974); and 12.5% (Craig-Holmes *et al.*, 1973). This surprising frequency variability is probably due to several reasons. In some cases, only 9qh+ has been recorded; the studies have been carried out on different selected and unselected populations as well as on different races; and, probably most important, the criteria for variant identification may vary from author to author. In any case, the frequency of variants of secondary constrictions of chromosome 9 are, in general, higher than previously estimated, although large scale studies should be carried out under well-defined conditions to determine the precise frequency and variability among populations.

Most of the variants have been found to be transmitted according to simple Mendelian patterns (Craig-Holmes *et al.*, 1975). However, in some cases, the variants observed were not present in either parent, while other variants appeared to have been changed, possibly by a mechanism of somatic crossing-over (Sekhon and Sly, 1975; Craig-Holmes *et al.*, 1975). The presence of variant secondary constrictions of chromosome 9 does not seem to produce phenotypic effects in the carrier, but may predispose to nondisjunction as suggested by the increased frequency of 9qh+ variant in parents of patients affected by chromosomal anomalies (Nielsen *et al.*, 1974a).

Although pericentrically inverted chromosomes may lead to duplication deficiencies as a result of meiotic crossing-over in the inverted region, no such cases have been reported in families carrying this rearrangement in the secondary constriction of chromosome 9. The diminished frequency of crossing-over in heterochromatic regions already observed in other species (Yunis and Yasmineh, 1971, 1972) may explain this fact.

### *3. Other Chromosomes*

The secondary constriction of chromosome 1 has been observed to be enlarged in 4 of 208 (1.9%) normal unrelated adults. One of them also had a pericentric inversion of the same area (Tüür *et al.*, 1974). Craig-Holmes *et al.* (1973), however, found 5 individuals with a variant of chromosome 1 in a study of 20 normal subjects (25%), and only 1 of the variants corresponded to an enlargement of that area, while the other 4 showed a small secondary constriction.

Chromosome 16 has also been the subject of studies on polymorphic variants. Nielsen *et al.* (1974b) found the frequency of 16qh+ to be 0.12% in the general population. In the 20 patients studied by Craig-Holmes *et al.* (1973), no 16qh+ variant was found.

Although chromosomes 1, 9, 16, and Y most frequently have identifiable variants (Fig. 6), polymorphisms are in no way limited to such chromosomes. Acrocentric chromosomes (13–15, 21–22) are well known to have variations in the size of their satellites which may have bright or dull fluorescence when stained with quinacrine mustard (Fig. 8). Indeed, Q-band polymorphism of chromosome 21 has been used to trace the paternal or maternal origin of the extra chromosome in patients with Down syndrome (Bott *et al.*, 1975). In 7 of 59 families, the extra chromosome had polymorphic characteristics that allowed the determination of its origin, and in 4 of the 7 cases, it was found to have come from the father. The authors also observed that, although the frequency of Q-band polymorphism of chromosome 21 in the general population is 7%, in these 59 families with patients affected by Down syndrome, 42.5% of the mothers had such polymorphisms, as well as 28.6% of the fathers and 33.3% of the patients. These frequencies, as already mentioned, suggest a possible association between polymorphic variants and nondisjunction phenomena. Additional information is needed to confirm or rule out this inference.

Chromosome 19 was recently subjected to a careful study for the presence of pericentromeric variants (Crossen, 1975). Chromosomes from 50 normal donors were stained with the G-banding technique. Four different heterochromatic patterns were found in the pericentromeric region. Twenty-three of the subjects (46%) studied showed a variant polymorphism in at least one of the two chromosomes 19. Chromosomal polymorphism has been found to be present in other chromosomes as well and to have a widespread occurrence. For example, Q- and C-band polymorphism were found on most chromosomes in a study of 69 individuals from 9 unrelated families (Sekhon and Sly, 1975). The authors reported a total of 352 Q- and C-band variants with an average of 5.1 per karyotype. On 11 different chromosomes, 209 Q-band polymorphisms were found, while 143 C-band variants were distributed on most chromosomes.

From the above-mentioned studies, it can be deduced that polymorphic variants are extensively distributed among all chromosomes of normal individuals. Whether these variants have any physiological effects is a question which remains unanswered.

The study of constitutive heterochromatin and its polymorphic variants has already been amply used in human cytogenetics, as markers in cell fusion studies (Yunis, 1977), to identify the chromosome involved in the origin of the Philadelphia chromosome (Rowley, 1975), for linkage studies in gene mapping (Donahue *et al.*, 1968), and to determine twin zygosity (Van Dyke *et al.*, 1974). Given the Mendelian inheritance and frequency of most of these polymorphisms, they may constitute a valuable aid in paternity studies. Since chromosomal variants are quite common, the possibility for two individuals having exactly the same variants on all chromosomes is small, and it would not be surprising if, in the future, chromosomal variants might be used for identification purposes.

## C. New Chromosomal Abnormalities and Birth Defects

Application of the banding techniques in the area of clinical cytogenetics has been most productive. The characteristic banding patterns of human chromosomes have allowed the precise identification of previously suspected trisomies, i.e., trisomies 8 and 22 (Caspersson *et al.*, 1972; Hirschhorn *et al.*, 1973), and monosomies such as −21 and −22 (De Cicco *et al.*, 1973; Kancko *et al.*, 1975). More valuable yet has been the application of the banding methods to cases of balanced or unbalanced translocations, peri- and paracentric inversions, ring, recombinant, and derivative chromosomes. In these cases, only segments of chromosomes are usually involved, and the rearranged chromosomes are so distorted from their natural configuration that identification without the help of the chromosomal bands would have been impossible. The results have been impressive, and the clinical cytogenetic knowledge has expanded considerably from the relatively few syndromes known until 5 years ago. Several new syndromes due to partial trisomies and deletions have been firmly established (Lewandowski and Yunis, 1975; see also Chapter 12, Fig. 1) and the large number of reports which have appeared dealing with chromosomal abnormalities involving all chromosomes of the human set will surely lead to the identification of more syndromes in the future. A detailed description of all the definitive and possible new syndromes identified falls beyond the purpose of this chapter. In the following pages of this book, the reader will find detailed reviews on the classical and new chromosomal syndromes (Chapter 3–11). However, an attempt has been made to illustrate in a concise form the available information regarding new chromosomal syndromes and chromosome defects recently described and the involvement of every human chromosome in patients affected by chromosomal diseases. An extensive search of the recent literature has been used to construct Table II. From this table, the relatively large number of new chromosomal syndromes that are known can be appreciated. It is also evident that several isolated cases of trisomies and monosomies for different chromosomal segments have been described. An analysis of the data contained in this table allows one to make certain deductions. For example, contrary to what was believed before the introduction of the banding techniques, partial monosomies are not necessarily lethal and are, in fact, relatively frequent and involve most of the human chromosomes (Table II; also Fig. 1 of Chapter 12). As in *Drosophila,* however, it is true that trisomies in general seem to be less deleterious and thus more frequent than monosomies. Kajii *et al.* (1973), for example, have

**TABLE II**
**Chromosomes or Chromosome Segments Involved in Duplication/Deficiencies in Man[a]**

| | |
|---|---|
| Chromosome 1 | |
| *Short arm* | |
| Partial deletion (1q24→1q32.1) | Turleau *et al.*, 1974 |
| Partial trisomy (1p22→1p36) | Gray *et al.*, 1972 |
| *Long arm* | |
| Partial deletion (1q44→1qter) | Laurent *et al.*, 1973a |
| Partial trisomy (1q24→1qter) | van der Berghe *et al.*, 1973 |
| Partial trisomy (1q32→1qter) | Seabright, 1972; Sanger *et al.*, 1974 |
| Partial trisomy (1q25→1qter) | Finley *et al.*, 1975 |
| Partial trisomy (1q23→1qter) | Norwood and Hoehn, 1974 |
| Chromosome 2 | |
| (Complete trisomy described in spontaneous abortuses, Lauritsen *et al.*, 1972; Kajii *et al.*, 1973) | |
| *Short arm* | |
| Partial deletion (2p23→2pter) | Ferguson-Smith *et al.*, 1973 |
| Partial trisomy (2p22→2pter) | Stoll *et al.*, 1974 |
| *Long arm* | |
| Partial deletion (2q14) | German and Chaganti, 1973 |
| Partial trisomy (2q31→2qter) | Warren *et al.*, 1975 |
| Partial trisomy (2q32→2qter) | Forabosco *et al.*, 1973 |
| Partial trisomy (2q33→2qter) | Shapiro and Warburton, 1972 |
| Partial trisomy (2q33→2qter) | Rosenthal *et al.*, 1974 |
| Chromosome 3 | |
| (Complete trisomy described in spontaneous abortuses, Kajii *et al.*, 1973) | |
| *Short arm* | |
| Partial deletion (3p25→3pter) | Allderdice *et al.*, 1975 |
| Partial trisomy (3p21→3p26) | Rethoré *et al.*, 1972 |
| Partial trisomy (3p23→3pter) | Ballesta and Vehi, 1974 |
| Partial trisomy (3p13→3pter) | Creasy *et al.*, 1974 |
| Partial trisomy (3p24→3pter) | Say *et al.*, 1976 |
| *Long arm* | |
| Partial deletion | |
| Partial trisomy (3q21→3qter) | Allderdice *et al.*, 1975 |
| Partial trisomy (3q25→3qter) | Lockhart and Meyne, 1975 |
| Chromosome 4 | |
| (Complete trisomy described in spontaneous abortuses, Kajii *et al.*, 1973) | |
| *Short arm* | |
| Partial deletion (4p– or Wolf-Hirschhorn syndrome: see Chapter 3) | |
| Partial trisomy (duplication 4p syndrome: Rethoré et al., 1974; also Chapter 4) | |
| *Long arm* | |
| Partial deletion (4q31→4qter) | van Kempen, 1975 |
| Partial deletion (4q33→4qter) | Dutrillaux *et al.*, 1973 |

*continued*

**TABLE II** (*continued*)

| | |
|---|---|
| Partial trisomy (4q21→4qter) | de la Chapelle *et al.*, 1973b |
| Partial trisomy (4q25→4q34) | Dutrillaux *et al.*, 1975 |
| Partial trisomy (4q26→4qter) | Schrott *et al.*, 1974 |
| Partial trisomy (4q28→4qter) | Dutrillaux *et al.*, 1975 |
| Partial trisomy (4q34→4qter) | Baccicheti *et al.*, 1975 |
| Chromosome 5 | |
| *Short arm* | |
| Partial deletion (5p– or cri-du-chat syndrome: see Chapter 3) | |
| Partial trisomy (5p11→5pter) | Opitz and Patau, 1975; Lewandowski and Yunis (Chapter 12) |
| *Long arm* | |
| Partial trisomy (5q11→5q23) | Jalbert *et al.*, 1975 |
| Partial trisomy (5q31→5qter) | Osztovics and Kiss, 1975; Ferguson-Smith *et al.*, 1973 |
| Chromosome 6 | |
| (Complete trisomy described in spontaneous abortuses, Kajii *et al.*, 1973) | |
| *Short arm* | |
| Partial deletion (6p25→6pter) | Borgaonkar *et al.*, 1973 |
| Partial trisomy (6p21→6pter) | Chiyo *et al.*, 1975a |
| Partial trisomy (6p22→6pter) | Gouw *et al.*, 1973 |
| *Long arm* | |
| Partial deletion (6q25→6qter) | Mikkelsen *et al.*, 1973 |
| Partial deletion (6q26→6qter) | Mikkelsen and Dyggve, 1973 |
| Partial trisomy (6q21→6qter) | Robertson *et al.*, 1975 |
| Chromosome 7 | |
| (Complete trisomy described in spontaneous abortuses, Boué *et al.*, 1975) | |
| *Short arm* | |
| Partial deletion (7p15→7p15) | Friedrich, 1975; Kajii *et al.*, 1973 |
| *Long arm* (see Chapter 5) | |
| Partial deletion (7q35→7qter) | de Grouchy and Turleau, 1974 |
| Partial deletion (7qter) | Shokeir *et al.*, 1973 |
| Partial deletion (7q32→7qter) | Bass *et al.*, 1973a |
| Partial trisomy (7q22→7qter) | Carpentier *et al.*, 1972 |
| Partial trisomy (7q22→7q31) | Serville *et al.*, 1975; Berger *et al.*, 1974 |
| Partial trisomy (7q22→7q32) | Grace *et al.*, 1973 |
| Partial trisomy (7q31→7qter) | Vogel *et al.*, 1973; Alfi *et al.*, 1973a |
| Partial trisomy (7q32→7qter) | Bass *et al.*, 1973a |
| Chromosome 8 | |
| (Trisomy 8 syndrome: Caspersson *et al.*, 1972; see Chapter 6) | |
| *Short arm* | |
| Partial deletion (8p12→8p22) | Ladda *et al.*, 1974 |
| Partial deletion (8p22→8p23) | Fujimoto *et al.*, 1975 |
| Partial deletion (8p21→8pter) | Lockhart and Meyne, 1975 |

*continued*

**TABLE II** (*continued*)

| | |
|---|---|
| Partial trisomy (8p11→8pter) | Chiyo *et al.*, 1975b |
| Partial trisomy (8pter→8q23) | Lejeune and Rethoré, 1973 |
| *Long arm* | |
| Partial trisomy (8q13→8qter) | Lockhart and Meyne, 1975 |
| Partial trisomy (8q21→8qter) | Fryns *et al.*, 1974a |
| Partial trisomy (8q22→8q24) | Fujimoto *et al.*, 1975 |
| Partial trisomy (8q22→8qter) | Laurent *et al.*, 1974 |
| Partial trisomy (8q24→8qter) | Sanchez and Yunis, 1974 |
| Chromosome 9 | |
| (Trisomy 9: Francke *et al.*, 1975a; Feingold and Stkins, 1973; Lewandowski and Yunis, 1977; see Chapter 4) | |
| *Short arm* | |
| Partial deletion (9p– syndrome) | Alfi *et al.*, 1973b; Chapter 4 |
| Partial trisomy (9p21→9pter) | Lewandowski *et al.*, 1976 |
| Partial trisomy (dup 9p syndrome) | Rethoré *et al.*, 1973a; Chapter 4 |
| *Long arm* | |
| Partial trisomy (9pter→9q22) | Centerwall *et al.*, 1975 |
| Partial trisomy (9q11→9q33) | Turleau *et al.*, 1975a |
| Partial trisomy (9q21→9qter) | Lockhart and Meyne, 1975 |
| Partial trisomy (9q33→9qter) | Turleau *et al.*, 1975a |
| Chromosome 10 | |
| (Complete trisomy described in spontaneous abortuses, Kajii *et al.*, 1973; Boué *et al.*, 1975) | |
| *Short arm* | |
| Partial deletion (10p13→10pter) | Francke *et al.*, 1975b; Shokeir *et al.*, 1975 |
| Partial trisomy (dup 10p syndrome: Cantu *et al.*, 1975; see Chapter 7) | |
| *Long arm* | |
| Partial deletion (10q24→10qter) | Robertson *et al.*, 1975 |
| Partial trisomy (dup 10q syndrome: Yunis and Sanchez, 1974; see Chapter 7) | |
| Chromosome 11 | |
| *Short arm* (see Chapter 8) | |
| Partial trisomy (11p11.2→11pter) | Francke, 1972; Falk *et al.*, 1973 |
| Partial trisomy (11p11→11pter) | Palmer *et al.*, 1976 |
| Partial trisomy (11p12→11pter) | Sanchez *et al.*, 1974 |
| *Long arm* (see Chapter 8) | |
| Partial deletion (11q14→11q22) | Taillemite *et al.*, 1975 |
| Partial deletion (11q21→11qter) | Faust *et al.*, 1974 |
| Partial deletion (11q23→11qter) | Jacobsen *et al.*, 1973 |
| Partial trisomy (dup 11q syndrome: see Chapter 4) | |
| Chromosome 12 | |
| (Complete trisomy described in abortuses, Boué *et al.*, 1975) | |
| *Short arm* (see Chapter 4) | |
| Partial deletion (12p12) | Orye and Craen, 1975 |

*continued*

TABLE II (*continued*)

| | |
|---|---|
| Partial deletion (12p11→12p12) | Tenconi *et al.*, 1975 |
| Partial deletion (12p11→12pter) | Friedrich and Nielsen, 1974 |
| Partial deletion (12p12→12pter) | Magnelli and Therman, 1975 |
| Partial trisomy (12q12→12pter) | Rethoré *et al.*, 1975 |
| Partial trisomy (12cent→12pter) | Fryns *et al.*, 1974b |
| Partial trisomy (12p11→12pter) | Armendarez *et al.*, 1975 |
| Partial trisomy (12p12→12pter) | Uchida and Lin, 1973 |
| *Long arm* | |
| Partial trisomy (12q14→12qter) | Mikkelsen, 1974 |
| Partial trisomy (12q24→12qter) | Holboth *et al.*, 1974 |
| Chromosome 13 | |
| (Trisomy 13 or Patau syndrome: see Chapter 3) | |
| *Long arm* (see Chapter 9) | |
| Partial deletion (13q12→13qter) | Kajii *et al.*, 1974 |
| Partial deletion (13q22→13qter) | Ikeuchi *et al.*, 1974 |
| Partial deletion (13q21) | Orye *et al.*, 1974 |
| Partial deletion (13q34→13qter) | Forabosco *et al.*, 1973 |
| Partial trisomy (13pter→13q12) | Escobar and Yunis, 1974 |
| Partial trisomy (13q12→13qter) | Kajii *et al.*, 1974 |
| Partial trisomy (13q14→13qter) | Schinzel *et al.*, 1974 |
| Partial trisomy (13q21→13qter) | Stoll and Halb, 1974 |
| Partial trisomy (13q22→13qter) | Escobar *et al.*, 1974 |
| Chromosome 14 | |
| (Complete trisomy described in abortuses, Kajii *et al.*, 1973) | |
| *Long arm* (see Chapter 10) | |
| Partial trisomy (14pter→14q12) | Fawcett *et al.*, 1975; Laurent *et al.*, 1973b |
| Partial trisomy (14pter→14q21) | Turleau *et al.*, 1975b |
| Partial trisomy (14pter→14q22) | Raoul *et al.*, 1975 |
| Partial trisomy (14pter→14q22 or 23) | Wahlström, 1974; Pfeiffer *et al.*, 1973 |
| Partial trisomy (14pter→14q23 or 24) | Muldal *et al.*, 1973 |
| Chromosome 15 | |
| (Complete trisomy described in abortuses, Lauritsen *et al.*, 1972; Kajii *et al.*, 1973) | |
| *Long arm* (see Chapter 10) | |
| Partial trisomy (15pter→15q13) | Rethoré *et al.*, 1973b |
| Partial trisomy (15pter→15q22) | Fujimoto *et al.*, 1974; Crandall *et al.*, 1973; Watson and Gordon, 1974; Cohen *et al.*, 1975 |
| Chromosome 16 | |
| (Complete trisomy described in abortuses, Lauritsen *et al.*, 1972; Kajii *et al.*, 1973) | |
| *Short arm* | |
| Partial trisomy (16p12→16pter) | Stern and Murch, 1975 |
| *Long arm* | |
| Partial trisomy (16cent→16qter) | Schmikel, 1975 |

*continued*

**TABLE II (*continued*)**

| | |
|---|---|
| Chromosome 17 | |
| *Short arm* | |
| Partial deletion (17p12→17pter) | Nakagome *et al.*, 1973 |
| Partial trisomy (17q1?→17pter) | Latta and Hoo, 1974 |
| Chromosome 18 | |
| (Trisomy 18 or Edward syndrome: see Chapter 3) | |
| *Short arm* | |
| Partial deletion (18cent→18pter) | Surana *et al.*, 1973; Malpuech *et al.*, 1971 |
| *Long arm* | |
| Partial deletion (18q11→18pter) | Stern and Murch, 1975 |
| Partial deletion (18q11→18q21) | Chudley *et al.*, 1974 |
| Partial deletion (18q12→18qter) | Gouw *et al.*, 1973 |
| Partial deletion (18q21→18qter) | Curtis, 1973; Faed *et al.*, 1972; Schinzel *et al.*, 1974 |
| Partial trisomy (18cent→18qter) | Surana *et al.*, 1973 |
| Partial trisomy (18q11→18q21) | Chudley *et al.*, 1974 |
| Partial trisomy (18q21→18qter) | Steele *et al.*, 1974 |
| Chromosome 19 | |
| *Long arm* | |
| Partial trisomy (19q13→19qter) | Lange and Alfi, 1976 |
| Chromosome 20 | |
| (One case described with possible trisomy 20, Krmpotic *et al.*, 1971) | |
| *Short arm* | |
| Partial deletion (20p11→20pter) | Lockhart and Meyne, 1975 |
| Partial trisomy (dup 20p syndrome: see Chapter 8) | |
| *Long arm* | |
| Partial deletion (20q13→20qter) | Dutrillaux *et al.*, 1973 |
| Partial trisomy (20qter) | Sanchez *et al.*, (1977); Fawcett *et al.*, 1975 |
| Chromosome 21 | |
| Trisomy 21 (Down Syndrome): see Chapter 3 | |
| Monosomy 21 (Antimongolism) | Gripenberg *et al.*, 1972; Halloran *et al.*, 1974; Kancko *et al.*, 1975 |
| *Long arm* | |
| Partial deletion (21pter→21q21.1) | Rethoré *et al.*, 1973b |
| Partial deletion (21pter→21q22?) | Wyandt *et al.*, 1971; Dutrillaux *et al.*, 1973; Cooksley *et al.*, 1973 |
| Partial trisomy (21pter→21q21) | Lewandowski *et al.*, 1976; Sanchez *et al.*, (1977) |
| Partial trisomy (21pter→21q22) | Mikkelsen, 1974 |
| Partial trisomy (21q11→21qter) | Borgaonkar *et al.*, 1973 |
| Partial trisomy (21q21→21qter) | Williams *et al.*, 1975 |
| Chromosome 22 | |
| (Trisomy 22 syndrome: Bass *et al.*, 1973b; see Chapter 11) | |
| Monosomy 22 | De Cicco *et al.*, 1973 |
| *Long arm* | |
| Partial deletion (22pter→22q11) | Nakagome *et al.*, 1973 |

*continued*

TABLE II (*continued*)

| | |
|---|---|
| Partial deletion (22q12→22qter) | Wurster-Hill and Hoefnagel, 1975; Wahlström, 1974 |
| Partial trisomy (22pter→22q12) | Bühler *et al.*, 1972; M. K. Bofinger and S. W. Soukup, personal communication, 1976; (see also Chapter 11) |
| Chromosomes X and Y (see Chapter 3) | |

[a]Numbers in parentheses denote chromosomal bands involved in each case. Generally, references have been limited to those cases in which a definitive identification of the bands involved was achieved. In some cases, there was a combination of a partial duplication–deficiency and a reference was given for only one of the two chromosome segments involved. Balanced translocations, ring chromosomes, and chromosomal mosaics have not been included. In most cases where a definitive syndrome has been recognized, the reader is referred to the article in which that syndrome was described or to the corresponding sections in this book. Selected cases primarily cover the literature of the 1973–1975 period, and no attempt has been made to be all inclusive. For possible additional chromosome defects, consult Borgaonkar (1975).

shown that in early abortions, trisomies for several chromosomes can be found but monosomies are rare except for the X and 21 chromosomes; this could indicate that full monosomies are highly lethal and lost very early.

A concept which has been reinforced over the years is the idea that duplications or deletions of late replicating or heterochromatic areas are less harmful than those involving earlier replicating regions (Yunis, 1965). Positive Q- and G-bands are now known to represent late replicating regions (Ganner and Evans, 1971; Calderon and Schnedl, 1973; Crossen *et al.*, 1975; Dutrillaux, 1977) and the chromosomal segments generally involved in the chromosomal syndromes are particularly rich in Q- and G-positive bands.

Another point which arises upon observation of Table II is that, although trisomies or monosomies may involve any chromosome, the amount of chromosomal material present in excess or missing does not generally exceed 5% of the human genome. Again, this seems to be a similar phenomenon to that observed in *Drosophila* (Lindsley and Sandler *et al.*, 1972).

Even though the classical chromosomal syndromes are relatively well-defined entities and can be recognized on clinical grounds alone, partial deletions and duplications may sometimes be difficult to diagnose because phenotypic findings may be few and largely nonspecific. Furthermore, many patients can be considered unique cases because of their different chromosomal content. Although more frequent at the levels of centromeres and telomeres (Jacobs *et al.*, 1974; Nakagome and Chiyo, 1976), break points in chromosomes are in essence randomly located. As a direct consequence, the chromosomal segments involved in partial trisomies and deletions may vary from patient to patient and, consequently, the phenotype may show considerable variation. For example, offspring from carriers of balanced translocations may receive recombinant chromosomes, leading not only to partial trisomy or deletion, but usually to a combination of both which can be phenotypically translated into various mixtures of two given syndromes Figs. 3 and 4).

The merits of the banding techniques are not questionable. However, one should recognize that these techniques are not a panacea that will solve all the problems confronted in clinical cytogenetics. Mosaicism, for example, is still a difficult subject, especially when prenatal diagnosis is involved (Bloom *et al.*, 1974). Another situation that may sometimes be difficult to resolve even with the application of the banding techniques is found in cases of derivative chromosomes present as a result of a "*de novo* translocation," i.e., derivative chromosomes not present or detectable in the parents that seem to have occurred early in embryonic life or in a germinal cell. In this situation, the chromosomal segments present in the abnormal chromosome 10 could not be identified. A similar situation was confronted recently in this laboratory when studying a patient with several congenital malformations whose metaphase chromosomes exhibited an abnormally long chromosome 1. Both parents presented normal karyotypes. The origin of the extra bands present in the abnormal chromosome 1 was not identified with certainty, although clinically the pateint showed the typical phenotype of the duplication 10q syndrome and early metaphases showed the typical subbanding pattern of the distal end of the long arm of chromosome 10 on the abnormally long chromosome 1 (cf. Fig. 8 of Chapter 7).

## D. Specific Chromosomal Defects in Cancer

For technical reasons, the study of chromosomal defects has been easier in hematopoeitic neoplasia than in solid tumors. Although usually of only mediocre quality, application of the banding techniques to such preparations has provided valuable information (Rowley, 1974a). Indeed, in some cases, some types of neoplasia (e.g., chronic myelogenous leukemia, retinoblastoma) have been found to have a consistent relationship with certain specific chromosomal anomalies.

### *1. Chronic Myelogenous Leukemia*

The most conspicuous and well-known example of a constant association between chromosomal anomalies and neoplasia is the presence of the $Ph^1$ chromosome in chronic myelogenous leukemia (CML) (Nowell and Hungerford, 1960; Whang-Peng *et al.*, 1968). With the use of banding techniques, the abnormal chromosome was identified as a No. 22 (Caspersson *et al.*, 1970b; O'Riordan *et al.*, 1971). Rowley, in 1973 (1973c) demonstrated that the $Ph^1$ chromosome does not represent a simple deletion of chromosome 22 but rather an apparent balanced translocation between chromosomes 9 and 22. The involvement of these two chromosomes in the genesis of the Ph[1] chromosome has been repeatedly confirmed (Berger, 1973; Gahrton *et al.,* 1974c; Prigogina and Fleischman, 1975a; Fleischman and Prigogina, 1975). However, chromosome 9 does not seem to be a *sine qua non* prerequisite or component, since translocation between chromosome 22 and other chromosomes (even the homologous 22) have been reported (Hayata *et al.,* 1973; Gahrton *et al.,* 1974c; Foerster *et al.,* 1974). The Ph[1] chromosome is known to be acquired, since it cannot be demonstrated in the cultured fibroblasts of affected patients, in cells of the parents (Gahrton *et al.,* 1973), or in the identical twins of affected patients (Goh *et al.,* 1967). Either the paternal or the maternal chromosome 22 may be involved, and there is some evidence that the Ph[1] chromosome is of monoclonal origin (Gahrton *et al.,* 1973, 1974a). This chromosome is known to be present in approximately 85% of the cases. The 15% of patients not showing the Ph[1] chromosome usually follow a more malignant course and present unusual characteristics (Sandberg and Hossfeld, 1970).

Blastic crisis in CML is usually accompanied by other chromosomal anomalies in addition to the presence of the Ph[1] chromosome. These changes do not follow random patterns, but frequently involve the presence of extra copies of the Ph[1] chromosome, isochromosomes for the long arm of chromosome 17, and trisomy 8 and/or 9 (Lobb *et al.,* 1972; Hossfeld, 1974; Gahrton *et al.,* 1974b; Engel *et al.,* 1975; Prigogina and Fleischman, 1975a; Rowley, 1975). The exact significance of these changes is not known.

### *2. Acute Leukemias and Other Myeloproliferative Diseases*

Acute leukemias, either myelo- or lymphoblastic, show chromosomal abnormalities in bone marrow cells in approximately 50% of the cases (Sandberg and Hossfeld, 1970). Acute lymphatic leukemia (ALL) tends to show hyperdiploidy, while in acute myelogenous leukemia (AML) hypodiploidy predominants. Chromosomal changes in bone marrow cells of acute leukemias were believed to be extremely diverse and thus probably unrelated to the cause or prognosis of these diseases (Sandberg and Hossfeld, 1970). Nevertheless, an extra C group chromosome has been repeatedly reported to be present in bone marrow cells of leukemic patients as well as in patients with other hematopoietic disorders (Philip, 1975). Application of the banding techniques led to the identification of the extra chromosome as 8 or 9 in most cases (Jonasson *et al.,* 1974; Ford *et al.,* 1974; Mitelman and Brandt, 1974; Philip, 1975), and sometimes as 10 or 11 (Philip, 1975; Rowley, 1975). A missing or deleted chromosome 7 has also been identified (Rowley, 1973a; Petit *et al.,* 1973; Zech *et al.,* 1975; Lawler *et al.,* 1975). Another relatively frequent finding has been the presence of an isochromosome for the long arm of chromosome 17 (Mitelman *et al.,* 1973, 1975; Fleischman and Prigogina, 1975).

Thus, when chromosomal abnormalities are present in patients suffering from acute leukemias and other related hematopoietic disorders, they show a preferential involvement of chromosomes 7, 8, 9, and 17. Further studies are needed before reaching definitive conclusions on the relationship between specific chromosomes and acute leukemias.

### *3. Lymphomas*

Burkitt lymphoma is a neoplastic process in which there is a close association with the Epstein-Barr (EB) virus, possibly as a cause-effect type of relationship (Epstein and Achong, 1973; Henle and Henle, 1973). Manolov and Manolova (1972) reported the presence of an abnormal chromosome 14 in 8 of 12 tumors examined. The abnormal chromosome showed a similar anomaly in all cases, involving an extra dark band at the distal end of the long arm of chromosome 14. This finding was later confirmed by Jarvis *et al.* (1974) who found the same marker chromosome in 7 cell lines derived from Burkitt lymphoma. The anomaly seems to be directly related to the neoplastic process itself and not to the virus, since in 31 cases of mononucleosis-derived cell lines (a disease closely associated with the EB virus) the marker chromosome 14 was not shown. Zech (1975) has recently provided evidence suggesting that the extra dark band present in chromosome 14 is the result of a balanced translocation from chromosome 8. This has been confirmed by McCaw *et al.* (1977) who have found that various types of non-Burkitt lymphoma also have this unusual translocation. The widespread finding of the abnormal chromosome 14 among various kinds of lymphoma has been reported by different investigators (Reeves, 1973; Prigogina and Fleischman, 1975b).

The similarities between this chromosomal abnormality in lymphomas and that found in CML are striking. Both are the result of acquired balanced translocations and are possibly specific for each respective neoplastic process.

### *4. Retinoblastoma*

Retinoblastoma has long been associated with a partial deletion of the long arm of a D (13–15) group chromosome (Lele *et al.*, 1963; Wilson *et al.*, 1969; Taylor, 1970). In 1971, Orye *et al.* applied the banding techniques to chromosomes of a patient with retinoblastoma and a deletion of a D group chromosome, and identified the chromosome involved as No. 13. This finding has been confirmed in several cases (Wilson *et al.*, 1973; Howard *et al.*, 1974; Orye *et al.*, 1974; O'Grady *et al.*, 1974). Lewandowski and Yunis (Chapter 12) have reexamined the available information and concluded that the specific segment involved probably includes band 13q14 and the proximal portion of band 13q21.*

*Most recently, Yunis (1977b) has sublocalized the responsible segment to a portion of band q14.

### 5. *Meningiomas*

Cytogenetic studies of human meningioma have shown that the loss of a G group chromosome is typical for that tumor (Zang and Singer, 1967). Banding techniques identified the missing chromosome as No. 22 (Mark *et al.*, 1972). In addition, loss of chromosomes 8, 9, X, and Y have also been reported in a few cases (Mark, 1973; Zankl *et al.*, 1975). In a patient with aniridia and Wilm's tumor, a partial deletion of chromosome 8 (8p12→8p22) has been reported (Ladda *et al.*, 1974). The question of whether these associations are due to chance remains to be determined.

In conclusion, the application of the banding techniques to the study of chromosomes in human tumors offers fruitful results. It is clear, however, that many more studies are needed to precisely clarify the role played by chromosomal abnormalities in the genesis of human neoplasias.

## ACKNOWLEDGMENTS

This work was supported in part by NIH Grant No. HD01962. The authors are grateful to Ms. Marylyn Hoglund for her editorial assistance.

## REFERENCES

Alfi, O. S., Donnell, G. N., and Kramer, S. L. (1973a). *J. Med. Genet.* **10,** 187–189.

Alfi, O. S., Donnell, G. N., Crandall, B. F., Derencsenyi, A., and Menon, R. (1973b). *Ann. Genet.* **15,** 17–22.

Allderdice, P. W., Browne, N., and Murphy, D. P. (1975). *Am. J. Hum. Genet.* **27,** 699–718.

Allen, F. S., Moen, R. P., and Hollstein, U. (1976). *J. Am. Chem. Soc.* **98,** 3–4.

Armendarez, S., Salamanca, F., Nava, S., Ramirez, S., and Cantu, J.-M. (1975). *Ann. Genet.* **18,** 89–94.

Arrighi, F. E., and Hsu, T. C. (1971). *Cytogenetics* **10,** 81–86.

Arrighi, F. E., and Hsu, T. C. (1965). *Exp. Cell Res.* **39,** 305–308.

Baccicheti, C., Tenconi, R., Anglani, F., and Zacchello, F. (1975). *J. Med. Genet.* **12,** 425–427.

Bahr, G. F. (1977). *In* "Molecular Structure of Human Chromosomes" (J. Yunis, ed.), Chromosomes in Biology and Medicine Ser., Academic Press, New York.

Bahr, G. F., and Larsen, P. M. (1974). *Adv. Cell Mol. Biol.* **3,** 191.

Bahr, G. F., Mikel, U., and Engler, W. F. (1973). *In* "Chromosome Identification" (T. Caspersson and L. Zech, eds.), p. 280. Academic Press, New York.

Ballesta, F., and Vehi, L. (1974). *Ann. Genet.* **17,** 287–290.

Baranovskaya, L. T., Zakharov, A. F., Dutrillaux, B., Carpentier, S., Prieur, M., and Lejeune, J. (1972). *Ann. Genet.* **15,** 271–274.

Bass, H. N., Crandall, B. F., and Marey, S. M. (1973a). *J. Pediatr.* **83,** 1034–1038.

Bass, H. N., Crandall, B. F., and Sparkes, R. S. (1973b). *Ann. Genet.* **16,** 189–192.

Berg, J. M., Smith, G. F., Ridler, M. A. C., McCreary, B. D., Faunch, J. A., Farnham, F. N., and Allen, M. L. (1967). *J. Med. Genet.* **4,** 184–189.

Berger, R. (1973). *Nouv. Presse Med.* **5,** 3121.
Berger, R., Derre, J., and Ortiz, M. A. (1974). *Nouv. Presse Med.* **3,** 1801–1804.
Bigger, T. R. L., and Savage, J. R. K. (1975). *Cytogenet. Cell Genet.* **15,** 112–121.
Bloom, A. D., Schmickel, R., Barr, M., and Burdi, A. R. (1974). *J. Pediatr.* **84,** 732–733.
Bobrow, M., Pearson, P. L., Pike, M. C., and El-Alfi, O. S. (1971). *Cytogenetics* **10,** 190–198.
Bobrow, M., Madan, K., and Pearson, P. L. (1972). *Nature (London) New Biol.* **238,** 122–124.
Borgaonkar, D. S. (1975). "Chromosomal Variations in Man. A Catalog of Chromosomal Variants and Anomalies." Johns Hopkins Press, Baltimore, Maryland.
Borganokar, D. S., and Hollander, D. H. (1971). *Nature (London)* **230,** 52.
Borganokar, D. S., Bias, W. B., Chase, G. A., Sadasivan, G., Herr, H. M., Golomb, H. M., Bahr, G. F., and Kunkel, L. M. (1973). *Clin. Genet.* **4,** 53–57.
Borganokar, D. S., and Hollander, D. H. (1971). *Nature (London)* **230,** 52.
Bott, C. E., Sekhon, G. S., and Lubs, H. A. (1975). *Am. J. Hum. Genet.* **27,** 20A.
Boué, J. G., Boué, A., Deluchat, C., Perraudin, N., and Yvert, F. (1975). *J. Med. Genet.* **12,** 165–168.
Bühler, E. M., Mèhes, K., Müller, H., and Stalder, G. R. (1972). *Humangenetik* **15,** 150–162.
Bühler, E. M., Tsuchimoto, T., Jurik, L. P., and Stalder, G. R. (1975). *Humangenetik* **26,** 329–333.
Calderon, D., and Schnedl, W. (1973). *Humangenetik* **18,** 63–70.
Cantu, J.-M., Salamanca, F., Buentello, L., Carnevale, A., and Armendarez, S. (1975). *Ann. Genet.* **18,** 5–11.
Carpentier, S., Rethoré, M. O., and Lejeune, J. (1972). *Ann. Genet.* **15,** 283–286.
Caspersson, T., and Zech, L., eds. (1973). "Chromosome Identification." Academic Press, New York.
Caspersson, T., Farber, S., Foley, G. E., Kudynowski, J., Modest, E. J., Simonsson, E., Wagh, E., and Zech, L. (1968). *Exp. Cell Res.* **49,** 219–222.
Caspersson, T., Zech, L., Johansson, C., and Modest, E. J. (1970a). *Chromosoma* **30,** 215–227.
Caspersson, T., Gahrton, G., Lindsten, J., and Zech, L. (1970b). *Exp. Cell Res.* **63,** 238–239.
Caspersson, T., Lindsten, J., Zech, L., Buckton, K. E., and Price, W. H. (1972). *J. Med. Genet.* **9,** 1–7.
Castoldi, G. L., Brusovin, G. D., Scapoli, G. L., and Spanedda, R. (1972). *Acta Genet. Med. Gemellol.* **21,** 319–326.
Centerwall, W. R., Mayeski, C. A., and Cha, C. C. (1975). *Humangenetik* **29,** 91–98.
Chaganti, R. S. K., Schonberg, S., and German, J. (1974). *Proc. Natl. Acad. Sci. U.S.A.* **71,** 4508–4512.
Chen, T. R., and Ruddle, F. H. (1971). *Chromosoma* **34,** 51–72.
Chicago Conference. (1966). Standardization in Human Cytogenetics. *Birth Defects, Orig. Artic. Ser.* **2,** No. 2, 1–21.
Chiyo, H., Kuroki, Y., Matsui, I., Yanagida, K., and Nakagome, Y. (1975a). *Humangenetik* **30,** 63–67.
Chiyo, H., Nakagome, Y., Matsui, I., Kuroki, Y., Kobayashi, H., and Ohno, K. (1975b). *Clin. Genet.* **7,** 328–333.
Chudley, A. E., Bauder, F., Ray, M., McAlpine, P. J., Pena, S. D. J., and Hamerton, J. L. (1974). *J. Med. Genet.* **11,** 353–363.
Cohen, M. M., Shaw, M. W., and MacCluer, J. W. (1966). *Cytogenetics* **5,** 34–52.
Cohen, M. M., Ornoy, A., Rosenmann, A., and Kohn, G. (1975). *Ann. Genet.* **18,** 99–103.

Cooksley, W. G. E., Firouz-Abadi, A., and Wallace, D. C. (1973). *Med. J. Aust.* **2,** 178–180.
Court-Brown, W. M. (1967). *In* "Human Population Cytogenetics" (A. Neuberger and E. L. Tatum, eds.), pp. 55–72. North-Holland Publ., Amsterdam.
Couturier, J., Dutrillaux, B., and Lejeune, J. (1973). *C. R. Hebd. Seances Acad. Sci., Ser. D.* **279,** 339–342.
Craig-Holmes, A. P., Moore, F. B., and Shaw, M. W. (1973). *Am. J. Hum. Genet.* **25,** 181–192.
Craig-Holmes, A. P., Moore, F. B., and Shaw, M. W. (1975). *Am. J. Hum. Genet.* **27,** 178–189.
Crandall, B. F., Muller, H. M., and Bass, H. (1973). *Am. J. Ment. Defic.* **77,** 571–578.
Creasy, M. R., Crolla, J. A., and Daker, M. G. (1974). *Humangenetik* **24,** 303–308.
Crossen, P. E. (1975). *Clin. Genet.* **8,** 218–222.
Crossen, P. E., Pathak, S., and Arrighi, F. E. (1975). *Chromosoma* **52,** 339–347.
Curtis, D. J. (1973). *Humangenetik* **18,** 273–277.
Dallapicola, B., and Ricci, N. (1975). *Humangenetik* **26,** 251–255.
De Cicco, F., Steele, M. W., Pan, S., and Park, S. C. (1973). *J. Pediatr.* **83,** 836–838.
de Grouchy, J., and Turleau, C. (1974). *Humangenetik* **24,** 197–200.
de la Chapelle, A., Schröder, J., Selander, R.-K., and Stenstrand, K. (1973a). *Chromosoma* **42,** 365–382.
de la Chapelle, A., Koivisto, M., and Schröder, J. (1973b). *J. Med. Genet.* **10,** 384–389.
Donahue, R. P., Bias, W. B., Renwick, J. H., and McKusick, V. A. (1968). *Proc. Natl. Acad. Sci. U.S.A.* **61,** 949–955.
Drets, M. E., and Shaw, M. W. (1971). *Proc. Natl. Acad. Sci. U.S.A.* **68,** 2073–2077.
Dutrillaux, B. (1973). *Chromosoma* **41,** 395–402.
Dutrillaux, B. (1975). *Chromosoma* **52,** 261–273.
Dutrillaux, B. (1977). *In* "Molecular Structure of Human Chromosomes" (J. Yunis, ed.), Chromosomes in Biology and Medicine Ser., Academic Press, New York.
Dutrillaux, B., and Lejeune, J. (1971). *C. R. Hebd. Seances Acad. Sci.* **272,** 2638–2640.
Dutrillaux, B., Jonasson, J., Laurén, K., Lejeune, J., Lindsten, J., Petersen, G. B., and Saldaña-García, P. (1973). *Ann. Genet.* **16,** 11–16.
Dutrillaux, B., Laurent, C., Forabosco, A., Noel, B., Suerinc, E., Biedmont, M.-C., and Cotton, J.-B. (1975). *Ann. Genet.* **18,** 21–27.
Edwards, J. H., Harnden, D., Cameron, A., Crosse, V., and Wolff, O. (1960). *Lancet 1,* 787–789.
Eiberg, H. (1973). *Clin. Genet.* **4,** 556–562.
Eiberg, H. (1974a). *Lancet* **2,** 836–837.
Eiberg, H. (1974b). *Nature (London)* **248,** 55.
Engel, E., McKee, L. C., Flexner, J. M., and McGee, B. J. (1975). *Ann. Genet.* **18,** 56–60.
Epstein, M. A., and Achong, B. E. (1973). *Annu. Rev. Microbiol.* **27,** 413–436.
Escobar, J. I., and Yunis, J. J. (1974). *Am. J. Dis. Child.* **128,** 221–222.
Escobar, J. I., Sanchez, O., and Yunis, J. J. (1974). *Am. J. Dis. Child.* **128,** 217–220.
Evans, H. J., Buckland, R. A., and Pardue, M. L. (1974). *Chromosoma* **48,** 405–426.
Faed, M. J. W., Whyte, R., Paterson, C. R., McCathie, M., and Robertson, J. (1972). *J. Med. Genet.* **9,** 102–105.
Falk, R. E., Carrel, R. E., Valente, M., Crandall, B. F., and Sparkes, R. (1973). *Am. J. Ment. Defic.* **77,** 383–388.
Faust, J., Vogel, W., and Löning, B. (1974). *Clin. Genet.* **6,** 90–97.
Fawcett, W. A., McCord, W. K., and Francke, U. (1975). *Birth Defects, Orig. Artic. Ser.* **11,** No. 5, 223.
Feingold, M., and Atkins, L. (1973). *J. Med. Genet.* **10,** 184–187.

Ferguson-Smith, M. A., and Handmaker, S. D. (1961). *Lancet* **1,** 638–640.

Ferguson-Smith, M. A., Ferguson-Smith, M. E., Ellis, P. M., and Dickson, M. (1962). *Cytogenetics* **1,** 325–343.

Ferguson-Smith, M. A., Newman, B. F., Ellis, P. M., and Thomson, D. M. G. (1973). *Nature (London), New Biol.* **243,** 271–273.

Finley, W. H., Garrett, J. H., and Finley, S. C. (1975). *Am. J. Hum. Genet.* **27,** 35A.

Fleischman, E. W., and Prigogina, E. L. (1975). *Humangenetik* **26,** 335–342.

Foerster, W., Medau, H. J., and Löffler, H. (1974). *Klin. Wochenschr.* **42,** 123–126.

Forabosco, A., Dutrillaux, B., Toni, G., Tamborino, G., and Cavazzuti, G. (1973). *Ann. Genet.* **16,** 255–258.

Ford, J. H., Pittman, S. M., and Gunz, F. W. (1974). *Br. Med. J.* **4,** 227–228.

Francke, U. (1972). *Am. J. Hum. Genet.* **24,** 189–213.

Francke, U., Benirschke, K., and Jones, O. W. (1975a). *Humangenetik* **29,** 243–250.

Francke, U., Mahan, G. M., Dixson, K., and Jones, O. W. (1975b). *Birth Defects, Orig. Artic. Ser.* **11,** No. 5, 207.

Friedrich, U. (1975). *Humangenetik* **26,** 161–165.

Friedrich, U., and Nielsen, J. (1974). *Humangenetik* **21,** 127–132.

Fryns, J. P., Verresen, H., van der Berghe, H., and Van Kerchkvoorde, J. (1974a). *Humangenetik* **24,** 241–246.

Fryns, J. P., van der Berghe, H., and Van Herck, G. (1974b). *Humangenetik* **24,** 247–252.

Fujimoto, A., Towner, J. W., Ebbin, A. J., Kahlstrom, E. J., and Wilson, M. G. (1974). *J. Med. Genet.* **11,** 287–291.

Fujimoto, A., Wilson, M. G., and Towner, J. W. (1975). *Humangenetik* **27,** 67–73.

Funaki, K., Matsui, S., and Sasaki, M. (1975). *Chromosoma* **49,** 357–370.

Gahrton, G., Lindsten, J., and Zech, L. (1973). *Exp. Cell Res.* **79,** 246–247.

Gahrton, G., Lindsten, J., and Zech, L. (1974a). *Blood* **43,** 837–840.

Gahrton, G., Lindsten, J., and Zech, L. (1974b). *Acta Med. Scand.* **196,** 355–360.

Gahrton, G., Zech, L., and Lindsten, J. (1974c). *Exp. Cell Res.* **86,** 214–216.

Galloway, S. M., and Evans, H. J. (1975). *Cytogenet. Cell Genet.* **15,** 17–29.

Ganner, E., and Evans, H. J. (1971). *Chromosoma* **35,** 326–341.

Genest, P. (1973). *Ann. Genet.* **16,** 35–38.

Geraedts, J., and Pearson, P. (1973). *Humangenetik* **20,** 171–173.

German, J., and Chaganti, R. S. K. (1973). *Science* **182,** 1261–1262.

Gilgenkrantz, S., Mauuary, G., Dutrillaux, B., and Masocco, G. (1975). *Humangenetik* **26,** 25–34.

Goh, K., Swisher, S. N., and Herman, E. C. (1967). *Arch. Intern. Med.* **120,** 214–219.

Goodpasture, C., and Bloom, S. E. (1975). *Chromosoma* **53,** 37–50.

Goodpasture, C., Bloom, S. E., Hsu, T. C., and Arrighi, F. E. (1976). *Am. J. Hum. Genet.* **28,** 559–566.

Gouw, W. L., ten Kate, L. P., and Anders, G. J. P. A. (1973). *Humangenetik* **19,** 123–126.

Grace, E., Sutherland, G. R., Stark, G. D., and Bain, A. D. (1973). *Ann. Genet.* **16,** 51–54.

Gray, J. E. Syrett, J. E., Ritchie, K. M., and Elliot, W. D. (1972). *Lancet* **2,** 92–93.

Gripenberg, U., Elfving, J., and Gripenberg, L. (1972). *J. Med. Genet.* **9,** 110–115.

Halloran, K. H., Roy Breg, W., and Mahoney, M. J. (1974). *J. Med. Genet.* **11,** 299–303.

Hamerton, J. L. (1971). *Hum. Cytogenet.* **1,** 71–100.

Hayashi, K., and Schmid, W. (1975a). *Humangenetik* **29,** 201–206.

Hayashi, K., and Schmid, W. (1975b). *Humangenetik* **30,** 135–141.

Hayata, I., Kakati, S., and Sandberg, A. A. (1973). *Lancet* **2,** 1385.

Henderson, A. S., Warburton, D., and Atwood, K. C. (1972). *Proc. Natl. Acad. Sci. U.S.A.* **69,** 3394–3398.

Henle, W., and Henle, G. (1973). *Cancer Res.* **33,** 1419–1423.
Hirschhorn, K., Lucas, M., and Wallace, I. (1973). *Ann. Hum. Genet.* **36,** 375–379.
Holboth, N., Jacobsen, P., and Mikkelsen, M. (1974). *J. Med. Genet.* **11,** 299–303.
Hossfeld, D. K. (1974). *Humangenetik* **23,** 111–118.
Howard, R. O., Roy Breg, W., Albert, D. M., and Lesser, R. L. (1974). *Arch. Opthalmol.* **92,** 490–493.
Howell, W. M., and Denton, T. E. (1974). *Experientia* **30,** 1364–1366.
Howell, W. M., Denton, T. E., and Diamond, J. R. (1975). *Experientia* **31,** 260–262.
Hsu, T. C., Pathak, S., and Shafer, D. A. (1973). *Exp. Cell Res.* **79,** 484–487.
Hsu, T. C., Spirito, S. E., and Pardue, M. L. (1975). *Chromosoma* **43,** 25–36.
Hungerford, D. A., Donnelly, A. J., Nowell, P. C., and Beck, S. (1959). *Am. J. Hum. Genet.* **11,** 215–236.
Ikeuchi, T., Sonta, S., Sasaki, M., Hujita, M., and Tsunematsu, K. (1974). *Humangenetik* **21,** 309–314.
Ikushima, T., and Wolff, S. (1974). *Exp. Cell Res.* **87,** 15–19.
Jacobs, P. A. (1972). *Clin. Genet.* **3,** 226–248.
Jacobs, P. A., Buckton, K. E., Cunningham, C., and Newton, M. (1974). *J. Med. Genet.* **11,** 50–64.
Jacobsen, P., Hauge, M., Henningsen, K., Holboth, N., Mikkelsen, M., and Philip, J. (1973). *Hum. Hered.* **23,** 568–585.
Jalbert, P., Jalbert, H., Sele, B., Mouriquand, C., Malka, J., Boucharlat, J., and Pison, H. (1975). *J. Med. Genet.* **12,** 418–423.
Jarvis, J., Ball, G., Rickinson, A. B., and Epstein, M. D. (1974). *Int. J. Cancer* **14,** 716–721.
Jonasson, J., Gahrton, G., Lindsten, J., Simonsson-Lindelman, C., and Zech, L. (1974). *Blood* **43,** 557–563.
Jones, K. W., (1977). *In* "Molecular Structure of Human Chromosomes" (J. Yunis, ed.), Chromosomes in Biology and Medicine Ser., Academic Press, New York.
Kajii, T., Ohama, K., Niikawa, N., Ferrier, A., and Auirachan, S. (1973). *Am. J. Hum. Genet.* **25,** 539–547.
Kajii, T., Meylan, J., and Mikamo, K. (1974). *Cytogenet. Cell Genet.* **12,** 426–436.
Kancko, Y., Ikeuchi, T., Sasaki, Y., Satake, Y., and Kuwajima, S. (1975). *Humangenetik* **29,** 1–7.
Kato, H., and Moriwaki, K. (1972). *Chromosoma* **38,** 105–120.
Kleiman, L., and Huang, R. C. (1971). *J. Mol. Biol.* **55,** 503–521.
Koulischer, L., and Lambrotto, C. (1974). *Ann. Genet.* **17,** 189–192.
Krmpotic, E., Rosenthal, I. M., Szego, K., and Bocian, M. (1971). *Ann. Genet.* **14,** 291–299.
Ladda, R., Atkins, L., Littlefield, J., Neurath, P., and Marimuthu, K. (1974). *Science* **185,** 784–787.
Lange, M., and Alfi, O. S. (1976). *Ann. Genet.* **19,** 17–21.
Latt, S. A. (1973). *Proc. Natl. Acad. Sci. U.S.A.* **70,** 3395–3399.
Latt, S. A., and Gerald, P. S. (1973). *Exp. Cell Res.* **81,** 401–406.
Latta, E., and Hoo, J. J. (1974). *Humangenetik* **23,** 213–217.
Laurent, C., Bovier-Lapierre, M., and Dutrillaux, B. (1973a). *Humangenetik* **18,** 321–327.
Laurent, C., Dutrillaux, B., Biemont, M.-Cl., Genood, J., and Bethenod, M. (1973b). *Ann. Genet.* **16,** 281–284.
Laurent, C., Biemont, M.-Cl., Midenet, M., Couturier, P., and Dutrillaux, B. (1974). *Lyon Med.* **232,** 609–615.
Laurent, C., Biemont, M.-Cl., and Dutrillaux, B. (1975). *Humangenetik* **26,** 35–46.
Lauritsen, J. G., Jonasson, J., Therkelsen, A. J., Lass, F., Lindstein, J., and Petersen, G. B. (1972). *Hereditas* **71,** 160–163.

Lawler, S. D., Secker Walker, L. M., Summersgill, B. M., Reeves, B. R., Lewis, J., Kay, H. E. M., and Hardisty, R. M. (1975). *Scand. J. Haematol.* **14,** 312–320.
Lee, C. L. Y., Welch, J. P., and Lee, S. H. S. (1973). *Nature (London) New Biol.* **241,** 142–143.
Lejeune, J. (1959). *Ann. Genet.* **1,** 41–49.
Lejeune, J., and Rethoré, M. O. (1973). *In* "Chromosome Identification" (T. Caspersson and L. Zech, eds.), p. 214. Academic Press, New York.
Lejeune, J., Lafourcade, J., Berger, R., Vialatte, J., Boeswillwald, M., Seringe, P., and Turpin, R. (1963). *C. R. Hebd. Seances Acad. Sci.* **257,** 3098–3102.
Lele, K. P., Penrose, L. S., and Stallard, H. B. (1963). *Ann. Hum. Genet.* **27,** 171–174.
Levan, G., and Mitelman, F. (1975). *Hereditas* **79,** 156–160.
Lewandowski, R. C., and Yunis, J. J. (1975). *Am. J. Dis. Child.* **129,** 515–529.
Lewandowski, R. C., and Yunis, J. J. (1977). *Clin. Genet.* **11** (in press).
Lewandowski, R. C., Yunis, J. J., Lehrke, R., O'Leary, J., Sawiman, K. F., and Sanchez, O. (1976). *Am. J. Dis. Child.* **130,** 663–667.
Lindsley, D. L., Sandler, L., Baker, B. S., Carpenter, A. T. C., Denell, R. E., Hall, J. C., Jacobs, P. A., Miklos, G. L. G., Davis, B. K., Gethmann, R. C., Hardy, R. W., Hessler, A., Miller, S. M., Nogawa, A., Parry, D. M., and Gould-Somero, M. (1972). *Genetics* **71,** 157–184.
Lobb, D. S., Reeves, B. R., and Lawler, S. D. (1972). *Lancet* **1,** 849–850.
Lockhart, L. H., and Meyne, J. (1975). *Am. J. Hum. Genet.* **27,** 59A.
Lubs, H. A., and Ruddle, F. H. (1970). *In* "Human Population Cytogenetics" (P. A. Jacobs, W. H. Price, and P. Larv, eds.), p. 110. University Press, Edinburgh.
McCaw, B. K., Epstein, A. L., Kaplan, H. S., and Hecht, F. (1977). *Int. J. Cancer* (in press).
McKenzie, W. H., and Lubs, H. A. (1975). *Cytogenet. Cell Genet.* **14,** 97–115.
Madan, K., and Bobrow, M. (1974). *Ann. Genet.* **17,** 81–86.
Magnelli, N. C., and Therman, E. (1975). *J. Med. Genet.* **12,** 105–108.
Malpuech, G., Raynaud, E. J., Belin, J., Godeneche, P., and de Grouchy, J. (1971). *Ann. Genet.* **14,** 213–218.
Manolov, G., and Manolova, Y. (1972). *Nature (London)* **237,** 33–34.
Mark, J. (1973). *Hereditas* **75,** 213–220.
Mark, J. Levan, G., and Mitelman, F. (1972). *Hereditas* **71,** 163–168.
Matsui, S., and Sasaki, M. (1973). *Nature (London)* **246,** 148–150.
Meisner, L. F., and Inhorn, S. (1972). *J. Med. Genet.* **9,** 373–377.
Mikelsaar, A. V. N., Tüür, S. J., and Känosaar, M. E. (1973). *Humangenetik* **20,** 89–101.
Mikkelsen, M. (1974). *Hum. Hered.* **24,** 160–166.
Mikkelsen, M., and Dyggve, H. (1973). *Humangenetik* **18,** 195–202.
Mitelman, F., and Brandt, L. (1974). *Scand. J. Haematol.* **13,** 321–330.
Mitelman, F., Brandt, L., and Levan, G. (1973). *Lancet* **2,** 972.
Mitelman, F., Panai, A., and Brandt, L. (1975). *Scand. J. Haematol.* **14,** 308–312.
Mittwoch, U. (1974). *In* "Human Chromosome Methodology" (J. J. Yunis, ed.), 2nd ed., pp. 73–93. Academic Press, New York.
Moorhead, P. S., Nowell, P. C., Mellman, W. J., Battips, D. M., and Hungerford, D. A. (1960). *Exp. Cell Res.* **20,** 613–616.
Muldal, S., Enoch, B. A., Ahmed, A., and Harris, R. (1973). *Clin. Genet.* **4,** 480–489.
Muller, W., and Crothers, D. M. (1968). *J. Mol. Biol.* **35,** 251–290.
Mutton, D. E., and Daker, M. G. (1973). *Nature (London) New Biol.* **241,** 80.
Nakagome, Y., and Chiyo, H. (1976). *Am. J. Hum. Genet.* **28,** 31–41.

Nakagome, Y., Iinoma, K., and Matsui, I. (1973). *J. Med. Genet.* **10,** 174–176.
Niebuhr, E. (1971). *J. Ment. Defic. Res.* **15,** 277–291.
Nielsen, J., and Friedrich, U. (1972). *Clin. Genet.* **3,** 281–285.
Nielsen, J., Friedrich, U., Hreidarsson, A. B., and Zeuthen, E. (1974a). *Humangenetik* **21,** 211–216.
Nielsen, J., Friedrich, U., Hreidarsson, A. B., and Zeuthen, E. (1974b). *Clin. Genet.* **5,** 316–321.
Norwood, T. H., and Hoehn, H. (1974). *Humangenetik* **25,** 79–82.
Nowell, P. C., and Hungerford, D. A. (1960). *J. Natl. Cancer Inst.* **25,** 85–108.
O'Grady, R. B., Rothstein, T. B., and Romano, P. E. (1974). *Am. J. Opthalmol.* **77,** 40–45.
Opitz, J. M., and Patau, K. (1975). *Birth Defects, Orig. Artic. Ser.* **11,** No. 5, 191.
O'Riordan, M. L., Robinson, J. A., Buckton, K. E., and Evans, H. J. (1971). *Nature (London)* **230,** 167–168.
Orye, E., and Craen, M. (1975). *Humangenetik* **28,** 335–342.
Orye, E., Delbeke, M. J., and Vandenabeele, B. (1971). *Lancet* **2,** 1376.
Orye, E., Delbeke, M. H., and Vandenabeele, B. (1974). *Clin. Genet.* **5,** 457–464.
Orye, E., Verhaaren, H., Samüel, K., and van Mele, B. (1975). *Humangenetik* **28,** 1–8.
Ostovics, M., and Kiss, P. (1975). *Clin. Genet.* **8,** 112–116.
Palmer, C. G., Pland, C., Reed, T., and Kojetin, J. (1976). *Humangenetik* (in press).
Pardo, D., Luciani, J. M., and Stahl, A. (1975). *Ann. Genet.* **18,** 105–109.
Paris Conference, 1971, (1972). *Birth Defects, Orig. Artic. Ser.* **3,** No. 7.
Park, J., and Antley, R. M. (1975). *Am. J. Hum. Genet.* **27,** 70A.
Pasquali, F. (1973). *Ann. Genet.* **16,** 47–50.
Patau, K. (1965). *In* "Human Chromosome Methodology" (J. J. Yunis, ed.), 1st ed., pp. 155–186. Academic Press, New York.
Patau, K., Smith, D., Therman, E. M., Inhorn, S. L., and Wagner, H. P. (1960). *Lancet* **1,** 790–793.
Patil, S. R., and Lubs, H. A. (1975). *Am. J. Hum. Genet.* **16,** 66A.
Patil, S. R., Merrick, S., and Lubs, H. A. (1971). *Science* **173,** 821–822.
Pawlowitzki, I. H., and Pearson, P. L. (1972). *Humangenetik* **16,** 119–122.
Pearson, P. (1972). *J. Med. Genet.* **9,** 264–275.
Pearson, P. (1973). *In* "Chromosome Identification" (T. Caspersson and L. Zech, eds.), p. 145. Academic Press, New York.
Pearson, P. L., and Bobrow, M. (1970). *J. Reprod. Fertil.* **22,** 177–179.
Pearson, P. L., Bobrow, M., and Vosa, C. G. (1970). *Nature (London)* **226,** 78–80.
Pearson, P. L., Geraedts, J. P. M., and van der Linden, A. G. J. M. (1973). *In* "Symposia Medica Hoechst 6, Modern Aspects of Cytogenetics: Constitutive Heterochromatin in Man." (R. A. Pfeiffer, ed.), pp. 201–213. F. K. Schattauer, Verlag, Stuttgart, Germany.
Perry, P., and Wolff, S. (1974). *Nature (London)* **251,** 156–158.
Petit, P., Alexander, M., and Findu, P. (1973). *Lancet* **2,** 1326–1327.
Pfeiffer, R. A., Buttinghaus, K., and Struck, H. (1973). *Humangenetik* **20,** 187–189.
Philip, P. (1975). *Scand. J. Haematol.* **14,** 140–147.
Prieur, M., Dutrillaux, B., and Lejeune, J. (1973). *Ann. Genet.* **16,** 39–46.
Prigogina, E. L., and Fleischman, E. W. (1975a). *Humangenetik* **30,** 113–119.
Prigogina, E. L., and Fleischman, E. W. (1975b). *Humangenetik* **30,** 109–112.
Raoul, O., Rethoré, M. O., Dutrillaux, B., Michon, L., and Lejeune, J. (1975). *Ann. Genet.* **18,** 35–39.
Reeves, B. R. (1973). *Humangenetik* **20,** 231–250.
Reich, E., and Goldberg, I. H. (1964). *Prog. Nucleic Acid Res. Mol. Biol.* **3,** 184.

Rethoré, M. O., Lejeune, J., Carpentier, S., Prieur, M., Dutrillaux, B., Seringe, P., and Job, J.-C. (1972). *Ann. Genet.* **15,** 159–165.
Rethoré, M. O., Hoehn, H., Rott, H. D., Couturier, J. Dutrillaux, B., and Lejeune, J. (1973a). *Humangenetik* **18,** 129–138.
Rethoré, M. O., Dutrillaux, B., and Lejeune, J. (1973b). *Ann. Genet.* **16,** 271–275.
Rethoré, M. O., Dutrillaux, B., Giovannelli, G., Forabosco, A., Dallapicola, B., and Lejeune, J. (1974). *Ann. Genet.* **17,** 125–128.
Rethoré, M. O., Kaplan, J.-C., Junien, C., Cruveiller, J., Dutrillaux, B., Aurias, A., Carpentier, S., Lafourcade, J., and Lejeune, J. (1975). *Ann. Genet.* **18,** 81–87.
Robertson, K. P., Thurmon, T. F., and Tracy, M. C. (1975). *Birth Defects, Orig. Artic Ser.* **11,** No. 5, 267.
Rodman, T. C. (1974). *Science* **184,** 171–173.
Rosenthal, I. M., Beligere, N., Thompson, F., Pruzansky, S., and Reinglass, H. (1974). *Am. J. Hum. Genet.* **26,** 73A.
Rotterdam Conference, 1974 (1975). Second International Workshop on Human Gene Mapping. *Birth Defects, Orig. Artic. Ser.,* **11,** No. 3.
Rowley, J. D. (1973a). *Lancet* **2,** 1385–1386.
Rowley, J. D. (1973b). *Ann. Genet.* **16,** 109–112.
Rowley, J. D. (1973c). *Nature (London)* **243,** 290–293.
Rowley, J. D. (1974a). *In* "Human Chromosome Methodology" (J. J. Yunis, ed.), 2nd ed., pp. 17–46. Academic Press, New York.
Rowley, J. D. (1974b). *Lancet* **2,** 835.
Rowley, J. D. (1975). *Proc. Natl. Acad. Sci. U.S.A.,* **72,** 152–156.
Sanchez, O., and Yunis, J. J. (1974). *Humangenetik* **23,** 297–303.
Sanchez, O., Escobar, J. I., and Yunis, J. J. (1973). *Lancet* **2,** 269.
Sanchez, O., Yunis, J. J., and Escobar, J. I. (1974). *Humangenetik* **22,** 59–65.
Sanchez, O., Mamunes, P., and Yunis, J. J. (1977). *J. Med. Genet.* (in press).
Sandberg, A. A., and Hossfeld, D. K. (1970). *Annu. Rev. Med.* **21,** 379–408.
Sanger, R., Alfi, O., and Donnell, G. (1974). *Am. J. Hum. Genet.* **26,** 75A.
Say, B., Barber, N., Bobrow, M., Jones, K., and Coldwell, J. G. (1976). *J. Pediatr.* **88,** 447–450.
Schinzel, A., Schmid, W., and Mürset, G. (1974). *Humangenetik* **22,** 287–298.
Schmikel, R. (1975). *Birth Defects, Orig. Artic. Ser.* **11,** No. 5, 229.
Schnedl, W. (1973). *In* "Chromosome Identification" (T. Caspersson and L. Zech, eds.), p. 34. Academic Press, New York.
Schrott, H. G. Sakaguchi, S., Francke, U., Luzzatti, L., and Failkow, P. J. (1974). *J. Med. Genet.* **11,** 201–205.
Schwinger, E., and Wild, P. (1974). *Humangenetik* **22,** 67–69.
Seabright, M. (1971). *Lancet* **2,** 971–972.
Seabright, M. (1972). *Chromosoma* **36,** 204–210.
Sehested, J. (1974). *Humangenetik* **21,** 55–58.
Sekhon, G. S., and Sly, W. S. (1975). *Am. J. Hum. Genet.* **27,** 79A.
Shapiro, L. R., and Warburton, D. (1972). *Lancet* **2,** 712–713.
Shiraishi, Y., and Yosida, T. H. (1972). *Chromosoma* **37,** 75–83.
Shokeir, M. H. K., Ying, K. L., and Pabello, P. (1973). *Clin. Genet.* **4,** 360–368.
Shokeir, M. H. K., Ray, M., Hamerton, J. L., Bauder, F., and O'Brien, H. (1975). *J. Med. Genet.* **12,** 99–103.
Skovby, F. (1975). *Clin. Genet.* **7,** 21–28.
Soudek, D., and Laraya, P. (1974). *Clin. Genet.* **6,** 225–229.

Soudek, D., Langmuir, V., and Steward, D. J. (1973). *Humangenetik* **18**, 285–290.
Sperling, K., Wegner, R. d., Riehm, H., and Obe, G. (1975). *Humangenetik* **27**, 227–230.
Steele, M. W., Pan, S., Mickell, J., and Senders, V. (1974). *J. Pediatr.* **85**, 827–829.
Stern, L. M., and Murch, A. R. (1975). *J. Med. Genet.* **12**, 305–307.
Stoll, C., and Halb, A. (1974). *Pediatrie* **29**, 725–729.
Stoll, C., Messer, J., and Vors, J. (1974). *Ann. Genet.* **17**, 193–196.
Sumner, A. T., Robinson, J. A., and Evans, H. J. (1971a). *Nature (London) New Biol.* **229**, 231–233.
Sumner, A. T., Evans, H. J., and Buckland, R. A. (1971b). *Nature (London) New Biol.* **232**, 31–32.
Surana, R. B., McKendry, J. B., Bailey, J. D., and Conen, P. E. (1973). *Am. J. Hum. Genet.* **25**, 77A.
Taillemite, J.-L., Baheux-Morlier, G., and Roux, Ch. (1975). *Ann. Genet.* **18**, 61–63.
Taylor, A. I. (1970). *Humangenetik* **10**, 209–217.
Tenconi, R., Baccicheti, C., Anglani, F., Dellegrino, P.-A., Kaplan, J.-C., and Junien, C. (1975). *Ann. Genet.* **18**, 95–98.
Tjio, J. H., and Levan, A. (1956). *Hereditas* **42**, 1–6.
Turleau, C., Roubin, M., Chavin-Colin, F., Satge, M., and de Grouchy, J. (1974). *Ann. Genet.* **17**, 291–294.
Turleau, C., de Grouchy, J., Chavin-Colin, F., Roubin, M., Brissaud, P. E., Repessé, G., Safar, A., and Borniche, P. (1975a). *Humangenetik* **29**, 233–241.
Turleau, C., de Grouchy, J., Bocquentin, F., Roubin, M., and Chavin-Colin, F. (1975b). *Ann. Genet.* **18**, 41–44.
Tüür, S., Känosaar, M., and Mikelsaar, A. V. (1974). *Humangenetik* **24**, 217–220.
Uchida, I. A., and Lin, C. C. (1973). *J. Pediatr.* **82**, 269–272.
Uchida, I. A., and Lin, C. C. (1974). *In* "Human Chromosome Methodology" (J. J. Yunis, ed.), 2nd ed., pp. 47–58. Academic Press, New York.
van der Berghe, H., van Eygen, M., Fryns, J. P., Tanghe, W., and Verresen, H. (1973). *Humangenetik* **18**, 225–230.
Van Dyke, D. L., Palmer, C. G., and Nance, W. E. (1974). *Am. J. Hum. Genet.* **26**, 88A.
Van Kempen, C. (1975). *J. Med. Genet.* **12**, 204–207.
Vogel, W., Siebers, J.-W., and Reinwein, H. (1973). *Ann. Genet.* **16**, 277–280.
Wahlström, J. (1974). *Hereditas* **78**, 251–254.
Warren, R. J., Panizales, E. G., and Cantwell, R. J. (1975). *Birth Defects, Orig. Artic. Ser.* **11**, No. 5, 177.
Watson, E. J., and Gordon, R. R. (1974). *J. Med. Genet.* **11**, 400–402.
Whang-Peng, J., Canellos, G. P., Carbone, P. P., and Tjio, J. H. (1968). *Blood* **32**, 755–766.
Williams, J. D., Summitt, R. L., Martens, P. R., and Kimbrell, R. A. (1975). *Am. J. Hum. Genet.* **27**, 478–485.
Wilson, M. G., Melnyk, J., and Towner, J. W. (1969). *J. Med. Genet.* **6**, 322–327.
Wilson, M. G., Towner, J. W., and Fujimoto, A. (1973). *Am. J. Hum. Genet.* **25**, 57–61.
Wurster-Hill, D. H., and Hoefnagel, D. (1975). *J. Ment. Defic. Res.* **19**, 145–150.
Wyandt, H. E., Hecht, F., Lovrien, E. W., and Stewart, R. E. (1971). *Cytogenetics* **10**, 413–416.
Wyandt, H. E., Vlietinck, R. F., Magenis, R. E., and Hecht, F. (1974). *Humangenetik* **23**, 119–130.
Yasmineh, W. G., and Yunis, J. J. (1969). *Biochem. Biophys. Res. Commun.* **35**, 779–782.
Yunis, J. J. (1964). *In* "Human Chromosome Methodology" J. J. Yunis, ed.), 1st ed., pp. 187–242. Academic Press, New York.

Yunis, J. J. (1976). *Science* **191,** 1268–1270.
Yunis, J. J., ed. (1977a). "Molecular Structure of Human Chromosomes." Chromosomes in Biology and Medicine Ser., Academic Press, New York.
Yunis, J. J. (1977b). *Am. J. Dis. Child.* (in press).
Yunis, J. J., and Chandler, M. E. (1977a). *Am. J. Path.* (in press).
Yunis, J. J., and Chandler, M. E. (1977b). In: "Progress in Clinical Pathology" (M. Stefanini and A. Hossaini, eds), Vol. **7,** Grune & Stratton, New York (in press).
Yunis, J. J., and Sanchez, O. (1973). *Chromosoma* **44,** 15–23.
Yunis, J. J., and Sanchez, O. (1974). *J. Pediatr.* **84,** 567–570.
Yunis, J. J., and Sanchez, O. (1975). *Humangenetik* **27,** 167–172.
Yunis, J. J., and Yasmineh, W. G. (1971). *Science* **174,** 1200–1209.
Yunis, J. J., and Yasmineh, W. G. (1972). *Adv. Cell Mol. Biol.* **2,** 1.
Yunis, J. J., Roldan, L., Yasmineh, W. G., and Lee, J. C. (1971). *Nature (London)* **231,**
Zakharov, A. F., and Egolina, N. A. (1972). *Chromosoma* **38,** 341–365.
Zakharov, A. F., Seleznev, J. V., Benjusch, V. A., Baranovskaya, L. I., and Demintseva, V. S. (1971). *Excerpta Med. Found. Int. Cong. Ser.* **233,** 193.
Zang, K. D., and Singer, H. (1967). *Nature (London)* **216,** 84–85.
Zankl, H., Seidel, H., and Zang, K. D. (1975). *Humangenetik* **27,** 119–128.
Zech, L. (1969). *Exp. Cell Res.* **58,** 463.
Zech, L. (1975). *Proc. Int. Cancer Cong., 11th, 1974* Abstract, p. 644 (cited by Levan and Mitelman, 1975).
Zech, L., Lindsten, J., Uden, A.-M., and Gahrton, G. (1975). *Scand. J. Haematol.* **15,** 251–155.

# 2

# Clinical Diagnosis and Nature of Chromosomal Abnormalities

DAVID W. SMITH

A chromosome abnormality does not generally imply a defect in the quality of genes but rather a *quantitative* abnormality in gene functions. Thus, the basic problem is a genetic imbalance, an aberration in gene dosage. Though the fundamental control of morphogenesis remains a mystery, the processes depend on both the quality and quantity of genes. When there is a genetic imbalance, the control of growth and morphogenesis is liable to be interfered with, most commonly giving rise to multiple problems in morphogenesis. Missing genetic material appears to be more deleterious than extra genes. The great majority of chromosomal abnormalities which result in a whole extra chromosome, or one missing, are early lethals. Thus Boué (1974; Boué *et al.*, 1971) has estimated that as many as one-half of fertilized ova may have a chromosomal abnormality, the majority being lethal in early gestation. She found about 60% chromosomal abnormalities among over 900 spontaneously aborted conceptuses. At the usual time of birth, about 1 in 200 babies have a gross chromosomal abnormality. Of these, about half are sex chromosome abnormalities, especially XXY and XYY, and the other half are autosomal trisomies, predominantly 21 trisomy, 18 trisomy and 13 trisomy. The impact of the sex chromosome abnormalities is generally not as severe as that of the autosomal abnormalities. However, the greater the imbalance, the more serious are the consequences. For example, XXX usually does not give rise to obvious physical abnormalities in the XXX female, whereas XXXX often gives rise to some growth and mental deficiency, and XXXXX usually causes serious growth and mental deficiency, plus multiple other defects.

With the advent of various chromosomal banding techniques, it has been possible to discriminate chromosomal abnormalities which were previously difficult to detect (Yunis, 1974). These include small extra pieces of chromosomal material, pieces missing, and rearrangements within chromosomes; predominantly the consequence of chromosomal breakage. A large number of clinical disorders have been recognized which are due to only a part of a chromosome being missing or extra. Basically, the nature of the problem is the same, namely genetic imbalance.

The type of disorders in development which result from genetic imbalance are predominantly problems in morphogenesis. There would appear to be an imbalance in the delicate control mechanisms which manage the timeliness and sequential nature of morphogenesis as well as the control over cell number in particular tissues. It has been rare to detect any succinct biochemical abnormality in a patient with a chromosomal abnormality. There are certain generalities which can be set forth about the effects of genetic imbalance in the human. These are as follows:

1. Most autosomal abnormalities adversely affect prenatal growth, both in terms of linear size and brain growth. Thus growth deficiency and mental deficiency are frequent and nonspecific features. This is generally due to hypoplasia, there being a diminished number of cells. The growth deficiency is frequently malproportionate and many of the "malformations" may simply represent dysharmonic growth deficiency of particular tissues. For example, 18 trisomy syndrome infants tend to have disproportionately small facies, short sternum and narrow pelvis (Smith, 1970). The low arch dermal ridge patterning on the fingertips in this syndrome could simply be secondary to hypoplasia of the fetal fingertip pads, resulting in low arch fingertip dermal ridge patterning. At a more speculative level, the ventricular septal defect in the 18 trisomy syndrome might simply be the consequence of growth deficiency in the cardiac septum with resultant failure of its closure.

Some anomalies which are frequent features in a variety of chromosomal disorders may also be a consequence of problems in the control systems responsible for synchronous and harmonious morphogenesis. These include: inner epicanthal folds, usually the result of redundant skin due to a low nasal bridge; short and inturned 5th finger, the result of undue hypoplasia of the middle phalanx of this finger; simian crease, the consequence of a mild difference in form of the hand resulting in the development of but a single flexural folding crease in the upper palm.

2. No individual feature has been shown to be pathognomonic for a particular chromosomal abnormality. Rather the total pattern of abnormalities tends to have diagnostic specificity. However, for each of the more common chromosomal abnormality disorders, there have been instances in which the clinical diagnosis was not validated by extensive chromosomal studies. Hence it is obviously possible to have close phenocopies of the chromosomal abnormality disorders, and it is therefore generally worthwhile to obtain chromosomal

studies for confirmation of the clinical diagnosis as well as for counseling purposes.

3. Certain anomalies besides prenatal onset growth deficiency and mental deficiency are somewhat more likely to occur in chromosomal abnormality conditions than in multiple defect disorders due to other modes of etiology. Among those which may be suggestive of a chromosome imbalance are redundant folds of skin in the posterior neck at birth, upslanting palpebral fissures, low set and malformed auricles, and the combination of simian crease plus distal axial triradius in the palms.

Most recognized chromosomal abnormalities give rise to multiple defects in morphogenesis: chromosomal studies are rarely merited in the otherwise normal individual with a single malformation.

Knowledge of the nature of a particular chromosomal abnormality disorder and its natural history may be of great value in the counseling of the family and the management for the patient. For example, XXY syndrome patients generally do not produce adequate testosterone, and most of the features of the Klinefelter syndrome are the consequence of hypogonadism. Early detection of the XXY syndrome allows for prospective testosterone therapy at the age of adolescence and thereafter, thus preventing the signs of hypogonadism (Caldwell and Smith, 1972). Children with the Down syndrome, who formerly received no special training during early childhood, are performing better at the age of 5 years following early special training. To what extent this will result in a significant difference in the adaption of the individual with Down syndrome in society remains to be determined. At the other end of the spectrum are those newborn babies with disorders such as 18 trisomy syndrome and 13 trisomy syndrome for which the defect of development of the brain is quite severe and the capacity for survival limited without medical intervention (Smith, 1970). For such problems, the author gives the parents the option of no medical intervention. The situation is explained as one in which the genetic imbalance—such as 18 trisomy—usually does not allow for survival and results in miscarriage. The parents are told "you must have a very good genetic background" and the mother is told "you must be very good in carrying a fetus, to have had a fetus with this problem survive to be born. Having survived to be born, the capacity for postnatal survival is quite limited. If we intervene medically and are successful in maintaining survival, the baby will be severely limited in brain function. We feel the kindest approach for all concerned is no medical intervention." We then ask the parents if this is their wish. Thus the spectrum extends from XXY individuals who can be helped to adapt into normal society, to Down syndrome children who are handicapped but can apparently function better with special training, to those severe problems for which the interpretation is toward that of a late miscarriage.

The capacity for early chromosomal studies on a developing fetus has changed the genetic counsel relative to future pregnancies into a positive situation of prevention. Most women who have had a seriously affected child would desire amniocentesis in future pregnancies. Ideally the evaluation may eventually become available for monitoring of all pregnancies, thus extending nature's role from the early lethality of most chromosomal abnormalities to the early lethality of almost all chromosomal abnormalities.

## REFERENCES

Boué, J. G. (1974). Chromosomal studies in more than 900 spontaneous abortuses. *Teratol. Soc. Meet., 1974.*

Boué, J. G., Boué, A., Lazar, P., and Guéguen, S. (1971). Sur les durées de développement et de retention de 716 zygotes, produits d'avortements spontanés precoces. *C. R. Hebd. Seances Acad. Sci., Ser. D* **272,** 2992.

Caldwell, P. D., and Smith, D. W. (1972). The XXY (Klinefelter's) syndrome in childhood: Detection and treatment. *J. Pediatr.* **80,** 250.

Smith, D. W. (1970). "Recognizable Patterns of Human Malformation." Saunders, Philadelphia, Pennsylvania.

Yunis, J. J., ed. (1974). "Human Chromosome Methodology," 2nd ed. Academic Press, New York.

# 3

# Classical Chromosome Disorders

ROBERT J. GORLIN

The conditions discussed in this chapter are with rare exception those initially described within the first decade of chromosome study employing standard chromosome techniques described by Tjio and Levan in 1956.

## I. 4p− SYNDROME (WOLF–HIRSCHHORN SYNDROME) (FIG. 1)

Wolf *et al.* (1965) and Hirschhorn *et al.* (1965) independently reported partial deletion of the short arm of one of the chromosomes 4. It is much less common than the 5p− syndrome. About 45 sporadic cases have been described to date (Johnson *et al.*, 1976). Translocation has rarely been demonstrated (Wilson *et al.*, 1970). Parental age has been somewhat advanced (Fryns *et al.*, 1973). A 4r syndrome has been described (Carter *et al.*, 1969).

Psychomotor and growth retardation are marked. Birth weight is usually about 2000 gm in spite of normal gestation time. Fetal activity is diminished, and most affected infants are hypotonic. Seizures occur in about 80% of patients (Leão *et al.*, 1967; Pfeiffer, 1968; Passarge *et al.*, 1970; Judge *et al.*, 1974).

The skull is uniformly microcephalic with frequent cranial asymmetry. Occasionally, midline scalp defects have been noted (Hirschhorn *et al.*, 1965; Wolf *et al.*, 1965; Miller *et al.*, 1970). Hemangioma on the brow is frequent. A prominent glabella and ocular hypertelorism are almost constant features. Divergent strabismus, eyelid ptosis, and antimongoloid obliquity of the palpebral fissures have been noted in about half the cases. Iris coloboma has been found in over 30% (Judge *et al.*, 1974).

The ears have narrow external canals, are low set and simplified in form. A preauricular dimple or sinus has been present in 50% (Leão *et al.*, 1967). The nose is misshapen or beaked with a broad base. The philtrum is short with a down-turned mouth. Cleft lip or, especially, cleft palate and micrognathia have been noted in most cases (Miller *et al.*, 1970; Taylor *et al.*, 1970; Arias *et al.*, 1970).

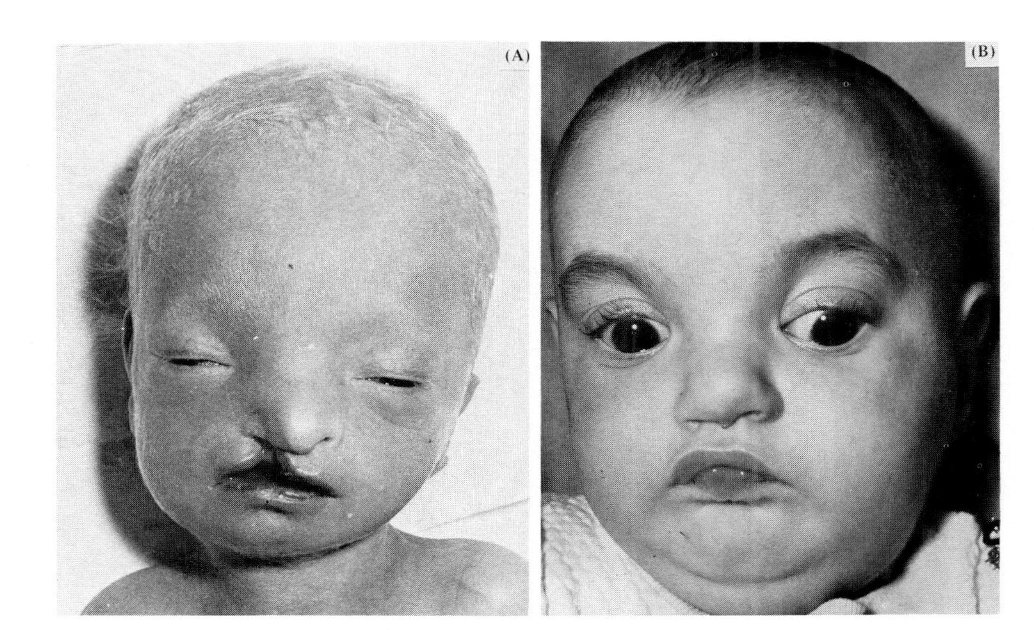


**Fig. 1.** 4p– (Wolf-Hirschhorn) syndrome (A) microcephaly, ocular hypertelorism, flattened nose, cleft lip with short philtrum (from Taylor, 1968). (B) Microcephaly, cranial asymmetry, ocular hypertelorism, strabismus, broad based nose, short philtrum, and down turned mouth (from Arias *et al.*, 1970).

Cryptorchidism and, especially, hypospadias are common. Absent uterus and streak gonad have been described (Wilcock *et al.*, 1970). Congenital heart malformations, most often atrial or ventricular septal defects, have been noted in about 60% (Wolf *et al.*, 1965; Taylor *et al.*, 1970; Arias *et al.*, 1970; Guthrie *et al.*, 1971; Judge *et al.*, 1974). About one-third die within the first 2 years of life.

Several patients have exhibited dimpling of the skin over the sacrum and elsewhere, such as shoulders, elbows, or knuckles. The pelvic and carpal bones are late in ossification. Pseudoepiphyses may be seen in the phalanges and at the base of each metacarpal. Various bony anomalies have included microcephaly, hypertelorism, prominent glabella, internal hydrocephalus, proximal radioulnar synostosis, anterior fusion of vertebrae, fused ribs, cervical ribs, dislocation of hips and club feet (Franceschini *et al.*, 1971; Dunbar *et al.*, 1975).

Simian palmar creases have been present in over 20%. Dermal ridges are frequently hypoplastic with low arch pattern and low total ridge count on fingers (Fryns *et al.*, 1973).

## II. 5p− SYNDROME (CRI-DU-CHAT SYNDROME) (FIG. 2)

Since Lejeune *et al.* (1963) first reported the cri-du-chat syndrome, over 150 examples have been documented to date. The syndrome is present in about 1%

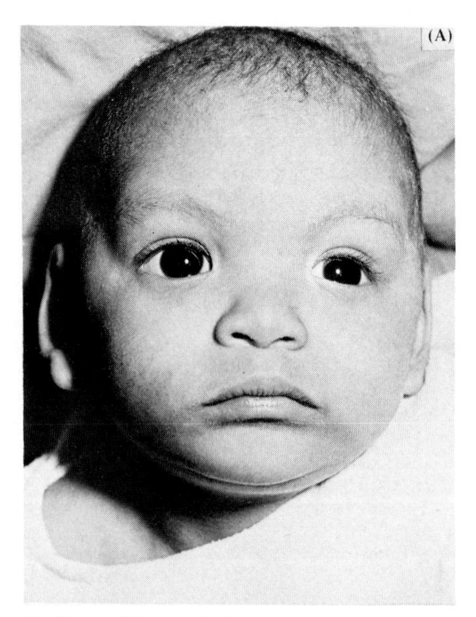


**Fig. 2.** 5p− (cri-du-chat). (A) Microcephaly, round face, ocular hypertelorism with broad nasal bridge, and malformed pinnas (from Weinkove and McDonald, 1969). (B) Compare facies in this child with that of child in A (from Gordon and Cooke 1968). (C) As child ages, round face disappears. Note preauricular tag (from Dyggve and Mikkelsen, 1965)).

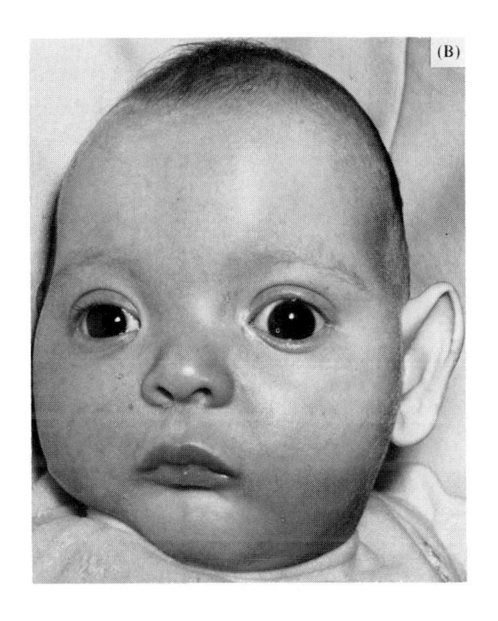


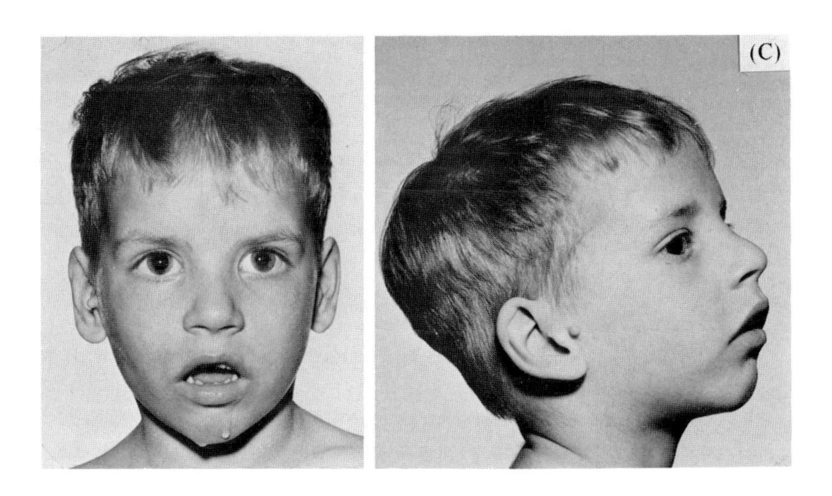


**Fig. 2 (*Continued*)**

of institutionalized individuals with intelligence quotients less than 35. The condition results from deletion of 15–80% of the short arm of one of the chromosomes 5 (German *et al.*, 1964; Miller *et al.*, 1969; Niebuhr, 1972; Singh *et al.*, 1973). Bands 5p14 and 5p15 are missing in all cases of the syndrome. Theoretically, this could arise from at least four different mechanisms: terminal deletion, interstitial deletion, translocation, and unequal interchange of the short arm. Most deletions are thought to occur as a result of two breaks. If these occur in the short arm, an interstitial deletion results (Berger *et al.*, 1974). If deletion occurs in both arms, a ring chromosome is produced (Rohde and Tompkins, 1965). Maternal age is not advanced. About 70% of those identified at birth are females; however, most older patients have been male (Breg *et al.*, 1970). The reason for this discrepancy is not evident. Mosaicism has also been described, patients having all the stigmata of the full-blown syndrome (Zellweger, 1966; Neuhäuser *et al.*, 1968; Mennicken *et al.*, 1968). About 10–15% result from translocation (de Capoa *et al.*, 1967; Warburton and Miller, 1967). The chromosome to which the segment is translocated has been quite variable (Singh *et al.*, 1973). Pericentric inversion has also been described (Faed *et al.*, 1972).

As the name implies, the cri-du-chat syndrome is characterized by a catlike, weak, shrill cry in infancy caused by hypoplasia of the larynx (Ward *et al.*, 1968). However, the cry usually disappears with time, even within a few weeks of age (Gordon and Cooke, 1968; Berg *et al.*, 1970). The cry, almost one octave higher than normal, is quite monotone in quality (Schroeder *et al.*, 1967).

The infant face is characterized by microcephaly, round form, hypertelorism, antimongoloid obliquity of palpebral fissures, epicanthus, bilateral alternating strabismus, broad nasal bones, and low-set ears. Howard (1972) described a bizarre pupillary response to methacholine. Preauricular tags are occasionally noted. Most patients have mild micrognathia. However, the roundness of the face and the ocular hypertelorism disappear with age. The face becomes thin and the philtrum short. Premature graying of the hair has been noted in about 30%. Dental malocclusion is common (Taylor, 1967; Mennicken *et al.*, 1968; Breg *et al.*, 1970; Gordon and Cooke, 1968; Niebuhr, 1971).

There is usually severe mental retardation (IQ less than 25), failure to thrive, and hypotonia in infancy. Birth weight is usually less than 2500 gm in spite of normal gestation time. Adult height usually ranges from 124 to 168 cm (49 to 66 inches). Various musculoskeletal anomalies have included hypotonia, flat feet, mild scoliosis, large frontal sinuses, small ilia, syndactyly, and short metacarpals and metatarsals (Neuhäuser and Lother, 1966; Mennicken *et al.*, 1968).

Dermatoglyphic alterations include simian creases in about 35%, a high frequency of thenar patterns (about 50%) and t′ axial triradii and a deficiency of ulnar loops (Warburton and Miller, 1967; Mennicken *et al.*, 1968). Eight or more whorls have been noted in about 30%.

## III. TRISOMY 13 SYNDROME (PATAU SYNDROME) (FIG. 3)

Patau (Patau *et al.,* 1960) is credited for first identifying trisomy 13, although Bartholin in 1657 may have given the first description of the clinical features (Warburg, 1960). The incidence has been estimated to be about 1 per 6000 births (Conen and Erkman, 1966a).

The phenotype is striking. Mean birth weight is about 2500 gm; often there is a single umbilical artery. About 45% die within the first month, 70% by the sixth month, and less than 5% survive more than 3 years (Magenis *et al.,* 1968). The oldest known child with the disorder was 10 years old (Marden and Yunis, 1967). The mean maternal age is about 31 years. As in the case of 21 and 18 trisomics, most examples result from nondisjunction.

Moderate microcephaly with sloping forehead and wide sagittal suture and fontanels have been noted in over 60%. Arhinencephaly, apneic spells, seizures, feeding difficulties, severe mental retardation, and deafness are common. Any of the holoprosencephalic states (cyclopia, ethmocephaly, cebocephaly, and premaxillary agenesis) may be associated with trisomy 13 (Conen *et al.,* 1966; Halbrecht *et al.,* 1971; Fujimoto *et al.,* 1973). The inner ear anomalies are of the Mondini or Scheibe types (Kos *et al.,* 1966; Maniglia *et al.,* 1970).

About 80% exhibit microphthalmia or iris coloboma with retinal dysplasia, ocular hypertelorism, and malformed pinnas (Cogan and Kuwabara, 1964). The retinal dysplasia is characterized by intraocular cartilage extending from the retrolental region to the sclera at the site of the iris coloboma (Ginsberg and Perrin, 1965).

Capillary hemangiomas in the glabellar region and localized scalp defects in the parietooccipital area have been described in about 75%. Cleft lip and/or cleft palate and micrognathia have been noted in 60–70% (Conen *et al.,* 1966; Taylor, 1968).

Musculoskeletal abnormalities include postaxial polydactyly of the hands or feet with overlapping flexed fingers (about 75%). The calcaneus is often prominent, and frequently there are rocker-bottom feet. The fingernails are hyperconvex and narrow.

At least 80% have congenital heart defects (atrial septal defect, patent ductus arteriosus, ventricular septal defect, and dextroposition) (Smith, 1969). There may be accessory spleens or splenic tissue in the pancreas. Genital anomalies include cryptorchidism (over 90%) in males, and bicornuate uterus (about 50%) and hypoplastic ovaries in females.

Dermatoglyphic alterations include simian palmar crease (60%), distal palmar axial triradius (80%), and hallucal arch fibular (40%) or loop tibial (35%) (Preus and Fraser, 1972). Also frequent are thenar exit of the A line (80%) and a radial loop on other than the index finger (50%).

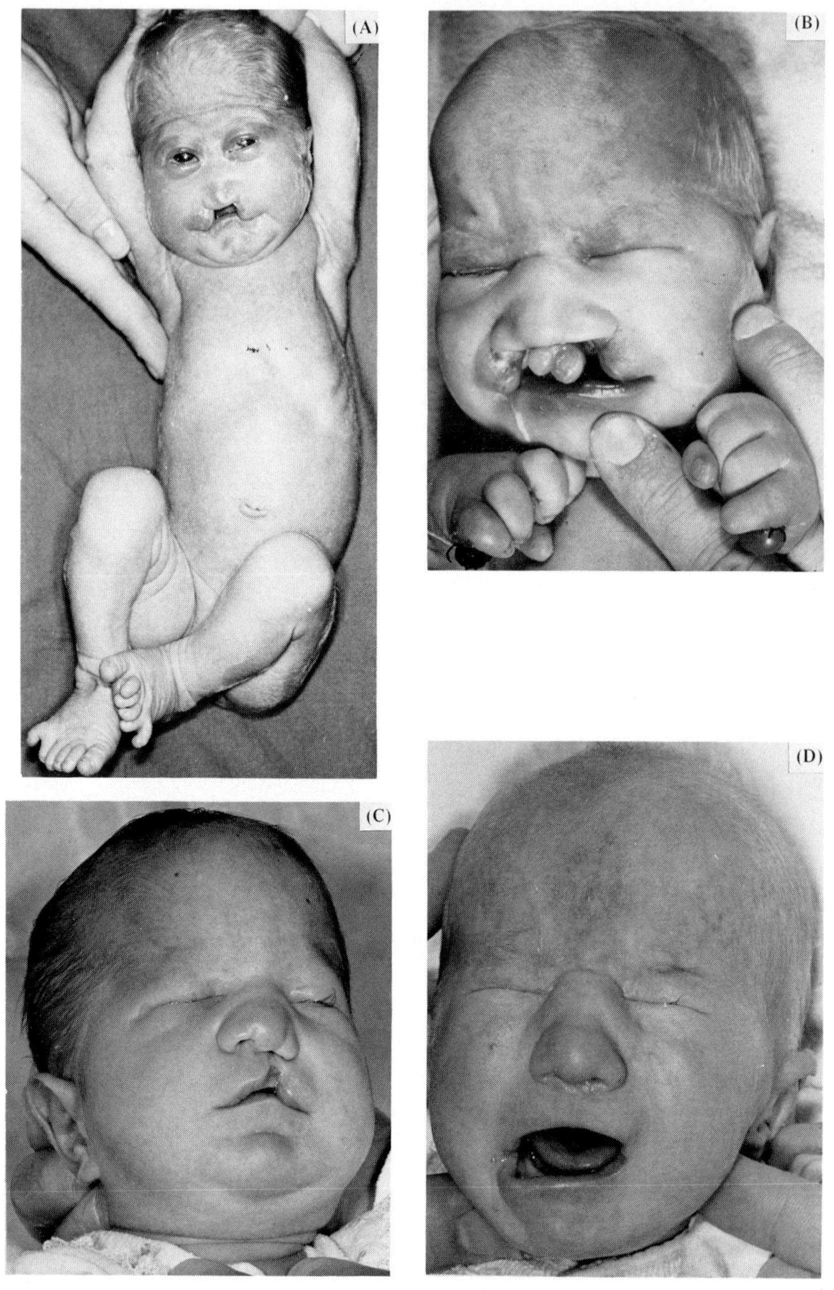

**Fig. 3.** Trisomy 13 syndrome. (A) Premaxillary agenesis form of holoprosencephaly with trisomy 13. Note ocular hypotelorism, lack of nasal bones, extra digit (from Conen *et al.*, 1966). (B,C,D) Compare facies of three infants with trisomy 13. Two have cleft lip, and palate, the third has cleft palate. All have microphthalmus and unusually bulbous nose. (E) Cutaneous defect of occipitoparietal area. (F) Malformed pinna. (G) Hyperconvex fingernails. (H) Hypoplastic penis.

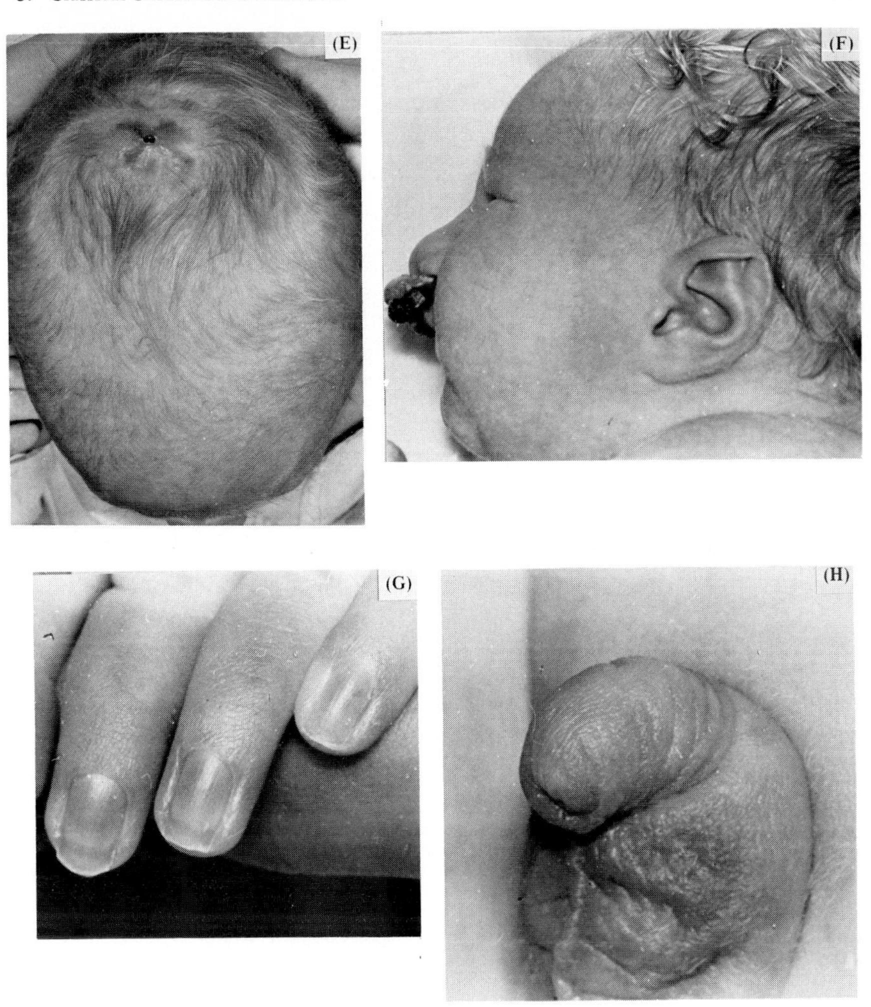


Fig. 3. (*continued*)

Polymorphonuclear neutrophils often (25–80%) have nuclear projections in cases of trisomy 13, due to primary nondisjunction. Excellent ultrastructural study of the projections has been carried out (Waltzer *et al.*, 1966; Lutzner and Hecht, 1966). Fetal hemoglobin, Hb-Gower, and other hemoglobins have been elevated, but there is good evidence that these changes disappear with age and may represent general delayed maturity (Marden and Yunis, 1967).

DNA replication studies and various banding techniques have demonstrated that the D-group chromosome involved is number 13, which is the longest and the latest of the pairs to replicate (Yunis and Hook, 1966).

## A. Trisomy 13 Caused by Primary Nondisjunction

About 75% of cases of trisomy 13 are caused by primary nondisjunction. There is no sex predilection. The mean age for mothers of infants with trisomy 13 caused by this type is advanced (32.4 years), far higher than for cases caused by translocation or mosaicism (Magenis *et al.*, 1968; Taylor *et al.*, 1970).

There have been several examples of trisomy 13 occurring with other chromosomal abnormalities in the same sibship (Klinefelter syndrome, Turner syndrome, Down Syndrome, and triploidy), but this may be chance association (Visfeldt, 1969).

## B. Translocation Trisomy 13

About 20% of the cases of trisomy 13 are caused by translocation, far more frequently than occurs in Down syndrome (Magenis *et al.*, 1968; Taylor *et al.*, 1970). In at least 85%, the translocation occurred between two D-group chromosomes. Maternal age is not advanced (25.6 years). A definite male predilection has been established. Fertility and intelligence in balanced carriers are quite variable (Wilson, 1971).

In most cases, a chromosome 13 will translocate to a chromosome 14 (Krmpotic *et al.*, 1970; Cohen, 1971). Cohen *et al.* (1968) reported trisomy 13 caused by two D/D translocations, 13 to 14, and 13 to 15. Rarely (about 5%) is the translocation familial, and then through the maternal side (Taylor *et al.*, 1970). Clinical findings are the same as those noted for trisomy 13 syndrome caused by primary nondisjunction, except for the lower frequency of nuclear projections in polymorphonuclear neutrophils (Waltzer *et al.*, 1966). Female carriers of the translocation may be more prone to miscarry. Male carriers may be more likely to give rise to balanced carriers than are female carriers (Neu *et al.*, 1973).

## C. Trisomy 13 Mosaicism

About 5% of the cases of trisomy 13 are caused by mosaicism. About half of these examples are caused by an extra chromosome 13 in a proportion of the cells. The remainder result from a complex assortment of chromosomal abnormalities (Magenis *et al.*, 1968; Taylor *et al.*, 1970).

As in translocation trisomy 13, the age of the mother of a trisomy 13 mosaic is not advanced (25.4 years), in contrast to mothers of trisomy 21 mosaics.

The clinical stigmata, as expected, are less severe than in those children with classic trisomy 13 (Bain *et al.*, 1965).

## IV. TRISOMY 18 SYNDROME (FIG. 4)

Edwards *et al.* (1960) described a new syndrome associated with the presence of an extra chromosome in the E group which was subsequently shown to be chromosome 18 (Patau *et al.*, 1961; Yunis *et al.*, 1964b).

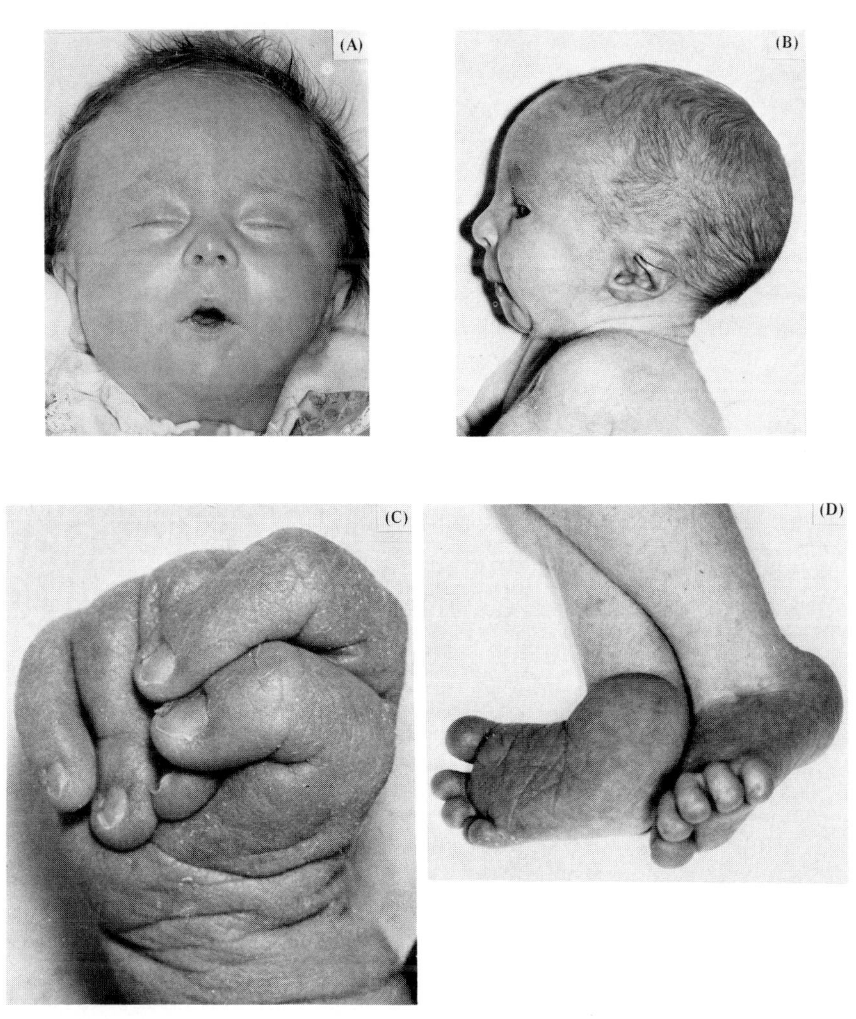


**Fig. 4.** Trisomy 18 syndrome. (A) Narrow bifrontal diameter. (B) Small mandible, prominent occiput, malformed pinna. (C) Overlapping fingers with contractures. (D) Talipes. (B,C from Paerregård, 1966).

Features of the syndrome noted in over 75% of the cases include failure to thrive, developmental retardation, feeding difficulties, hypertonia, limited hip abduction, flexion deformities (usually ulnar deviation) of fingers, short sternum, congenital heart disease (ventricular septal defect–90%, patent ductus arteriosus–70%, and atrial septal defect–20%), short dorsiflexed halluces, rockerbottom feet, calcaneovalgus deformity of feet, and cryptorchidism (Butler *et al.,* 1965; Kurien and Duke, 1968; Taylor, 1968; Weber and Sparkes, 1970).

Craniofacial anomalies almost always present include prominent occiput, low-set malformed pinnas, and micrognathia. Less frequent are extra skin at nape (25–50%) and cleft lip and/or palate (15%) (Butler *et al.,* 1965; Taylor, 1968; Weber and Sparkes, 1970; Schinzel and Schmid, 1971). Ocular anomalies, although frequent, are relatively minor (Ginsberg *et al.,* 1968; Keith, 1968).

Severe anomalies found at autopsy, apart from the cardiac anomalies noted above, include Meckel diverticulum, heterotopic pancreatic tissue, thin diaphragm with eventration, and various renal anomalies.

Dermatoglyphic alterations are frequent. Over 85% of finger prints are simple arches. More than 30% have a simian palmar crease, and greater than 40% have a single flexion crease in the fifth finger. The atd angle is somewhat increased (73°), although by no means as great as in trisomy 21 or in trisomy 13. There is also an increased frequency (about 15%) of radial loops on thumbs (Ross, 1968; Penrose, 1969). Hallucal arches have been noted in 40%. Dermal ridges not uncommonly are hypoplastic.

Trisomy 18 has an uncommon but yet definite association with aplasia of the radius and thrombocytopenia (Rabinowitz *et al.,* 1967; Schinzel and Schmid, 1971; Stoll *et al.,* 1972). We have also seen such an example.

## A. Trisomy 18

The incidence of trisomy 18 in the more recent surveys has varied from 1 per 3500 to 1 per 7000 births (Taylor, 1968; Benady and Harris, 1969; Garfinkel and Porter, 1971). Mean maternal age is advanced, 32 years (Taylor, 1968).

The mother often exhibits small weight gain during pregnancy and indicates that fetal movements were feeble. Most examples are postmature. Mean birth weight is less than 2300 gm. The placenta is often small with the single umbilical artery, and hydramnios has been noted in over 50%.

30% fail to survive more than 1 month, 50% succumb by 2 months, and less than 10% live more than 1 year. A few examples have lived to at least 15 years (Stoll *et al.,* 1974). Mean survival time is about 70 days (females–134 days; males–15 days). Taylor (1968) suggested that two-thirds of all infants with trisomy 18 were conceived between September and February and that the maternal age in the summer-conceived group was considerably lower than those of the winter-conceived group.

### B. Double Trisomies

Double primary nondisjunction has been observed in 5–10% of cases (Hamerton, 1971). There are a few examples of 48,XXX,18+ and 48,XXY,18+ patients who have, as may be expected, principally exhibited the stigmata of trisomy 18 (Taylor, 1968; Cohen and Bumbalo, 1967; Bach *et al.*, 1973). A single example of 48,XY,18+,21+ (Gagnon *et al.*, 1961) expressed the stigmata of both syndromes. Mean survival time for double trisomics has been 3 weeks. Maternal age is markedly advanced in this group.

### C. Translocation Trisomy 18

Translocation is usually sporadic, but examples of familial translocation have been recorded (Hamerton, 1971). Mean maternal age is lower than for trisomy 18 caused by nondisjunction.

### D. Trisomy 18 Mosaicism

Possibly 10% of trisomy 18 cases have exhibited mosaicism. These cases have been elegantly tabulated by Hamerton (1971). As expected, they exhibit milder manifestations of the disorder and survive for a longer period of time (Shih *et al.*, 1974). Several examples of asymmetry have been recorded (Hook and Yunis, 1965; Backus and Darien, 1968; Pavone *et al.*, 1972). Double mosaicism has also been described (Bodensteiner and Zellweger, 1971).

## V. 18p− SYNDROME

Deletion of the short arm of chromosome 18 is associated with a variable phenotype. There appears to be a two peak distribution curve, one between 25 and 29 years, the other between 40–45 years (Aksu *et al.*, 1976). There is a 2:1 female sex predilection (Parker *et al.*, 1973). About 85 cases have been reported to date (Aksu *et al.*, 1976).

Mental retardation is a constant feature but of variable degree. Birth weight is low, somatic growth retarded, and often there is hypotonia. Several infants have exhibited some features of Turner syndrome (pterygium colli, lymphedema of hands and feet, and shield chest with widespread nipples). The disorder is infrequently caused by translocation (Jacobsen and Mikkelsen, 1968).

There is no characteristic facial dysmorphia. Frequently, however, microcephaly, hypertelorism, epicanthal folds, strabismus, and ptosis of lids are noted. The ears are low-set, large, floppy, outstanding and poorly formed. The mandible is generally small (Pfeiffer, 1966; Reinwein *et al.*, 1968; de Grouchy, 1969; Lurie and Lazjuk, 1972). Dental caries is marked.

Within the group of 18p− and 18r cases, there is a distinct group exhibiting various degrees of holoprosencephaly (Uchida *et al.*, 1965; McDermott *et al.*, 1968; Nitowsky *et al.*, 1966; Gorlin *et al.*, 1968; Dumars *et al.*, 1970; Sabater *et al.*, 1972). These cases have differed in no way from other examples of holoprosencephaly associated with a normal karyotype. Possibly these cases arise from deletion of the normal allele in the heterozygote to allow expression of a recessive holoprosencephalic gene (Gorlin *et al.*, 1968). Serum IgA has been absent in some cases of 18p−, 18q−, and 18r, probably being nonspecific (Ruvalcaba and Thuline, 1969; Fischer *et al.*, 1970; Stewart *et al.*, 1970; Aksu *et al.*, 1976).

## VI. 18q− SYNDROME (FIG. 5)

The disorder was first described by de Grouchy in 1964. Birth weight is generally below 2700 gm. Maternal age is not advanced. Rarely is the condition caused by translocation (Aarskog, 1969; Law and Masterson, 1969; Šubrt and Pokorny, 1970). De Grouchy (1969) found 4 of 24 cases to be associated with mosaicism.

Mental retardation is profound, few having an IQ over 30. Somatic growth is also retarded. Hypotonia and seizures are frequent. The voice is often low-pitched. Skin dimples may be present over the subacromion and epitrochlear areas, lateral to the patellae, and over the knuckles. The fingers are long and tapered. Skeletal anomalies are limited to supernumerary ribs. Congenital heart anomalies are present in over 65%. The genitalia are hypoplastic in both sexes, the labia, clitoris, and penis being small (Parker *et al.*, 1972).

Characteristic are midfacial hypoplasia and mild microcephaly. The eyes are deeply set, and there are frequent ophthalmological defects: glaucoma, strabismus, nystagmus, tapetoretinal degeneration, and optic atrophy. The nose is short. A small subcutaneous nodule may be present at the site of cheek dimples. The mouth is carp-shaped in 75%. The pinnae are somewhat unusual, the antitragus and antihelix being especially prominent. The ear canals are atretic in over 50%. In 40% of the cases, cleft lip and/or cleft palate have been noted (de Grouchy, 1969; Law and Masterson, 1969; Gorlin *et al.*, 1971; Kunze *et al.*, 1972; Aksu *et al.*, 1976). As noted above, IgA has been diminished in some cases.

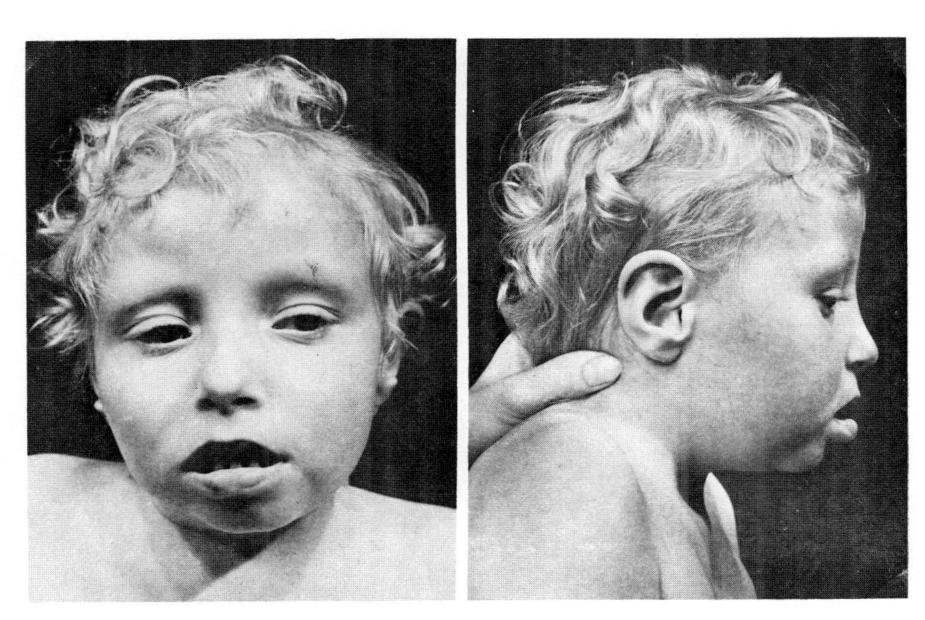

**Fig. 5.** 18q− syndrome. Midface hypoplasia, deepset eyes, antitragus and antihelix prominent (from Murken *et al.*, 1970).

Fingerprint whorls characteristically exceed 5 in number (Wolf *et al.*, 1967; Law and Masterson, 1969; Lurie and Lazjuk, 1972), and there may be a high frequency of large composite patterns (Mavalwala *et al.*, 1970).

## VII. TRISOMY 21 (DOWN SYNDROME) (FIG. 6)

Langdon Down (1866) first extensively described the syndrome which has received his name, calling it "Mongoloid idiocy" or "mongolism." Lejeune (1959) demonstrated that the disorder was associated with an extra chromosome in the G group. Polani *et al.* (1960) described translocation Down syndrome, and Clarke *et al.* (1961) discovered mosaicism for an extra G-group chromosome. Yunis *et al.* (1965), by means of autoradiography, identified the chromosome as one of the chromosomes 22, although by this time the term trisomy 21 had been so extensively employed that it has remained. Comprehensive reviews are by Penrose and Smith (1966) and Benda (1969).

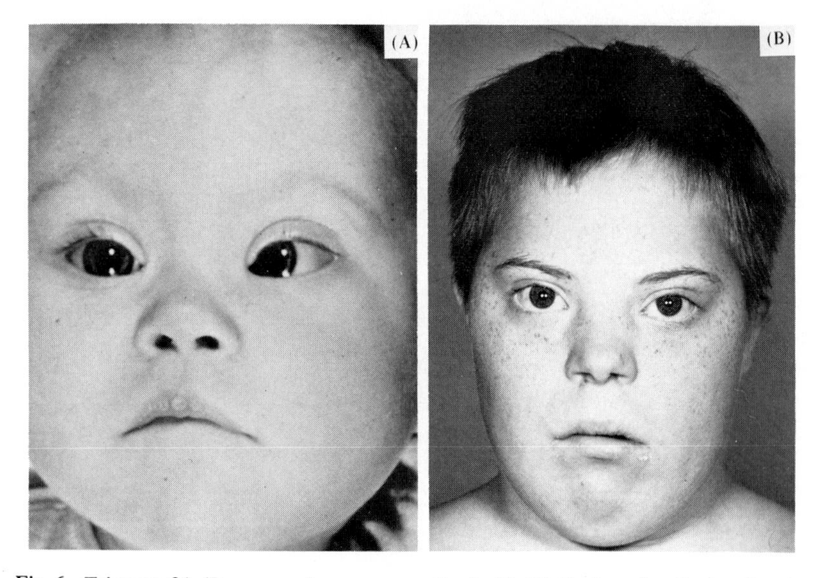


**Fig. 6.** Trisomy 21 (Down syndrome, mongolism). (A–D) Facies of patients of various ages (from R. Goodman, Tel Hashomer, Israel). (E) Brushfield spots of iris. (F) Serrated vermilion of lips. (G) Plicated tongue. (H) Missing teeth.

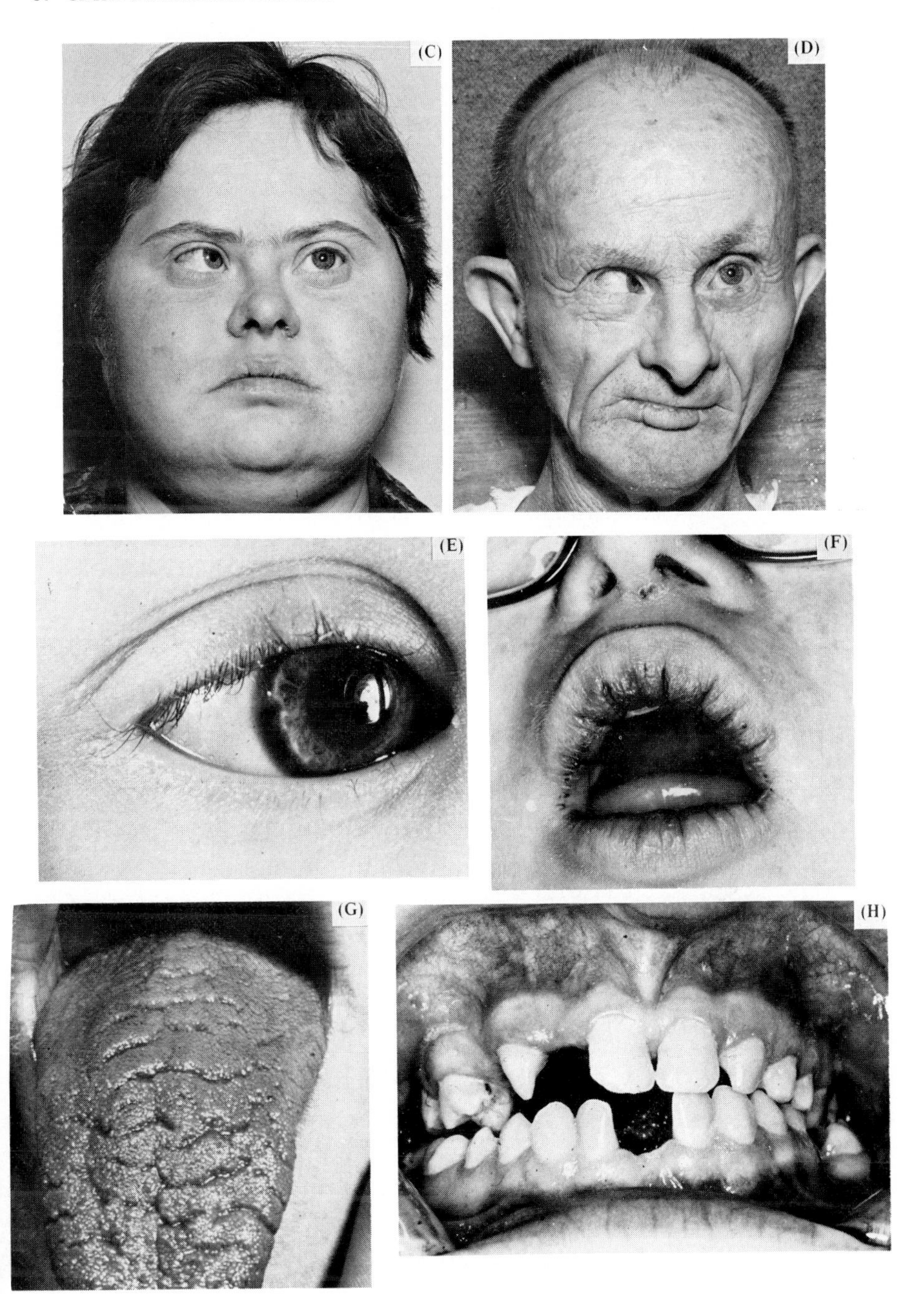


Fig. 6. (*continued*)

The incidence of trisomy 21 is between 1 and 2 per 1000 live births among various populations (Mikkelsen, 1971a). Over 95% of the cases are caused by nondisjunction, the remainder resulting from translocation. Clinical features cannot be utilized to differentiate trisomic and translocation forms (Ong *et al.*, 1967).

The skull is brachycephalic with shortening of the anteroposterior diameter and flattening of the occiput in about 75%. The cephalic index (normally 0.75–0.80) is usually greater than 0.80 and may exceed 1.00 (Roche *et al.*, 1961). In infants with trisomy 21, the fontanels are larger than normal, and closure is late. A third fontanel is present in over 50% of infants with Down syndrome (Tan, 1971). In patients over 10 years of age, a patent metopic suture is found in 65% of males (normal, 9%) and in 40% of females (normal, 12%). Absence of frontal and sphenoid sinuses and hypoplasia of the maxillary sinuses have been found in over 90% (Levinson *et al.*, 1955; Spitzer *et al.*, 1961; Betlejewski *et al.*, 1964). The bones of the middle face are poorly developed, producing a relative prognathism and ocular hypotelorism (Gerald and Silverman, 1965).

The profile is flattened, due to hypoplasia of the nasal bones. The palpebral fissures are oblique, the outer canthus being slightly higher than the inner. Epicanthal folds are extremely common. Speckled iris (Brushfield spots) and lens opacity are present in about 85% and 60% of patients, respectively. Interpupillary distance is reduced by about 5 mm, and convergent strabismus and nystagmus are common.

Ear anomalies are present in at least 60%, especially angular overlapping helix, prominent antihelix, and small or absent ear lobe. Hearing loss is common, probably as a result of frequent otitis media (Brooks *et al.*, 1972). Excessive skin on the nape has been noted in about 80% (Hall, 1966).

Open mouth occurs in over 60%, with the tongue often thrust beyond the lips. The tongue, however, is not enlarged (Ardran *et al.*, 1972). The lips are usually broad, irregular, fissured, and dry. Fissuring and furrowing of the tongue are seen in at least 30–40% (normal, 5%). The hard palate is narrower and shorter but not higher than normal (Shapiro *et al.*, 1967).

Various other anomalies include missing or malformed teeth (especially maxillary lateral incisors and mandibular second premolars), delayed eruption, increased periodontal destruction, and malocclusion (Cohen and Cohen, 1971).

The hands are characteristically short and broad, the fifth finger usually being abbreviated and clinodactylous, and having a single flexion crease in about 20% of the cases. There is usually greater space than normal between the hallux and the rest of the toes.

Hypotonia, especially marked in infancy, improves with age. Joints are usually hyperextensible. The Moro reflex is absent in over 80%.

The penis and scrotum are usually small, and about 25% have cryptorchidism. Pubic hair is straight.

Congenital cardiac anomalies, present in about 40%, in decreasing order of frequency are: ventricular septal defect, AV communis, atrial septal defect, and patent ductus arteriosus (Cullum and Liebman, 1969).

Diastasis recti, duodenal atresia, or umbilical hernia occur in about 10% (Penrose and Smith, 1966). Syringoma has been described in almost 20% (Butterworth *et al.*, 1964).

Radiographic changes include reduced iliac and acetabular angles in the young infant, and hypoplastic middle phalanx of the fifth finger. Skeletal alterations were found in 80% of 21 trisomics, and in 55% with translocation or mosaicism. Also noted with increased frequency were retarded maturation of the manubrium sterni and aplasia of the 12th rib (Nicolis and Sacchetti, 1963; Willich *et al.*, 1975).

Intelligence quotients range from 25 to 70, most Down syndrome patients 3 years of age or less having IQs of 50–59 but slipping with increasing age to 25–49 (Penrose and Smith, 1966).

Dermatoglyphic anomalies include distal axial triradius in the palm (over 80%), bilateral simian creases (30%), single flexion crease in fifth finger (20%), 10 ulnar loops (30%), and hallucal arch tibial (70%) or small loop distal (30%) patterns (Preus and Fraser, 1972). There are no dermatoglyphic differences between trisomic and translocation Down syndrome patients (Rosner and Ong, 1967). Reed *et al.* (1970) described a helpful dermatoglyphic nomogram.

Because of susceptibility to respiratory infection, early mortality used to be great. Low IgE levels have been found (Lopez, 1974), but these findings have not been supported (Kadowaki *et al.*, 1976). With the introduction of antibiotics, the mean survival age is almost 20 years. There is a twentyfold increased association with acute leukemia (Conen and Erkman, 1966b).

Numerous attempts have been made to establish specific biochemical alterations. Rosner *et al.* (1965) found decreased blood serotonin and increased galactose-1-phosphate uridyltransferase, leukocyte alkaline phosphatase, and galactose-6-phosphate dehydrogenase levels in patients with Down syndrome. Miller and Hyde (1966) confirmed elevated leukocyte alkaline phosphate levels in primary trisomics but not in translocation patients.

## A. Trisomy 21 Caused by Primary Nondisjunction

As discussed above, about 95% of cases of Down syndrome are sporadic primary trisomics, resulting from nondisjunction which is age-dependent. It probably occurs as often in the male as in the female parent and arises in both the first and second meiotic divisions (Uchida, 1973; Wagenbichler *et al.*, 1976). If the mother is less than 20 years of age at the time of conception, the risk of producing a child with trisomy 21 is about 1 per 2500 live births (Robinson, 1973). This risk gradually increases until 35 years, after which there is a more marked increase in frequency such that a mother over 45 years has about 1 chance in 50 or less of having a child with Down syndrome (Penrose and Smith, 1966; Stene, 1970a).

There are less than two dozen published cases in which a female with trisomy 21 has given birth. In about half the cases, the offspring had Down syndrome, while in the remainder the child was normal. There is no known example of a male with trisomy 21 having sired a child.

### B. Association of Down Syndrome with Other Primary Nondisjunctions

Individuals with trisomy 21 have been occasionally (about 1 per 200) found to have another extra chromosome (double primary nondisjunction), the most frequent type being 48,XXY,G+ (Hamerton *et al.,* 1965; Taylor and Moores, 1967). Other forms such as 48,XXX,G+ and 48,XYY,G+ have also been described (Yunis *et al.,* 1964a; Uchida *et al.,* 1966). This association is much higher than might be expected by chance.

### C. Translocation Down Syndrome

Down syndrome patients born to young mothers as well as those with affected relatives often have the extra 21 chromosome attached to another chromosome. This comprises about 3.5% of cases of Down syndrome. Translocation Down syndrome is not age-dependent. It may be sporadic or familial. About 8% of Down syndrome patients born to mothers less than 30 years of age have exhibited translocation as opposed to 1.5% born to mothers over 30 years old. It is widely accepted that the short arms of acrocentric chromosomes have nucleolar organizers and that these points are likely to break, producing a high frequency of structural chromosome aberrations.

In familial translocation Down syndrome, one of the parents has 45 chromosomes instead of the normal 46. One of the small G-group chromosomes is "missing" since it has been translocated to another chromosome. The parent carrying the translocation chromosome is phenotypically normal, since no significant amount of genetic material has been lost in the translocation process. In most cases, chromosome 21 has been translocated onto an acrocentric chromosome: to a D chromosome (46,XX or XY,–D,+t [DqGq]) in about half the cases, or to another G chromosome (46,XX or XY,–G,+t [GqGq]) in the remainder. There are rare examples arising from an isochromosome for the long arm of a chromosome 21 (46,XX or XY,–G,Gqi).

Among cases of t(DqGq), about half are inherited and half are sporadic. Hecht *et al.* (1968) demonstrated that the D chromosome is usually 14, rarely 15, and almost never 13 (see Nagel and Hoehn, 1971). The preferential involvement of chromosome 14 in both t(DqGq) and t(DqDq) may be related to repetitive DNA sequences in the subcentromeric position, conceivably making the chromosome more liable to breakage (Nagel and Hoehn, 1971).

If the translocation is of the DqGq type and the mother bears the translocation, approximately 10% of the offspring have Down syndrome. However, only about 2.5% born to carrier fathers have the DqGq translocation (Hamerton, 1970).

When translocation occurred between two G chromosomes, in prebanding years, it was morphologically impossible to tell whether the translocation was t(21q22q) or t(21q21q). Pedigree study at times was helpful since a t(21q21q) carrier, that is not a mosaic, would produce only trisomy 21 or monosomic (lethal) offspring. On the other hand, birth of normal individuals or of balanced translocation carriers to a patient with t(GqGq) would indicate that the translocation is t(21q22q). However, distinction between these types can be made by means of banding techniques (Mikkelsen, 1971b) and should *always* be carried out for purposes of genetic counseling.

Although data are not abundant, in the case of a balanced t(21q22q) state a female carrier has about a 9% chance of producing a child with Down syndrome. Reliable data are not available for carrier fathers but the risk appears to be small (Mikkelsen and Stene, 1970; Stene, 1970b; Mikkelsen, 1971b). In 21q21q translocation, the risk is 100%.

21qi results from an error in centromeric division at the second meiotic division or during the first division by formation of an unstable telocentric chromosome. The 21qi heterozygote can produce only offspring with Down syndrome.

Rarely, noncentric translocation has been reported (Aarskog, 1966; Laurent and Robert, 1966; Cohen and Davidson, 1967; Orye *et al.*, 1969).

### D. Down Syndrome Mosaicism

Patients having two different cell populations, one trisomic for chromosome 21 and another normal, constitute about 2–3% of patients with Down syndrome (Chitham and MacIver, 1965; Richards, 1969, 1974; Mikkelsen, 1971a; Sutherland and Wiener, 1972). This condition is usually suspected when the phenotypic expression of trisomy 21 is not fully expressed, or when the intelligence of the patient is higher than expected. In addition, they may have children with Down syndrome (Weinstein and Warkany, 1963). Individuals having trisomy 21 mosaicism may vary in phenotype from typical trisomy 21 to normal. There is no age dependency (Richards, 1969). One cannot correlate the percentage of trisomic lymphocytes with intelligence. Richards (1969) found about 20% more trisomic cells in fibroblasts than in lymphocytes.

Mosaicism involving more than two stem cell lines has also been described (Reinwein *et al.*, 1966; Richards, 1969). If mosaicism is found in one of the parents of a child with Down syndrome, meiotic study of ovary or testis should be carried out. There is evidence that if half the cells are abnormal, about 25% of the children will have Down syndrome (Mikkelsen, 1971a). Detection of mosaicism in one of the parents may be extremely difficult (Sutherland *et al.*, 1972).

The dermatoglyphic findings in mosaic patients are more similar to those with pure trisomy 21 than to normal controls (Loesch, 1974). Intellectual capacity and verbal facility are greater than with pure trisomy 21 patients (Fishler *et al.*, 1976).

## VIII. 21q− SYNDROME (ANTIMONGOLISM) (FIG. 7)

This syndrome consists of mental and growth retardation, hypertonia, nail anomalies, skeletal malformations, cryptorchidism, hypospadias, inguinal hernia, pyloric stenosis, thrombocytopenia, eosinophilia, and hypogammaglobulinemia. It should be borne in mind that most of the cases cited were identified prior to the time of banding and that some may represent examples of the 22q−

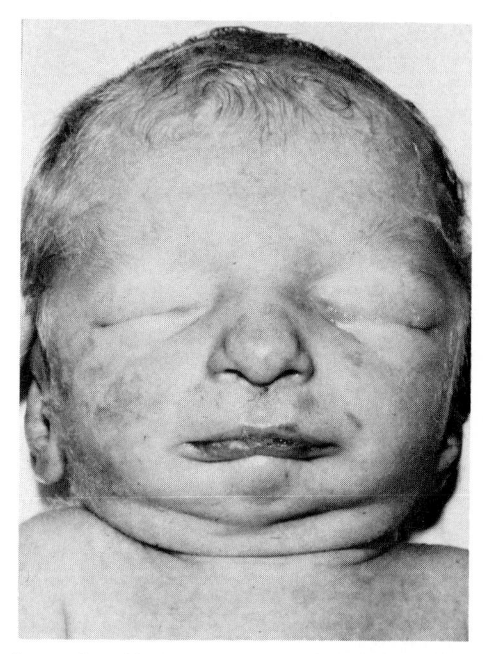

Fig. 7. 21q− syndrome. In addition to severe mental retardation, infant had antimongoloid obliquity of palpebral fissures and broad-based nose (from Reisman *et al.*, 1966).

syndrome or other disorders (Warren *et al.*, 1973). Banded cases have been cited by Mikkelsen and Vestermark (1974). Facial and oral manifestations include microcephaly, large, poorly-modeled low-set ears, antimongoloid obliquity of palpebral fissures, blepharochalasis, highly arched or cleft palate, cleft lip-palate, and micrognathia (Lejeune *et al.*, 1964; Reisman *et al.*, 1966; Penrose, 1966; Thorburn and Johnson, 1966; Hall *et al.*, 1967; Schultz and Krmpotic, 1968; Böhm and Fuhrmann, 1969; Endo *et al.*, 1969; Challacombe and Taylor, 1969; Emberger *et al.*, 1970; Kelch *et al.*, 1971; Crandall *et al.*, 1972; Magenis *et al.*, 1972; Richmond *et al.*, 1973; Shibata *et al.*, 1973; Mikkelsen and Vestermark, 1974). Dermatoglyphic analysis has shown a marked increase in radial loops (Schindeler and Warren, 1973).

## IX. TRIPLOIDY AND TETRAPLOIDY (FIG. 8)

Triploidy is a frequent cause of fetal wastage prior to the eighth intrauterine week (Jonasson *et al.*, 1972). About 1% of conceptuses and 20% of chromosomally abnormal abortuses are triploid (Carr, 1971). Diploid/triploid mosaicism is occasionally compatible with survival, and there have been several examples of pure triploidy (Niebuhr, 1974). The supernumerary haploid set may be of maternal (digyny) or paternal (diandry) origin. Theoretically, triploid zygotes may arise from mitotic anomalies in germ cell precursors, from failure in the first or second meiotic division of oocytes and spermatocytes or dispermy. The majority (60%) of examples have been 69,XXY. These and various other possibilities are discussed by Niebuhr (1974). Animal experiments suggest that delayed fertilization results in triploidy (Yamamoto and Ingalls, 1972).

Most patients with triploidy or triploid/diploid mosaicism have been mentally and physically retarded (Schindeler and Mikamo, 1970; Prats *et al.*, 1971; Schmickel *et al.*, 1971; Simpson *et al.*, 1972; David *et al.*, 1975; Leisti *et al.*, 1974; Chambon *et al.*, 1975, Al Saadi *et al.*, 1976; Wertelecki *et al.*, 1976). Cranial or craniofacial asymmetry has been noted in over half the cases. The posterior fontanel is often open. Hypotonia may be marked. Microphthalmia and/or colobomata of the iris and choroid and mild hypertelorism have been common features. Holoprosencephaly has been noted in one infant (Zergollern *et al.*, 1972) and myelomeningocoele is not uncommon (David *et al.*, 1975). The pinna may be low-set or malformed, and the mandible is usually small. Cleft lip and/or palate has been noted in about 30% of cases (Keutel *et al.*, 1970; Finley *et al.*, 1972; David *et al.*, 1975). Several infants had hydrocephalus with hypoplasia or aplasia of the falx cerebri and corpus callosum (Zergollern *et al.*, 1972). In most 69,XXY cases, the genitals have been ambiguous with hypospadias, bifid scrotum, cryptorchidism and Leydig cell hyperplasia. Soft tissue syndactyly of the third and fourth fingers and, less often, toes, and simian

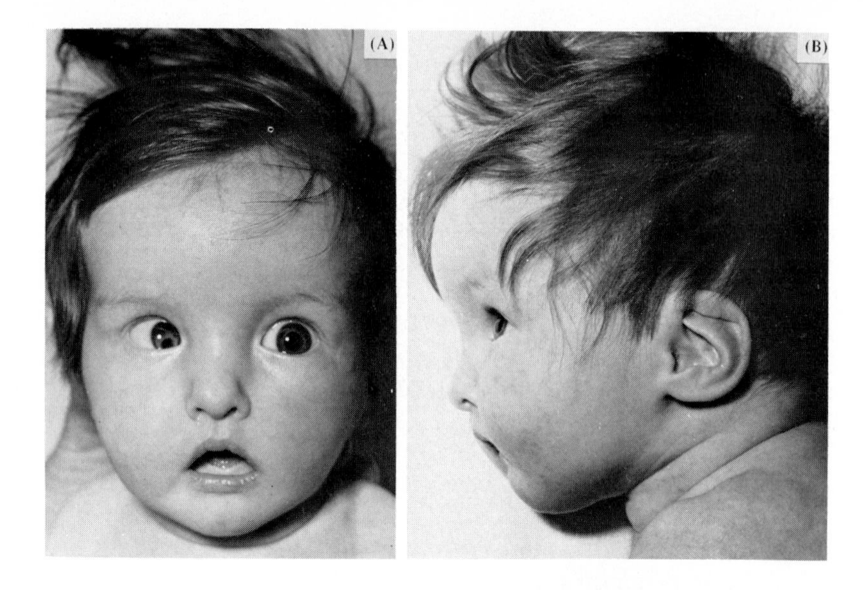


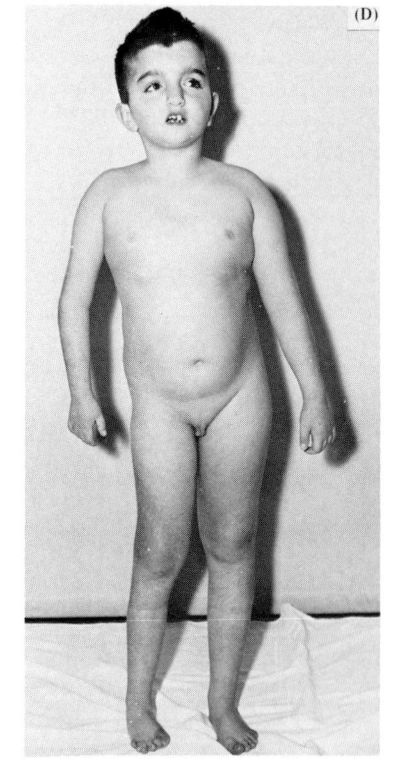


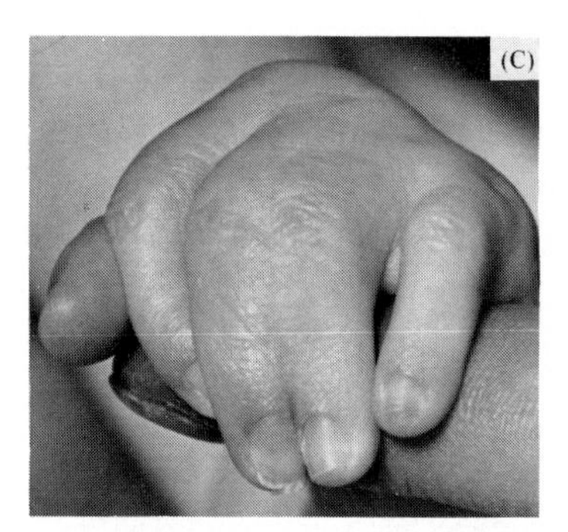


**Fig. 8.** Triploidy. (A–C) Frontal bossing, coloboma of iris, strabismus, malformed pinna. Soft tissue syndactyly of third and fourth fingers (from Schmid and Vischer, 1967). (D) Congenital asymmetry associated with diploid-triploid mosaicism (from Ferrier *et al.*, 1964). (E–G) Large placenta with hydatidiform degeneration and 69/XXY karyotype (from Walker *et al.*, 1973).

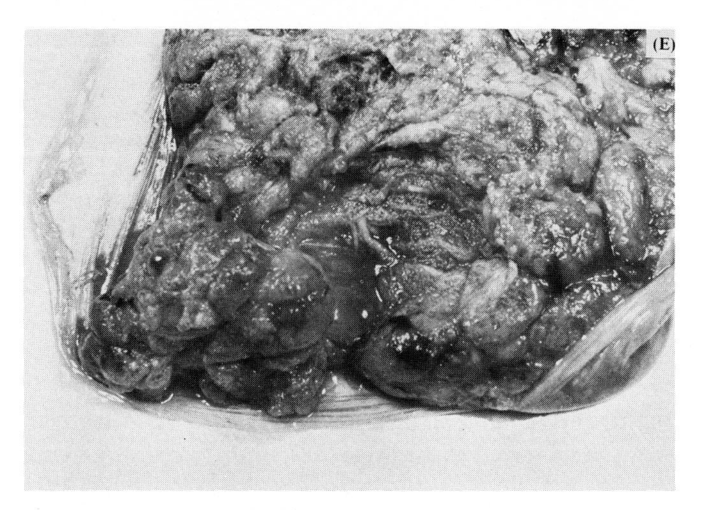


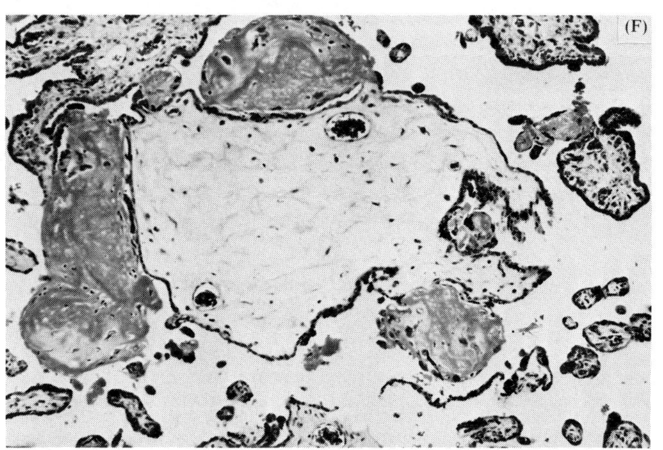


(G)

A B

C + X

D E

F G Y

Fig. 8. (*continued*)

creases have been frequent findings. An increased number of digital whorls has been noted (Butler *et al.*, 1969; Zergollern *et al.*, 1972). Cystic degeneration of the kidneys and aplasia of the adrenals have also been found. Present in nearly all cases has been hydatidiform degeneration of a large placenta (Walker *et al.*, 1973).

Two cases of tetraploid/diploid mosaicism have been reported (Kohn *et al.*, 1967; Kelly and Rary, 1974). The infant described by Kohn *et al.* had microcephaly and closed fontanels and overriding sutures, iris coloboma, aphakia, bilateral retinal detachment, two phalanges in each finger, and oligodactyly and syndactyly of toes. The pathological changes have been reviewed by Yanoff and Rorke (1973). The child noted by Kelly and Rary exhibited mental retardation, a catlike cry, hypotonia, microcephaly, delayed closure of anterior fontanel, microstomia, partial syndactyly of toes 2, 3, and 4 bilaterally, talipes equinovarus, and simian creases. A less severely affected child having tetraploidy and 18 trisomy was reported by Atnip and Summitt (1971).

An infant, completely tetraploid, who survived to 1 year of age was documented by Golbus *et al.* (1976). The boy was small with microcephaly, narrow bifrontal diameter, sparse blond hair, poorly modeled pinnae, preauricular ear tags, microphthalmia, short philtrum, beaked nose, bifid uvula, hyperconvex fingernails and hypertonia.

## X. KLINEFELTER SYNDROME (FIG. 9)

Klinefelter *et al.* (1942) described postpubertal males having small firm testes with tubular hyalinization but with a normal number of Leydig cells, azospermia, gynecomastia, elevated urinary gonadotropins, and decreased urinary 17-ketosteroids. Bradbury *et al.* (1956) and Plunkett and Barr (1956) noted chromatin-positive nuclei (Barr bodies) in the tissues of such patients, and Jacobs and Strong (1959) described an XXY sex chromosome complement in chromatin-positive Klinefelter syndrome.*

Chromatin-positive males comprise about 2 per 1000 live male births. However, this group contains XXY, XXYY, XY/XXY, and other rarer forms of Klinefelter syndrome. About 80% are XXY, 10% are mosaics, and the rest represent the more unusual types.

Klinefelter syndrome has been found in about 0.5–1% of males institutionalized either for mental retardation, epilepsy, or mental illness (Hambert, 1967; Anders *et al.*, 1968; Maclean *et al.*, 1968) and in about 10% of males manifesting sterility (Williams and Runyan, 1966). As mentioned earlier, there is a greater than chance association of Klinefelter syndrome and Down syndrome.

---

*We would prefer to limit use of the term Klinefelter syndrome to chromatin-positive males. However, XY/XXY mosaics may be chromatin-negative, thus creating problems (Yunis, 1965).

## A. XXY Klinefelter Syndrome (See Figs. 9 A and B)

The clinical features of XXY Klinefelter syndrome do not become apparent until after puberty (Puck *et al.,* 1975). Body proportions usually do not appear remarkably abnormal. However, the lower extremities tend to be long and about 60% have a span that exceeds their height by 3 cm or more (Becker *et al.,* 1966). The prepubertal testes are of normal size and microscopic appearance, but during adolescence they fail to enlarge and remain small and firm, averaging less than 2 cm in length. The seminiferous tubules are usually shrunken, hyalinized, and irregularly arranged. Those tubules which are not sclerotic are immature and lined exclusively with Sertoli cells. Elastic fibers are absent around the tunica propria of the tubules. Leydig cells are clumped. Rarely, spermatogenesis can be demonstrated (Steinberger *et al.,* 1965). In nearly all cases the testes descend. The penis is usually of normal size but may be somewhat shorter than normal. The prostate is smaller than normal (Frøland, 1969; Gordon *et al.,* 1972).

Gynecomastia develops after puberty in about 50%, and facial hair is sparse in about 60–75%. Axillary hair may also be deficient. About 50% have a female pubic escutcheon. Libido and potency are usually decreased and probably all legitimate examples of nonmosaic patients are sterile. Gonadotropins have been elevated in about 75%.

There is some evidence of increased tendency to pulmonary disorders, varicose veins, and, possibly, breast cancer (Rohde, 1964; Cuenca and Becker, 1968; Scheike *et al.,* 1973).

There is the same frequency of color blindness among XXY patients as in normal females.

Although intelligence may be reduced, at least 75% of XXY males have normal intelligence. Personality is usually passive (Nielsen, 1971; Theilgaard *et al.,* 1971).

The incidence of XXY Klinefelter syndrome is about 1.3 per 1000 live male births. Maternal age is significantly advanced for XXY but not for XXYY, XXXY, or XXXXY patients (Court-Brown *et al.,* 1969). About 60% of XXY males are $X^mX^mY$, while 40% are $X^mX^pY$. The $X^mX^mY$ state arises from nondisjunction either during oogenesis or at an early postzygotic division. The $X^mX^pY$ condition probably has its origin in nondisjunction during the first meiotic division (Frøland *et al.,* 1968; Race and Sanger, 1969).

## B. XXYY Klinefelter Syndrome (Figs. 9 C and D)

Patients with the XXYY variant tend to be about 4 cm taller, more aggressive, and more mentally retarded than those with XXY Klinefelter syndrome (Schlegel *et al.,* 1965; Parker *et al.,* 1970; Borgaonkar *et al.,* 1970). Otherwise the phenotype is quite similar; small firm testes, eunuchoid body build, sparse

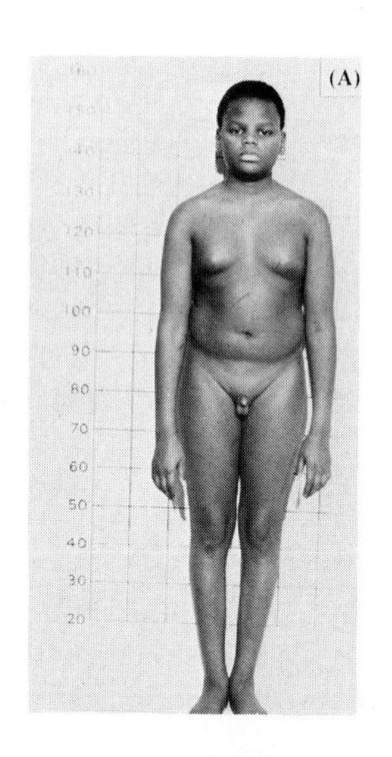
(A)

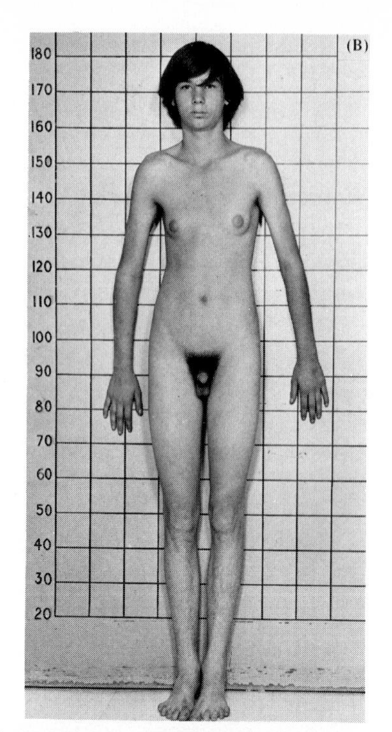
(B)
180
170
160
150
140
130
120
110
100
90
80
70
60
50
40
30
20

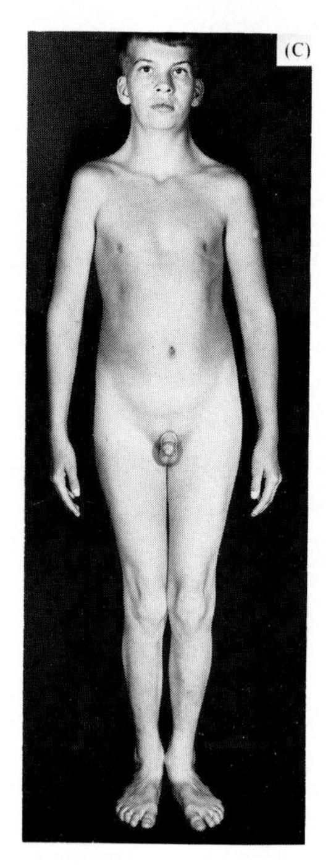
(C)

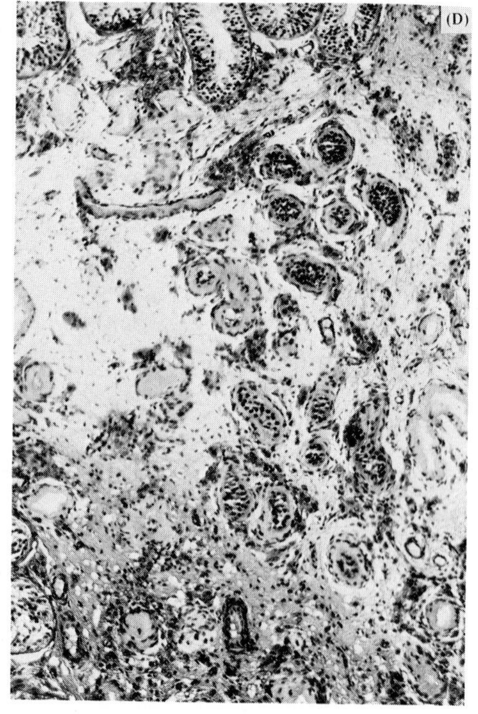
(D)

body hair, gynecomastia, and elevated gonadotropins. Almost all XXYY males described to date have been mentally retarded and many have been aggressive (Schlegel *et al.*, 1965).

As mentioned above, there is no advanced parental age in contrast to XXY Klinefelter syndrome. The disorder is most likely due to nondisjunction in both the first and second meiotic divisions during spermatogenesis with production of an XYY sperm. The less likely possibility of nondisjunction at the second meiotic division in both parents cannot be ruled out.

Dermatoglyphic studies have shown that digital arch patterns are more common in the XXYY patient than in the XXY individual who, in turn, has more than the normal male. Thus, the total ridge count is low. About 80% of XXYY patients have hypothenar patterns with attendant triradii on the ulnar border (Alter *et al.*, 1966; Penrose, 1967; Hunter, 1968; Cushman and Soltan, 1969).

A child with an XXYYY sex chromosome complement was noted to have mental retardation, lordosis, flexed index and fifth fingers, pes planus, and aggressive personality (Gracey and Fitzgerald, 1967).

## C. XXXY Klinefelter Syndrome

Over 30 cases of XXXY Klinefelter syndrome have been published (Vormittag and Weninger, 1972; Simpson *et al.*, 1974). All have been mentally retarded. The phenotype is similar to that of the XXY male but the penis is hypoplastic in at least 50% of the cases (Close *et al.*, 1968; Zollinger, 1969; McGann *et al.*, 1970; Simpson *et al.*, 1974). Gynecomastia has been present in about 35%, epicanthal folds in 25%.

Radioulnar synostosis has been noted in 10% of the cases (Greenstein *et al.*, 1970; McGann *et al.*, 1970), less frequently than in XXXXY Klinefelter syndrome. Clinodactyly of the fifth finger has been observed in about 30% (Simpson *et al.*, 1974).

Two late-labeling X chromosomes have been demonstrated. However, two Barr bodies are seen in only a fraction of the cells (Vormittag and Weninger, 1972). The condition may arise from successive nondisjunction in either the maternal or paternal meiotic divisions (Pfeiffer and Sanger, 1973).

---

**Fig. 9.** Klinefelter syndrome. (A) Prepubescent boy with XXY Klinefelter syndrome. Note gynecomastia. (Courtesy of D. Rimoin, Torrance, California.) (B) Postpubescent male. Note gynecomastia and female pubic escutcheon. (C) 16 year-old male with XXYY Klinefelter syndrome. Note pterygium colli, sparse pubic hair. (D) A large proportion of testicular tubules are small and exhibit partial to complete hyalinization. Many areas are devoid of tubules. (C & D from Schlegel *et al.*, 1965).

Dermatoglyphic findings have not been consistent. Greenstein *et al.* (1970) found numerous arches, but Borgaonkar and Mules (1970) and Vormittag and Weninger (1972) did not. However, finger ridge counts have been low. Hypothenar patterns with an ulnar displaced triradius but no axial triradii are probably common (Vormittag and Weninger, 1972).

XXXYY males have been described by Bray and Josephine (1963) and Lecluse-van der Bilt *et al.* (1974). Both had mental retardation, delayed bone age and mandibular prognathism.

## D. XXXXY Klinefelter Syndrome (Fig. 10)

There have been over 90 cases of 49,XXXXY males published since Fraccaro and Lindsten documented the first example in 1960. Nearly all have been severely mentally retarded, intelligence quotients usually ranging from 20 to 60 (Terheggen *et al.*, 1973). Exceptions have been noted by Shapiro *et al.* (1970) and Assemany *et al.* (1971). A marked difference between the XXY and XXXXY male is poor development of the external genitalia in the latter. The penis is always minute, and the testes very small and undescended with hypoplastic Leydig cells and absence of germ cells. The scrotum is usually hypoplastic.

Average birth weight is about 2500 gm. Height is often below the 3rd percentile.

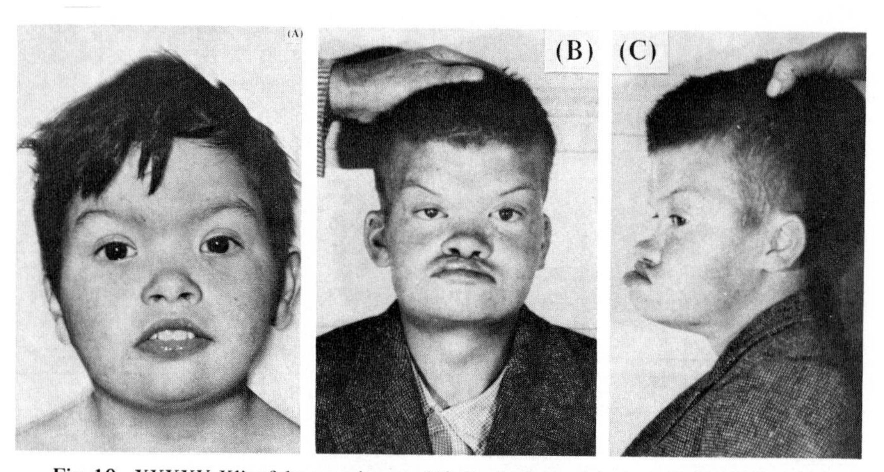


**Fig. 10.** XXXXY Klinefelter syndrome. (A) Ocular hypertelorism. (B,C) Midfacial hypoplasia. (D) Radiograph showing relative mandibular prognathism. (E) Micropenis. (F) Radioulnar synostosis. (Figs. A and E from Joseph *et al.*, 1964. Figs. B, C, and D from H. Schade, Münster. Fig. F from Zaleski *et al.*, 1966.)

Mild microcephaly, ocular hypertelorism (90%), myopia (25%), strabismus (50%), mild mongoloid obliquity of palpebral fissures (35%), epicanthus (80%), and short neck and redundant skin on the nape have been noted. About 15% have cleft palate (Sacrez *et al.*, 1965). In infancy, the face is often rounded. This, however, disappears with age, and midfacial growth is retarded with relative mandibular prognathism, especially after puberty. The pinnae are poorly modeled. The neck is short (80%) and mildly webbed (20%). Gynecomastia is not a feature of the XXXXY syndrome (Tumba, 1972).

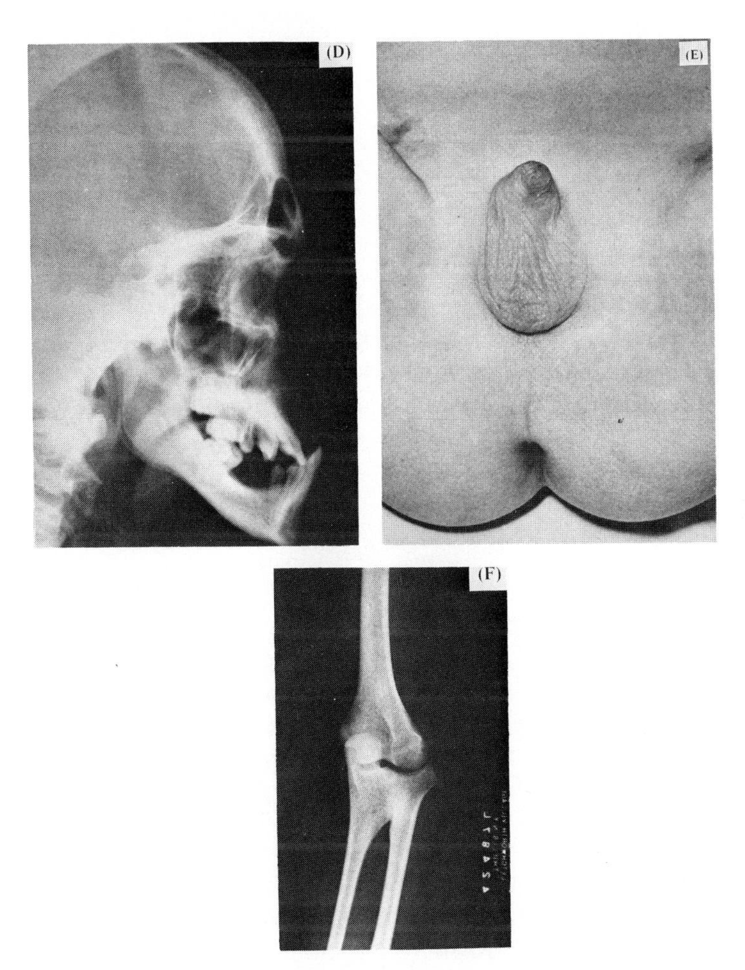


**Fig. 10.** (*continued*)

Skeletal anomalies present in over half the cases include radioulnar synostosis with reduced ability to pronate the forearm, cubitus valgus, retarded bone age, coxa valga, genua valga, elongation of distal ulna and proximal radius, wide proximal ulna, hypoplasia of middle phalanx of fifth digit, malformed cervical vertebrae, pseudoepiphyses of metacarpals and metatarsals, and pes planus (Lauritzen *et al.*, 1965; Sacrez *et al.*, 1965; Zaleski *et al.*, 1966; Houston, 1967; Christensen and Therkelsen, 1970; Terheggen *et al.*, 1973). Taurodontism is a frequent finding (Stewart, 1974).

Congenital heart disorders, most frequently patent ductus arteriosus, have been noted in about 15% of cases (Assemany *et al.*, 1971; Karsh *et al.*, 1975). Gonadotropins have not been elevated.

Autoradiographic evidence has shown three heavily labeled X chromosomes (Hsu and Lockhart, 1965). Three Barr bodies may be found in some interphase nuclei (Miller and Warburton, 1968).

Parental age is not advanced. Postzygotic nondisjunction in an XXY zygote appears to be the cause for the XXXXY state, all the X chromosomes coming from the mother (Murken and Scholz, 1967; Race and Sanger, 1969).

## E. XX Klinefelter Syndrome

Less than 60 cases have been published of males who have 46,XX karyotypes. They exhibit many of the stigmata of Klinefelter syndrome and hence will be considered here (Anderson *et al.*, 1972).

All have small testes, are infertile, and rarely shave; some have hypoplastic genitalia prior to puberty (Laurance *et al.*, 1976). About 70% have gynecomastia and elevated gonadotropin levels. Plasma testosterone levels are very low (Neuwirth *et al.*, 1972). The penis and scrotum have been small in about half the cases. All are of normal intelligence and have normal skeletal proportions. Histopathological examination of the testes reveals a picture similar to that seen in Klinefelter syndrome, but tubular hyalinization is usually absent (Lindsten *et al.*, 1966; de Grouchy *et al.*, 1967).

These patients defy the thesis that a Y chromosome is necessary for testicular differentiation unless there has been X-Y chromosomal interchange in the father (Ferguson-Smith, 1966). However, recent data suggest that both X chromosomes are maternal in origin (de la Chapelle, 1972; Sanger *et al.*, 1971). Furthermore, fluorescent staining has failed to demonstrate any marked fluorescent region in either X chromosome (George and Polani, 1970). It is possible that these individuals were originally XXY to allow for male differentiation but that the Y chromosome became subsequently lost (de la Chapelle, 1972; Laurance *et al.*, 1976). Testicular differentiation depends on material located in the short arms of the Y chromosome in the centromeric region (Böök *et al.*, 1973). Since this material is not fluorescent, it conceivably is translocated to another chromosome, a hypothesis we would favor (Krmpotik *et al.*, 1972). Palutke *et al.* (1973) found brightly fluorescent material in the Sertoli cells of an otherwise XX male. Bartsch-Sandhoff *et al.* (1974) supported the mosaicism theory for the origin of the disorder.

### F. Klinefelter Syndrome Mosaicism

About 15% of patients with Klinefelter syndrome have been found to have two or more chromosomally distinct cell populations. In each of these individuals, one of the cell populations generally has an XXY, XXXY, or XXXXY sex chromosome constitution while the other is XX or XY. The clinical expression of mosaicism for Klinefelter syndrome depends on the type of sex constitution present at a critical time of development. Thus one can find, for example, an XY/XXY mosaic who is phenotypically normal, provided the XY cells exerted the predominant genetic effect. It is important to bear this possibility in mind when trying to predict the development of testicular dysgenesis and the mental status of newborn males with sex chromatin-positive buccal smears. This is particularly noteworthy because many newborn males with chromatin-positive smears are mosaics (Maclean *et al.*, 1964).

In a study of XY/XXY mosaics, Gordon *et al.* (1972) found that only one-half exhibited azospermia, and about one-third had gynecomastia and elevated gonadotropins. About one-quarter had germinal epithelium. Among 6 patients with XX/XXY mosaicism, Ferguson-Smith (1969) noted comparable findings.

## XI. XYY SYNDROME

Although Sandberg *et al.* (1961) reported an extra Y chromosome in a male individual as early as 1961, interest was markedly aroused when a disproportionately high percentage (usually 2–4%) of such males was found in prisons and mental hospitals (Casey *et al.*, 1968; Jacobs *et al.*, 1968; Marinello *et al.*, 1969; Casey and Blank, 1972; Hook, 1973; Noël *et al.*, 1974). The frequency of the condition among newborn male infants is about 1 per 700 births (Ratcliffe *et al.*, 1970) and probably few of these individuals lead other than quite routine lives. The association of this chromosome state with criminality has been seriously questioned (Noël *et al.*, 1974). Personality has been extensively studied in XYY patients by Money *et al.* (1974). The adult height of an XYY individual is usually over 180 cm, while XYY children are usually above the ninetieth percentile in height by 6 years of age (Valentine *et al.*, 1971). Leg length and trunk length are increased, but the leg/trunk ratio is normal (Keutel and Dauner, 1969). Muscle weakness (especially of the pectoralis major) and poor coordination are commonly noted. Phenotypical alterations are subtle: mild facial asymmetry, mild pectus excavatum, and mild scapular winging. The ears tend to be long, and often there is a bony chin point. Most have exhibited normal sexual development (Court-Brown, 1969). There are no characteristic dermatoglyphic alterations (Hubbell *et al.*, 1973).

The disorder probably arises from paternal nondisjunction during the second meiosis. Sumner *et al.* (1971), utilizing fluorescent technique, showed that over 1% of sperm from normal males contained two Y chromosomes, implying marked selection against such sperm. Parental age is not advanced above the norm for the general population. There is no evidence that the extra Y chromosome is transmitted to the progeny of XYY males (Parker *et al.*, 1969).

Extensive psychiatric study has been carried out on XYY prisoners (Hope *et al.*, 1967; Price and Whatmore, 1967; Nielsen, 1969; Nielsen *et al.*, 1973). In general, explosive behavior and a propensity to destroy property rather than to display violence to individuals have been noted. Deviant behavior was exhibited quite early, in most cases soon after puberty.

## XYYY and XYYYY Syndromes

There have been but few documented examples of the XYYY and XYYYY syndromes (Sele *et al.*, 1975). Presumably the XYYY condition arises from nondisjunction in spermatologic mitosis followed by a second nondisjunction in one of the Y chromosomes in meiosis, resulting in a sperm bearing three Y chromosomes.

Townes *et al.* (1965) described a 5-year-old male with mild mental retardation (IQ=80), inguinal hernia, undescended testes, valvular pulmonary stenosis, and simian creases.

Retarded intelligence (IQ=70), impulsive aggressive behavior, bilateral simian creases, clinodactyly of fifth fingers, retarded bone age with pseudoepiphyses at the bases of the metacarpals and metatarsals, and lack of patellar epiphyseal calcification were described by Schoepflin and Centerwall (1972).

Ridler *et al.* (1973) noted low normal intelligence, behavior problems with aggressive outbursts, repeated pulmonary infections, hypotrophic testes, sparse body hair, and acne in a 48,XYYY patient. Conversely, Hunter and Quaife (1973) described no stigmata other than sterility.

A most remarkable mosaic 45,XO/49,XYYYY was documented by van den Berghe *et al.* (1968). The boy was mentally retarded with bilateral cataracts, facial asymmetry, clinodactyly of fifth fingers, and brachymesophalangy.

## XII. TURNER SYNDROME (FIG. 11)

Turner (1938) described postpubertal females exhibiting sexual infantilism, short stature, short and webbed neck, and cubitus valgus. Albright *et al.* (1942) showed that these patients had elevated urinary gonadotropins, and Wilkins and Fleischmann (1944) described "streak" gonads devoid of ovarian follicles in such cases. Polani *et al.* (1954) and Wilkins *et al.* (1954) demonstrated that most cases are chromatin-negative, and Ford *et al.* (1959) first described the XO karyotype.

Turner syndrome has been estimated to occur 1 per 2500 female births (Maclean *et al.,* 1964; Mikamo, 1968) and has been noted frequently in abortuses. Parental age is not advanced.

Variation in phenotype has led to some confusion concerning nomenclature. Since the most common features are short stature, streak gonads, and X monosomy or short arm loss of X chromosomal material, all patients with these features are classified here as examples of Turner syndrome (Yunis, 1965). Deletion of the long arm of the Y chromosome has occasionally been associated with the Turner syndrome. Cases with streak gonads and sexual infantilism but normal or increased stature and normal female or male sex chromosome complement are referred to as having "pure gonadal dysgenesis" or, more accurately, XY or XX gonadal dysgenesis.

Primary amenorrhea and sterility are almost constant features of the XO Turner syndrome although exceptions have been noted (Ferguson-Smith, 1965; Greenblatt *et al.,* 1967; Hausmann and Goebel, 1972). Breast development is poor, the chest is broad with seemingly widely spaced, hypoplastic, at times inverted, nipples. The external genitalia are infantile, and pubic hair is sparse.

The histological pattern of the dysgenetic gonad found in Turner syndrome consists of long streaks of white wavy connective tissue stroma without follicles. Follicles are present, however, in fetal and infantile ovaries of patients with Turner syndrome (Gordon and O'Neill, 1969).

Adult height is usually less than 57 inches (144 cm). Various skeletal anomalies include cubitus valgus (about 75%), short fourth metacarpals (about 65%), deformity of medial tibial condyle (about 65%), osteoporosis (about 50%), hypoplasia of cervical vertebrae (about 80%), and small carpal angle (de la Chapelle, 1962; Lemli and Smith, 1963; Finby and Archibald, 1963; Kosowicz, 1965; Engel and Forbes, 1965; Felix *et al.,* 1974).

Birth weight is below the 3rd percentile in about half the cases. In infants, excess skin on the nape and peripheral lymphedema have been noted in 15–50% of the cases. During embryonic life, neck blebs or cystic hygroma are common (Singh and Carr, 1966; Rushton *et al.,* 1969). Toenails are frequently hypoplastic and deepset. With age, the excess skin on the nape metamorphoses into pterygium colli and, with improvement in deep lymphatic circulation, the peripheral lymphedema gradually disappears. Increased numbers of cutaneous nevi are found in about 60%. There is also an increased tendency to keloid formation.

Epicanthal folds, ptosis of upper eyelids, prominent ears, and micrognathia are common facial features. The hairline is low at the nape, and the neck is often short.

Coarctation of the aorta and idiopathic hypertension occur in about 25%. Various renal anomalies, especially horseshoe kidney, can be found in over 80%

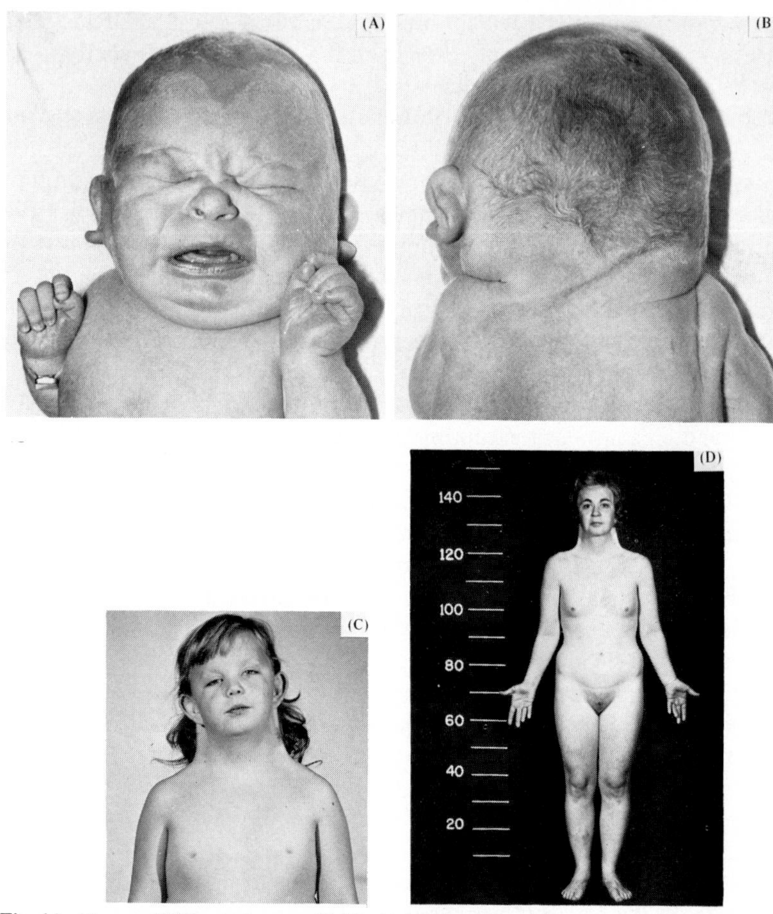


**Fig. 11.** Turner (XO) syndrome. (A,B) At birth, excess skin is present at nape rather than pterygium which develops later. Note protruding pinnas, widely spaced nipples (from Gordon and O'Neill, 1969). (C) Pterygium colli, eyelid ptosis, protruding ears and broad shieldlike chest with small nipples and areolae. (D) Similar phenotype in adult (from G. Tagatz, Minneapolis, Minn.). (E) Low posterior hairline, numerous nevi (from Schönenberg *et al.*, 1957). (F) Short fourth metacarpals. (G) Lymphangiectatic edema of toes, hypoplastic nails. (H) Deepset toenails. (Figs. F and H from G. Tagatz, Minneapolis, Minnesota.)

(Hung and Lo Presti, 1965). Telangiectasia of the small bowel occurs in about 5% (Rosen *et al.*, 1968).

Thyroid antibodies are elevated in XO Turner syndrome but less frequently than in the X-iso X mosaic. Glucose intolerance occurs with greater frequency in patients with Turner syndrome and in their parents than in the normal population (Rimoin, 1975). Many patients with Turner syndrome exhibit a neurocognitional defect in space for perception and orientation (Money and Alexander, 1966).

### A. X Monosomy

Patients with an XO sex complement appear to comprise about 60% of the cases of Turner syndrome. About 20% of spontaneous abortuses lost during the

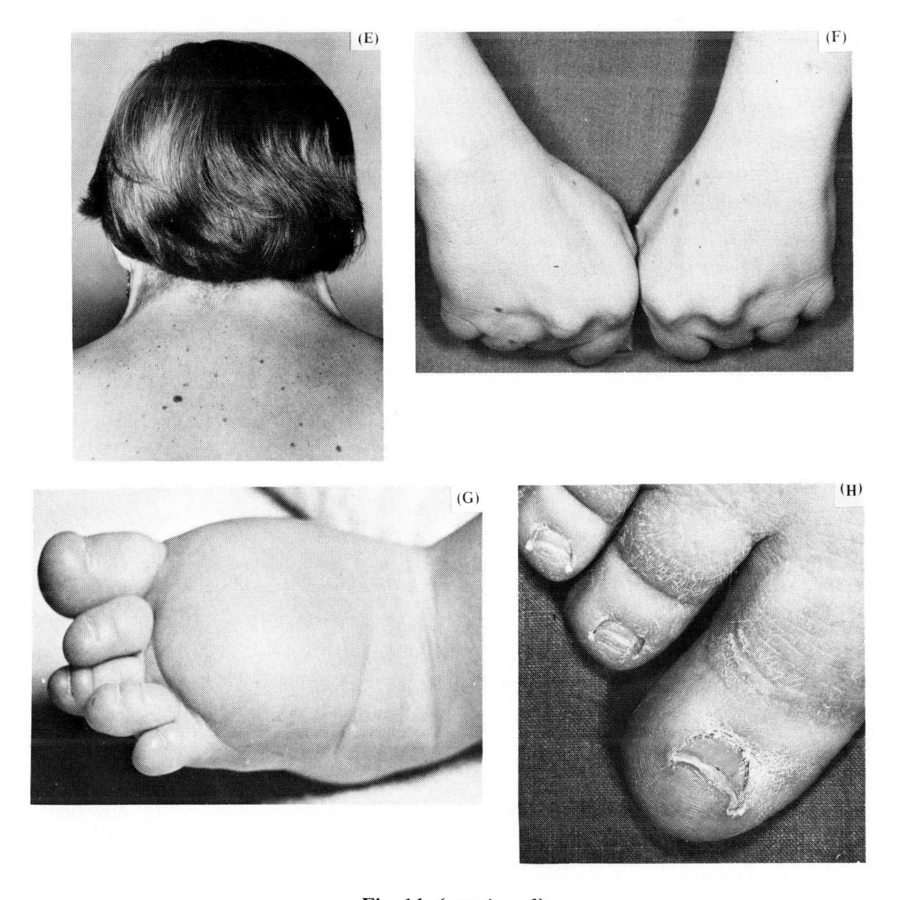


**Fig. 11.** (*continued*)

first eight weeks of pregnancy have an XO karyotype (Larson and Titus, 1970). The apparent high lethality of the 45,XO embryo cannot be explained. XO Turner syndrome patients appear to be more severely affected clinically than other forms of the disorder. Coarctation of the aorta, neck webbing, epicanthal folds, and congenital lymphedema occur more frequently in this group (Ferguson-Smith, 1965). Menstruation has been noted in less than 10% of the cases. Race and Sanger (1969) estimated that in about 75% of the cases, the X was of maternal origin ($X^m$) while 25% were paternal ($X^p$). Dermatoglyphic studies have shown a markedly elevated total finger ridge count (Kerr *et al.*, 1974). There is a coarse reticular pattern of the carpal bones (Bercu *et al.*, 1976).

## B. XO/XX and XO/XXX Mosaicism

Patients with Turner syndrome may have two different cell populations, one having an XO sex constitution, the other a normal XX sex complement. Such individuals are called XO/XX mosaics and constitute about 7–10% of the cases of Turner syndrome (Schmid *et al.*, 1974). The two cell population types may appear in every tissue of the body or only in certain ones. Because of this difference in cell population distribution, study of chromosomes or sex chromatin in a single tissue does not necessarily give an accurate picture.

The usual accepted explanation for XO/XX mosaicism is loss of an X chromosome during cleavage in the early embryo. Difference in clinical patterns may be related to the time of loss and the particular tissues involved.

About 5% of the cases of Turner syndrome are XO/XXX mosaics. Clinically they resemble the XO/XX mosaic. Patients having three stem lines XO/XX/XXX have been reported but are quite similar phenotypically to XO/XX mosaics.

Presumptive evidence for mosaicism lies in a discrepancy between sex chromatin pattern and karyotype, or through observing a low percentage of chromatin-positive nuclei (5–15%) in phenotypic females. The clinical spectrum of XO/XX mosaicism is wide and may vary from cases quite typical of Turner syndrome with many associated anomalies to cases with normal gonads and normal stature. About 20% menstruate (Ferguson-Smith, 1969) and fertility is markedly reduced (Reyes *et al.*, 1976). In contrast to patients with XO Turner syndrome who are prone to aortic coarctation, those with XO/XX karyotypes are likely to have pulmonary stenosis with or without atrial septal defect (Nora *et al.*, 1970), being similar to patients with Noonan syndrome. Total finger ridge counts are also elevated in XO/XX mosaicism (Kerr *et al.*, 1974).

Cases reported as having a typical XO sex complement with the characteristic Turner phenotype might actually be XO/XX mosaics. This state has been described in monozygotic twins, only one of which had the stigmata of Turner syndrome (Potter and Taitz, 1972).

## C. Isochromosome X (XXqi)

About 15–20% of patients with Turner syndrome have an X isochromosome, i.e., replication of the long arm of the late replicating X chromosome (de la Chapelle *et al.*, 1966; Ockey *et al.*, 1966; Sparkes and Motulsky, 1967; Schmid *et al.*, 1974). While they exhibit many of the stigmata seen in those with XO Turner syndrome, i.e., short stature, sexual infantilism, primary amenorrhea, and skeletal anomalies, they are less likely to have pterygium colli and only very rarely have aortic coarctation (Senzer *et al.*, 1973). More than two-thirds of patients having an iso-X chromosome are mosaics (XO/XXqi), i.e., only about 5% of those with Turner syndrome are monosomic XXqi examples. Xg blood group evidence suggests that the isochromosome X arises in the male during meiosis (Race and Sanger, 1969).

The Barr body and polymorphonuclear neutrophil drumsticks are larger than normal. Drumsticks are also more numerous (Taft *et al.*, 1965; Sparkes and Motulsky, 1967).

About 20% of XXqi patients have Hashimoto thyroiditis (Sparkes and Motulsky, 1967; Milet *et al.*, 1967), a finding not observed in those mosaic for XO and isochromosome X (XO/XXqi). Diabetes mellitus probably has a higher frequency in XXqi individuals. Dermatoglyphic studies have demonstrated high fingerprint ridge counts but normal axial triradius placement (Milet *et al.*, 1967).

## D. Short and Long Arm Deletion of an X Chromosome

Deletion of the long arm of an X chromosome (XXq–) is far less likely to be associated with short stature. Perhaps 5% of cases of Turner syndrome have this genotype (Schmid *et al.*, 1974). Ferguson-Smith's (1969) analysis of 14 cases (of which 7 were mosaic) showed that over 60% had normal stature, 55% menstruated, and only 30% had associated anomalies. De la Chapelle *et al.* (1972) described isochromosome for the short arm of an X chromosome, i.e., 46,XXpi. The girl was 159 cm tall and exhibited no stigmata of Turner syndrome. She never menstruated and her ovaries were not palpable. As expected, her Barr bodies were smaller than normal. $Xg^a$ studies showed that the Xpi was of maternal origin. These cases suggest that gonadal differentiation depends in part on material located on the long arm of the X chromosome. However, Turner syndrome has been reported in association with XXq–. Possible explanations have been discussed by Lippe and Crandall (1973).

Ring chromosome X cases (XXr) differ phenotypically according to the size of the ring, reflecting the amount of deletion of both short and long arms (Pfeiffer *et al.*, 1968). The smaller the ring, the greater the deletion and the closer the resemblance to classic XO Turner phenotype (Ferguson-Smith, 1969). XXp–, XXq–, and XXr cases together comprise no more than 5% of cases of Turner syndrome.

### E. Y Deletion

At least a dozen cases of Turner syndrome associated with dicentric Y chromosome have been published (Armendares *et al.*, 1972; Cohen *et al.*, 1973). All have short stature, female phenotype, and most have associated anomalies. A patient with long arm isochromosome Y had Turner syndrome but she was mosaic with a 45,XO cell line (Jacobs and Ross, 1966; Ferguson-Smith, 1969).

### F. XO/XY Mosaicism

About 15% of patients with XO/XY mosaicism exhibit Turner syndrome (Pfeiffer *et al.*, 1968). Internal examination has demonstrated that virtually all patients with an XO/XY karyotype have a uterus. Over 40% have one streak gonad and one testis. 25% have bilateral testes, 15% have two streak gonads, and 10% have no demonstrable gonads. Seminoma or gonadoblastoma develops after puberty in about 20% of the cases and is bilateral in about one-fourth of the patients.

## XIII. XO/XY SYNDROME

The phenotype of the XO/XY individual is quite variable. About 80% have ambiguous genitalia. Approximately 15% exhibit some stigmata of Turner syndrome. Of the 20% with unambiguous genitalia, 15% have a female phenotype (with streak gonads) and 5% have a male phenotype (with bilateral undescended testes).

Virtually all those with ambiguous genitalia have a uterus. About 40% have one streak gonad and one testis, and approximately 15% have two streak gonads. 25% have bilateral testes, while 10% have no gonads. Only 2% are true hermaphrodites.

A serious complication occurring in one-fourth of those with ambiguous genitalia is the development of seminoma or gonadoblastoma. The tumor has bilateral occurrence in about 25%. In no known case has the tumor presented before 11 years of age (Van Campenhout *et al.*, 1969).

## XIV. TRISOMY, TETRASOMY, AND PENTASOMY X

### A. Trisomy X

Jacobs *et al.* (1959) first reported trisomy X. XXX females, in contrast to other human trisomies, do not exhibit a distinctive phenotype (Anderson, 1965;

Pfeiffer *et al.,* 1967). About one-fourth have some type of congenital deformity but there is no constant pattern (Barr *et al.,* 1969); about one-fourth are essentially normal; about one-fourth have mild developmental lags; and one-fourth are borderline between normal and defective, either intellectually or emotionally (Tennes *et al.,* 1975). There is probably also a predisposition to psychosis (Kidd *et al.,* 1963; Raphael and Shaw, 1963; Barr *et al.,* 1969). The sexual disturbances vary from amenorrhea and sterility to menstrual disorders of milder type. Abnormal hormonal levels or deficient ovarian follicles have been found in less than 10% of the cases, but at least 75% have a normal menstrual history and breast development, and several have had children (Barr *et al.,* 1969). Almost without exception the offspring of XXX females have no chromosomal abnormalities. The lack of clinical or chromosomal abnormalities in these children suggests that gametes with two X chromosomes are selected against. The mothers of XXX individuals tend to be older (Kohn *et al.,* 1968; Barr *et al.,* 1969).

Most adult patients with XXX karyotype have been found by screening mental institutions so that an observational bias has been introduced in most reports. Approximately 4–5 per 1000 institutionalized females have trisomy X (Day *et al.,* 1964; Barr *et al.,* 1969), whereas the incidence in newborns has been reported to be as high as 1.2 per 1000 births (Maclean *et al.,* 1964). Barr *et al.* (1969) suggested a general population frequency of 0.56 per 1000. The difference between these rates is said to be statistically significant, so there actually is a true increased frequency of mental deficiency in trisomy X.

About 10% of buccal smear cells show two Barr bodies. An increased number of drumsticks is not seen in polymorphonuclear leukocytes.

### *1. Mosaicism and Trisomy X.*

XXX/XX, XXX/XO, XXX/XX/XO, and other cases of mosaicism have been reported (Day *et al.,* 1964; Barr *et al.,* 1969). Patients with XXX/XX, like those with trisomy X, have no definite phenotype. Those having XO cells have tended to exhibit some stigmata of the Turner syndrome. Several patients with XXX/XX syndrome have had children with chromosome abnormalities (Barr *et al.,* 1969).

### *2. Double Trisomies*

Some individuals with trisomy X have also been found with an extra autosome. Both XXX-trisomy 18 (Uchida and Bowman, 1961; Ricci and Borgatti, 1963; Kohn *et al.,* 1968; Barr *et al.,* 1969) and XXX-trisomy 21 (Yunis *et al.,* 1964a; Day *et al.,* 1964) have been described. In these cases, the phenotype has reflected the extra autosome involved. This is in accordance with the mild and variable expression of the extra X chromosome found in cases of simple trisomy X.

## B. Tetrasomy and Pentasomy X

About two dozen examples of females with more than three X chromosomes have been reported. Mental retardation has been a common feature.

Tetrasomy X has been described by Carr *et al.*, (1961), di Cagno and Franceschini (1968), de Grouchy *et al.* (1968), Park *et al.* (1970), Blackston and Chen (1972), Larget-Piet *et al.* (1972a), Rerrick (1972), Witkowski and Zabel (1974), and others cited by the last authors. All the X chromosomes have been of maternal origin.

Clinical findings have included mental retardation (IQ 30–80) and frequent ocular hypertelorism, epicanthal folds, and mild relative mandibular prognathism. Normal intelligence has, however, been described (Blackston and Chen, 1972). Dermatoglyphic findings have not been remarkable but have included low ridge counts (Telfer *et al.*, 1970). Three Barr bodies have been found in less than 20% of buccal mucosal epithelial cells. Fertility is probably reduced (Gardner *et al.*, 1973).

Pentasomy X is equally rare (Fig. 12). There is some resemblance in phenotype to the XXXXY male (Kesaree and Woolley, 1963; Sergovich *et al.*, 1971; Larget-Piet *et al.*, 1972b). These include severe mental retardation (IQ 20–40),

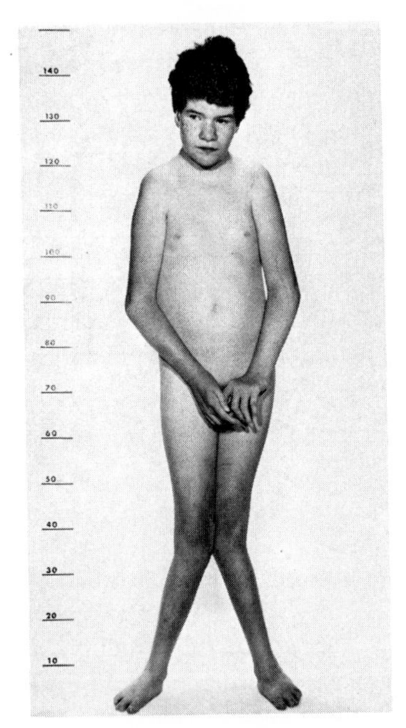


**Fig. 12.** XXXXX syndrome. Phenotype has certain similarity to XXXXY Klinefelter syndrome. Patient has mental retardation, genua valga, and pes planus (from F. Sergovich *et al.*, 1971).

ocular hypertelorism with uncoordinated eye movements, broad flat nose, everted furrowed lips, short neck, low hairline, infantile breasts, normal external genitalia, scanty pubic hair, infantile uterus, and an increased number of digital arches. Nuclei of buccal and vaginal epithelial cells have exhibited one to four Barr bodies, and up to three drumsticks have been noted in polymorphonuclear leukocytes. Tritiated thymidine studies have shown four of the five X's to be late-replicating. A host of skeletal anomalies have included sclerotic cranial sutures, relative mandibular prognathism, retarded bone age, clinodactyly of fifth fingers, pseudoepiphyses of metacarpals, radioulnar synostosis, scoliosis, thick sternum with abnormal segmentation, coxa valga, genua valga, and pes planus. Present in XXXXX females but absent in XXXXY males have been hypoplastic frontal sinuses, microbrachycephaly, and malformed sacral vertebrae. All the X chromosomes have been of maternal origin (Gardner *et al.*, 1973).

## REFERENCES


Aarskog, D. (1966). A new cytogenetic variant of translocation Down's syndrome. *Cytogenetics* **5**, 82–87.

Aarskog, D. (1969). A familial 3/18 reciprocal translocation resulting in duplication-deficiency (3?+, 18q–). *Acta Paediatr. Scand.* **58**, 397–406.

Aksu, F., Mietens, C., and Scholz, W. (1976). Numerische und strukturelle Aberrationen des Chromosoms Nr. 18. *Klin. Pädiat.* **188**, 220–232.

Albright, F., Smith, P. H., and Fraser, R. (1942). A syndrome characterized by primary ovarian insufficiency and decreased stature. Am. J. Med. Sci. **204**, 625–648.

Al Saadi, A. Jubar, J. F., Harm. J., Brough, A. J., Perrin, E. V., and Chen, H. (1976). Triploidy syndrome. *Clin. Genet.* **9**, 43–50.

Alter, M., Gorlin, R., Yunis, J., Peagler, F., and Bruhl, H. (1966). Dermatoglyphics in XXYY Klinefelter's syndrome. *Am. J. Dis. Child.* **111**, 421–424.

Anders, J. M., Jagiello, G., Polani, P., Giannelli, F., Hamerton, J., and Lieberman, D. (1968). Chromosome findings in chronic psychotic patients. *Br. J. Psychiatry* **114**, 1167–1174.

Anderson, I. F. (1965). The triple X syndrome. Clinical and cytological features. *S. Afr. Med. J.* **39**, 841–846.

Anderson, L., Bergman, S., Reitalu, J., and Ansehn, S. (1972). A case of XX male. Cytogenetic findings by autoradiography and fluorescence microscopy. *Hereditas* **70**, 311–314.

Ardran, G. M., Harker, P., and Kemp, F. H. (1972). Tongue size in Down's syndrome. *J. Ment. Defic. Res.* **16**, 160–166.

Arias, D., Passarge, E., Engle, M. A., and German, J. (1970). Human chromosomal deletion–two patients with the 4p– syndrome. *J. Pediatr.* **76**, 82–88.

Armendares, S., Buentello, L., Salamanca, F., and Cantu-Garza, J. (1972). A dicentric Y chromosome without evident sex chromosomal mosaicism, 46,XYq dic, in a patient with features of Turner's syndrome. *J. Med. Genet,* **9**, 96–100.

Assemany, S. R., Neu, R. L., and Gardner, L. I. (1971). XXXXY syndrome in a phenotypic male infant with cardiac abnormalities. *Humangenetik* **12**, 101–104.

Atnip, R., and Summitt, R. (1971). Tetraploidy and 18 trisomy in a 6-year-old triple mosaic boy. *Cytogenetics* **10,** 305–317.

Bach, C., Toublanc, J., and Gautier, M. (1973). Une observation de double aneuploidie chromosomique; trisomie 18 et XXY. *Ann. Genet.* **16,** 61–66.

Backus, J. A. and Darien, G. G. (1968). Group E triple cell line mosaicism with Sprengel's and other asymmetrical congenital abnormalities. *Am. J. Clin. Pathol.* **49,** 527–538.

Bain, A. D., Insley, J., Douglas, D. M., Gould, T. K., and Scott, H. A. (1965). Normal/ trisomy 13–15 mosaicism in two infants. *Arch. Dis. Child.* **40,** 442–445.

Barr, M. L., Sergovich, F. R., Carr, D. H., and Shaver, E. L. (1969). The triple X female: an appraisal based on a study of 12 cases and a review of the literature. *Can. Med. Assoc. J.* **101,** 247–258.

Bartsch-Sandhoff, M., Schade, H., Wiegelmann, W., Solbach, H. G., and Scholz, W. (1974). Ein Beitrag zur Genese von XX– Männern. *Humangenetik* **21,** 245–253.

Becker, K. L., Hoffman, D., Albert, A., Underdahl, L., and Mason, H. (1966). Klinefelter's syndrome. Clinical and laboratory findings in 50 patients. *Arch. Intern. Med.* **118,** 314–321.

Benady, S. G., and Harris, R. J. (1969). Trisomy 18. *Acta Paediatr. Scand.* **58,** 445–448.

Benda, C. E. (1969). "Down's Syndrome. Mongolism and its Management." Grune & Stratton, New York.

Bercu, B. B., Kramer, S., and Bode, H. (1976). A useful radiologic sign for the diagnosis of Turner's syndrome. *Pediatrics* **58,** 737–739.

Berger, R., Tauati, G., Derre, J., Ortiz, M. A., and Martinetti, J. (1974). "Cri-du-chat" syndrome with maternal insertional translocation. *Clin. Genet.* **5,** 428–432.

Betlejewski, S., Klajman, S., and Walczynski, Z. (1964). Radiologische Untersuchungen der Entwicklung der Nasennebenhöhlen im Down-Syndrom. *Ann. Paediatr.* **203,** 355–362.

Blackston, R. D., and Chen, A. T. (1972). A case of 48, XXXX female with normal intelligence. *J. Med. Genet.* **19,** 230–232.

Bodensteiner, J., and Zellweger, H. (1971). Trisomy E/trisomy G mosaicism. *Helv. Paediatr. Acta* **26,** 63–70.

Böhm, R., and Fuhrmann, W. (1969). Lebensfähigkeit bei Monosomie G. *Monatsschr. Kinderheilkd.* **117,** 184–187.

Böök, J. A., Eilon, B., Halbrecht, I., Komlos, L., and Shabatary, F. (1973). Isochromosome Y [46,X,i (Yq)] and female phenotype. *Clin. Genet.* **4,** 410–414.

Borgaonkar, D., and Mules, E. (1970). Comments on patients with sex chromosome aneuploidy. Dermatoglyphics, parental ages, $Xg^a$ blood group. *J. Med. Genet.* **7,** 345–350.

Borgaonkar, D., Mules, E., and Char, F. (1970). Do the 48, XXYY males have a characteristic phenotype? *Clin. Genet.* **1,** 272–275.

Bradbury, J. T., Bunge, R. G., and Boccabella, R. (1956). Chromatin test in Klinefelter's syndrome. *J. Clin. Endocrinol. Metab.* **16,** 689.

Bray, P., and Josephine, A. (1963). An XXXYY sex-chromosome anomaly. *J. Am. Med. Assoc.* **184,** 179–182.

Breg, W. R., Steele, M. W., Miller, O. J., Warburton, D., de Capoa, A., and Allerdice, P. W. (1970). The cri du chat syndrome in adolescents and adults: clinical findings in 13 older patients with partial deletion of the short arms of chromosome no. 5 (5p–). *J. Pediatr.* **77,** 782–791.

Brooks, D. N., Wooley, H., and Kanjilal, G. C. (1972). Hearing loss and middle ear disorders in patients with Down's syndrome (mongolism). *J. Ment. Defic. Res.* **16,** 21–29.

Butler, L. J., Snodgrass, G., France, N., Sinclair, L., and Russell, A. (1965). E (16–18) trisomy syndrome. Analysis of 13 cases. *Arch. Dis. Child.* **40,** 600–611.

Butler, L. J., Chantler, C., France, N., and Keith, C. (1969). A liveborn infant with complete triploidy (69, XXX). *J. Med. Genet.* **6**, 413–421.

Butterworth, T., Strean, L., Beerman, H., and Wood, M. (1964). Syringoma and mongolism. *Arch. Dermatol.* **90**, 483–487.

Carr, D. H. (1971). Chromosome studies in selected spontaneous abortions. Polyploidy in man. *J. Med. Genet.* **8**, 164–174.

Carr, D. H., Barr, M. L., and Plunkett, E. R. (1961). An XXXX sex chromosome complex in two mentally defective females. *Can. Med. Assoc. J.* **84**, 131.

Carter, R., Baker, E., and Hayman, D. (1969). Congenital malformations associated with a ring 4 chromosome. *J. Med. Genet.* **6**, 224–227.

Casey, M. D., and Blank, C. E. (1972). Male patients with chromosome abnormality in two state hospitals. *J. Ment. Defic. Res.* **16**, 215–256.

Casey, M. D., Street, D. R. K., Segall, L. J., and Blank, C. E. (1968). Patients with sex chromatin abnormality at two state hospitals. *Ann. Hum. Genet.* **32**, 53–63.

Challacombe, D. M., and Taylor, A. (1969). Monosomy for a G autosome. *Arch. Dis. Child.* **44**, 113–119.

Chambon, A., David, M., Laurent, C., and Plauchu, H. (1975). La triploïdie chez l'enfant. *Pediatrie* **30**, 371–389.

Chitham, R. G., and MacIver, E. (1965). A cytogenetic statistical survey of 105 cases of mongolism. *Ann. Hum. Genet.* **28**, 309–316.

Christensen, M. F., and Therkelsen, A. J. (1970). A case of XXXXY chromosome anomaly with 4 maternal X chromosomes and diabetic glucose tolerance. *Acta Paediatr. Scand.* **59**, 706–710.

Clarke, C. M., Edwards, J. H., and Smallpeice, V. (1961). 21-trisomy/normal mosaicism in an intelligent child with some mongoloid characters. *Lancet* **1**, 1028–1030.

Close, H., Goonetilleke, A., Jacobs, P., and Price, W. (1968). The incidence of sex chromosome abnormalities in mentally subnormal males. *Cytogenetics* **7**, 277–285.

Cogan, D. G., and Kuwabara, T. (1964). Ocular pathology of 13–15 trisomy syndrome. *Arch. Ophthalmol.* **72**, 246–253.

Cohen, M. M. (1971). The chromosomal constitution of 165 human translocations involving G group chromosomes identified by autoradiography. *Ann. Genet.* **14**, 87–96.

Cohen, M. M., and Bumbalo, T. S. (1967). Double aneuploidy, trisomy 18 and Klinefelter's syndrome. *Am. J. Dis. Child.* **113**, 483–486.

Cohen, M. M., and Davidson, R. G. (1967). Down's syndrome associated with a familial (21q–; 22q+) translocation. *Cytogenetics* **6**, 321–330.

Cohen, M. M., Takagi, N., and Harrod, E. K. (1968). Trisomy D with two D/D translocation chromosomes. *Amer. I. Dis. Child.* **115**, 185–190.

Cohen, M. M., Macgillivray, M., Capraro, V., and Aceto, R. (1973). Human dicentric Y chromosomes. *J. Med. Genet.* **10**, 74–79.

Cohen, M. M., Sr., and Cohen, M. M., Jr. (1971). The oral manifestations of trisomy $G_1$ (Down's syndrome). *Birth Defects, Orig. Artic. Ser.* **7**, 241–251.

Conen, P. E., and Erkman, G. (1966a). Frequency and occurrence of chromosomal syndromes. I. D-trisomy. *Am. J. Hum. Genet.* **18**, 374–386.

Conen, P. E., and Erkman, B. (1966b). Combined mongolism and leukemia. *Am. J. Dis. Child.* **112**, 429–443.

Conen, P. E., Erkman, B., and Metaxotou, C. (1966). The "D" syndrome. *Am. J. Dis. Child.* **111**, 236–247.

Court-Brown, W. M. (1969), Males with an XYY sex chromosome complement. *J. Med. Genet.* **5**, 341–359.

Court-Brown, W. M., Law, P., and Smith, P. G. (1969). Sex chromosome aneuploidy and parental age. *Ann. Hum. Genet.* **33,** 1–14.

Crandall, B. F., Weber, F., Muller, H. M., and Burnwell, J. K. (1972). Identification of the 21r and 22r chromosomes by quinacrine fluorescence, *Clin. Genet.* **3,** 264–270.

Cuenca, C. R., and Becker, K. L. (1968). Klinefelter's syndrome and carcinoma of the breast. *Arch. Intern. Med.* **121,** 159–162.

Cullum, L., and Liebman, J. (1969). The association of congenital heart disease with Down's syndrome (mongolism). *Am. J. Cardiol.* **24,** 354–357.

Cushman, C. J., and Soltan, H. C. (1969). Dermatoglyphics in Klinefelter's syndrome (47,XXY). *Hum. Hered.* **19,** 641–653.

David, M., Chambon, A., Laurent, C., Plauchu, H., Lindler, D., Rouchon, A., de Peretti, E., Genoud, J., and Jeune, M. (1975). La triploidie chez l'enfant. Etude du phenotype. *Pediatrie* **30,** 281–298.

Day, R. W., Larson, W., and Wright, S. W. (1964). Clinical and cytogenetic studies on a group of females with XXX sex chromosome complements. *J. Pediatr.* **64,** 24–33.

de Capoa, A., Warburton, D., Breg, W. R., Miller, D. A., and Miller, O. J. (1967). Translocation heterozygosis: a cause of five cases of the cri du chat syndrome and two cases with a duplication of chromosome number five in three families. *Am. J. Hum. Genet.* **19,** 586–603.

de Grouchy, J. (1969). The 18p–, 18q–, and 18r syndromes. *Birth Defects, Orig. Artic. Ser.* **5,** 74–87.

de Grouchy, J., Royer, P., Salmon, C., and Lamy, M. (1964). Délétion partielle des bras longs du chromosome 18. *Path. Biol.* **12,** 579–581.

de Grouchy, J., Canivet, J., Canlorbe, P., Mantel, O., Borniche, P., and Poitout, M. (1967). Deux observations d'hommes 46,XX. *Ann. Genet.* **10,** 193–200.

de Grouchy, J., Brissaud, H., Richardet, J., Repesee, G., Sanger, R., Race, R., Salmon, C., and Salmon, D. (1968). Syndrome 48,XXXX chez une enfant de six ans. Transmission anormale de groupe Xg. *Ann. Genet.* **11,** 120–124.

de la Chapelle, A. (1962). Cytogenetical and clinical observations in female gonadal dysgenesis. *Acta Endocrinol. (Copenhagen), Suppl.* **65,** 1–65.

de la Chapelle, A. (1972). Nature and origin of males with XX sex chromosomes. *Am. J. Hum. Genet.* **24,** 71–105.

de la Chapelle, A., Wennström, J., Hortling, H., and Ockey, C. H. (1966). Isochromosome X in man. I. *Hereditas* **54,** 260–267.

de la Chapelle, A., Schröder, J., and Pernu, M. (1972). Isochromosomes for the short arm of X, a human 46,XXpi syndrome. *Ann. Hum. Genet.* **36,** 79–87.

di Cagno, L., and Franceschini, P. (1968). Feeblemindedness and XXXX karyotype. *J. Ment. Defic. Res.* **12,** 226–236.

Dumars, K. W., Carnahan, L. G., and Barrett, R. V. (1970). Median facial cleft associated with ring E chromosome. *J. Med. Genet.* **7,** 86–90.

Dunbar, R. D., Toomey, F. B., and Centerwall, W. R. (1975). Radiologic signs of the 4p– (Wolf) syndrome. *Radiology* **117,** 395–396.

Dyggve, H. G., and Mikkelsen, M. (1965). Partial deletion of the short arms of a chromosome of the 4–5 group (Denver), *Arch. Dis. Child.* **40,** 82–85.

Edwards, J. H., Harnden, D., Cameron, A., Crosse, V., and Wolff, O. (1960). A new trisomic syndrome. *Lancet,* **1,** 787–789.

Emberger, J. M., Rey, J., Rieu, D., Dossa, D., Bonnet, H., and Jean, R. (1970). Monosomie 21 avec mosaique 45,XX,21-/46,XX,21 pi. *Arch. Fr. Pediatr.* **27,** 1069–1080.

Endo, A., Yamamoto, M., Watanabe, G., Suzuki, Y., and Sakai, K. (1969). Antimongolism syndrome. *Br. Med. J.* **4,** 148–149.

Engel, E., and Forbes, A. P. (1965). Cytogenetic and clinical findings in 48 patients with congenitally defective or absent ovaries. *Medicine (Baltimore)* **44,** 135–164.

Faed, M., Marrian, V., Robertson, J., Robson, E., and Cook, P. (1972). Inherited pericentric inversion of chromosome 5: a family with history of neonatal death and a case of the "cri du chat" syndrome. *Cytogenetics* **11,** 400–411.

Felix, A., Capek, V., and Pashayan, H. M. (1974). The neck in the XO and XX/XO mosaic Turner's syndrome. *Clin. Genet.* **5,** 77–80.

Ferguson-Smith, M. A. (1965). Karyotype-phenotype correlations in gonadal dysgenesis and their bearing on the pathogenesis of malformations. *J. Med. Genet.* **2,** 142–155.

Ferguson-Smith, M. A. (1966). X-Y chromosomal interchange in the aetiology of true hermaphroditism and XX Klinefelter's syndrome. *Lancet* **2,** 475–476.

Ferguson-Smith, M. A. (1969). Phenotypic aspects of sex chromosome aberrations. *Birth Defects, Orig. Artic. Ser.* **5,** 3–9.

Ferrier, P., Stalder, G., Bamatter, F., Ferrier, S., Büchler, E., and Klein, D. (1964). Congenital asymmetry associated with diploid–triploid mosaicism. *Lancet,* **1,** 80–82.

Finby, N., and Archibald, R. M. (1963). Skeletal abnormalities associated with gonadal dysgenesis. *Am. J. Roentgenol., Radium Ther. Nucl. Med.* [N.S.] **89,** 1222–1235.

Finley, W. H., Finley, S. C., Green, M., and Bush, S. (1972). Triploidy in a liveborn male infant. *J. Pediatr.* **81,** 885–856.

Fischer, P., Glob, E., Friedrich, F., Kunze-Mühl, E. Doleschel, W., and Aichmair, H. (1970). Autosomal deletion syndrome: 46,XX,18q–: A new case report with absence of IgA in serum. *J. Med. Genet.* **7,** 91–98.

Fishler, K., Koch, R., and Donnell, G. N. (1976). Comparison of mental development in individuals with mosaic and trisomy 21 Down's syndrome. *Pediatrics* **58,** 744–748.

Ford, C. E., Jones, K., Polani, P., de Almeida, J., and Briggs, J. (1959). A sex chromosome anomaly in a case of gonadal dysgenesis (Turner's syndrome). *Lancet* **1,** 711–713.

Fraccaro, M., and Lindsten, J. (1960). A child with 49 chromosomes. *Lancet* **2,** 1303.

Franceschini, P., Grassi, E., and Marchese, G. S. (1971). Les principaux signes radiologiques du syndrome 4p–. *Ann. Radiol.* (Paris) **14,** 335–340.

Frøland, A. (1969). Klinefelter's syndrome: clinical, endocrinological, and cytogenetical studies. *Dan. Med. Bull.* **16,** Suppl. 6, 1–108.

Frøland, A., Sanger, R., and Race, R. R. (1968). Xg blood groups in 78 patients with Klinefelter's syndrome and some of their parents. *J. Med. Genet.* **5,** 161–164.

Fryns, J. P., Eggermont, E., Verresen, H., and van den Berghe, H. (1973). The 4p– syndrome, with a report of two new cases. *Humangenetik* **19,** 99–109.

Fujimoto, A., Ebbin, A. J., Towner, J., and Wilson, M. G. (1973). Trisomy 13 in two infants with cyclops. *J. Med. Genet.* **10,** 294–296.

Gagnon, J., Katyk-Longtin, N., de Groot, J., and Barbeau, A. (1961). Double trisomie autosomique à 48 chromosomes (21 + 18). *Union Med. Can.* **90,** 1220–1226.

Gardner, R. J., Veale, A., Sands, V., and Holdaway, M. (1973). XXXX syndrome: case report, and a note on genetic counselling and fertility. *Humangenetik* **17,** 323–330.

Garfinkel, J., and Porter, I. H. (1971). Trisomy 18 in New York state. *Lancet* **2,** 1421–1422.

George, K. P., and Polani, P. E. (1970). Y heterochromatin and XX males. *Nature (London)* **228,** 1215–1216.

Gerald, B. E., and Silverman, F. N. (1965). Normal and abnormal interorbital distances, with specific reference to mongolism. *Am. J. Roentgenol., Radium Ther. Nucl. Med.* [N.S.] **95,** 154–161.

German, J., Lejeune, J., Macintyre, M. N., and de Grouchy, J. (1964). Chromosomal autoradiography in the cri du chat syndrome. *Cytogenetics* **3,** 347–352.

Ginsberg, J., and Perrin, E. V. D. (1965). Ocular manifestations of 13–15 trisomy. *Arch. Ophthalmol.* **74,** 487–495.
Ginsberg, J., Perrin, E. V., and Sueoka, W. T. (1968). Trisomy 18. *Am. J. Ophthalmol.* [3] **66,** 59–67.
Golbus, M. S., Bachman, R., Wiltse, S., and Hall, B. D. (1976). Tetraploidy in a liveborn infant. *J. Med. Genet.* **13,** 329–332.
Gordon, D. L., Krmpotic, E., Thomas, W., Gandy, H., and Paulsen, C. (1972). Pathologic findings in Klinefelter's syndrome. 47,XXY vs. 46 XY/47,XXY. *Arch. Intern. Med.* **130,** 726–729.
Gordon, R. R., and Cooke, P. (1968). Facial appearance in cri du chat syndrome. *Dev. Med. Child Neurol.* **10,** 69–76.
Gordon, R. R., and O'Neill, E. M. (1969). Turner's infantile phenotype. *Br. Med. J.* **1,** 483–485.
Gorlin, R. J., Yunis, J., and Anderson, V. E. (1968). Short arm deletion of chromosome 18. *Am. J. Dis. Child.* **115,** 453–476.
Gorlin, R. J., Cervenka, J., and Pruzansky, S. (1971). Facial clefting and its syndromes. *Birth Defects, Orig. Artic. Ser.* **7,** 3–49.
Gracey, M., and Fitzgerald. M. G. (1967). The XXYYY sex chromosome complement in a mentally retarded child. *Aust. Paediatr. J.* **3,** 119–121.
Greenblatt, R. B., Byrd, J. R., McDonough, P. G., and Mahesh, V. (1967). The spectrum of gonadal dysgenesis. *Am. J. Obstet. Gynecol.* **98,** 151–172.
Greenstein, R., Harris, D., Luzzatti, L., and Cann, H. (1970). Cytogenetic analysis of a boy with XXXY syndrome: origin of the X chromosomes. *Pediatrics* **45,** 677–686.
Guthrie, R. D., Aase, J. M., Asper, A. C., and Smith, D. W. (1971). The 4p– syndrome. *Am. J. Dis. Child.* **122,** 421–425.
Halbrecht, I., Kletzky, O., Komlos, L., Lotker, M., and Gersht, N. (1971). Trisomy D in cyclops. *Obstet. Gynecol.* **37,** 391–393.
Hall, B. (1966). Mongolism in newborn infants. *Clin. Pediatr.* **5,** 4–12.
Hall, B., Fredga, K., and Svenningsen, N. (1967). A case of monosomy G? *Hereditas* **57,** 356–366.
Hambert, G. (1967). Positive sex chromatin in men with epilepsy. *Acta Med. Scand.* **175,** 663–665.
Hamerton, J. L., Giannelli, F., and Plani, P. E. (1965). Cytogenetics of Down's syndrome (mongolism). *Cytogenetics* **4,** 171–185.
Hamerton, J. L. (1970). Fetal sex. *Lancet* **1,** 516–517.
Hamerton, J. L. (1971). "Human Cytogenetics," Vol. 2. Academic Press, New York.
Hausmann, L., and Goebel, K. (1972). Turner's syndrome with menstruation. *J. Med. Genet.* **9,** 100–101.
Hecht, F., Case, M. P., Lovrien, E., Higgins, J., Thuline, H., and Melnyk, J. (1968). Nonrandomness of translocations in man. *Science* **161,** 371–372.
Hirschhorn, K., Cooper, H. L., and Firschein, I. (1965). Deletion of short arms of chromosome 4–5 in a child with defects of midline fusion. *Humangenetik* **1,** 479–482.
Hook, E. B. (1973). Behavioral implications of the human XYY genotype. *Science* **179,** 139–150.
Hook, E. B., and Yunis, J. J. (1965). Congenital asymmetry associated with trisomy 18 mosaicism. *Am. J. Dis. Child.* **110,** 551–555.
Hope, K., Philip, A. E., and Loughrin, J. M. (1967). Psychological characteristics associated with XYY sex-chromosone complement in a state mental hospital. *Br. J. Psychiatry* **113,** 495–498.

Houston, C. S. (1967). Roentgen findings in the XXXXY chromosome anomaly. *J. Can. Assoc. Radiol.* **18,** 258–267.

Howard, R. O. (1972). Ocular abnormalities in the cri-du-chat syndrome. *Am. J. Dis. Child.* **73,** 949–954.

Hsu, T. C., and Lockhart, L. H. (1965). The beginning of the terminal stages of DNA synthesis of human cells with a XXXXY constitution. *Hereditas* **52,** 320–324.

Hubbell, H., Borgaonkar, D., and Bolling, D. (1973). Dermatoglyphic studies of the 47,XYY male. *Clin. Genet.* **4,** 145–157.

Hung, W., and LoPresti, J. M. (1965). Urinary tract anomalies in gonadal dysgenesis. *Am. J. Roentgenol., Radium Ther. Nucl. Med.* [N.S.] **95,** 439–441.

Hunter, H. (1968). Finger and palm prints in chromatin positive males. *J. Med. Genet.* **5,** 112–117.

Hunter, H., and Quaife, R. (1973). A 48,XYYY male: A somatic and psychiatric description. *J. Med. Genet.* **10,** 80–96.

Jacobs, P. A., and Ross, A. (1966). Structural abnormalities of the Y chromosome in man. *Nature (London)* **210,** 352–354.

Jacobs, P. A., and Strong, J. A. (1959). A case of human intersexuality having a possible XXY sex-determining mechanism. *Nature (London)* **183,** 302–303.

Jacobs, P. A., Baikie, A. G., Court-Brown, W. M., MacGregor, T., Maclean, N., and Harnden, D. (1959). Evidence on the existence of the human "super female." *Lancet* **2,** 423–425.

Jacobs, P. A., Price, W. H., Court-Brown, W. M., Brittain, R. P., and Whatmore, P. B. (1968). Chromosome studies on men in a maximum security hospital. *Ann. Hum. Genet.* **31,** 339–358.

Jacobsen, P., and Mikkelsen, M. (1968). Chromosome 18 abnormalities in a family with a translocation t(18p–, 21p+). *J. Ment. Defic. Res.* **12,** 144–157.

Johnson, V. P., Mulder, R., and Hosen, R. (1976). The Wolf-Hirschhorn (4p–) syndrome. *Clin. Genet.* **10,** 104–112.

Jonasson, J., Therkelsen, A., Lauritsen, J., and Lindsten, J. (1972). Origin of triploidy in human abortuses. *Hereditas* **71,** 168–171.

Joseph, M. C., Anders, J. M., and Taylor, A. I. (1964). A boy with XXXXY sex chromosomes. *J. Med. Genet.* **1,** 95–101.

Judge, C. G., Garson, O. M., Pitt, D. B., and Sutherland, G. R. (1974). A girl with Wolf-Hirschhorn syndrome and mosaicism 46,XX/46,XX,4p–. *J. Ment. Defic. Res.* **18,** 79–85.

Kadowaki, J., Shiono, H., Umetso, A., and Ohnishi, M. (1976). Serum IgE concentration in Down's syndrome. *J. Ment. Defic. Res.* **20,** 105–108.

Karsh, R. B. Kanpp, R. F., Nora, J. J., Wolfe, R. R., and Robinson, A. (1975). Congenital heart disease in 49,XXXXY syndrome. *Pediatrics* **56,** 462–464.

Keith, C. G. (1968). The ocular findings in the trisomy syndromes. *Proc. R. Soc. Med.* **61,** 251–253.

Kelch, R. P., Franklin, M., and Schmickel, R. D. (1971). Group G deletion syndromes. *J. Med. Genet.* **8,** 341–345.

Kelly, T., and Rary, J. M. (1974). Mosaicism tetraploidy in a two-year-old female. *Clin. Genet.* **6,** 221–224.

Kerr, J. M., Hsu, L. Y., Workman, P., and Hirschhorn, K. (1974). Total finger ridge count and 45,X mosaicism (with and without Y chromosome). *Clin. Genet.* **5,** 68–71.

Kesaree, N., and Woolley, P. V. (1963). A phenotypic female with 49 chromosomes presumably XXXXX. *J. Pediatr.* **63,** 1099–1103.

Keutel, J., and Dauner, I. (1969). XYY Status bei Kindern. *Z. Kinderheilkd.* **106,** 314–332.

Keutel, J., Dollman, A., and Münster, W. (1970). Triploidie (69,XXY) bei einem lebend geborenen Kind. *Z. Kinderheilkd.* **109,** 104–117.

Kidd, C. B., Knox, R. S., and Mantle, D. J. (1963). A psychiatric investigation of triple X chromosome females. *Br. J. Psychiatry* **109,** 90–94.

Klinefelter, H. F., Jr., Reifenstein, E. C., Jr., and Albright, F. (1942). Gynecomastia, aspermatogenesis without aLeydigism and increased excretion of follicle-stimulating hormone. *J. Clin. Endocrinol.* **2,** 615–627.

Kohn, G., Mayall, B., Miller, M., and Mellman. W. (1967). Tetraploid-diploid mosaicism in a surviving infant. *Pediatr. Res.* **1,** 461–469.

Kohn, G., Winter, J. S. D., and Mellman, W. J. (1968). Trisomy X in three children. *J. Pediatr.* **72,** 248–254.

Kos, A. O., Schuknecht, H. F., and Singer, J. D. (1966). Temporal bone studies in 13–15 and 18 trisomy syndromes. *Arch. Otolaryngol.* **83,** 439–445.

Kosowicz, J. (1965). The roentgen appearance of the hand and wrist in gonadal dysgenesis. *Am. J. Roentgenol., Radium Ther. Nucl. Med.* [N.S.] **93,** 354–361.

Krmpotic, E., Choi, S. Y., and Grossman, A. (1970). Nonrandomness in D- group chromosomes involved in centric-fusion translocation. *Clin. Genet.* **1,** 232–243.

Krmpotic, E., Szego, K., Modestas, R., and Molabola, G. (1972). Localization of male determining factor on short arms of Y chromosome. *Clin. Genet.* **3,** 381–387.

Kunze, J., Stephan, E., and Tolksdorf, M. (1972). Ring-Chromosom 18. *Humangenetik* **15,** 289–318.

Kurien, V. A., and Duke, M. (1968). Trisomy 17–18 syndrome. *Am. J. Cardiol.* **21,** 431–435.

Langdon Down, J. (1866). Observations in an ethnic classification of idiots. *London Hosp. Clin. Lect. Rep.* **3,** 259–262.

Larget-Piet, L., Pignier, J., Bertholet, J., Ayache, P., Bourdon, P., and Larget-Piet, A. (1972a). Syndrome 48,XXXX chez une enfant de 5 ans. *Pediatrie* **27,** 433–443.

Larget-Piet, L., Rivron, J., Baillif, P., Dugay, J., Emerit, L., Larget-Piet, A., and Berthelot, J. (1972b). Syndrome 49,XXXXX chez une fille de 5 ans. *Ann. Genet.* **15,** 115–119.

Larson, S. L., and Titus, J. L. (1970). Chromosomes and abortions. *Mayo Clin. Proc.* **45,** 60–72.

Laurance, B. M., Darby, C. W., and Vanderschueren-Lodeweyckx, M. (1976). Two XX males diagnosed in childhood. *Arch. Dis. Childh.* **51,** 144–148.

Laurent, C., and Robert, J. M. (1966). Segregation d'une translocation D/G "en tandem" sur trois generations. *Ann. Genet.* **9,** 134–136.

Lauritzen, J., Frøland, A., and Johnsen, S. G. (1965). Sex differentiation in the XXXXY chromosome constitution. *Acta Pathol. Microbiol. Scand.* **65,** 321–328.

Law, E. M., and Masterson, J. G. (1969). Familial 18q–syndromes. *Ann. Genet.* **12,** 215–222.

Leão, J. C., Bargman, G. J., Neu, R., Kajii, T., and Gardner, L. (1967). New syndrome associated with partial deletion of short arms of chromosome No. 4. *J. Am. Med. Assoc.* **202,** 434–437.

Lecluse-van der Bilt, F. A., Hagemeijer, A., Smit, E. M., Visser, H. K., and Vaandreger, G. J. (1974). An infant with an XXXYY karyotype. *Clin. Genet.* **5,** 263–270.

Leisti, J. T. Raivio, K. O., Rapola, M., Saksela, E., and Aula, P. (1974). The phenotype of human triploidy. *Birth Defects, Orig. Artic. Ser.* **10,** 248–253.

Lejeune, J. (1959). Le mongolisme. Premier exemple d'aberration autosomique humaine. *Ann. Genet.* **1,** 41–49.

Lejeune, J., Lafourcade, J., Berger, R., Vialatte, J., Boeswillwald, M., Seringe, P., and

Turpin, R. (1963). Trois cas de délétion partielle du bras court d'un chromosome 5. *C. R. Hebd. Seances Acad. Sci.* **257**, 3098–3102.

Lejeune, J., Berger, R., Rethoré, M. *et al.* (1964). Monosomie partielle pour un petit acrocentrique. *C. R. Hebd. Seances Acad. Sci., Ser. D* **259**, 4187–4190.

Lemli, L., and Smith, D. W. (1963). The XO syndrome. A study of the differential phenotype in 25 patients. *J. Pediatr.* **63**, 577–588.

Levinson, A., Friedman, A., and Stamps, F. (1955). Variability in mongolism. *Pediatrics* **16**, 43–54.

Lindsten, J., Bergstrand, C., Tillinger, K., Schwarzacher, H., Tiepolo, L., Muldal, S., and Hökfelt, B. (1966). A clinical and cytogenetical study of three patients with male phenotype and apparent XX sex chromosome constitution. *Acta Endocrinol. (Copenhagen)* **52**, 91–112.

Lippe, B., and Crandall, B. (1973). Turner syndrome with partial deletion of the X chromosome long arm. *Am. J. Dis. Child.* **126**, 222–224.

Loesch, D. (1974). Dermatoglyphic characteristics of 21-trisomy mosaicism in relation to the fully developed syndrome and normal controls. *J. Ment. Defic. Res.* **18**, 209–269.

Lopez, V. (1974). Serum IgE concentration in trisomy 21. *J. Ment. Defic. Res.* **18**, 111–114.

Lurie, I., and Lazjuk, G. (1972). Partial monosomies 18. *Humangenetik* **15**, 203–222.

Lutzner, M. A., and Hecht, F. (1966). Nuclear anomalies of the neutrophil in a chromosomal triplication. The $D_1$ (13–15) trisomy syndrome. *Lab. Invest.* **15**, 597–605.

McDermott, A., Insley, J., Barton, M. E., Rowe, P., Edwards, J. H., and Cameron, A. (1968). Arrhinencephaly associated with a deficiency involving chromosome 18. *J. Med. Genet.* **5**, 60–67.

McGann, B., Alexander, M., and Fox, F. (1970). XXXY chromosomal abnormality in a child. *Calif. Med.* **112**, 30–32.

Maclean, N., Harnden, D., Court-Brown, W., Bond, J., and Mantle, D. (1964). Sex-chromosome abnormalities in newborn babies. *Lancet* **1**, 286–290.

Maclean, N., Court-Brown, W. M., Jacobs, P., Mantle, D., and Strong, J. (1968). A survey of sex chromosome abnormalities in mental hospitals. *J. Med. Genet.* **5**, 165–172.

Magenis, R. E., Hecht, R., and Mulham, S., Jr. (1968). Trisomy 13 ($D_1$) syndrome. Studies on parental age, sex ratio, and survival. *J. Pediatr.* **73**, 222–228.

Magenis, R. E., Armendares, S., Hecht, F., Webber, R. G., and Overton, K. (1972). Identification by fluorescence of two G rings. *Ann. Genet.* **15**, 265–266.

Maniglia, A. J., Wolff, D., and Herques, A. J. (1970). Congenital deafness in 13–15 trisomy syndrome. *Arch. Otolaryngol.* **92**, 181–188.

Marden, P. M., and Yunis, J. J. (1967). Trisomy $D_1$ in a 10-year-old girl. *Am. J. Dis. Child.* **114**, 662–664.

Marinello, M. J., Berkson, R. A., Edwards, J. A., and Bannerman, R. M. (1969). The study of the XYY syndrome in tall men and juvenile delinquents. *J. Am. Med. Assoc.* **208**, 321–325.

Mavalwala, J., Wilson, M. G., and Parker, C. E. (1970). The dermatoglyphics of the 18q– syndrome. *Am. J. Phys. Anthropol.* **32**, 443–450.

Mennicken, U., Pfeiffer, R., Puyn, U., Worbes, H., and Wagener, A. (1968). Klinische und cytogenetische Befunde von 7 Patienten mit Cri-du-chat-Syndrom. *Z. Kinderheilkd.* **104**, 230–256.

Mikamo, K. (1968). Sex chromosomal anomalies in newborn infants. *Obstet. Gynecol.* **32**, 688–699.

Mikkelsen, M. (1971a). Down's syndrome. Current stage of cytogenetic research. *Humangenetik* **12,** 1–28.
Mikkelsen, M. (1971b). Identification of G group anomalies in Down's syndrome by quinacrine dihydrochloride fluorescence staining. *Humangenetik* **12,** 67–73.
Mikkelsen, M., and Stene, J. (1970). Genetic counseling in Down's syndrome. *Hum. Hered.* **20,** 457–464.
Mikkelsen, M., and Vestermark, S. (1974). Karyotype 45,XX,–21/46,XX,21q– in an infant with symptoms of the G-deletion syndrome. *J. Med. Genet.* **11,** 389–393.
Milet, R. G., Plunkett, E. R., and Carr, D. (1967). Gonadal dysgenesis with XX-iso-chromosome constitution and abnormal thyroid patterns. *Acta Endocrinol. (Copenhagen)* **54,** 609–617.
Miller, D. A., Warburton, D., and Miller, O. J. (1969). Clustering in deleted short arm length among 25 cases with a Bp– chromosome. *Cytogenetics* **8,** 109–116.
Miller, J. Q., and Hyde, M. S. (1966). Leukocyte alkaline phosphatase in mongolism. *Neurology* **16,** 577–580.
Miller, O. J., and Warburton, D. (1968). The control of sex chromatin. *Cytogenetics* **7,** 58–77.
Miller, O. J., Breg, W. R., Warburton, D., Miller, D. A., de Capoa, A., Allerdice, P. W., Davis, J., Klinger, H. P., McGilvray, P., and Allen, F. (1970). Partial deletion of the short arm of chromosome No. 4 (4p–). Clinical studies in five unrelated patients. *J. Pediatr.* **77,** 792–801.
Money, J., and Alexander, D. (1966). Turner's syndrome. Further demonstration of the presence of specific cognitial deficiencies. *J. Med. Genet.* **3,** 47–48.
Money, J., Annecillo, C., Van Orman, C., and Borgaonkar, D. S. (1974). Cytogenetics, hormones and behavior disability: comparison of XYY and XXY syndromes. *Clin. Genet.* **6,** 370–382.
Murken, J. D., and Scholz, W. (1967). Serologische Klärung der Herkunft der überzähligen X-Chromosomen beim XXXXY-Syndrom. *Blut* **16,** 164–168.
Murken, J. D., Salzer, G., and Kunze, D. (1970). Ringchromosom Nr. 18 und fehlendes IgA bei einem 6 jährigen Mädchen (46,XX,18r). *Z. Kinderheilkd.* **109,** 1–10.
Nagel, M., and Hoehn, H. (1971). On the non-random involvement of D-group chromosomes in centric fusion translocations in man. *Humangenetik* **11,** 351–354.
Neu, R. L., Gardner, L. I., Williams, M., and Barlow, M. (1973). Three generations and six family members with a t(13q 15q) chromosome. *J. Med. Genet.* **10,** 94–95.
Neuhäuser, G., and Lother, K. (1966). Das Katzenschrei Syndrom. *Monatsschr. Kinderheilkd.* **114,** 278–281.
Neuhäuser, G., Singer, H., and Zang, K. (1968). Cri du chat Syndrom mit Chromosomen-mosaik 46,XY/46,XY,5p–. *Humangenetik* **5,** 315–320.
Neuwirth, J., Starka, L., and Raboch, J. (1972). Different variants of Klinefelter's syndrome and plasma testosterone. *Humangenetik* **15,** 93–95.
Nicolis, F., and Sacchetti, G. (1973). X-ray evaluation of pelvis in mongolism. *Pediatrics* **32,** 1074–1077.
Niebuhr, E. (1971). The cat cry syndrome (5p–) in adolescents and adults. *J. Ment. Defic. Res.* **15,** 277–291.
Niebuhr, E. (1972). Localization of the deleted segment in the cri-du-chat syndrome. *Humangenetik* **16,** 357–358.
Niebuhr, E. (1974). Triploidy in man. *Humangenetik* **21,** 103–125.
Nielsen, J. (1969). Klinefelter's syndrome and the XYY syndrome. Genetical, endocrinological and psychiatric-psychological study of thirty-three severely hypogonadal male pa-

tients and two patients with karyotype, 47,XYY. *Acta Psychiatr. Scand., Suppl.* **209,** 1–353.

Nielsen, J., Christensen, A. L., Schultz-Larsen, J., and Yde, H. (1973). A psychiatric–psychological study of patients with XYY syndrome found outside of institutions. *Acta Psyciat. (Scandinavia)* **49,** 159–168.

Nitowsky, H. M., Sindhavanada, N., Konigsberg, U., and Weinberg, T. (1966). Partial 18 monosomy in the cyclops malformation. *Pediatrics* **37,** 260–269.

Noël, B., Duport, J. P., Revil, D., Dussuyer, I., and Quack, B. (1974). The XYY syndrome: Reality or myth. *Clin. Genet.* **5,** 387–394.

Nora, J. J., Torres, F., Sinha, A., and McNamara, D. (1970). Anomalies of XO Turner syndrome, XX and XY phenotype, and XO/XX Turner mosaic. *Am. J. Cardiol.* **25,** 639–641.

Ockey, C. H., Wennström, J., and de la Chapelle, A. (1966). Isochromosome X in man. II. *Hereditas* **54,** 277–292.

Ong, B. H., Rosner, F., Mahanand, D., Houck, J., and Paine, R. (1967). Clinical, psychological, and radiological comparisons of trisomic and translocation Down's syndrome. *Dev. Med. Child Neurol.* **9,** 307–312.

Orye, E., Coetsier, H., and Hooft, C. (1969). A probably pericentric inversion of a G/C translocation in a mentally retarded child with mongoloid traits. *Hum. Hered.* **19,** 288–298.

Paerregård, P., Mikkelsen, M., Frøland, P., and Anderson, H., (1966). Trisomy No. 17–18. *Acta Pathol. Microbiol. Scand.* **67,** 479–487.

Palutke, W., Chen, Y., and Chen, H. (1973). Presence of brightly fluroescent material in testes of XX males. *J. Med. Genet.* **10,** 170–173.

Park, I. J., Tyson, J. E., and Jones, H. W. (1970). A 48,XXXX female with mental retardation. *Obstet. Gynecol.* **35,** 248–252.

Parker, C. E., Melnyk, J., and Fish, C. H. (1969). The XYY syndrome. *Am. J. Med.* **47,** 801–808.

Parker, C. E., Mavalwala, J., Melnyk, J., and Fish, C. (1970). The 48,XXYY syndrome. *Am. J. Med.* **48,** 777–781.

Parker, C. E., Mavalwala, J., Koch, R., Hatashita, A., and Derencsenyi, A. (1972). The syndrome associated with the partial deletion of the long arms of chromosome 18 (18q–). *Calif. Med.* **117,** 65–71.

Parker, C. E., Donnell, G. N., Mavalwala, J., Hurst, N., and Derencsenyi, A. (1973). A short retarded child with a deletion of the short arm of chromosome 18 (18p–). *Clin. Pediatr.* **12,** 42–46.

Passarge, E., Altrogge, H. C., and Rüdiger, R. A. (1970). Human chromosomal deficiency. The 4p– syndrome. *Humangenetik* **10,** 51–57.

Patau, K., Smith, D., Therman, E. M., Inhorn, S. L., and Wagner, H. P. (1960). Multiple congential anomaly caused by an extra autosome. *Lancet* **1,** 790–793.

Pavone, L., Zellweger, H., Abbo, G., Gauchat, R., and Knecht, B. (1971). A case of trisomy 18 mosaicism with peculiar features. *Humangenetik* **11,** 29–34.

Penrose, L. S. (1966). Anti-mongolism. *Lancet* **1,** 497.

Penrose, L. S. (1967). Finger-print pattern and the sex chromosomes. *Lancet* **1,** 298–300.

Penrose, L. S. (1969). Dermatoglyphics in trisomy 17 or 18. *J. Ment. Defic. Res.* **13,** 44–59.

Penrose, L. S., and Smith, G. F. (1966). "Down's Anomaly." Little, Brown, Boston, Massachusetts.

Pfeiffer, R. A. (1966). Deletion der kurzen Arme des Chromosoms Nr. 18. *Humangenetik* **2,** 178–185.

Pfeiffer, R. A. (1968). Neue Dokumentation zur Abgrenzung eines Syndroms der Deletion des kurzen Arms eines Chromosoms Nr. 4. *Z. Kinderheilkd.* **102,** 49–61.

Pfeiffer, R. A., and Sanger, R. (1973). Origin of 48 XXXY: the evidence of the Xg blood groups. *J. Med. Genet.* **10,** 142–143.

Pfeiffer, R. A., Palm, D., and Jochmus, J. (1967). Das Erscheinungsbild der Trisomie des X–Chromosoms bei Jugendlichen. *Monatsschr. Kinderheilkd.* **115,** 9–18.

Pfeiffer, R. A., Scharfenberg, W., Büchner, R., and Stolecke, H. (1968). Ringchromosomen und zentrische Fragmente bei Turner-Syndrom. *Geburtshilfe Frauenheilkd.* **28,** 11–25.

Plunkett, E. R., and Barr, M. L. (1956). Testicular dysgenesis affecting the seminiferous tubules principally, with chromatin positive males. *Lancet* **2,** 853–857.

Polani, P. E., Hunter, W. F., and Lennox, B. (1954). Chromosomal sex in Turner's syndrome with coarctation of the aorta. *Lancet* **2,** 120–121.

Polani, P. E., Briggs, J. H., Ford, C. E., Clarke, C. M., and Berg, J. M. (1960). A mongol girl with 46 chromosomes. *Lancet* **1,** 721–724.

Potter, A. M., and Taitz, L. S. (1972). Turner's syndrome in one of monozygotic twins with mosaicism. *Acta Paediatr. Scand.* **61,** 413–416.

Prats, J., Sarret, E., Moragas, A., and Martin, C. (1971). Triploid live full term infant. *Helv. Paediatr. Acta* **26,** 164–167.

Preus, M., and Fraser, F. C. (1972). Dermatoglyphics and syndromes. *Am. J. Dis. Child.* **124,** 933–943.

Price, W., and Whatmore, P. B. (1967). Behavior disorders and pattern of crime among XYY males identified at a maximum security hospital. *Br. Med. J.* **1,** 533–536.

Puck, M., Tennes, K., Frankenburg, W., Bryant, K., and Robinson, A. (1975). Early childhood development of four boys with 47,XXY karyotype. *Clin. Genet.* **7,** 8–20.

Rabinowitz, J., Moseley, J. E., Mitty, H., and Hirschhorn, K. (1967). Trisomy 18, esophageal atresia, anomalies of the radius, and congenital hypoplastic thrombocytopenia. *Radiology* **89,** 488–491.

Race, R. R., and Sanger, R. (1969). Xg and sex chromosome abnormalities. *Br. Med. Bull.* **25,** 99–103.

Raphael, T., and Shaw, M. W. (1963). Chromosome studies in schizophrenia. *J. Am. Med. Assoc.* **183,** 1022–1028.

Ratcliffe, S. G., Stewart, A. L., Melville, M. M., Jacobs, P., and Keay, A. J. (1970). Chromosome studies on 3,500 newborn male infants. *Lancet* **1,** 121–122.

Reed, T. E., Borgaonkar, D., Conneally, P. M., Yu, P., Nance, W. E., and Christian, J. (1970). Dermatoglyphic nomogram for the diagnosis of Down's syndrome. *J. Pediatr.* **77,** 1024–1032.

Reinwein, H., Wolf, U., and Ising, H. (1966). Bericht über 3 Mosaikfälle mit $G_1$ Trisomie (Mongolismus). *Helv. Paediatr. Acta* **21,** 300–314.

Reinwein, H., Struwe, F., Bettecken, F., and Wolf, U. (1968). Defizienz am kurzen Arm eines Chromosoms Nr. 18 (46,XX,18p–). Ein einheitliches Missbildungssyndrom. *Monatsschr. Kinderheilkd.* **116,** 511–514.

Reisman, L. E., Kasahara, S., Chung, C., Darnell, A., and Hall, B. (1966). Anti-mongolism: studies in an infant with a partial monosomy of the 21 chromosome. *Lancet* **1,** 394–397.

Rerrick, E. G. (1972). A female with XXXX sex chromosome complement. *J. Ment. Defic. Res.* **16,** 84–89.

Reyes, F., Koh, K. S., and Faiman, C. (1976). Fertility in women with gonadal dysgenesis. *Am. J. Obstet. Gynecol.* **126,** 668–670.

Ricci, N., and Borgatti, L. (1963). XXX-18 trisomy. *Lancet* **2,** 1276–1277.

Richards, B. W. (1969). Mosaic mongolism. *J. Ment. Defic. Res.* **13,** 66–83.

Richards, B. W. (1974). Investigation of 142 mongols and mosaic parents of mongols: cytogenetic analysis and maternal age at birth. *J. Ment. Defic. Res.* **18,** 199–208.

Richmond, H. G., Macarthur, P., and Hunter, D. (1973). A "G" deletion syndrome: antimongolism. *Acta Paediatr. Scand.* **62,** 216–220.

Ridler, M., Lax, R., Mitchell, M., Shapiro, A., and Saldaña-Garcia, P. (1973). An adult with XYYY sex chromosomes. *Clin. Genet.* **4,** 69–71.

Rimoin, D. L. (1975). Genetic syndromes associated with abnormal glucose tolerance in childhood and adolescence. *Modern Problems Pediat.* **12,** 403–408.

Robinson, J. A. (1973). Origin of extra chromosome in trisomy 21. *Lancet* **1,** 131–133.

Roche, A. F., Seward, F., and Sunderland, S. (1961). Nonmetrical observations on cranial roentgenograms in mongolism. *Am. J. Roentgenol., Radium Ther. Nucl. Med.* [N.S.] **85,** 659–661.

Rohde, R. A. (1964). Klinefelter's syndrome with pulmonary disease and other disorders. *Lancet* **2,** 149–150.

Rohde, R. A., and Tompkins, R. (1965). Cri-du-chat due to a ring B (5) chromosome. *Lancet* **2,** 1075–1076.

Rosen, K. M., Sirota, D. K., and Marinoff, S. (1968). Gastrointestinal bleeding in Turner's syndrome. *Ann. Intern. Med.* **67,** 145–150.

Rosner, F., and Ong, B. H. (1967). Dermatoglyphic patterns in trisomic and translocation Down's syndrome (mongolism). *Am. J. Med. Sci.* **253,** 566–570.

Rosner, F., Ong, B., Paine, R., and Mahanand, D. (1965). Biochemical differentiation of trisomic Down's syndrome (mongolism) from that due to translocation. *N. Engl. J. Med.* **273,** 1356–1361.

Ross, L. J. (1968). Dermatoglyphic observation in a patient with trisomy 18. *J. Pediatr.* **72,** 862–863.

Rushton, D. I., Faed, M., Richards, S., and Bain, A. D. (1969). The fetal manifestations of the 45XO karyotype. *J. Obstet. Gynaecol. Br. Commonw.* **76,** 266–272.

Ruvalcaba, R., and Thuline, H. (1969). IgA absence associated with short arm deletion of chromosome No. 18. *J. Pediatr.* **74,** 964–965.

Sabater, J., Antich, J., Lluch, M., and Pérez del Pulgar, J. (1972). Deletion of short arm of chromosome 18 with normal levels of IgA. *J. Ment. Defic. Res.* **16,** 103–111.

Sacrez, R., Clevert, J., Klein, M., Paira, M., Rumpler, J., Mandry, J., and Meyer, R. (1965). Dysgénésie gonadosomatique XXXXY. *Arch. Fr. Pediatr.* **22,** 41–52.

Sandberg, A. A., Koepf, G., Ishihara, T., and Hauschka, T. (1961). An XYY human male *Lancet* **2,** 488.

Sanger, R., Tippett, P., and Gavin, J. (1971). Xg groups and sex chromosome abnormalities in people of northern European ancestry. *J. Med. Genet.* **8,** 417–426.

Scheike, O., Visfeldt, J., and Petersen, B. (1973). Male breast cancer. *Acta Pathol. Microbiol. Scand., Sect. A* **81,** 352–358.

Schindler, A. M., and Mikamo, K. (1970). Triploidy in man. Report of a case and a discussion on etiology. *Cytogenetics* **9,** 116–130.

Schindeler, J. D., and Warren, R. J. (1973). Dermatoglyphics in the G deletion syndromes. *J. Ment. Defic. Res.* **17,** 149–159.

Schinzel, A., and Schmid, W. (1971). Trisomie 18. *Helv. Paediatr. Acta* **26,** 673–685.

Schlegel, R. J., Aspigalla, M., Neu, R., and Gardner, L. (1965). A boy with XXYY chromosome constitution. *Pediatrics* **36,** 113–119.

Schmickel, R. D., Silverman, E. M. Floyd, A., Payne, F., Pooley, J., and Beck, M. (1971). A live born infant with 69 chromosomes. *J. Pediatr.* **79,** 97–103.

Schmid, W., and Vischer, D. (1967). A malformed boy with double aneuploidy and diploid-triploid mosaicism 48,XXYY/71,XXXYY. *Cytogenetics* **6,** 145–155.

Schmid, W., Naef, E., Mürset, G., and Prader, A. (1974). Cytogenetic findings in 89 cases of Turner's syndrome with abnormal karyotypes. *Humangenetik* **24,** 93–104.

Schoepflin, G. S., and Centerwall, W. R. (1972). 48,XYYYY: a new syndrome. *J. Med. Genet.* **9,** 356–359.

Schönenberg, H., Hollstein, K., and Kosenow, W. (1957). Das klinische Bild und das chromosomale Geschlecht der Gonadendysgenese. *Z. Kinderheilkd.* **79,** 383–412.

Schroeder, H., Schleiermacher, E., Schroeder, T., Bauer, H., Richter, C., and Schwenk, J. (1967). Zur klinischen Differentialdiagnose des Cri du Chat Syndroms. *Humangenetik* **4,** 294–304.

Schultz, J., and Krmpotic, E. (1968). Monosomy G mosaicism in two unrelated children. *J. Ment. Defic. Res.* **12,** 255–268. (Case 1-type I, case 2-type II.)

Sele, B., Bachelot, Y., Richard, J., Muller, J., Jalbert, P., and Berthet, J. (1975). Les hommes 48,XYYY. *Pediatrie* **30,** 601–607.

Senzer, N., Aceto, T., and Cohen, M. M. (1973). Isochromosome X. Clinical and psychological findings. *Am. J. Dis. Child.* **126,** 312–316.

Sergovich, F., Uilenberg, C., and Pozsonyi, J. (1971). The 49,XXXXX chromosome constitution: similarities to the 49 XXXXY condition. *J. Pediatr.* **78,** 285–290.

Shapiro, B. L., Gorlin, R. J., Redman, R. S., and Bruhl, H. H. (1967). The palate and Down's syndrome. *N. Engl. J. Med.* **276,** 1460–1463.

Shapiro, L., Hsu, L., Calvin, M., and Hirschhorn, K. (1970). XXXXY boy. A 15-month-old child with normal intelligence development. *Am. J. Dis. Child.* **119,** 79–81.

Shibata, K., Waldenmaier, C., and Hirsch, W. (1973). A child with a 21-ring chromosome, 45,XX,21–/46,XX, 21r investigated with the banding technique. *Humangenetik* **18,** 315–319.

Shih, L., Hsu, L. Y., Sujansky, E., and Kushnick, T. (1974). Trisomy 18 mosaicism in two siblings. *Clin. Genet.* **5,** 420–427.

Simpson, J., Dische, R., Morillo-Cucci, G., and Connolly, C. (1972). Triploidy (69,XXY) in a liveborn infant. *Ann. Genet.* **15,** 103–106.

Simpson, J. L., Morillo-Cucci, G., Horwith, M., Stiefel, F. H., Feldman, F., and German, J. (1974). Monozygotic twins with the complement 48,XXXY. *Humangenetik* **21,** 301–308.

Singh, D. N., Osborne, R. A., and Wiscovitch, R. A. (1973). Transmission of the cri-du-chat syndrome from a maternal balanced translocation carrier, t(5p–; 11q+). *Humangenetik* **20,** 361–365.

Singh, R. P., and Carr, D. H. (1966). The anatomy and histology of XO human embryos and fetuses. *Anat. Rec.* **155,** 369–384.

Smith, D. W. (1969). The 18 trisomy and 13 trisomy syndromes. *Birth Defects, Orig. Artic. Ser.* **5,** 67–71.

Sparkes, R. S., and Motulsky, A. G. (1967). The Turner syndrome with isochromosome X and Hashimoto's thyroiditis. *Ann. Intern. Med.* **67,** 132–144.

Spitzer, R., Rabinowitch, J., and Wybar, K. C. (1961). A study of the abnormalities of the skull, teeth and lenses in mongolism. *Can. Med. Assoc. J.* **84,** 567–572.

Steinberger, E., Smith, K. D., and Perloff, W. (1965). Spermatogenesis in Klinefelter's syndrome. *J. Clin. Endocrinol. Metab.* **25,** 1325–1330.

Stene, J. (1970a). Detection of higher recurrence risk for age-dependent chromosome abnormalities with an application to trisomy $G_1$ (Down's syndrome). *Hum. Hered.* **20,** 112–122.

Stene, J. (1970b). A statistical segregation analysis of (21q 22q) translocations. *Hum. Hered.* **20,** 465–472.

Stewart, J. M., Go, S., Ellis, E., and Robinson, A. (1970). Absent IgA and deletion of chromosome 18. *J. Med. Genet.* **7**, 11–19.

Stewart, R. E. (1974). Taurodontism in X chromosome aneuploid syndromes. *Clin Genet.* **6**, 341–344.

Stoll, C., Sacrez, R., Willard, D., and Freysz, H. (1972). Un cas de trisomie 18 avec aplasie bilaterale du radius et thrombopenie. *Pediatrie* **27**, 537–542.

Stoll, C., Levy, J. M., and Terrade, E. (1974). Les survies prolongées dans la trisomie 18. *Ann. Pediatr.* **21**, 185–190.

Šubrt, I., and Pokorny, J. (1970). Familial occurrence of 18q–. *Humangenetik* **10**, 181–187.

Sumner, A. T., Robinson, J. A., and Evans, H. J. (1971). Distinguishing between X,Y, and YY-bearing human spermatozoa by fluorescence and DNA content. *Nature (London), New Biol.* **229**, 231–233.

Sutherland, G. R., and Wiener, S. (1972). Cytogenetics of 271 mongols. *Aust. Paediatr. J.* **8**, 90–91.

Sutherland, G. R., Fitzgerald, M. G., and Danks, D. (1972). Difficulty in showing mosaicism in the mother of three mongols. *Arch. Dis. Child.* **47**, 970–971.

Taft, P. D., Dalal, K. P., McArthur, J. W., and Worcester, J. (1965). Sex chromatin body size and its relation to X chromosomes structure. *Cytogenetics* **4**, 87–95.

Tan, K. L. (1971). The third fontanelle. *Acta Paediatr. Scand.* **60**, 329–332.

Taylor, A. I. (1967). Patau's, Edwards' and cri du chat syndromes: a tabulated summary of current findings. *Dev. Med. Child Neurol.* **9**, 76–86.

Taylor, A. I. (1968). Autosomal trisomy syndromes. A detailed study of 27 cases of Edwards' syndrome and 27 cases of Patau's syndrome. *J. Med. Genet.* **5**, 227–252.

Taylor, A. I., and Moores, E. C. (1967). A sex chromatin survey of newborn children in two London hospitals. *J. Med. Genet.* **4**, 258–259.

Taylor, A. I., Challacombe, D. N., and Howlett, R. M. (1970). Short arm deletion, chromosome 4, (4p–), a syndrome? *Ann. Hum. Genet.* **34**, 137–144.

Taylor, M. B., Juberg, R., Jones, B., and Johnson, W. A. (1970). Chromosomal variability in the $D_1$ trisomy syndrome. *Am. J. Dis. Child.* **120**, 374–381.

Telfer, M. A., Richardson, C. E., Helmken, J., and Smith, G. F. (1970). Divergent phenotypes among 48,XXXX and 47,XXX females. *Am. J. Hum. Genet.* **22**, 326–335.

Tennes, K., Puck, M., Bryant, K., Frankenburg, W., and Robinson, A. (1975). A developmental study of girls with trisomy X. *Am. J. Hum. Genet.* **27**, 71–86.

Terheggen, H. G., Pfeiffer, R. A., Hang, H., Hertl, M., Diggins, A., and Schünke, W. (1973). Das XXXXY Syndrom. *Z. Kinderheilkd.* **115**, 209–234.

Theilgaard, A., Nielsen, J., Sorensen, A., Frøland, A., and Johnsen, A. (1971). "A Psychological-Psychiatric Study of Patients with Klinefelter's Syndrome." Munksgaard, Copenhagen.

Thorburn, M. J., and Johnson, B. E. (1966). Apparent monosomy of a G autosome in a Jamaican infant. *J. Med. Genet.* **3**, 290–292.

Townes, P. L., Ziegler, N. A., and Lenhard, L. W. (1965). A patient with 48 chromosomes (XYYYY). *Lancet* **1**, 1041–1045.

Tumba, A. (1972). Le phénotype XXXXY. Etude analytique et synthétique. À-propos de 3 cas personnels et de 67 autres cas de la literature. *J. Genet. Hum.* **20**, 9–48.

Turner, H. H. (1938). A syndrome of infantilism, congenital webbed neck, and cubitus valgus. *Endocrinology* **23**, 566–574.

Uchida, I. A. (1973). Paternal origin of the extra chromosome in Down's syndrome. *Lancet* **2**, 1258.

Uchida, I. A., and Bowman, L. (1961). XXX-18 trisomy. *Lancet* **2**, 1094.

Uchida, I. A., McRae, K., Wang, A., and Roy, M. (1965). Familial short arm deficiency of chromosome 18 concomitant with arhinencephaly and alopecia congenita. *Am. J. Hum. Genet.* **17,** 410–419.

Uchida, I. A., Ray, M., and Duncan, B. P. (1966). 21 trisomy with a XYY set chromosome complement. *J. Pediatr.* **69,** 295–298.

Valentine, C. H., McClelland, M. A., and Sergovich, F. R. (1971). The growth and development of four XYY infants. *Pediatrics* **48,** 583–594.

Van Campenhout, J., Lord, J., Vauclair, R., Lanthier, A., and Berard, M. (1969). The phenotype and gonadal histology in XO/XY mosaic individuals: report of two personal cases. *J. Obstet. Gynaecol. Br. Commonw.* **76,** 631–639.

van den Berghe, H., Verresen, H., and Cassiman, J. J. (1968). A male with 4 Y chromosomes. *J. Clin. Endocrinol. Metab.* **28,** 1370–1372.

Visfeldt, J. (1969). Familial D/D translocation. *Acta Pathol. Microbiol. Scand.* **75,** 545–554.

Vormittag, W., and Weninger, M. (1972). XXXY Klinefelter-Syndrom. *Humangenetik* **15,** 327–333.

Wagenbichler, P., Killian, W., Rett, A., and Schnedl, W. (1976). Origin of the extra chromosome No. 21 in Down's syndrome. *Humangenetik* **32,** 13–16.

Walker, S., Andrews, J., Gregson, N. M., and Gault, W. (1973). Three further cases of triploidy in man surviving to birth. *J. Med. Genet.* **10,** 135–141.

Waltzer, S., Gerald, P. S., Breau, G., O'Neill, D., and Diamond, L. K. (1966). Hematologic changes in the $D_1$ trisomy syndrome. *Pediatrics* **38,** 419–429.

Warburg, M. (1960). Anophthalmos complicated by mental retardation and cleft palate. *Acta Ophthalmol.* **38,** 395–404.

Warburton, D., and Miller, O. J. (1967). Dermatoglyphic features of patients with a partial short arm deletion of a B group chromosome. *Ann. Hum. Genet.* **31,** 189–208.

Ward, P. H., Engel, E., and Nance, W. E. (1968). The larynx in the cri du chat (cat cry) syndrome. *Trans. Am. Acad. Ophthalmol. Otolaryngol.* **72,** 90–102.

Warren, R., Rimoin, D., and Summitt, R. (1973). Identification by fluorescent microscopy of the abnormal chromosomes associated with the G-deletion syndromes. *Am. J. Hum. Genet.* **25,** 77–81.

Weber, F. M., and Sparkes, R. S. (1970). Trisomy E (18) syndrome. Clinical spectrum in 12 new cases including chromosomal autoradiography in 4. *J. Med. Genet.* **7,** 363–366.

Weinkove, C., Webb, P. M., and McDonald, R. (1969). The cat-cry syndrome. *S. Afr. Med. J.* **43,** 218–220.

Weinstein, E. D., and Warkany, J. (1963). Maternal mosaicism and Down's syndrome. *J. Pediatr.* **63,** 599–604.

Wertelecki, W., Graham, J. M. Jr., and Sergovich, F. R. (1976). The clinical syndrome of triploidy. *Obstet. Gynecol.* **47,** 69–76.

Wilcock, A. R., Adams, F. G., Cooke, P., and Gordan, R. R. (1970). Deletion of short arm of No. 4 (4p–). *J. Med. Genet.* **7,** 171–176.

Wilkins, L., and Fleischmann, W. (1944). Ovarian agenesis. Pathology, associated clinical symptoms, and the bearing on the theories of sex differentiation. *J. Clin. Endocrinol.* **4,** 357–375.

Wilkins, L., Grumbach, M., and Van Wyk, J. (1954). Chromosome sex in "ovarian dysgenesis." *J. Clin. Endocrinol. Metab.* **14,** 1270–1271.

Williams, D. L., and Runyan, J. W., Jr. (1966). Sex chromatin and chromosome analysis in the diagnosis of sex anomalies. *Ann. Intern. Med.* **64,** 422–459.

Willich, E., Fuhr, U., and Kroll. W. (1975). Skeletal manifestations in Down's syndrome. Correlation between roentgenologic and cytogenetic findings. *Ann. Radiol.* **18,** 355–358.

Wilson, J. A. (1971). Fertility in balanced heterozygotes for a familial centric fusion translocation, t (Dq Dq). *J. Med. Genet.* **8,** 175–178.

Wilson, M. G., Towner, J. W., and Negus, L. (1970). Wolf-Hirschhorn syndrome associated with an unusual abnormality of chromosome No. 4. *J. Med. Genet.* **7,** 164–170.

Witkowski, R., and Zabel, R. (1974). Tetrasomie X: Ursache für primäre Amennorrhoe bei einer hochwuchsigen Frau. *Deut. Gesundheitsw.* **29,** 793–795.

Wolf, U., Reinwein, H., Porsch, R., Schröter, R., and Baitsch, H. (1965). Deficienz an den kurzen Armen eines Chromosoms Nr. 4. *Humangenetik* **1,** 397–413.

Wolf, U., Reinwein, H., Gorman, L., and Kunzer, W. (1967). Deletion on long arm of chromosome 18 (46,XX,18q–). *Humangenetik* **5,** 70–71.

Yamamoto, M., and Ingalls, T. H. (1972). Delayed fertilization and chromosome anomalies in the hamster embryo. *Science* **176,** 518–519.

Yanoff, M., and Rorke, L. (1973). Ocular and central nervous system findings in tetraploid-diploid mosaicism. *Am. J. Ophthalmol.* **75,** 1036–1042.

Yunis, J. J. (1965). Human chromosomes in disease. *In* "Human Chromosome Methodology" (J. J. Yunis, ed.), 1st ed., p. 187. Academic Press, New York.

Yunis, J. J., and Hook, E. B. (1966). Desoxyribonucleic acid replication and mapping of the $D_1$ chromosome. *Am. J. Dis. Child.* **111,** 83–89.

Yunis, J. J., Hook, E. B., and Alter, M. (1964a). XXX 21-trisomy. *Lancet* **1,** 437–438.

Yunis, J. J., Hook, E. B., and Mayer, M. (1964b). Desoxyribosenucleic acid replication pattern of trisomy 18. *Lancet* **2,** 286–287.

Yunis, J. J., Hook, E., and Mayer, M. (1965). Identification of the mongolism chromosome by DNA replication analysis. *Am. J. Hum. Genet.* **17,** 191–201.

Zaleski, W. A., Houston, C., Pozsonyi, J., and Ying, K. (1966). The XXXXY chromosome anomaly. Report of three new cases and review of 30 cases from the literature. *Can. Med. Assoc. J.* **94,** 1143–1154.

Zellweger, H. (1966). Cri du chat with chromosomal mosaicism. *Lancet* **2,** 57.

Zergollern, L., Drazancic, A., Damjanov, I., Hitrec, V., and Gorecam, V. (1972). A liveborn infant with triploidy (69,XXX). *Z. Kinderheilkd.* **112,** 293–300.

Zollinger, H. (1969). Das XXXY Syndrom. *Helv. Paediatr. Acta.* **24,** 589–599.

# 4

# Syndromes Involving Chromosomes 4, 9, and 12

MARIE-ODILE RETHORÉ



## I. ABNORMALITIES OF CHROMOSOME 4

### A. Introduction

In 1965, Wolf *et al.* (1965) described the first abnormality of chromosome 4, which involved partial monosomy of the short arm of this chromosome. After this initial identification, a great number of publications (Rethoré, 1976) permitted precise description of the cytological and clinical features of the 4p– syndrome (see Chapter 3).

The availability of techniques for revealing the structural details of chromatids has very recently permitted identification of two new syndromes: partial trisomy of the short arm (40 trisomy) and partial trisomy of the distal segment of the long arm (partial 4q trisomy). It has also permitted analysis of three observations of partial monosomy of the long arm (partial 4q monosomy).

## B. 4p Trisomy

Trisomy of the short arm of chromosome 4 is manifested by features which are sufficiently strong to permit clinical diagnosis at birth (Rethoré *et al.,* 1974b).

Wilson *et al.,* in 1970, offered the first cytological demonstration of the syndrome, and 21 cases have been reported since (Table I).

The patient whom Gustavson *et al.* (1964) reported as being affected by trisomy of the short arm of a B chromosome had malformation phenomena fully comparable to those of the 4p trisomy syndrome. Although this case was not the subject of an analysis relying on banding techniques, the similarity of phenotypes justifies the opinion that this was actually the first observation of a 4p trisomy.

### *1. Phenotype (Figs. 1 and 2)*

In the newborn, the face is rounded. The forehead, which is low and flat, is abundantly covered with lanugo. The eyebrows, which are low set and are very developed in their innermost portions, intersect above the nose. Aplasia of the nasal bones exists in contrast to the glabella, which forms a protrusion extending over the long and flat nasal saddle. The nose is small at the tip, and the tip is round-shaped. The nostril openings, which are very dilated, are bordered by thick wings.

In older patients, protrusion of the glabella and the globular shape of the tip of the nose persist, although nose length becomes normal, thereby accounting for the semblance of a "boxer's nose". On the forehead, the metopic suture sometimes protrudes.

Generally, the eyelid slits are horizontal, although they are sometimes affected by blepharophimosis. The eyeballs are small, and convergent or divergent strabismus is frequent.

The upper lip is elongated and it projects slightly, while the philtrum is barely visible. In one case (Rethoré *et al.,* 1974a), there was a unilateral harelip. The mouth is excessively wide, and there is a gothic palate, while the chin is pointed and recessed. The teeth are unevenly arranged, of diverse sizes, and subject to decay.

**TABLE I**
**Observations of Trisomy 4p**

| Authors | Parents | | Patients | | |
|---|---|---|---|---|---|
| | Maternal age | Parental translocation | Karyotype | Birth | Death |
| Gustavson *et al*. (1964) | 33 | t(Bp–;Gp+) mat. | 46,XY,Gp+ | ? | 3½ months |
| Wilson *et al*. (1970) | 20 | Inv.(4)(p–;q+) mat. | 46,XY,4q+ | ? | 18 months |
| Gouw *et al*. (1972) | 34 | t(4p–;Dp+) pat. | 46,XY,Dp+ | ? | – |
| Schinzel and Schmid (1972) | 27 | t(4p–;18q+) pat. | 46,XY,18q+ | 1970 | 7 months |
| Metz *et al*. (1973) | 36 | t(4p–;22p+) mat. | 46,XX,22p+ | 1969 | – |
| Schwanitz and Grosse (1973) | | | | | |
| Case 1 | 28 | t(4;22)(p14;p11) pat. | 46,XX,22p+ | 1956 | – |
| Case 2 | 27 | | 46,XY,22p+ | 1955 | – |
| Rethoré *et al*. (1974a) | 20 | Inv.(4)(p14;q35) pat. | 46,XY,4q+ | 1971 | – |
| Dallapiccola *et al*. (1974) | 27 | Inv.(4)(p13q;35) pat. | 46,XX,4q+ | 1970 | – |
| Giovannelli *et al*. (1974) | | | | | |
| Case 1 | 20 | t(4;22)(p11;p12) mat. | 46,XX,22p+ | 1960 | – |
| Case 2 | 27 | | 46,XX,22p+ | 1967 | – |
| Sartori *et al*. (1974) | | | | | |
| Case 1 | 27 | t(4;22)(p14;q13) pat. | 46,XX,22q+ | 1968 | – |
| Case 2 | 30 | | 46,XY,22q+ | 1971 | – |

*continued*

**TABLE I (*continued*)**

| Authors | Parents | | Patients | | |
|---|---|---|---|---|---|
| | Maternal age | Parental translocation | Karyotype | Birth | Death |
| Owen *et al*. (1974) | 24 | t(4;21)(p14;p11) pat. or t(4;21)(q13;q21) pat. | 46,XX,21p+ | ? | 3 months |
| Hustinx *et al*. (1975) | 31 | t(4;15)(p12;q11) mat. | 47,XX,+der(15),t(4;15)(p12;q11) | 1962 | 11 years |
| Furbetta *et al*. (1975) | 26 | t(4p 21p)(4q 21q) mat. | 47,XX,+der(21),der(4),der(21) t(4p21p;4q21q) | ? | – |
| Schröcksnadel *et al*. (1975) | 27 | t(4;15)(p14;p12) pat. | 46,XX,15p+ | 1959 | – |
| Giraud *et al*. (1975) | | | | | |
| Case 1 | 28 | t(4;15)(q12;q13) mat. | 46,XY,15p+ | ? | 26 days |
| Case 2 | ? | | 46,XX,15p+ | ? | 2½ months |
| Case 3 | 30 | – | 46,XX,4p+ | ? | – |
| André *et al*. (1976) | 26 | – | 46,XY,–4,t(1;4)(p36;q11)+i(4) pter→c→pter) | 1972 | – |
| Darmady and Seabright (1975) | | | | | |
| Case 1 | 28 | t(4;21)(p11;p12) mat. | 46,XY,21p+ | 1971 | 15 months |
| Case 2 | 27 | | ? | 1970 | 13 days |

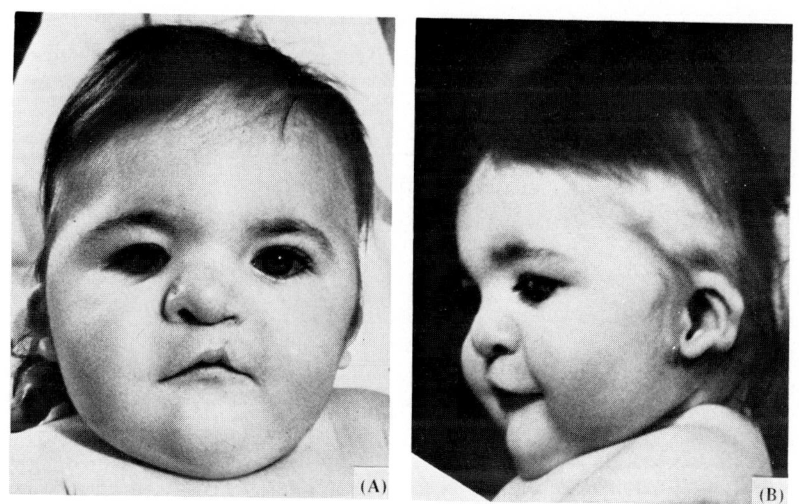

**Fig. 1.** Thirteen-month old male child with 4p trisomy (Rethoré *et al.,* 1974a. (A) Notice aplasia of the nasal root, short nose, and the roundness of the face. The cleft lip has been operated upon. (B) Notice protrusion of the glabella above the nasal saddle, low placement of the ear, the hem on the helix, protrusion of the antihelix, and the appearance of the concha.

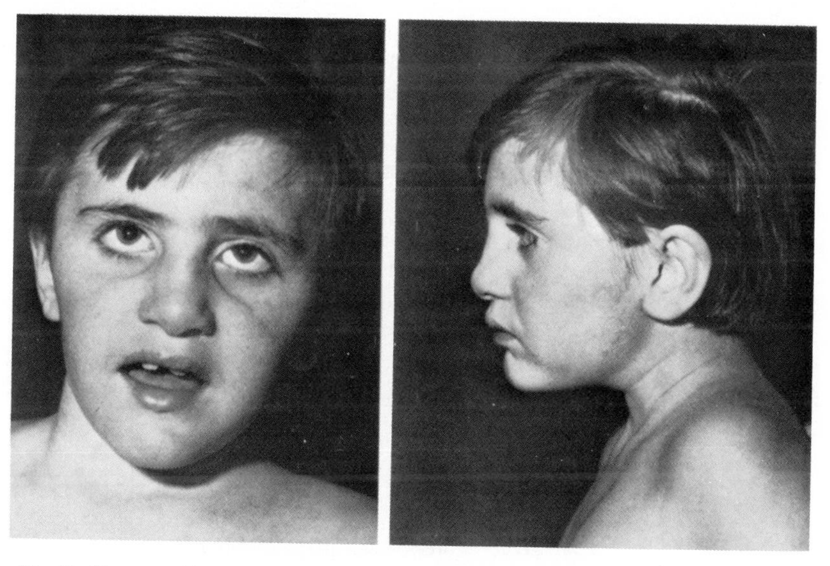

**Fig. 2.** Six-year-old female child with 4p trisomy (Giovannelli *et al.,* 1974). Notice the eyebrows, forehead, "boxer's" nose, and pointed chin.

The auricles of the ears are low set. With the exception of the subject studied by Wilson *et al.* (1970), whose auricles were truncated and deformed, other patients have conchae that are greater in length than in height. The antihelixes protrude and the upper parts of the helixes, which are horizontal, have very pronounced borders.

The cranium is small and round, but is sometimes asymmetrical as a result of plagiocephaly. Hair grows low along the back of the neck, as well as along the forehead.

The torso is distinguished by shortness of the neck, exaggerated spacing of the nipples, and the presence of scoliosis.

In female patients, the external genital organs have a normal appearance. Among males, a small penis was observed on two occasions (Rethoré *et al.*, 1974a; Giraud *et al.*, case 1, 1975). The presence of hypospadias (Wilson *et al.*, 1970; Gouw *et al.*, 1972; Darmady and Seabright, case 1, 1975) or cryptorchidism (Schinzel and Schmid, 1972; Giraud *et al.*, 1975; Darmady and Seabright, case 1, 1975) was observed three times.

The extremities are frequently deformed: bifurcation of the thumb (Wilson *et al.*, 1970); an extra finger (Darmady and Seabright, case 2, 1975); camptodactyly (Darmady and Seabright, cases 1 and 2, 1975; Schinzel and Schmid, 1972); clubbed hands and feet (Rethoré *et al.*, 1974a); rocker-bottom feet (Schinzel and Schmid, 1972; Giraud *et al.*, cases 1 and 2, 1975); hallux valgus (Sartori *et al.*, case 1, 1974; Schröcksnadel *et al.*, 1975); inward bending of the feet (Darmady and Seabright, 1975); and talus valgus (Giovannelli *et al.*, 1974).

Skeletal deformities, which were recently studied by Dallapiccola *et al.* (1974), are the most frequently encountered features of the 4p trisomy syndrome. In addition to the anomalous features of the extremities cited above, one can observe the following:

(*a*) *Face and cranium:* microcephaly; finger-shaped imprints in the temporal and occipital areas; small and closed sella turcica; hypertelorism; gothic palate; protruding upper incisors; elongation of the angle of the jawbone.

(*b*) *Spinal column:* kyphosis and scoliosis; Y-shaped clavicles (Giovannelli *et al.*, cases 1 and 2, 1974); hypoplasia or aplasia of the first pair of ribs (Giovannelli *et al.*, cases 1 and 2, 1974) and of the twelfth pair (Schinzel and Schmid, 1972; Sartori *et al.*, cases 1 and 2, 1974; Dallapiccola *et al.*, 1974; Metz *et al.*, 1973); hemivertebra at S1 (Schinzel and Schmid, 1972); hemisacralization at L5 (Furbetta *et al.*, 1975); hypoplasia of the fourth segment of the sacrum (Giraud *et al.*, case 2, 1975).

(*c*) *Pelvis and hips:* poorly developed and narrow iliac wings; enlarged iliac angles; reduced acetabular angles; dislocation of the hips (Schinzel and Schmid, 1972; Sartori *et al.*, cases 1 and 2, 1974; Darmady and Seabright, cases 1 and 2, 1975).

Bone development was very slow except in one patient (Furbetta *et al.*, 1975).

Among other deformities, and aside from the genital anomalies which have already been cited, the following have also been observed:

(*a*) *Brain:* hydrocephaly (Gustavson *et al.*, 1964); cistern of the velum interpositum (Rethoré *et al.*, 1974a); agenesis of the corpus callosum and dilation of the ventricle (Hustinx *et al.*, 1975); microencephalism (Wilson *et al.*, 1970).

(*b*) *Eyes:* coloboma (Gustavson *et al.*, 1964; Wilson *et al.*, 1970).

(*c*) *Heart and blood vessels:* ventricular septum defect (VSD) (Rethoré *et al.*, 1974a); VSD, hypertrophy of left ventricle, rudimentary tricuspid valve, single umbilical artery (Owen *et al.*, 1974); aortic hypoplasia (Giraud *et al.*, case 2, 1975; Darmady and Seabright, case 1, 1975).

(*d*) *Kidneys:* hydronephrosis (Hustinx *et al.*, 1975; Gustavson *et al.*, 1964); prolapsed bladder (Giovannelli *et all*, case 2, 1974).

(*e*) *Abdomen:* hiatus hernia, improper intestinal malrotation, anal atresia, omphalocele (Gustavson *et al.*, 1964); pyloric stenosis (Metz *et al.*, 1973); diaphragmatic hernia (Darmady and Seabright, case 1, case 2, 1975).

### 2. *Dermatoglyphic Features*

Among the eighteen palms which were analyzed (6 females and 3 males), the main lines had a normal arrangement, except for five hands where there was a transverse palmar crease or its equivalent. The axial triradius was in position t six times, t′ eight times, and in an ulnar position four times.

On the fingers, the flexion creases were normal. On the fingertips, there was an excess of whorls, and a low proportion of arches: whorls, 54.1%; ulnar loops, 35.8%; radial loops, 8.3%; arches, 1.8%.

### 3. *Growth*

***a. Height and Weight.*** At birth, the average weight is 2680 gm (for 20 cases), and the average height is 47.7 cm (for 8 cases). Overweight persists in many patients, but for all of them height remained considerably below normal.

***b. Psychomotor Development.*** Mental retardation, which is severe, was complicated in five cases by seizures. The intelligence quotient, which is close to 50 at the age of two years, declines to 30 or even to 10 in older subjects.

***c. Life-span.*** Mortality is relatively high. Nine patients died, usually during the first weeks of life, while another died at the age of eleven. Among these subjects, four (Gustavson *et al.*, 1964; Owen *et al.*, 1974; Darmady and Seabright, cases 1 and 2, 1975) had a pure 4p trisomy. Death usually occurs as a result of lung disease.

### 4. *Family data*

Among the twenty-three cases studied, the sex ratio was normal (11 males; 12 females). Duration of pregnancy, which was generally normal, was sometimes shorter in seven cases and longer in three cases. In several cases bleeding occurred during the first trimester. Two cases were the result of *de novo* chromosome defect (Giraud *et al.*, case 3, 1975; André *et al.*, 1976). The parents had not been exposed to radiation and were very young.

Among families in which there was a balanced chromosome translocation, five had two trisomic children (Table I). In one family (Schwanitz and Grosse, 1973), it appeared that a child from a protracted pregnancy, who died very soon after, was afflicted with the 4p– syndrome, but a chromosome analysis had not been performed. Among the other families, there was a relatively high incidence of premature deaths, still births, and spontaneous abortions.

### 5. *Cytogenetic data*

Twenty-three cases were studied (Table I). Twenty-one were the result of balanced translocation transmitted by one of the parents, and two arose *de novo*.

***a. Thirteen Cases of Pure 4p Trisomy (Figs. 3a and b).*** Ten of these cases resulted from translocation of the short arm of chromosome 4 to the centromere of an acrocentric chromosome (Figs. 3a and b), such as chromosome 15 (Schröcksnadel *et al.*, 1975), chromosome 14 or 15 (Gouw *et al.*, 1972), chromosome 21 (Darmady and Seabright, cases 1 and 2, 1975), chromosome 22 (Metz *et al.*, 1973; Schwanitz and Grosse, cases 1 and 2, 1973; Giovannelli *et al.*, cases 1 and 2, 1974), or a G chromosome (Gustavson *et al.*, 1964).

One case resulted from a *de novo* short arm isochromosome (Giraud *et al.*, case 3, 1975) and another was accompanied by a *de novo* balanced translocation (André *et al.*, 1976).

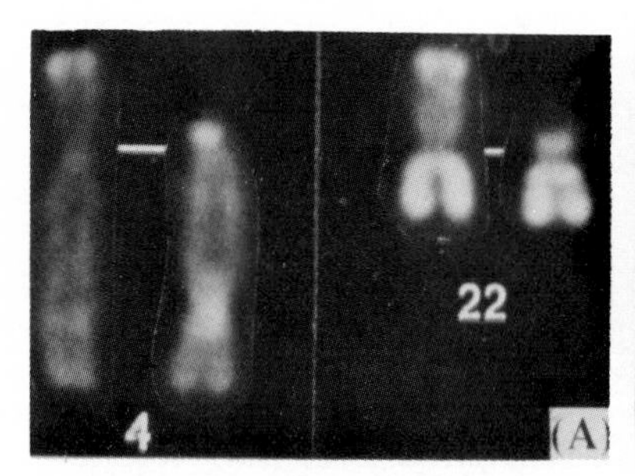


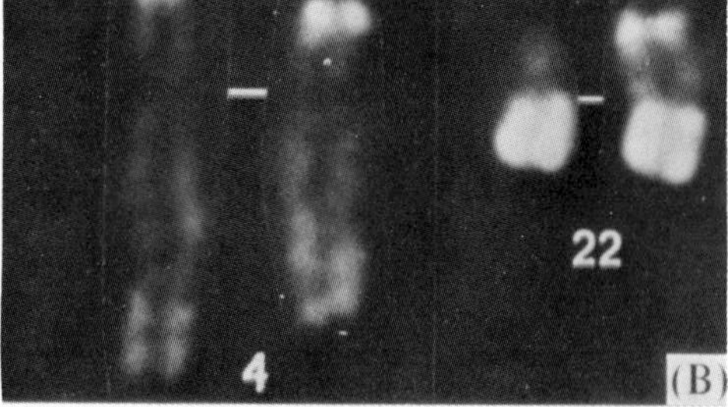


**Fig. 3.** Chromosomes 4 and 22 with T bands (Giovannelli *et al.*, 1974). (A) Maternal translocation t(4;22)(p11;p12). (B) 4p Trisomy in a patient who received a normal 4 chromosome and a 22p+ from his mother.

Another case resulted from nondisjunction of a maternal translocation (Furbetta *et al.*, 1975). The patient had 47 chromosomes and the extra component consisted of the short arm of chromosome 4 and the short arm of chromosome 21.

*b. Nine Cases of Associated 4p Trisomy (Fig. 4).* Three of the cases (Wilson *et al.*, 1970; Rethoré *et al.*, 1974a; Dallapiccola *et al.*, 1974) resulted from "aneusomy by recombination" a process described by Lejeune in 1965, which corresponds to an exchange of chromatids inside an inversion loop (Fig. 4). In three of the patients, 4p trisomy was accompanied by partial monosomy of the long arm of chromosome 4.

In five cases which resulted from a balanced parental translocation, 4p trisomy was accompanied by partial monosomy of another chromosome: the telomeric end of the long arm of chromosome 18 (Schinzel and Schmid, 1972) or of chromosome 22 (Sartori *et al.*, cases 1 and 2, 1974), or the proximal portion of the long arm of chromosome 15 (Giraud *et al.*, cases 1 and 2, 1975).

In one patient with 47 chromosomes (Hustinx *et al.*, 1975), 4p trisomy was accompanied by trisomy of the pericentromeric region on the long arm of chromosome 15.

Some uncertainty persists as to whether the case reported by Owen *et al.* (1974) corresponded to pure 4p trisomy or to a 4pter→4q13 trisomy. It was accompanied by partial monosomy of the long arm of chromosome 21.

Whether trisomy is pure or combined, it involves, according to the particular case, either the entire short arm of chromosome 4 (4p11→4pter) or only one portion (4p13 or 4p14→4pter).

## C. Partial 4q Trisomy

In 1972 Surana and Conen reported the first cytological demonstration of partial trisomy of the long arm of a 4 chromosome. 3 years later, Dutrillaux *et*

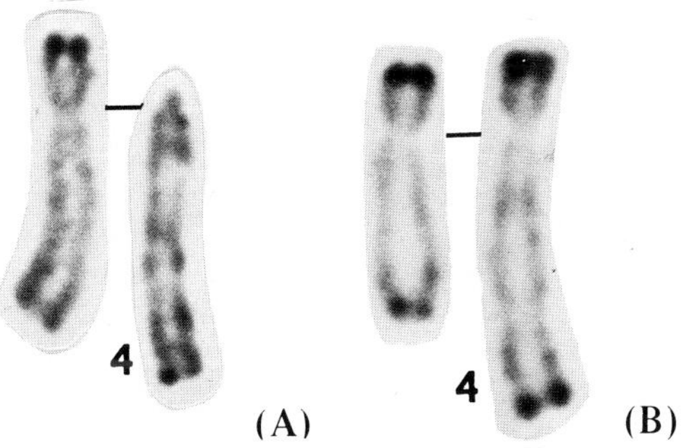


**Fig. 4.** Chromosomes 4 with R banding (Rethoré *et al.*, 1974a). (A) Pericentric paternal inversion: Inv(4)(p14q35). (B) 4p trisomy in the patient, duplication of p, inversion of (4)(p14q35).

*al.* (1975) furnished a precise description of the clinical syndrome for this affliction. The first observation was probably provided by Shaw *et al.* In 1965, they reported a case of partial trisomy resulting from maternal translocation t(Bp+; Bq–). The patient's overall deformities, which were comparable to those of the trisomy 4q syndrome, leads us to believe that the extra segment, which could not be identified at that time, may have been the distal portion of the long arm of chromosome 4. Thirteen cases of partial 4q trisomy were subsequently reported (Table II).

### *1. Phenotype (Fig. 5)*

Microcephaly is constant. At birth, the cranial perimeter is 32.3 (average for 6 cases), and the cranial sutures are usually elongated and open. In older children, the metopic suture sometimes protrudes.

The bridge of the nose, which is large and prominent, obscures the root of the nose and forehead. The crest of the nose is continuous with the forehead, which recedes. The eyelid slits are horizontal or slightly oblique; they are also narrow. Ptosis was observed in one case, and epicanthus in two. The eyeballs are small but not deformed. In one case (Shaw *et al.,* 1965), there were corneal leukomas.

The upper lip is extremely short. The philtrum, which is elongated and is bordered by protruding folds, is often intersected by a midseam which begins beneath the nasal septum and extends to the middle of the upper lip. In a resting position, the corners of the mouth droop, and the lower lip is accentuated by a horizontal dimple. When the child's mouth is closed, both lips project forward so as to give the mouth a very distinct appearance. The jawbone is straight and receding, the chin is pointed, and the palate is gothic.

The ears are low set and posteriorly rotated. In its upper portion, the helix is full of seams and is flattened. The antihelix, which is very prominent, extends as far as the lobe insertion, having a continuous crest which borders the concha. The concha itself is long and deep. The tragi are hypoplasic.

The neck is short and is sometimes webbed. In three cases, the presence of umbilical hernia, diastasis of the right side, or an inguinal hernia was observed. In all of the males, the testicles were undescended; among the females, the external genitals were normal. Some bone deformities were observed: abbreviation of the first rib and absence of the twelfth pair (Dutrillaux *et al.,* case 1, 1975), presence of thirteen ribs (Baccichetti *et al.,* 1975), dorsal scoliosis (Dutrillaux *et al.,* case 2, 1975), dislocation of the hip (Dutrillaux *et al.,* case 3, 1975).

The extremities are frequently deformed: bifurcated thumbs (Surana and Conen, 1972), absence of the thumb and bifurcation of the right index finger, digitalization of the left thumb (Dutrillaux *et al.,* case 3, 1975; Vogel *et al.,* case 3, 1975), or pointed feet (Shaw *et al.,* 1965; Surana and Conen, 1972).

Among the visceral deformities observed, those affecting the kidneys deserve special mention because of their frequency: horseshoe-shaped kidneys (Shaw *et*

**TABLE II**
**Observations of Partial Trisomy 4q**

| | Parents | | Patients | | |
|---|---|---|---|---|---|
| Authors | Maternal age | Parental translocation | Karyotype | Birth | Death |
| Shaw *et al.* (1965) | 27 | t(Bp+;Bq–) mat. | 46,XY,5?p+ | 1964 | 3 days |
| Surana and Conen (1972) | 23 | t(4q–;18q+) mat. | 46,XY,18q+ | ? | 7 months |
| Francke (1972), Case 5; Sparkes and Francke (1973) | ? | t(4q–;20q+) mat. | 46,XY,20q+ | ? | – |
| de la Chapelle *et al.* (1973) | 19 | t(4;21)(q21;q21) mat. | 47,XX,+der(21) | 1970 | – |
| Schwingshackl and Ganner (1973), Case 2 | ? | 3t(4q–;5p+) pat. | 46,XY,5p+ | 1970 | 2½ months |
| Fonatsch *et al.* (1975) | 20 | t(4;18)(q31;q21) pat. | 46,XX,18q+ | 1972 | – |
| Schrott *et al.* (1974) | 25 | t(4;13)(q27;q34) mat. | 46,XY,13q+ | ? | – |
| Dutrillaux *et al.* (1975) | | | | | |
| Case 1 | 22 | t(4;18)(q28;q22) mat. | 46,XY,18q+ | ? | – |
| Case 2 | 34 | – | 46,XY,der(2),t(2;4)(q37;q28) | ? | – |
| Case 3 (see also Vogel *et al.*, 1975) | 31 | – | 46,XX,ins (4) (pter→q34::q34→q25::q34→qter) | ? | – |
| Vogel *et al.* (1975) | | | | | |
| Case 1 | 30 | t(3;4)(p27;q27) pat. | 46,XX,3p+ | 1972 | – |
| Case 2 (see also Knörr-Gärtner *et al.*, 1974) | 21 | t(4q–;18q+) mat. | 46,XY,18q+ | 1972 | 7 days |
| Baccichetti *et al.* (1975) | 40 | t(4;21)(q32;q22) mat. | 46,XX,22q+ | 1966 | – |

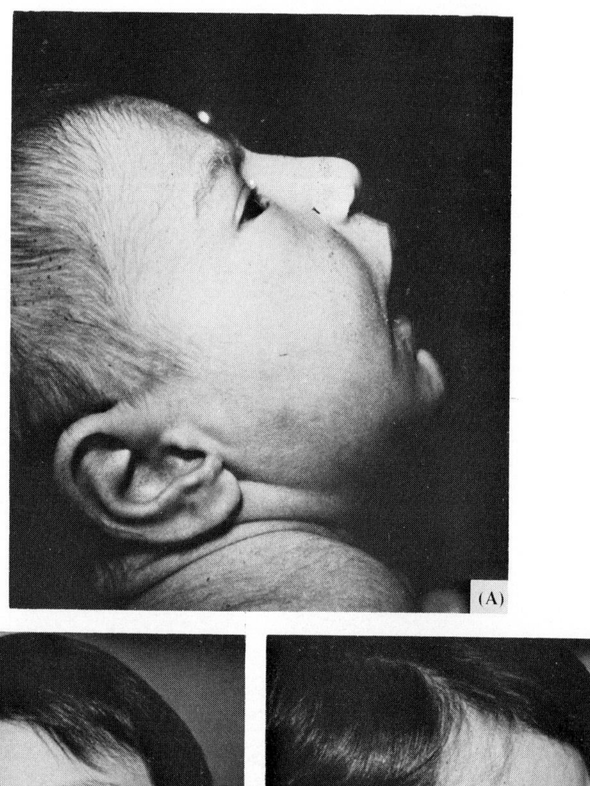

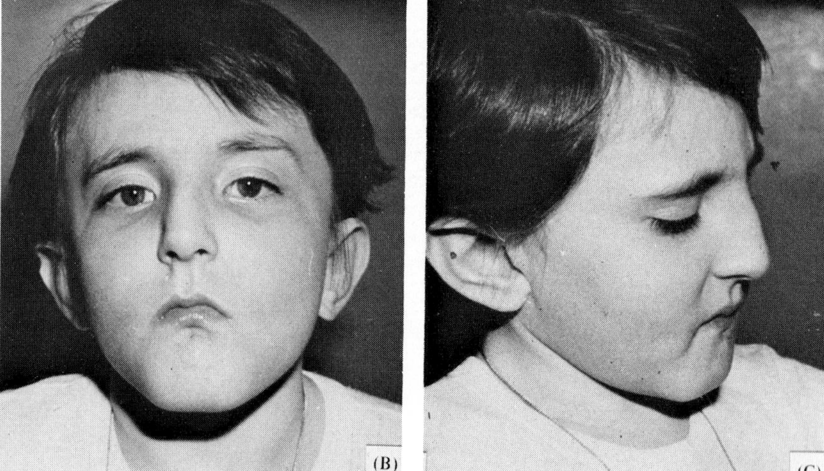

**Fig. 5.** Subjects afflicted with 4q trisomy. (A) Three-week-old female child (Dutrillaux *et al.,* case number 3, 1975). Notice protrusion of the nasal root, which obscures the nasofrontal saddle and the prolongation of the helix up to the beginning of the lobe. (B) Female child, 6 years, 9 months (Dutrillaux *et al.,* case no. 2, 1975). Notice projection of the philtrum pillars, and the horizontal dimple beneath the lower lip. (C) Same subject. Notice the position of the mouth, angling of the auricle, and prolongation of the antihelix.

*al.*, 1965); renal hypoplasia, either single or bilateral (Francke, 1972; Dutrillaux *et al.*, case 3, 1975; Vogel *et al.*, case 3, 1975; Vogel *et al.*, case 2, 1975); hydronephrosis (Surana and Conen, 1972; Schrott *et al.*, 1974); urethrovesicular reflux (Dutrillaux *et al.*, case 3, 1975).

Other deformities were revealed during anatomical examinations: agenesis of the corpus callosum, connection of the lateral ventricles, incomplete lung segmentation, ectopic pancreas (Shaw *et al.*, 1965); Fallot's tetralogy (Surana and Conen, 1972); large foramen ovale (Vogel *et al.*, case 2, 1975).

### *2. Growth and Family Data*

Weight and height retardation at birth is severe: 2720 gm (average for 10 cases) and 45.2 cm (average for 7 cases). Underdeveloped stature persists, whereas weight usually becomes normal for the particular age.

The level of mental retardation is profound. The intelligence quotient is close to 50. Seizures were cited in one case. Among the newborn, axial hypotonia exists, and, among older children, segmentary hypotonia has been frequently noted.

Three of the eleven children studied have died. They were afflicted with multiple deformities.

In two cases, 4q trisomy was the result of a *de novo* phenomenon (Dutrillaux *et al.*, case 2, 1975; Vogel *et al.*, case 3, 1975), and in eight cases it resulted from a maternal balanced translocation. In three cases, it resulted from a paternal balanced translocation. In one family (Sparkes and Francke, 1973), two sisters of the patient were affected with 4q trisomy. The older sister died from renal failure, and the younger one, whom the parents described as resembling the patient, died 3 weeks after birth.

### *3. Cytogenetic Data*

Thirteen cases were studied (Table II). In only one case (Dutrillaux *et al.*, case 3, 1975; Vogel *et al.*, case 3, 1975), was partial 4q trisomy truly a pure 4q trisomy. This was an instance of "mirror" duplication of segment 4q25 → 4q34.

In seven cases observed (Francke, 1972; Schrott *et al.*, 1974; Dutrillaux *et al.*, cases 1 and 2, 1975; Vogel *et al.*, case 1 and, perhaps, case 2, 1975; Baccichetti *et al.*, 1975), 4q trisomy was accompanied by such a small monosomy that it is permissible to consider the patients' phenotypes as essentially reflecting 4q trisomy (Fig. 6).

In one of the cases observed (de la Chapelle *et al.*, 1973), 4q trisomy was accompanied by partial 21 trisomy affecting the short arm and the proximal portion of the long arm of chromosome 21, but not affecting the 21q22 segment. According to several authors (Aula *et al.*, 1973; Niebuhr, 1974; Wahrman *et al.*, 1974; Williams *et al.*, 1975; Sinet *et al.*, 1976; Lewandowski *et al.*, 1976), the latter is responsible for the trisomy 21 syndrome. The patient did

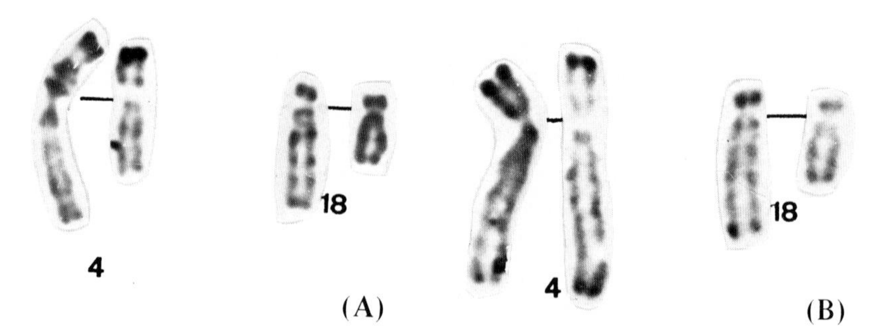


**Fig. 6.** Chromosomes 4 and 18 with R-bands (Dutrillaux *et al.*, case 1, 1975). (A) Maternal translocation t(4;18)(q28;q22). (B) 4q trisomy in a child who received a normal 4 chromosome and an 18q+ from his mother.

not display any signs of 21 trisomy and was not different from other 4q trisomy patients.

In two cases observed the phenotype was significantly changed by combined monosomy. In one case (Schwingshackl and Ganner, 1973), the patient also had a 5p monosomy and was afflicted with the cat cry syndrome. In the other case (Fonatsch *et al.*, 1975), the patient had monosomy 18q21 → 18qter, and the appearance was similar to that for the 18q– syndrome. In both of these cases, it was hardly possible to distinguish features directly attributable to 4q trisomy, and, hence, they were not considered in the clinical delineation of the syndrome.

Although the break points were not precisely identified in the observations furnished by Shaw *et al.* (1965) or by Surana and Conen (1972), one can assume that combined monosomies are minimal. In the first case, it involved the short arm of chromosome 5, and the patient did not display any sign of the cat cry syndrome. In the second case, it involved the long arm of chromosome 18, and the patient did not display any sign of the 18q– syndrome.

The segment of the long arm of chromosome 4 which is present in triplicate was only defined in eight of the cases. In seven cases, the distal segment was involved with break points at 4q21 (de la Chapelle *et al.*, 1973), 4q27 (Schrott *et al.*, 1974; Vogel *et al.*, case 1, 1975); 4q28 (Dutrillaux *et al.*, cases 1 and 2, 1975), 4q31 (Fonatsch *et al.*, 1975), and 4q32 (Baccichetti *et al.*, 1975).

## D. Partial 4q Monosomy

Depending upon the chromosome segment involved, cases of partial monosomy of the long arm of chromosome 4 can be classified into two groups.

In the first group (Table III), monosomy results from deletion and is limited to a portion of the long arm. The first observation of this type was by Ockey *et al.*, in 1967.

**TABLE III**
**Observations of Partial Monosomy 4q**

| | Parents | | Patients | | |
|---|---|---|---|---|---|
| Authors | Maternal age | Parental translocation | Karyotype | Birth | Death |
| Ockey *et al.* (1967) | 19 | ? | 46,XY, 4q− | ? | 3½ months |
| Golbus *et al.* (1973) | 22 | – | 46,XX, 4q− | ? | – |
| Van Kempen (1975) | 26 | – | 46,XY,del(4)(q31) | ? | – |

In the second type, one of the 4 chromosomes is replaced by a ring, and partial monosomy of the long arm is accompanied by partial monosomy of the short arm. Moreover, instability of the ring during successive mitoses accounts for the presence of various clones. Some of these are trisomic for a great portion of chromosome 4, while others are monosomic for 4. Because of this complexity, we shall not consider such observations (Carter *et al.*, 1969; Hecht, 1969; Faed *et al.*, 1969; Dallaire, 1969; Bobrow *et al.*, 1971; Surana *et al.*, 1971) in the clinical analysis of 4q monosomy.

### *1. Phenotype*

It is very early to define the syndrome for partial 4q monosomy since only three definite cases have been observed and it is not certain whether the deleted segment is the same for all three.

Analysis of the patients' phenotypes, however, has, at this point, permitted identification of a certain number of common symptoms. In two of the three subjects, the nasal saddle was long and flat, and the nostril openings were turned forward. The upper portion of the auricles was pointed; the jawbone was small and receding. The patient studied by Golbus *et al.* (1973) had a cleft palate and a bifurcated uvula. The patient studied by Van Kempen (1975) was microcephalic. In the patient studied by Ockey *et al.* (1967), the auricles were low set.

Cardiac malformations were observed in each of the three children (ASD and VSD, Ockey *et al.*, 1967; Golbus *et al.*, 1973). A right bundle branch block was observed by Van Kempen (1975). Kidney deformities were observed by Ockey *et al.* (1967) (dilation of the calyces) and by Golbus *et al.* (1973) (duplication of the left kidney). The patient described by Carnevale and De los Cobos (1973; see p.

3), also had a duplication of the left kidney, while the patient observed by Carter *et al.* (1969) had renal hypoplasia.

Only one of the three patients (Ockey *et al.*, 1967) displayed agenesis of the radius, the carpus, and four metacarpals. Among the patients who were not included for the reasons cited earlier, five displayed severe deformities of the extremities: Holt-Oram syndrome (Rybak *et al.*, all three cases, 1971); agenesis of the cubitus of the fourth and fifth fingers, hypoplasia of the fifth right metacarpal (Carnevale and de los Cobos, 1973); bilateral absence of the radius and the thumb (Faed *et al.*, 1969); absence of the right thumb (Dallaire, 1969); hypoplasia of the first metacarpal and the phalanges of the thumb (Carter *et al.*, 1969).

### 2. *Growth and Family Data*

Only one of the three patients died. As for the other two, mental retardation is severe. Seizures appeared in one case (Van Kempen, 1975) during the first year, and, in the other case, electroencephalograms are abnormal.

The birth weight, which was normal for two of the children, was low for the third, who has remained underdeveloped.

In the three cases, examination of the parents' chromosomes appear to show normal results, but in two of them (Ockey *et al.*, 1967; Golbus *et al.*, 1973) one cannot exclude the possibility of a balanced translocation because of technical difficulties.

### 3. *Cytogenetic Data*

The patient in the first case observed (Ockey *et al.*, 1967) has a B chromosome which lacks a portion of the long arm. Autoradiographic analysis and measurements which were performed show that it is a 4 chromosome whose long arm has been reduced by almost 23%.

In the second patient observed (Golbus *et al.*, 1973), analysis of the karyotype for G-bands (Sumner *et al.*, 1971) demonstrated that the missing segment involves the end portion of the long arm of chromosome 4, although the break points of separation were not indicated.

In the third patient observed (Van Kempen, 1975), analysis of G–bands showed pure monosomy of the segment 4q31 → 4qter.

For the other two cases observed, the B chromosomes which had a partial deletion of the long arm was not identified. Among the three males examined by Rybak *et al.* in 1971, it was possibly a case of deletion resulting from parental translocation, but a definite conclusion cannot be reached because of the inadequate techniques used. The case reported by Carnevale and de los Cobos (1973) corresponded to a mosaic composition in which the majority of the clones were monosomic for a portion of the long arm of a B chromosome and monosomic for the centromeric portion of a D chromosome.

## E. Conclusion

Among the features observed in patients affected by abnormalities of chromosome 4, there were those common for most syndromes involving chromosome defects, such as growth and mental retardation, and microcephaly. There were also other features which were characteristic in a given disorder. Their value was even more obvious if one compared the dysmorphology of the same portion of the anatomy of a trisomic patient to that of a monosomic patient. Thus, in 4p trisomy and 4p monosomy, the defects in facial development essentially involve the midportion, the forehead, the glabella, and the nose:

*For trisomy:* flat forehead, protruding glabella, aplasic nasal bones, and the tip of the nose is round.

*For monosomy:* recessed forehead, aplasic glabella, and a nose in the shape of a rectangular parallelogram.

Among patients who had a partial 4q trisomy, the most characteristic signs appeared to be attenuation of the nasofrontal saddle, shortness of the upper lip and a philtrum with seams, as well as compression of the lips. There is also flattening of the upper portion of the helix and prolongation of the antihelix as far as the beginning of the lobe. To these deformities, it is necessary to add, because of their frequency, renal defects and bone deformities.

Among the three patients who had a partial 4q monosomy, two had an aplasic nasal saddle, out-turned nostril openings, and angling of the upper portion of the helix. Other observations are undoubtedly necessary in order to identify more precisely the syndrome for monosomy of the long arm of chromosome 4, but even now it is possible to note the frequency of visceral, cardiac, and renal deformities, as well as anomalies affecting the extremities.

# II. ABNORMALITIES OF CHROMOSOME 9

## A. Introduction

Within group C, which was defined by the 1963 International Conference in London (London Conference, 1963), chromosome 9 is characterized using classic Giemsa staining, by secondary constriction opposite the centromere (Fig. 7). It is this feature which, in 1970, prior to availability of the banding techniques, permitted definition of the syndrome involving trisomy of the short arm of chromosome 9, the first example of a syndrome involving trisomy of an autosomal chromosome arm.

Use of techniques which permit analysis of the detailed structure of chromatids not only allowed confirmation of the distinct identity of 9p trisomy, but recognition of other abnormalities of chromosome 9.

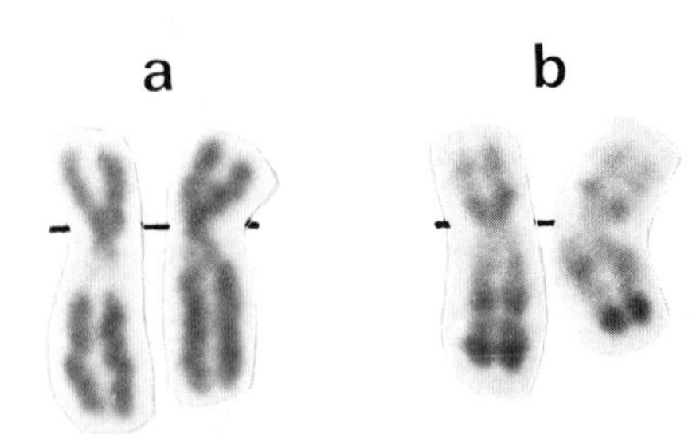


**Fig. 7.** Chromosome 9. (A) Conventional Giemsa staining; notice constriction on the long arm close to the centromere. (B) R-bands.

First, we will consider defects which only involve the short arm: 9p trisomy, 9p tetrasomy, and 9p monosomy. Subsequently, we will consider full 9 trisomy, trisomy of the distal end of the long arm, and, finally, ring 9.

## B. 9p Trisomy

Observation of the secondary constriction of the long arm of chromosome 9 in four parental translocations made it possible for Rethoré *et al.*, in 1970, to describe the syndrome for 9p trisomy. Later, with the aid of the banding techniques, it was possible to demonstrate the presence of triplication of the short arm of chromosome 9 in 59 cases. In five other observations of trisomy of the short arm of a group C chromosome (Table IV), the overall deformities of the patients

**TABLE IV**
**Observations of ?9p Trisomy**

| | Parents | | Patients | | |
|---|---|---|---|---|---|
| Authors | Maternal age | Parental translocation | Karyotype | Birth | Death |
| Edwards *et al.* (1962) | | | | | |
| Case 1 | 29 | t(9?p–;Bq+)pat | 46,XY,Bq+ | 1956 | – |
| Case 2 | 33 | | 46,XX,Bq+ | 1960 | – |
| Butler *et al.* (1969) | 28 | t(9?p+;15q–)mat | 47,XY,+15q– | ? | 13½ months |
| Cantu *et al.* (1971) | | | | | |
| Case 1 | 34 | ? | 46,XY,Dp+ | 1960 | – |
| Case 2 | 35 | | 46,XY,Dp+ | 1961 | – |

were entirely comparable to those of the 9p trisomy syndrome. Although these cases were not subjected to cytological confirmation by means of the banding techniques, phenotype convergence enabled one to conclude that these patients were trisomic with respect to the short arm of chromosome 9.

Analysis of the detailed structure of the chromosome defect permits identification of the following cases:

(a) *Pure 9p trisomy* (Table V) in which the defect consists of trisomy of the entire short arm of chromosome 9

(b) *9p Trisomy associated with proximal 9q trisomy* (Table VI) in which the defect consists of trisomy of the short arm and of the proximal portion of the long arm of chromosome 9

(c) *Associated 9p trisomy* (Table VII) in which trisomy of the short arm of chromosome 9 is accompanied by a trisomy or a monosomy of another chromosome segment

(d) *Partial 9p trisomy* (Table VIII) in which 9p trisomy is limited to the distal portion of the short arm of chromosome 9.

## 1. Phenotype

Morphological abnormalities described here are those which were observed in patients in whom only the short arm of chromosome 9 (Table V) was affected.

In the newborn child, the cranium is small and round. The anterior fontanel, which is large, extends forward into the metopic suture. In older children, the cranial perimeter is considerably below average; in three subjects who were more than 20 years of age, the cranial perimeter was 49,51, and 52 cm.

The eyeballs are sometimes small and sunken. The pupils are not centered, being displaced toward the inner edge of the irises. Strabismus is frequent, and it is often accompanied by refraction difficulties. The eyelid slits, usually, are slightly oblique. The space between the carunculae is exaggerated, but this form of hypertelorism is moderate.

The nose is large, especially at the tip. The nasal crest, which is initially normal, thickens into a protrusion which extends as a thin, protruding nasal subseptum. The nostril openings open downward.

In a resting position, the upper lip, which is very short, permits the incisors and upper canine teeth to be seen. The lower lip is turned outward. When the mouth is open, the resulting appearance is asymmetrical, and a very distinct unilateral rictus appears (Fig. 8c and d).

The auricles of the ears are situated normally, and are large and, especially, truncated. The jawbone is normal.

The neck is short and sometimes webbed. The thorax is funnel-shaped, and the nipples are wide-spaced. Subacromial and coccygeal dimples, diastasis of the right side, and umbilical hernias occur frequently.

**TABLE V**
**Observations of Pure 9p Trisomy**

| | Parents | | Patients | | |
|---|---|---|---|---|---|
| Authors | Maternal age | Parental rearrangement | Karyotype | Birth | Death |
| Rethoré *et al*. (1970, 1973) | | | | | |
| Case 3 | 25 | t(9;22)(p11;p11) pat. | 46,XY,22p+ | 1952 | – |
| Case 4 | 28 | | 46,XY,22p+ | 1955 | – |
| Hoehn *et al*. (1971); Rethoré *et al*. (1973), Case 6 | 32 | – | 47,XY,+9q– | ? | – |
| Newton *et al*. (1972) | 29 | t(9;22)(q11;p11) mat. | 47,XY,+9q– | 1926 | – |
| Rethoré *et al*. (1973), Case 10 | 28 | – | 46,XX,t(9p;15q) | 1964 | – |
| Baccichetti and Tenconi (1973) | 26 | – | 46,XY,t(9p;15q) | ? | – |
| Podruch and Weisskopf (1973) | | | | | |
| Case 1 | 40 | t(9;15)(q11;p11) mat. | 47,XY,+9q– | 1965 | – |
| Case 2 | ? | t(9;15)(q11;p11) mat. | 47,XX,+9q– | 1926 | – |

| | | | | | |
|---|---|---|---|---|---|
| Canki *et al*. (1975) | 36 | – | 46,XX,t(9p;14q) | 1972 | – |
| Jacobsen *et al*. (1975) | 29 | – | 46,XX,t(9p;15q) | 1962 | – |
| Turleau *et al*. (1975a) | 26 | 46,XY, del (9)(q11) 1 cell from the father | 47,XX,+9q– | 1968 | – |
| A. M. Potter (personal letter, 1975) | | | | | |
| Case 3 | ? | – | 46,XY,14p+ | ? | – |
| Case 4 | ? | – | 46,XX,14p+ | ? | – |
| Case 5 | ? | – | 46,XX,i(9p) | ? | – |
| F. Skovby, E. Niebuhr, and B. Møller (personal letter, 1976) | 39 | ? | 46,XY,t(9;19)(p11;q11), t(19;21)(q11;p11) | 1965 | – |
| T. W. J. Hustinx (personal letter, 1975) | | | | | |
| Case 1 | ? | – | 46,XX,i(9p) | 1962 | – |
| Case 2 | ? | Inv. (9) mat. | 46,XX,inv(9),i(9p) | 1945 | – |
| Rethore *et al.* (1977) | 23 | – | 46,XX,i(9p) | 1969 | – |

**TABLE VI**
**Observations of +9q− Trisomy**

| Authors | Parents | | Patients | | |
|---|---|---|---|---|---|
| | Maternal age | Parental translocation | Karyotype | Birth | Death |
| Zaremba *et al*. (1974) | | | | | |
| Case 1 | ? | t(9;15)(q13;q11) pat. | 46,XX,15p+ | 1963 | – |
| Case 2 | ? | | 46,XX,15p+ | 1967 | – |
| Case 3 | 28 | t(9;15)(q13;q11) mat. | 46,XX,15p+ | 1952 | – |
| Case 4 | 35 | | 46,XX,15p+ | 1959 | – |
| Blank *et al*. (1975) | | | | | |
| Case 1 | 19 | t(9;22)(q13;q11) mat. | 46,XX,22p+ | 1937 | – |
| Case 2 | 23 | t(9;22)(q13;q11) mat. | 46,XX,22p+ | 1950 | – |
| Case 3 | 27 | | 46,XX,22p+ | 1954 | – |
| Case 4 | 23 | t(9;22)(q13;q11) mat. | 46,XY,22p+ | 1967 | – |
| Balicek *et al*. (1975) | 22 | t(9;15)(q12;p11) mat. | 47,XX,+q9– | ? | – |
| Lord *et al*. (1967); A. M. Potter (personal letter, 1975) | | | | | |
| Case 1 | 21 | t(9;22)(q13;q11) mat. | 46,XX,22p+ | 1960 | – |
| Case 2 | 23 | | 46,XY,22p+ | 1962 | – |
| Case 3 | ? | | 46,XX,22p+ | ? | Fc |
| Case 4 | ? | | 46,XY,22p+ | ? | Fc |
| Dallapiccola (1975) | 37 | 45,XX,–9,–21,–22, t(9;22)(q12;q11), t(9;21)(q12;p13) | 46,XY,22p+ | 1967 | – |

**TABLE VII**
**Observations of Associated 9p Trisomy**

| | Parents | | Patients | | |
|---|---|---|---|---|---|
| Authors | Maternal age | Parental translocation | Karyotype | Birth | Death |
| Rethoré et al. (1970, 1973) | | | | | |
| Case 1 | 27 | t(6;9)(p25;q11) mat. | 47,XX,+9q– | 1967 | – |
| Case 2 | 24 | t(9;22)(q13;q13) mat. | 46,XX,22q+ | 1967 | – |
| Case 5 (see also Rethoré *et al.*, 1966) | 24 | t(9;19)(p12;p133) mat. | 46,XY,19p+ | 1957 | – |
| Rott *et al*. (1971, 1973 case 7) | 38 | t(4;9)(p15;q21) mat. | 47,XX,+9q– | 1968 | 5 months |
| Ebbin *et al*. (1973) | | | | | |
| Case 1 | 36 | t(9;18)(p?;p?) mat. | 46,XX,18p+ | ? | 20 days |
| Case 2 | 38 | | 46,XX,18p+ | ? | Induced abortion |
| Turleau *et al*. (1974) | | | | | |
| Case 1 | 33 | t(9;15)(q13;q25) mat. | 47,XX,+9q– | 1973 | – |
| Case 2 | 24 | t(7;9)(q35;p13) mat. | 46,XX,7q+ | 1973 | 3 months |
| Rethoré *et al*. (1974b) | 27 | t(4;9)(q34;q21) mat. | 47,XX,+9q– | 1973 | – |
| Schwanitz *et al*. (1974) | 20 | t(8;9)(q24;q22) mat. | 47,XX,+9q– | 1970 | 17 months |
| Lin *et al*. (1974) | | | | | |
| Case 1 | ? | | ? | ? | ? |
| Case 2 | ? | t(9;11)(p13;p15) mat. | ? | ? | ? |
| Case 3 | ? | | ? | ? | ? |

*continued*

TABLE VII (*continued*)

| Authors | Parents | | Patients | | |
|---|---|---|---|---|---|
| | Maternal age | Parental translocation | Karyotype | Birth | Death |
| Fujita *et al*. (1974) | 34 | t(9;13)(q12;q31), t(13;14)(q31;p1), t(9;14)(q12;p1) mat. | 47,XY,+ t(9;13) (q12;q31), 13q– | 1957 | – |
| Mulcahy and Jenkyn (1975) | | | | | |
| Case 1 | ? | t(8;9)(p23;p12) mat. | 46,XX,8p+ | 1965 | – |
| Case | ? | t(9;21)(p13;q22) mat. | 46,XX,21q+ | 1966 | – |
| Penchaszadeh and Coco (1975) | 25 | t(7;9)(q36;q13) mat. | 47,XY,+9q– | ? | – |
| Mason *et al*. (1975) | 24 | t(1;9)(p36;q21) mat. | 47,XX,+9q– | 1951 | – |
| Centerwall *et al*. (1975) | 38 | t(8;9)(p23;q22) mat. | 47,XY,+9q– | ? | – |
| Stoll *et al*. (1976) | 31 | t(9;20)(q12;p13) mat. | 47,XX,+9q– | 1973 | – |
| A. M. Potter (personal letter, 1975) | | | | | |
| Case 6 | ? | t(6;9)(q27;p22) pat. | 46,XX,6q+ | ? | – |
| Case 7 | ? | | 46,XY,6q+ | ? | – |
| B. Dallapiccola (personal letter, 1975) | ? | t(6;9)(q27;p13) pat. | 46,XX,6q+ | 1965 | – |
| C. Baccichetti (personal letter, 1975) | ? | t(9;18)(p12;p11) mat. | 46,XX,18p+ | ? | – |

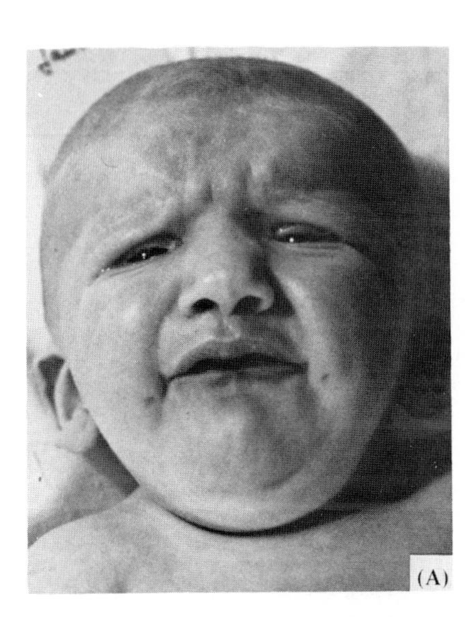


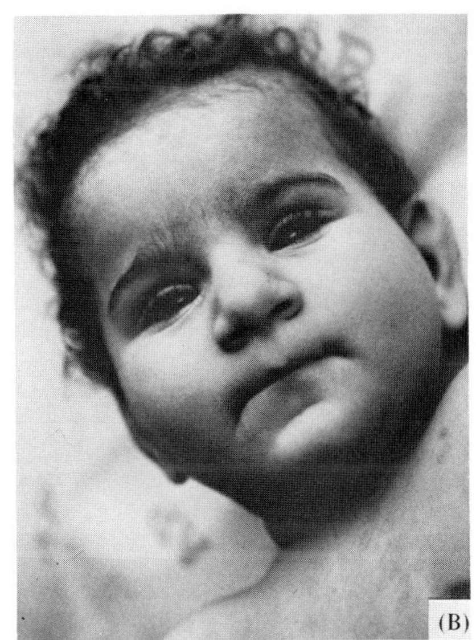


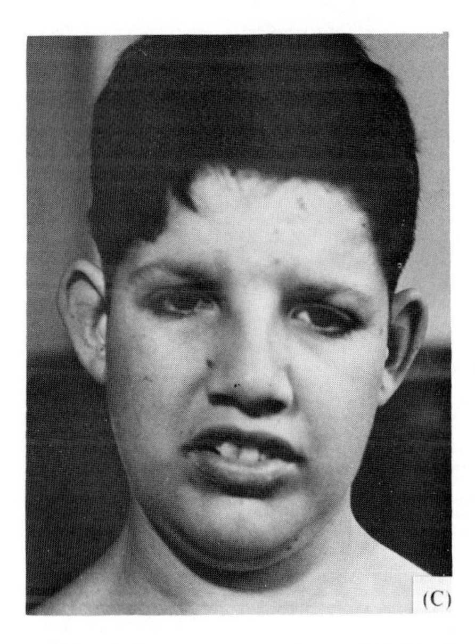


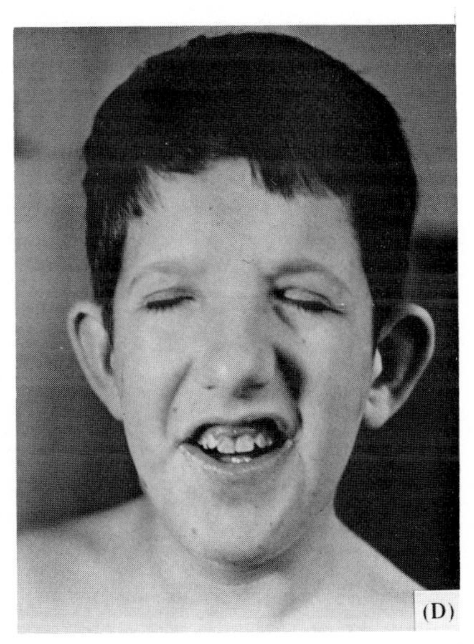


Fig. 8. 9p Trisomy patients. (A) Newborn in whom 9p trisomy is accompanied by 4qter trisomy (Rethoré *et al.*, 1974c). (B) Newborn in whom 9p trisomy is accompanied by 6pter trisomy (Rethoré *et al.*, case 1, 1970). (C) and (D) Two brothers with pure 9p trisomy (Rethoré *et al.*, case 3 and 4, 1970).

**TABLE VIII**
**Observations of Partial Trisomy 9p**

| | Parents | | Patients | | |
|---|---|---|---|---|---|
| Authors | Maternal age | Parental translocation | Karyotype | Birth | Death |
| Lewandowski *et al*. (1976) | | | | | |
| Case 1 | 30 | ins(18;9)(q11;p21p24)mat | 46,XY,18q+ | 1969 | – |
| Case 2 | 21 | t(9;21)(p21;q21)mat | 47,XY,+der(21) | 1959 | – |
| Case 3 | 34 | t(9;21)(p21;q21)mat | 47,XX,+der(21) | 1951 | – |

In older children, hypotrophy of the periscapular muscles and tendency toward hyperlordosis and scoliosis can be observed. Axillary hair is sparse and frequently absent. In two male patients who were 20 years of age, distribution of body fat was in conformity with female patterns, and the penis was at an infantile level of development. There was acrocyanosis of the extremities, and there were rod-shaped protrusions on the abdomen and the thighs. Body skin was very smooth. In female patients, the external genitals are normal (Fig. 8a–d).

The hands show a distinct morphology. The palms, which are exaggerated in length, display a single transverse crease. Although length of the fingers is normal, the segment situated between the two flexion creases is extremely small (see Fig. 9). Sometimes, especially on the joints, there is only one flexion crease. These abnormalities are very frequently accompanied by clinodactyly, which can affect all of the fingers except for the thumbs, which are too proximally implanted. Interphalangeal joints, which are hyperflexible in the youngest patients, permit dislocation of the phalanges. In older patients, there is sometimes an irreversible bending of the third phalanx. The fingernails are dysplasic, small, and sometimes claw-shaped, more frequently at the tips than in the bases.

Bone deformities essentially involve the phalanges of the fingers and the knuckles. The majority of observations indicate the existence of underdevelopment of the middle phalange on each digit except the thumb, and hypoplasia or even aplasia of the terminal phalanges.

Usually, there is a significant retardation of bone development; in one case (Centerwall *et al.*, 1975), bone development was advanced.

### 2. *Dermatoglyphic Features*

The very distinct dermatoglyphic patterns (Fig. 9 and 10) constitute a major element for diagnoses. In the great majority of cases, the two transverse creases

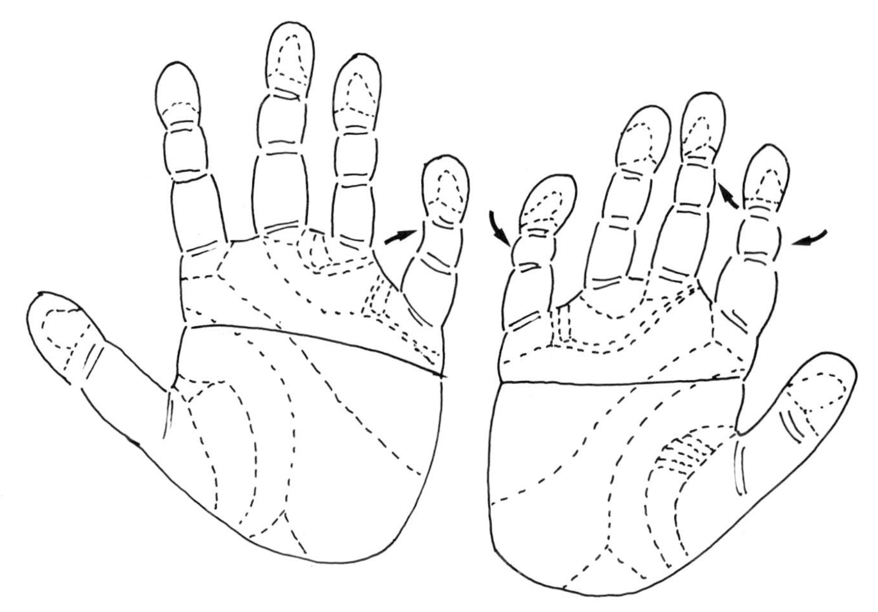

**Fig. 9.** Diagram of dermatoglyphics of a male child having pure 9p trisomy (Rethoré *et al.*, case 3, 1970). Note the transverse palmar crease, the patterns in the hypothenar area, and the brachymesophalangia.

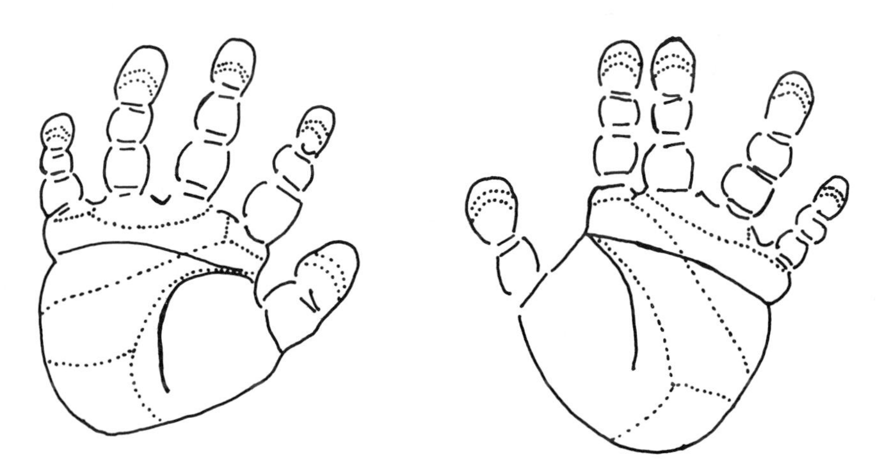

**Fig. 10.** Diagram of dermatoglyphics of a female child having 9p trisomy accompanied by 9q trisomy (Rethoré *et al.*, case 2, 1970).

of the palm are either replaced by a single transverse crease or they form the equivalent of a single transverse crease (Table IX). In 50.1% of the palms analyzed, the axial triradius t′ was observed (Table X).

Absence or fusion (Fig. 10) of the triradii located under the bases of the third (b) and fourth (c) fingers was observed in 66% of the palms studied (Table XI). This particular feature, which is very rare in the general population, was accompanied, in some cases, by a pronounced webbing between the third and fourth fingers. The main line D, of the subdigital triradius d, ends in 11 in 76.6% of the palms studied. This arrangement, which is rare in the general population, is frequently observed in patients with chromosome anomalies, particularly 21 trisomy. The frequency of interdigital patterns diminishes in relation to that which can be observed in the general population: 15% of the palms have an open loop between the fourth and fifth fingers (general population, 47.2%), and 21% have an open loop between the third and fourth fingers (general population, 41.2%).

Among 56 hands that were examined, 25 had a pattern on the thenar eminence. These included loops, whorls, and vestiges. In the general population, the frequency of such patterns is approximately 1%.

Frequency of hypothenar patterns was, on the other hand, very limited. They only appeared on 8 hands, whereas they are present in 25% of normal subjects.

**TABLE IX**
**Palm Creases in 9p Trisomy**

| Trisomy | Number of palms studied | Single transverse folds | |
|---|---|---|---|
| | | Observed | % |
| Pure 9p | | | |
| ♂ | 10 | 10 | 100.0 |
| ♀ | 12 | 11 | 91.6 |
| 9q− | | | |
| ♂ | 6 | 6 | 100.0 |
| ♀ | 18 | 8 | 44.4 |
| Associated 9p | | | |
| ♂ | 6 | 6 | 100.0 |
| ♀ | 14 | 14 | 100.0 |
| General population (France) | | | |
| ♂ | 200 | – | 0.75 |
| ♀ | 200 | – | 1.50 |

**TABLE X**
**Axial Triradius in 9p Trisomy**

| | | Position of axial triradius | | | | | | | | | |
|---|---|---|---|---|---|---|---|---|---|---|---|
| | | t | | t′ | | t and t′ | | t″ | | Cubital t | |
| Trisomy | Number of palms studied | Observed | % | Observed | % | Observed | % | Observed | % | Observed | % |
| Pure 9p | | | | | | | | | | | |
| ♂ | 10 | 5 | 36.3 | 2 | 45.4 | 1 | – | 1 | – | 1 | – |
| ♀ | 12 | 3 | – | 8 | – | 0 | – | 0 | – | 1 | – |
| 9q– | | | | | | | | | | | |
| ♂ | 4 | 2 | 36.3 | 1 | 50.0 | 1 | – | 0 | – | 0 | – |
| ♀ | 18 | 6 | – | 10 | – | 0 | – | 0 | – | 2 | – |
| Associated 9p | | | | | | | | | | | |
| ♂ | 6 | 1 | 45.0 | 5 | 55.0 | 0 | – | 0 | – | 0 | – |
| ♀ | 14 | 8 | – | 6 | – | 0 | – | 0 | – | 0 | – |
| General population (France) | | | | | | | | | | | |
| ♂ | 200 | | 70.5 | | 23.25 | | ? | | 6.25 | | ? |
| ♀ | 200 | | 65.0 | | 27.75 | | ? | | 7.25 | | ? |

TABLE XI
b,c Triradius in 9p Trisomy

| Trisomy | Number of palms | | | |
|---|---|---|---|---|
| | Studied | b and c Missing | b and c Fused | c Missing |
| Pure 9p | | | | |
| ♂ | 10 | 0 | 0 | 4 |
| ♀ | 12 | 0 | 0 | 5 |
| 9q– | | | | |
| ♂ | 4 | – | 2 | 2 |
| ♀ | 16 | 4 | 2 | 5 |
| Associated 9p | | | | |
| ♂ | 6 | 0 | 2 | 1 |
| ♀ | 14 | 6 | 0 | 8 |

On the fingertips, there is a noticeable excess of arches and a reduced number of whorls (Table XII). Deformities of the flexion creases of the fingers, which are consistent on the fifth digit, are very frequent on all other digits except for the thumbs: 20 digits had only one flexion crease, 4 index fingers, 1 middle finger, 3 ring fingers, and 12 little fingers; 87 digits displayed shortness of the middle phalanx, 27 index fingers, 12 middle fingers, 15 ring fingers, and 33 little fingers. It is not possible to estimate the frequency of this deformity of the flexion creases of the fingers since many authors do not furnish a precise description of this phenomenon.

### *3. Mental retardation*

Mental deficiency is constant, and it becomes noticeable during the first months of life.

(a) *Pure 9p trisomy.* The average intelligence quotient is 55 among 6 cases examined. The extreme levels are 30 for an adult and 75 for a 13-month-old child.

(b) *9p trisomy associated with proximal 9q trisomy.* The average intelligence quotient is 41 for 7 cases. The extreme levels are 16 in a 15-year-old patient and 60 in a 5-year-old patient.

(c) *Associated 9p trisomy.* The average intelligence quotient is 42 for 6 cases examined. The extreme levels are 30 for a 13-year-old child and 52 for an 11-month-old child.

**TABLE XII**
**Fingerprints in 9p Trisomy**

| Trisomy | Number of fingers studied | Fingerprints | | | | | | | |
|---|---|---|---|---|---|---|---|---|---|
| | | Arches | | Ulnar loops | | Radial loops | | Whorls | |
| | | Observed | % | Observed | % | Observed | % | Observed | % |
| Pure 9p | | | | | | | | | |
| ♂ | 50 | 4 | 8 | 41 | 82 | 4 | 8 | 1 | 2 |
| ♀ | 60 | 8 | 13.3 | 44 | 73.3 | 1 | 1.8 | 7 | 11.6 |
| 9q– | | | | | | | | | |
| ♂ | 30 | 7 | 23.3 | 23 | 76.6 | 0 | 0 | 0 | 0 |
| ♀ | 90 | 18 | 20.0 | 65 | 72.2 | 6 | 6.7 | 1 | 1.1 |
| Associated 9p | | | | | | | | | |
| ♂ | 30 | 10 | 33.3 | 18 | 60.0 | 1 | 3.35 | 1 | 3.35 |
| ♀ | 60 | 28 | 46.6 | 27 | 45.0 | 3 | 5.00 | 2 | 3.4 |
| both sexes | 120 | 67 | 55.8 | 46 | 38.3 | 4 | 3.3 | 3 | 2.6 |
| General population (France) | | | | | | | | | |
| ♂ | 1000 | | 6.74 | | 56.74 | | 4.70 | | 31.82 |
| ♀ | 1000 | | 11.30 | | 59.60 | | 3.50 | | 25.60 |

(d) *Partial 9p trisomy.* Intelligence quotients for three patients with trisomy for the distal half of the short arm 9 (Lewandowski *et al.*, 1976) were 90 (case 1), 46 (case 2), and 56 (case 3).

Speech difficulties are particularly severe; one of the patients displayed deafness and mutism (Rethoré *et al.*, 1977). Nervousness, poor balance, and coordination problems were observed in all of the patients who underwent psychological examination. Two of the adolescent patients were "filchers".

Convulsions were observed in two cases (Hoehn *et al.*, 1971b; Blank *et al.*, cases 1 and 2, 1975). Abnormal electroencephalograms were recorded for four of the patients observed, (Jacobsen *et al.*, 1975; Mason *et al.*, 1975; Turleau *et al.*, case 2, 1974; Zaremba *et al.*, case 3, 1974).

### 4. Growth

At birth, average weight was normal in pure 9p trisomy patients (3106 gm for 13 cases) and for patients with trisomy for the short arm and proximal segment of the long arm (3030 gm for 10 cases). However, for associated 9p trisomies, birth weight is low (2716 gm for 16 cases). At maturity, height varies from 160 cm to 175 cm for males, and from 149 cm to 159 cm for females. Weight is normal or above normal. The menses begin late for females, between 15 and 20 years of age.

Mortality from this syndrome appears very low. Ten patients reached adulthood, but none of them reproduced. Among the 53 patients examined, four females died during the first months of life. They were all afflicted with an associated 9p trisomy (Table VII). In addition to these cases, it is necessary to include the one reported by Butler *et al.* (1969) which probably corresponded to a 9p trisomy accompanied by partial trisomy of the long arm of chromosome 15.

### 5. Family data

Pregnancy progressed normally, and delivery occurred at full term or close to full term. The sex ratio was sharply altered in favor of females: pure 9p trisomy, 8 males and 10 females; 9p trisomy associated with trisomy for the proximal segment of the long arm, 4 males and 10 females; associated 9p trisomy, 5 males and 16 females; partial 9p trisomy, 2 males and 1 female.

In the twelve cases which resulted *de novo,* the average age of the mothers was 29 years ($\sigma = 4.56$), while the average age of the fathers was 30 years ($\sigma = 4.17$).

### 6. Cytogenetic data

The 9p trisomy syndrome displays highly diversified cytological features.

The short arm of a chromosome 9 may appear in the form of a small

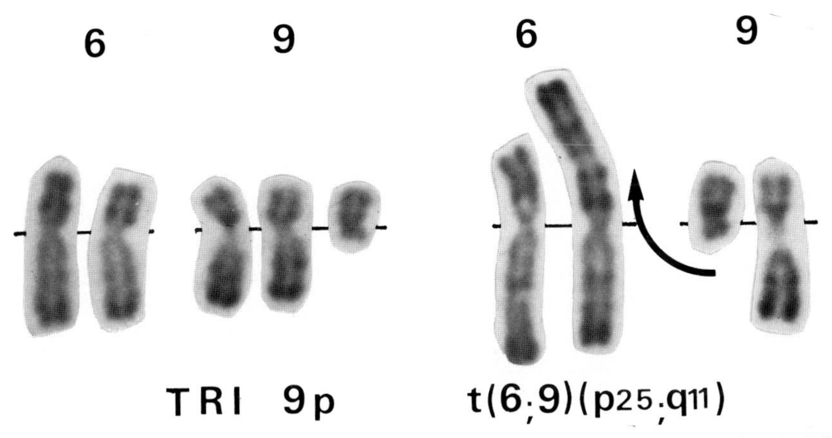


**Fig. 11.** View of 9p trisomy (Rethoré *et al.*, 1973). The extra chromosome (on the left) results from improper segregation of the maternal translocation (on the right).

supernumerary element (Fig. 11), or translocated into another chromosome (Fig. 12).

A parental chromosome abnormality was observed in 52 cases. Usually, it was a situation of reciprocal translocation, which occurred 39 times in the mothers and 9 times in the fathers. One of the cases of partial 9p trisomy (Lewandowski *et al.*, case 1, 1976) resulted from a 9p21p24 insertion within the long arm of one of the mother's 18 chromosomes. There was some uncertainty concerning the two cases reported by Cantu *et al.* (1971) (Table IV).

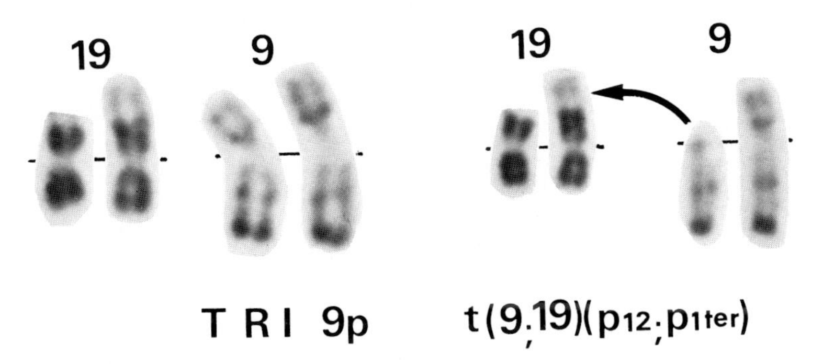


**Fig. 12.** View of 9p trisomy (Rethoré *et al.*, 1973). The extra short arm of the 9 chromosome is translocated to the short arm of chromosome 19. On the right, there is a balanced maternal translocation.

## C. 9p Tetrasomy

After the initial observation of tetrasomy of the short arm of chromosome 9 by Ghymers *et al.* (1973), two additional cases were reported (Table XIII).

### *1. Phenotype*

Facial appearance was described only for two of the patients observed (Rutten *et al.*, 1974; Orye *et al.*, 1975). For these two female patients, facial dysmorphology was relatively similar to that described for 9p trisomy. However, the eyelid slits were horizontal in one case (Rutten *et al.*, 1974), and the ears were low set in the other.

Analysis of dermatoglyphic patterns of these two female patients showed that there was a transverse palmar crease on only one hand. The axial triradii were in position t on two palms, and in t and t″ on the other two. The subdigital triradius, c, was absent on one hand. On the fingertips, there were 11 ulnar loops, 8 whorls, and 1 extended arch.

For the following three patients, there were multiple, severe deformities: hydrocephaly, enlarged heart, hypoplasia of the right kidney, thoracic hemivertebrae, camptodactyly, restricted bending of the knee, and thrombocytopenia (Ghymers *et al.*, 1973); hydrocephaly, myopia, pseudoarthrosis of the right clavicle, brachymesophalangia of the fifth fingers, and deformed feet (Rutten *et al.*, 1974); cleft lip on the right and a cleft palate, pulmonic stenosis, hypoplasia of the pulmonary artery, VSD, cubitus valgus, and brachymesophalangia of the fifth fingers (Orye *et al.*, 1975).

### *2. Growth and Family Data*

One of the patients (Ghymers *et al.*, 1973) died at 11 months. Two others displayed severe mental deficiency. At the age of five, the intelligence quotient of the patient studied by Rutten *et al.* (1974) was 48.

In one case (Ghymers *et al.*, 1973), the mother had undergone a complete ovariectomy on one side and a partial ovariectomy on the other side. In another case (Rutten *et al.*, 1974), profuse hemorrhages had occurred during the twelfth week of pregnancy.

### *3. Cytogenetic Data*

In three cases, 9p tetrasomy showed a mosaic pattern. Analysis of the detailed structure of the chromosomes showed, in the first case, that the supernumerary element consists of two short arms of chromosome 9 connected by a secondary constriction which is two times longer than that of the normal chromosome 9. In the other two cases, there was an isochromosome for the short arm of chromosome 9 without a secondary constriction: i(9)(pter→cen→pter).

**TABLE XIII**
**Observations of Mosaic 9p Tetrasomy**

| | Patient | | | | | | | Parents | | | |
|---|---|---|---|---|---|---|---|---|---|---|---|
| | Birth | | | | | | | Age at birth of child | | Karyotype | |
| Authors | Year | Term (weeks) | Weight | Height | Cranial perimeter | Sex | Death | Mother | Father | Mother | Father |
| Ghymers *et al.* (1973) | 1972 | ? | 2250 | 48 | ? | M | 11 months | ? | ? | ? | ? |
| Rutten *et al.* (1974) | 1968 | ? | 3400 | ? | ? | F | – | 31 | 32 | N | N |
| Orye *et al.* (1975) | 1972 | 40 | 2290 | 46.5 | 30.5 | F | – | 25 | 25 | N | N |

## D. 9p Monosomy

Since the initial description by Alfi *et al.* in 1973, nine other cases of partial monosomy of the short arm of chromosome 9 have been reported (Table XIV).

### *1. Phenotype (Figs. 13 and 14)*

In newborn infants, there is craniostenosis along with trigonocephaly. Diameter of the parietal bones is reduced, and the forehead is high. The eyebrows are very pointed, and the eyeballs protrude somewhat. The eye slits are oblique; their inner corner is concealed by an epicanthus. The nose is short and the nasal saddle is long and flat; its end does not extend beyond the area beneath the nasal septum. The nostril openings are pointed forward. The upper lip is long, and it hangs over the lower lip; the pillars of the philtrum are aplasic. The mouth is small, and the chin recedes. The auricles of the ears are very flat, and, frequently, the vertical portion does not have a pronounced border. The earlobes are joined to the cranium, or they bend backwards.

The neck is short and is sometimes webbed. The thorax is elongated, and the nipples are wide-spaced. Among female patients, the labia majora are hypoplas-

**TABLE XIV**
**Observations of 9p Monosomy**

| | Parents | | Patients | | |
|---|---|---|---|---|---|
| Authors | Maternal age | Parental translocation | Karyotype | Birth | Death |
| Alfi *et al.* (1973, 1976) | | | | | |
| Case 1 | 22 | t(9;16)(p22;q23) mat. | 46,XX,9p– | 1971 | – |
| Case 2 | 25 | – | 46,XY,9p– | 1973 | – |
| Case 3 | 27 | – | 46,XX,9p– | 1973 | – |
| Case 4 | 27 | – | 46,XY,9p– | 1974 | – |
| Case 5 | 45 | – | 46,XX,9p– | 1965 | – |
| Case 6 | 31 | – | 46,XX,9p– | 1955 | – |
| Orye *et al.* (1975) | 25 | t(9;15)(p22;q26) pat. | 46,XX,9p– | 1973 | – |
| Prieur *et al.* (1976a) | | | | | |
| Case 1 | 31 | – | 46,XX,9p– | 1970 | – |
| Case 2 | 26 | – | 46,XX,9p– | 1973 | – |
| Prieur *et al.* (1976b) | | | | | |
| Case 1 | 36 | t(5;9)(q32;p23) pat. | 46,XX,9p+ | 1971 | 4 months |
| Case 2 | 20 | t(9;12)(p21;q42) mat. | 46,XY,9p+ | 1972 | 6 months |

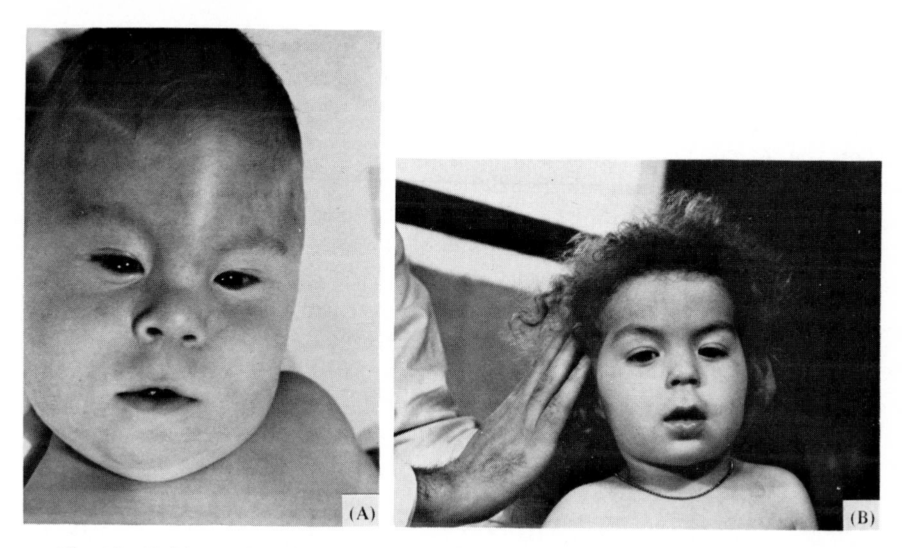


**Fig. 13.** Children afflicted with pure 9p monosomy. (A) Prieur *et al.* (1977a), case 1, (B) Prieur *et al.* (1976a), case 2.

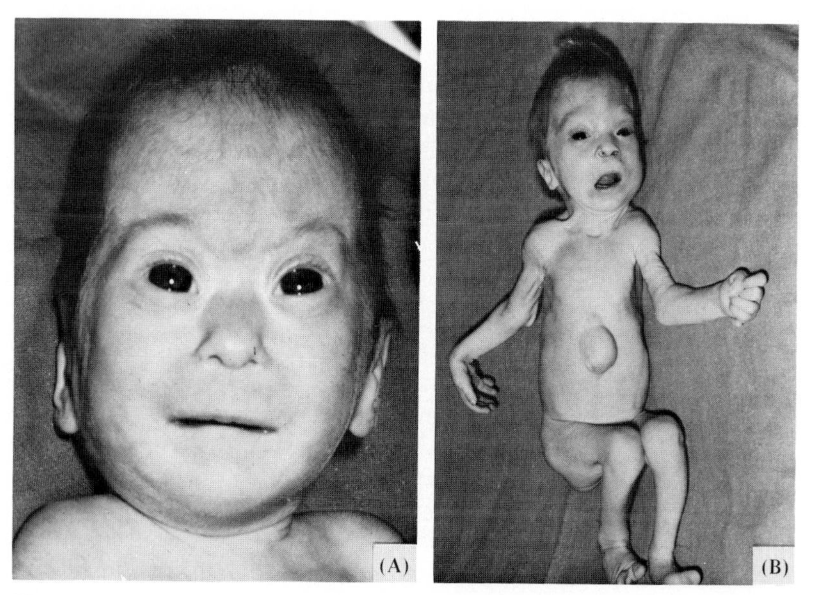


**Fig. 14.** Newborn children having associated 9p monosomy. (A) 5qter→5q32 trisomy (Prieur *et al,* case 1, 1976b). (B) 12qter→12q24.2 trisomy (Prieur *et al.*, case 2, 1977b).

tic, and the labia minora are hypertrophic. Two of the male patients displayed hypospadias.

The fingers are long, but, in particular, there is an exaggerated distance between the two flexion creases. On some fingers, one finds, in the middle of the second knuckle, a small additional flexion crease. The nails, which are well developed, are extremely convex, being of the "roman tile" type.

Among the seven patients manifesting pure monosomy, three were afflicted with cardiac deformities (Alfi *et al.*, case 2, 1973; Alfi *et al.*, cases 3 and 4, 1976). A pneumoencephalogram revealed ventricular dilatation in one patient and enlargement of the cistern in another (Prieur *et al.*, case 1, 1976a).

Deformities are most severe and most numerous among patients whose monosomy is accompanied by another chromosome imbalance: ductus arteriosus, omphalocele, diaphragmatic and inguinal hernia, pointed feet, and single umbilical artery (Alfi *et al.*, case 1, 1973); ductus arteriosus, perforation of the heart, bilateral hydronephrosis, and pointed feet (Orye *et al.*, 1975b); hypotelorism, diffused micropolygyria, and Fallot's tetrology (Prieur *et al.*, case 1, 1976b); adultlike coarctation of the aorta, patent foramen ovalis, ventricular hypertrophy, disc-shaped kidneys, diaphragm-like incisions on the convex portion of the liver, subumbilical hernia with agenesis of the muscles, hypospadias and bilateral undescended testes, and bilateral hexadactyly (Prieur *et al.*, case 2, 1976b).

### 2. *Dermatoglyphic Features*

Full analysis of the twelve hands showed that the flexion creases of the palms generally have a normal arrangement. There was a $t''$ elevation of the axial triradius in 9 palms, and $t'$ elevation in 2 palms. All except two of the subdigital triradii were in the proper positions: d was missing in one hand (Alfi *et al.*, case 1, 1973), and c was absent in another (Prieur *et al.*, case 2, 1976b).

There is a noticeable excess of whorls on the fingertips. Four patients had ten whorls, one had eight, five had six, and another had five. The distance between the flexion creases of the fingers is exaggerated.

### 3. *Growth and Family Data*

Among the eleven subjects, there were three males and eight females. Two died within the first months of life. Underdevelopment was severe, and psychomotor development was practically absent. Two of the children were afflicted with an associated 9p monosomy.

Among other patients, mental deficiency was severe. The intelligence quotient ranged from 33 for an infant of 21 months to 60 in an infant of 16 months. The newborn patients were hypertonic.

In the cases with a *de novo* defect, the average age of the mothers was 30.3 years and that of the fathers was 32.5 years. The parents had not been subjected to significant radiation prior to birth of the children.

Study of families where a translocation was present revealed a large number of subjects with deformities, as well as still births and spontaneous abortions.

### *4. Cytogenetic Data*

In seven cases (Table XIV), the patients were monosomic for the distal portion of the short arm of chromosome 9 (9p22→9pter).

In the four other cases, partial 9p monosomy resulted from parental translocation and, for that reason, was accompanied by another chromosome disorder: 9p22 monosomy→9pter and 16qter trisomy (Alfi *et al.*, case 1, 1973); 9p22 monosomy→9pter and 15q26 trisomy (Orye *et al.*, 1975); 9p23 monosomy→9pter and 5q32→5qter trisomy (Prieur *et al.*, case 1, 1977b); 9p22 monosomy→9pter and 12q24.2→12qter (Prieur *et al.*, case 2, 1977a,b) (Fig. 15).

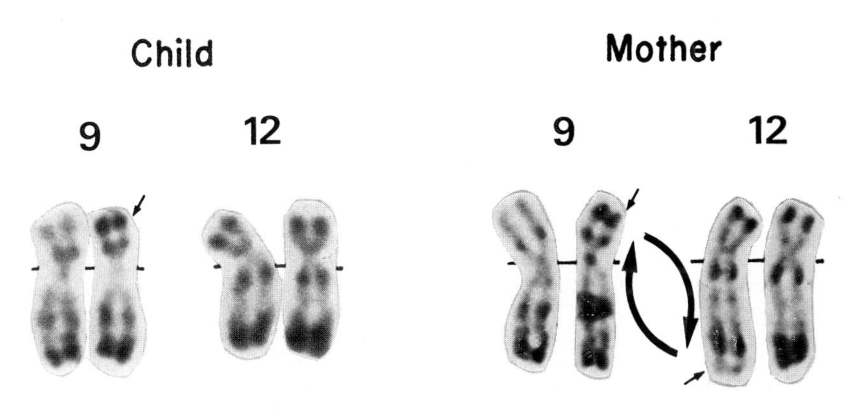


**Fig. 15.** Composite showing 9p monosomy (on the left) resulting from maternal translocation t(9;12)(p21;q42) (Prieur *et al.*, case 2, 1977b). Notice pericentric inversion of the secondary constriction on chromosome 9, which was not involved in the translocation.

## E. Trisomy 9

Since the original observation by Feingold and Atkins in 1973, the presence of a triplicated 9 chromosome has been demonstrated in four children that were born alive and in ten which were aborted.

Among the five children born alive (Table XVa), four were bearers of a mosaic pattern. In one case (Haslam *et al.*, 1973), the trisomy 9 clone was very infrequent,

**TABLE XV**
**Observations of 9 Trisomy**

(a) Children Born Alive

| Authors | Patient | | | | | | | | | | | Parents | | | |
|---|---|---|---|---|---|---|---|---|---|---|---|---|---|---|---|
| | Karyotype | | | | Birth | | | | | | | Age at birth of child | | Karyotype | |
| | Blood % | | Skin % | | | | | | | | | | | | |
| | +9 | N | +9 | N | Term (weeks) | Weight | Height | Cranial perimeter | Sex | Deces Death | | Mother | Father | Mother | Father |
| Feingold and Atkins (1973); Kurnick *et al.* (1974) | 100 | 0 | ? | ? | 40 | 3218 | 50.8 | 32.5 | M | 28 days | | 24 | 25 | N | N |
| Haslam *et al.* (1973) | 6 | 94 | 8 | 92 | 39 | 2700 | 48 | ? | M | 8 years | | 32 | 40 | N | N |
| Bowen *et al.* (1974) | 12 | 88 | 0 | 100 | 39 | 2211 | 49 | 32.5 | M | 42 days | | 17 | 19 | N | N |
| Schinzel *et al.* (1974) | 51 | 49 | 0 | 100 | 45 | 2420 | 51 | 35.6 | M | – | | 21 | 22 | N | N |
| Leisti *et al.* (1975) | 46,X,–X,+der(9), t(X;9)(q11;q32) | | | | ? | 2590 | ? | ? | F | – | | 27 | 33 | t(X;9) (q11;q32) | N |

(b) Abortuses

| | Fetus | | | | | Parents | | | |
|---|---|---|---|---|---|---|---|---|---|
| | | | | | | Age | | Karyotype | |
| | Tissue +9 % | AFC +9 % | Term (weeks) | Embryo development | Sex | Mother | Father | Mother | Father |
| Francke *et al*. (1975) | 100 | 100 | 17 | Interrupted pregnancy at 20 weeks | M | 40 | ? | N | N |
| Boué *et al*. (1975) | | | | | | | | | |
| Case 1 | 100 | – | 12 | 40 days | F | 43 | ? | ? | ? |
| Case 2 | 100 | – | 9 | 35 days | F | 21 | ? | ? | ? |
| Case 3 | 100 | – | 11 | 30 days | M | 33 | ? | ? | ? |
| Case 4 | 100 | – | 16 | 28 days | F | 25 | ? | ? | ? |
| Case 5 | 100 | – | 9 | 38 days | F | 25 | ? | ? | ? |

but it was present in both of the tissues which were examined. In two cases (Bowen *et al.*, 1974; Schinzel *et al.*, 1974) the trisomy 9 clone was only present in the blood. In the patient examined by Leisti *et al.* (1975), three 9 chromosomes were present in all of the cells, and one of those had undergone translocation. There was only one X chromosome due to improper separation of the maternal translocation t(9;X)(q11;q32). As is usually the case in unbalanced autosomal–X translocations (Laurent *et al.*, 1975), the abnormal X which contains an autosomal segment is inactive, and its inactivity is shared by the translocated autosome. Thus, it is a situation in which there is a functional mosaic which only affects a very low percentage of cells and which is accompanied by homogeneous monosomy of the short arm of the X chromosome. The patient's phenotype, moreover, is that which exists for Turner's syndrome. In the patient studied by Feingold and Atkins (1973), trisomy 9 was observed in the 63 cells which were examined. Likewise, one cannot formally exclude the presence of a mosaic, since only the blood was examined.

Among the fetuses which were examined (Table XV b), only one (Francke *et al.*, 1975) had been the subject of a prenatal examination, which was decided upon because of the mother's age. Abortion was at the twentieth week by saline injection. Chromosomal examination of the fetus and of its membranes confirmed the existence of full trisomy 9. In other cases, it was a matter of spontaneous abortion: five cases of full trisomy 9, among 12 C-trisomy specimens (Boué *et al.*, 1975); one case of full trisomy 9 and one case of trisomy 9 accompanied by Klinefelter's syndrome among 36 heteroploid specimens (Kajii *et al.*, 1973); three cases of full trisomy 9 among 10 specimens examined (Therkelsen *et al.*, 1973).

### *1. Phenotype*

If we set aside the particular case observed by Leisti *et al.* (1975), for the reasons previously cited, there remain only four patients, who are all males.

The same craniofacial deformities appeared in the four children: microcephaly and dolichocephaly; narrow eyelid slits that were pointed obliquely upward and outward; enophthalmos; hypertelorism; large nose; low-set auricles; retromicrognathism.

Severe and multiple deformities attest the lethal nature of the syndrome and include the following:

(a) *Brain:* dilatation of the fourth ventricle and nonfusion of the cerebellum along its median line (Feingold and Atkins, 1973); enlargement of the cisterna magna; hypoplasia of the frontal lobes; vertical orientation of the Sylvian cleft; dilatation of the third ventricle and the lateral ventricles (Haslam *et al.*, 1973).

(b) *Heart and blood vessels:* A-V canal and VSD (4/5); deformity of the tricuspid valve (Bowen *et al.*, 1974); coarctation of the aorta (Feingold and

Atkins, 1973); displacement to the right (Haslam *et al.*, 1973); a single umbilical artery (Francke *et al.*, 1975);

(c) *Genital organs:* cryptorchidism and a small penis (4/4); histological abnormalities in the testicles (Bowen *et al.*, 1974; Francke *et al.*, 1975).

(d) *Bone deformities:* hypoplasia of the twelfth rib, dislocation of the head of the radius, absence of the fibula and the joints (except the first) on the left leg, absence of the third phalanges in all toes (except the first) on the right foot (Feingold and Atkins, 1973); dislocation of the hips and the knees and scoliosis (Haslam *et al.*, 1973); dislocation of the hips and the knee and twisted feet (Bowen *et al.*, 1974); 13 ribs, clubbed hands, and twisted feet (Schinzel *et al.*, 1971).

### 2. *Growth*

All of the patients, except for one, died. Psychomotor development was practically nonexistent. The child who died in its ninth year could only say a few words.

Miscarriages occurred during the third month of pregnancy. Development of the fetus corresponded to that in a 1-month pregnancy.

### 3. *Family Data*

Pregnancies which came to term were characterized by hemorrhages at the end of the first trimester.

The parents were young except in four of the cases observed (Table XV). They had not been exposed to significant radiation. One mother who had epilepsy had been treated for ten years. The parents' karyotypes were normal. It should be noted, however, that, in two cases (Bowen *et al.*, 1974; Schinzel *et al.*, 1974) the mother had a pericentric inversion of the secondary constriction of chromosome 9, and that in the trisomal cells of the children the mother's chromosome 9 which had shown an inversion reappeared in duplicate form.

## F. Partial 9q Trisomy

In 1975, Turleau *et al.* (1975b) reported the first two cases of partial trisomy of the long arm of chromosome 9.

### 1. *Phenotype*

In both patients, there were common dysmorphologic characteristics: dolichocephaly; very sunken eyes, narrow eyelid slits; hypothelorism; flattened nose; small mouth; receding chin. The fingers were long and thin. The index finger overlapped the thumb and the third finger. The knuckles were long and poorly positioned.

Various deformities tended to appear in only one of the two patients–Case No. 1: microcephaly, epicanthus, strabismus, pyloric stenosis, and VSD; Case No. 2: low set ears, 13 dorsal vertebrae, and 13 ribs.

Mental deficiency is severe. At 12 months, the first child had an intelligence quotient of 55–60. At 5 years, his mental retardation worsened considerably. He became nervous and aggressive. His weight was 15 kg; his height was 99 cm. The heart was normal.

### *2. Cytogenetic data*

In one case, trisomy affected the segment, q11→q33; it was produced by a tandem duplication which arose *de novo.*

In the other case, trisomy affected the segment q31→qter and was accompanied by 4q34 monosomy. The defect resulted from improper segregation of a maternal translocation, t(4;9)(q33;q31).

## G. Ring 9

Since the original observation by Kistenmacher and Punnett in 1970, two additional cases have been reported (Table XVI).

### *1. Phenotype*

In both newborn children (Kistenmacher and Punnett, 1970; Fraisse *et al.*, 1974), craniofacial deformities were similar: microcephaly with trigonocephaly; enlarged and arched eyebrows; moderate exophthalmia; upward and outward obliqueness of the eye slits; epicanthus; inward curving nasal wings and outturned nostrils; a long upper lip; down-turned corners of the mouth and obscuring of the Cupid's bow; retrognathism; and significant protrusion of the antihelixes.

In the patient examined by Jacobsen *et al.* (1972), the cranial perimeter was small. The eyelid slits were moderately oblique upward and outward, and there were protuberant eyeballs. The eyebrows were horizontal. The nose was large at the tip, the upper lip was short, and the lower lip turned outward. The chin was small and receding.

In all three patients, the flexion creases of the palms were normal, except in one hand, where there was the equivalent of a transverse palmar crease. The axial triradius was in position t and the subdigital triradii were properly situated. Two patients had an excess of whorls.

### *2. Growth and Family Data*

One of the patients died as a result of cardiac arrest (Fraisse *et al.*, 1974). Mental deficiencies were severe. The newborn infants were hypotonic.

**TABLE XVI**
**Observations of Ring Chromosome 9**

| | Patient | | | | | | | Parents | | | |
|---|---|---|---|---|---|---|---|---|---|---|---|
| | | Birth | | | | | | Age at birth of child | | Karyotype | |
| Authors | Year | Term (weeks) | Weight | Height | Cranial perimeter | Sex | Death | Mother | Father | Mother | Father |
| Kistenmacher and Punnett (1970); Kistenmacher *et al.* (1975) | 1967 | 40 | 1700 | 47.5 | ? | M | – | 21 | 21 | N | N |
| Jacobsen *et al.* (1973) | 1951 | 40 | 3500 | 54 | ? | F | – | 38 | 42 | N | N |
| Fraisse *et al.* (1974) | 1973 | 40 | 2740 | ? | ? | M | 2 years | 29 | 36 | N | N |

The intelligence quotient for the patient examined by Jacobsen *et al.* (1975) was 50. At 22 years of age, she was 147 cm and weighed 52 kg. Her menses had occurred at 13. She was afflicted by hereditary diabetes.

The parents' karyotypes were normal. They had not been exposed to radiation prior to birth of the children.

### 3. *Cytogenetic Data*

As is usually the case when there is a ring chromosome (Lejeune, 1968), chromosome analysis revealed several cell categories in each of the three patients (Fig. 16). In two cases (Kistenmacher *et al.*, 1975; Fraisse *et al.*, 1974), analysis of the detailed structure of chromatids demonstrated that the break points were located at regions p2 and p3 (Fig. 16), corresponding to the telomeres of the short and long arms.

## H. Conclusion

As a result of this study, it appears that defects of chromosome 9 are relatively frequent and that they correspond to syndromes which are clinically recognizable from birth. Comparison of phenotypic changes observed for the various defects illustrates the concept of "type and countertype," which Lejeune introduced in 1968. If one takes trisomy as a model, the traits produced by monosomy of the same segment are opposite.

*9p Trisomy:* round cranium and extremely large anterior fontanel, enophthalmus, upwardly and outwardly oblique eyelid slits, large nose, short upper lip which exposes the upper incisors, truncated auricles of the ears, brachymesophalangia, excessive arches on the fingertips

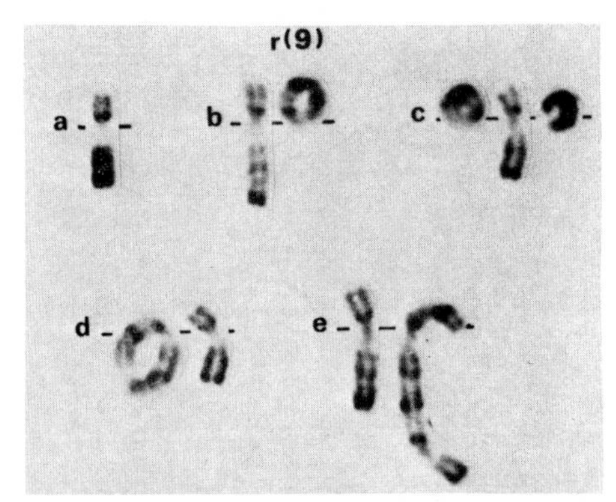


**Fig. 16.** Composite based upon five cells of a patient having a ring 9 chromosome (Fraisse *et al.*, 1974). (a) 9 monosomy; (b) r(9); (c) 2r(9); (d) dicentric r(9); (e) open dicentric r(9).

*9p Monosomy:* craniostenosis and trigonocephaly, exophthalmus, upwardly and outwardly oblique eyelid slits, short nose, upper lip that is long and overlaps the lower lip, very truncated auricles of the ears, elongated intermediate phalanges, excessive whorls.

Certain features of 9p trisomy recur in full 9 trisomy (enophthalmus, large nose), but they are accompanied by other traits that were observed for 9q trisomy (narrow eyelid slits, retromicrognathism). In addition, visceral deformities are more numerous and more severe.

Facial deformities in newborn children who have a ring 9 chromosome are comparable to those observed in children with 9p monosomy. On the other hand, the adolescent observed by Jacobsen *et al.* (1975) displayed certain features of trisomy. These phenotype variations reflect the coexistence of monosomic and trisomic cells in these patients.

## III. ABNORMALITIES OF CHROMOSOME 12

### A. Introduction

It was only after techniques permitting analysis of the detailed structure of chromosomes (Caspersson *et al.,* 1970; Drets and Shaw, 1971; Dutrillaux and Lejeune, 1971) became available that it was possible to distinguish defects involving chromosome 12. With conventional Giemsa staining, this chromosome was barely distinguishable from other submetacentric elements in group C, as defined by the 1963 International Conference in London.

It is certainly far too early to be able to furnish a full definition of syndromes corresponding to these defects, but, at this point, it is possible to analyze observations of six 12p monosomy cases and five 12p trisomy cases.

### B. Partial 12p Monosomy

Since the first observed case of partial deletion of the short arm of chromosome 12, described in 1974 by Mayeda *et al.,* five other cases have been reported (Table XVII).

#### *1. Phenotype (Figs. 17 and 18)*

It is not possible to develop an overview of deformities common to the six patients who were studied. However, phenotype analysis, which is summarized in Table XVII, shows that certain symptoms recurred in several patients: hypoplasia of the jaw 5/5; microcephaly 4/5; high and arched eyebrows 4/5;

**TABLE XVII**
**Observations of Partial Monosomy of the Short Arm of Chromosome 12**

| | Mayeda *et al.* (1974) | Malpuech *et al.* (1975) | Tenconi *et al.* (1975) | Orye and Craen (1975) | | Magnelli and Therman (1975) |
|---|---|---|---|---|---|---|
| | | | | Case 1 | Case 2 | |
| 12p segment deleted | End portion | 12p11→12p12.2 | 12p11→12p13 | 12p12.?→12p13 | 12p12.?→12p13 | ? |
| Sex | F | M | M | F | M | M |
| Birth | | | | | | |
| Term (weeks) | 41 | 40 | 38 | 36 | 40 | ? |
| Weight | 2,600 kg | 2,800 kg | 2,740 kg | 3,100 kg | 3,400 kg | ? |
| Age at time of diagnosis | 10 years | 3 years | 15 days | 13 months | 6½ years | 35 years |
| Death | – | – | 14 months; lung illness | – | – | – |
| Cranium | | | | | | |
| Occipital frontal Circumference | Reduced | Reduced | ? | Reduced; craniostenosis | Normal | Reduced |
| Occipital Bones | ? | Prominent | Prominent | ? | ? | ? |
| Forehead | ? | Normal | Straight and receding | Straight | High and broad | Straight and receding |
| Eyes | | | | | | |
| Eyebrows | ? | Arched | Arched | Arched | Normal | Arched |
| Epicanthus | ? | – | – | – | – | – |
| Eyelid slits | ? | Oblique downward | Horizontal | Horizontal | Horizontal | Oblique downward |

| | | | | | | |
|---|---|---|---|---|---|---|
| Nose | | | | | | |
| Saddle | ? | Protruding | Large and protruding | Broad and flat | Normal | ? |
| Crest | ? | Fine | Normal | Normal | Fine | Normal |
| Tip | ? | Pointed | Round | Fleshy | Pointed | Pointed |
| Lips | | | | | | |
| Upper | ? | Philtrum not distinct | Normal | Normal | Normal | ? |
| Lower | ? | Normal | Normal | Normal | Normal | ? |
| Mouth | ? | Small | Normal | Small | Small | Normal |
| Jaw | Hypoplastic | Hypoplastic | Hypoplastic | Hypoplastic | Hypoplastic | ? |
| Ears | | | | | | |
| Position | Low | Low | Low | Normal | Low | Normal |
| Helix | – | Broad and truncated | Normal | ? | ? | Hyperplastic |
| Antihelix | – | Hypoplastic | Normal | Hypoplastic | ? | ? |
| Neck | | Normal | Normal | Normal | Normal | Broad and short |
| Thorax | Keel-shaped | ? | ? | Normal | Broad | Asymmetrical |
| External Genitals | | Normal | ? | Small penis | Normal | Cryptorchidism, ♀Pubic hairiness |
| Hands | | | | | | |
| Palms | Hypoplastic metacarpals | Narrow | ? | ? | ? | Hypoplastic metacarpals |
| Fingers | | Clinodactyly of 5th | Long and slender Clinodactyly of 5th | Brachydactyly | ? | Short |
| Feet | | | | | | |
| Top of Foot | Hypoplastic metatarsals | Normal | ? | ? | ? | Hypoplastic metatarsals |
| Toes | | ? | Clinodactyly of 5th | Brachydactyly | ? | ? |
| LDH.B | Decreased | Decreased | Decreased | ? | ? | ? |

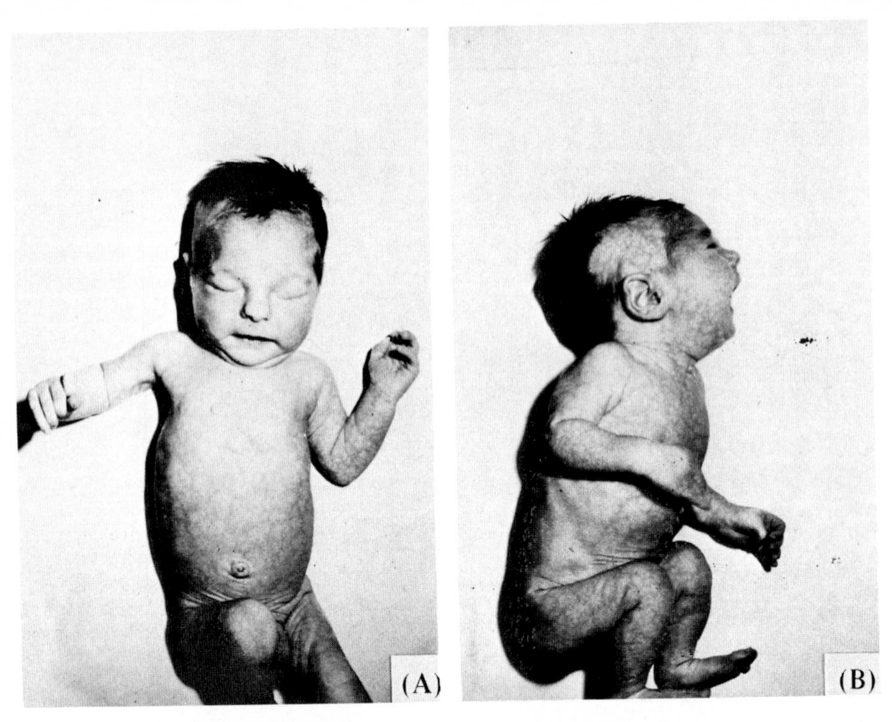


**Fig. 17.** Patient with 12p monosomy, 1-month-old (face and profile) (Tenconi *et al.*, 1975). Notice protrusion of occipital bones and retrognathism.

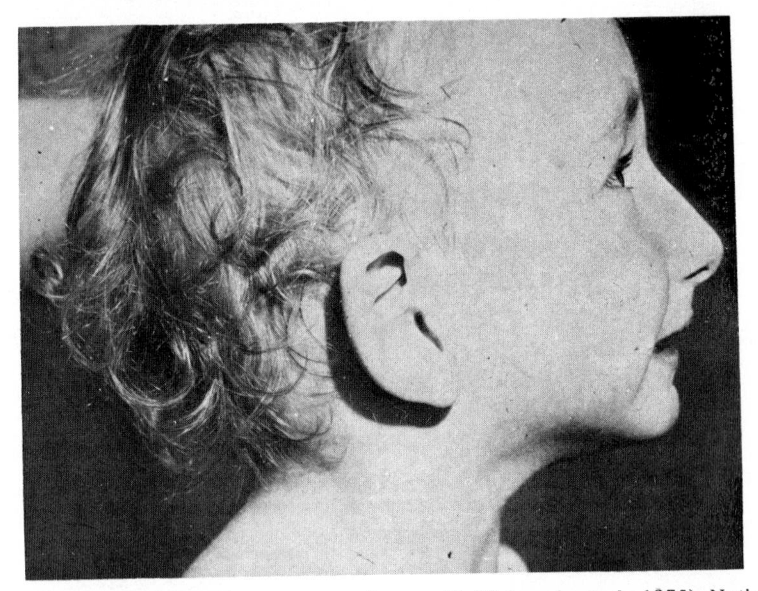

**Fig. 18.** Patient with 12p monosomy, 3-years-old (Malpuech *et al.*, 1975). Notice the low set auricle and aplasia of the antihelix.

low set auricles of the ears 4/5; narrow forehead 3/5; pointed tip of the nose 3/5; protruding occipital bone 2/2; downward and outward obliqueness of eyelid slits 2/5; hypoplasia of the antihelixes 2/3; and hypoplasia of the metacarpals and metatarsals 2/3.

In addition to the bone deformities cited in Table XVII which involve the metacarpals and metatarsals, one patient (Orye and Craen, case 2, 1975) displayed familial congenital osteogenesis. Bone age was normal except in one case (Orye and Craen, case 1, 1975), where it was advanced.

### 2. *Dermatoglyphic Features*

The data for monosomy 12p are summarized in Table XVIII. For four palms among eight, the axial triradius was in t or in t′ position. For three palms, the main line of the axial triradius ended in the first interdigital space, at 11. Flexion creases of the palms were normal.

### 3. *Gene Localization*

The relationship between the lactate dehydrogenase (LDH) gene and chromosome 12 was demonstrated in 1973 by Chen *et al.* using cell hybridization. Subsequently, Mayeda *et al.* (1974) and then Malpuech *et al.* (1975) and Tenconi *et al.* (1975) studied this enzyme in their patients. The three authors observed a reduction of nearly 40% in LDH activity in the red corpuscles. According to Vesel and Bearn (1961), subunit B is preponderant in the red corpuscles. These observations permit one to conclude that the locus of the *LDH.B* gene occurs along the short arm of chromosome 12, and, more precisely, in the midportion of band 12p12, since, among the three patients examined, it was only that region which had been deleted.

Routine tests gave normal results, and study of red cells did not reveal any abnormalities of blood groups.

### 4. *Growth and Psychomotor Development*

At birth, weight was slightly below average (2930 gm in five cases). Height and weight underdevelopment was observed by all authors. The patient of Magnelli and Therman (1975), at 35 years of age, weighed 45 kg and was 145 cm tall.

Severe mental retardation was always present. The intelligence quotient was 46 at 3 years (Malpuech *et al.*, 1975) and 20 during adulthood (Magnelli and Therman, 1975).

There were bilateral Babinski signs, as well as brisk deep tendon reflexes and generalized seizures, according to Mayeda *et al.* (1974). The patient observed by Magnelli and Therman (1975) was hypertonic, while those observed by Tenconi *et al.* (1975) and Orye and Craen (1975) were hypotonic.

**TABLE XVIII**
**Dermatoglyphic Features of Patients with 12p Monosomy**

| Authors | Subdigital triradius | | | | Axial triradius | Interdigital loops | | | Regions | | Main lines | | Fingerprints | | | | |
|---|---|---|---|---|---|---|---|---|---|---|---|---|---|---|---|---|---|
| | a | b | c | d | | 7 | 9 | 11 | Thenar | Hypothenar | Palms | Fingers | I | II | III | IV | V |
| Tenconi *et al.* (1975) | | | | | | | | | | | | | | | | | |
| R | 5 | 9 | 11 | 11 | t′→13 | ? | ? | ? | – | – | ? | Brachymesophalangy of 5th | W | UL | W | W | UL |
| L | 5 | 6 | 9 | 9 | t′→13 | ? | ? | ? | – | – | ? | Brachymesophalangy of 5th | W | UL | UL | W | UL |
| Case 1 | | | | | | | | | | | | | | | | | |
| R | 1 | 5 | 6 | 8 | t′→11 | + | ? | ? | ? | ? | ? | ? | UL | UL | UL | UL | UL |
| L | 1 | 5 | 5 | 5 | t→11 | ? | ? | ? | ? | ? | ? | ? | UL | UL | At | UL | UL |
| Orye and Craen (1975) | | | | | | | | | | | | | | | | | |
| R | 3 | 5 | 7 | 9 | t→13 | + | ? | ? | ? | ? | ? | ? | W | RL | UL | W | UL |
| L | 3 | 5 | 5 | 7 | t→13 | + | ? | ? | ? | ? | ? | ? | UL | UL | UL | UL | W |
| Malpuech *et al.* (1975) | | | | | | | | | | | | | | | | | |
| R | ? | ? | ? | ? | t′→13 | ? | ? | ? | ? | ? | N | ? | 7UL; 2W; 1A | | | | |
| L | ? | ? | ? | ? | t→11 | ? | ? | ? | ? | RL | N | ? | | | | | |

R = Right; L = Left; UL = Ulnar loop; RL = Radial loop; W = Whorl; A = Arch; t = Present; – = Missing.

### 5. *Family Data*

The average age of the parents, which was calculated in five cases, was somewhat high: 30.8 years for the mothers and 33.8 years for the fathers. There had been no particular exposure to radiation prior to birth of the patients.

Pregnancies were not characterized by any notable incidents, except for one case (Orye and Craen, case 1, 1975), where there had been a hemorrhage during the third month.

Births occurred at full term in two cases, prematurely in two cases, and one week after full term in another case.

### 6. *Cytogenetic Data*

In the six cases examined, the parents' karyotypes were normal, and the defects observed in the patients resulted from *de novo* phenomena.

Analysis of the detailed structure of chromatids showed, in each case, that only a portion of the short arm of chromosome 12 had been deleted. Thus, it was a situation of a pure partial 12p monosomy.

In each case, study of the karyotype with R bands permitted exact definition of the missing segment: band 12p11 and subband 12p12.1 (Malpuech *et al.*, 1975) (Fig. 19); all of bands 12p11 and 12p12 (Tenconi *et al.*, 1975).

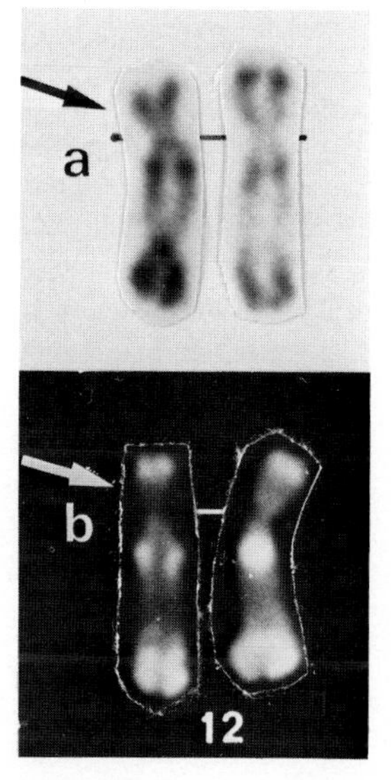


**Fig. 19.** View showing partial deletion of the short arm of chromosome 12: (p11→p12.2) (Malpuech *et al.*, 1975).

In two unrelated patients examined by Orye and Craen (1975), analysis according to Q- and G-bands showed that deletion had occurred in band 12p12, but the limits of the missing segment were not specified.

According to Mayeda *et al.* (1974), deletion in their patient was restricted to the distal portion of the short arm: that is, probable deletion of the 12p13-band and a portion of the 12p12-band.

Magnelli and Therman (1975) did not identify the missing segment and they only indicated that the length of the short arm of one of the 12 chromosomes was reduced by half.

Thus, from this analysis, the conclusion reached is that from one case to another the missing segments varied, although they did partially overlap.

## C. 12p Trisomy

Since the first case reported in 1973 by Uchida and Lin, four additional observations of patients have been provided (Table XIX).

### *1. Phenotype*

Although the cytological diversity of chromosome defects makes phenotype interpretation difficult, it is possible to distinguish a common set of deformities in these five very young children.

The face (Fig. 20), whose outlines are not very distinct, is flat and round. The forehead is high and the occipital bones protrude slightly.

The saddle of the nose is long and it projects slightly. Hypertelorism is less pronounced than in the cat-cry syndrome. The inner corners of the eyes are concealed by broad epicanthi. The eyelid slits are horizontal; the eyebrows are bushy and irregularly situated.

The nose is very short, and the nostril openings point forward. The upper lip is elongated and the philtrum is wide. The red edge is only visible at the level of the Cupid's bow. The lower lip turns outward.

The auricles are normally situated or are somewhat low set. They are only hemmed in the upper portion, except in one case (Fryns *et al.,* 1974) where the hem of the helix, which was rolled into a spiral, continued as far as the antitragus. The lower segment of the antihelix was absent in three cases.

The neck, which is very short, is covered with cutaneous folds. The abdomen is distended due to diastasis of the right side. The sacral dimples are pronounced. Outer genitals are normal.

The hands are large, covered with deep metacarpophalangial dimples. The fingers are spindly. The space separating the first and second knuckles is somewhat enlarged.

X-ray analyses that were performed did not reveal any skeletal deformities, except for the pointed appearance of the distal phalanges. In three cases, study of bone formation revealed retarded development of the head of the femur.

**TABLE XIX**
**Observations of Trisomy of the Short Arm of Chromosome 12**

| | Uchida and Lin (1973) | Fryns *et al.* (1974) | Rethoré *et al.* (1975) | Armendares *et al.* (1975) | |
|---|---|---|---|---|---|
| | | | | Case 1 | Case 2 |
| Chromosome study | | | | | |
| Banding techniques | Q | G | R | G | G |
| Parental translocation | t(8;12)(p23;p11) pat | t(6;12)(q26;p11) pat | t(12;14)(q12;p11) mat | t(12;21)(p11;p11) mat. | ? |
| Patient's karyotype | 46,XY,der(8),t(8;12)p23;p11) | 46,XY,der(6),t(6;12) (q26;p11) | 47,XY,+der(12),t(12;14) (q12;p11) | 46,XX,der(21),t(12;21) (p11;p11) | ? |
| Parents | | | | | |
| Paternal age (years) | 34 | 35 | 40 | 38 | ? |
| Maternal age (years) | 23 | 20 | 37 | 35 | ? |
| Patients | | | | | |
| Sex | M | M | M | F | F |
| Birth date | 2-6-1971 | ? | 11-13-1973 | 6-17-1971 | ? |
| Pregnancy: | | | | | |
| Term (weeks) | ? | ? | 37 | 40 | ? |
| Complications | Cesarean | Near miscarriage | – | – | ? |
| Birth | | | | | |
| Weight (kg) | 3930 | 3450 | 3350 | 4100 | ? |
| Height | ? | 48 cm | 50 cm | ? | ? |
| Cranial perimeter | ? | 33 cm | 34.5 cm | ? | ? |
| Age at diagnosis | 7 months | 1 month | at birth | 2 years 10 months | – |
| Death | | | | | |
| Age | – | 6 weeks | – | – | 3 years |
| Cause | – | Cardiac arrest | – | – | Cranial lesion |

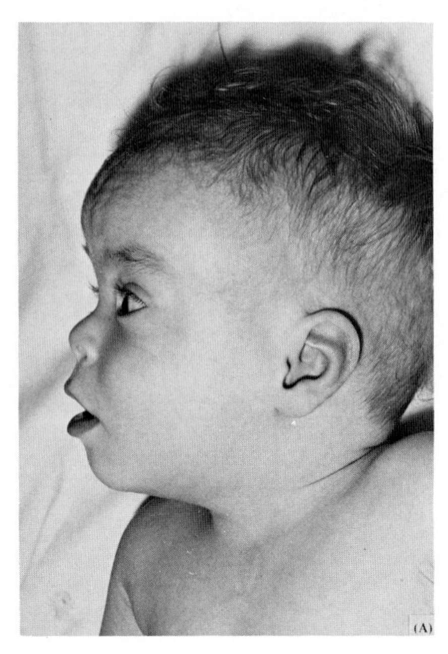


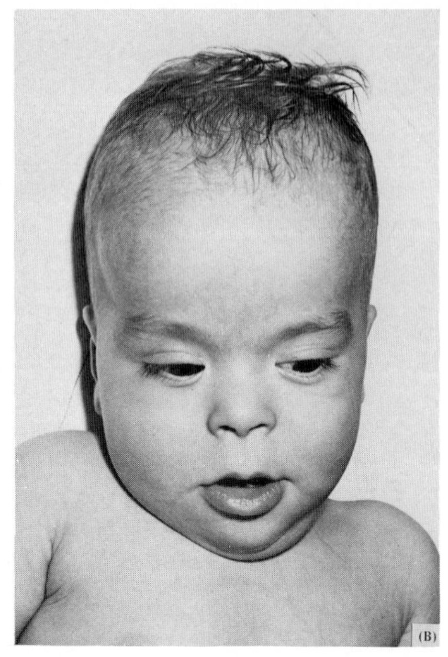


**Fig. 20.** Patient with 12p trisomy, 3-months-old (Face and profile) (Rethoré *et al.*, 1975). Notice the projecting forehead, and extremely short nose, the broad philtrum, the out-turned lower lip, and the appearance of the ears.

## *2. Dermatoglyphic Features (Table XX)*

Arrangement of the epidermal ridges varies considerably from one patient to another. It can be observed, however, that, among the six palms which were analyzed, four had a transverse palmar crease and two had an additional transverse crease which began at the base of the thumb and intersected the edge of the midportion of the hypothenar region. On all the palms, there was an open loop at the third interdigital space.

## *3. Metabolic Studies and Gene Localization*

Chromatography of amino acids from blood and urine, electrophoresis of serum proteins, muscle enzymes and tests for thyroid function gave normal results. Study of red cell blood groups did not reveal any abnormality in the transmission of blood groups.

Lactate dehydrogenase analysis was only performed in one case (Rethoré *et al.*, 1975). A 50% increase in LDH activity was observed in the red cells. Butyrate dehydrogenase activity increased in a parallel manner, and it was not possible to demonstrate an increase in the hydroxybutyrate dehydrogenase (HDBH/LDH) ratio. Electrophoretic analysis did not reveal significant modifica-

**TABLE XX**
**Dermatoglyphic Features of Three Patients with 12p Trisomy**

| Authors | Subdigital triradius | | | | Axial triradius | Interdigital loops | | | Regions | | Main lines | | Fingerprints | | | | |
|---|---|---|---|---|---|---|---|---|---|---|---|---|---|---|---|---|---|
| | a | b | c | d | | 7 | 9 | 11 | Thenar | Hypothenar | Palms | Fingers | I | II | III | IV | V |
| Rethoré *et al.* (1975) | | | | | | | | | | | | | | | | | |
| R | 4 | 7 | 9 | 11 | t | – | + | – | – | – | Additional transverse crease | N | A | A | A | A | A |
| L | 4 | 7 | 9 | 11 | t | – | + | – | – | – | Additional transverse crease | N | A | A | A | A | UL |
| Armendares *et al.* (1975) Case 1 | | | | | | | | | | | | | | | | | |
| R | 3 | 5 | 7 | 9 | t′ | + | – | – | – | – | Transverse palmar crease | N | W | RL | W | W | UL |
| L | 3 | 7 | 9 | 11 | t′ | – | + | – | Loop | – | Transverse palmar crease | N | UL | UL | RL | UL | UL |
| Uchida and Lin (1973) | | | | | | | | | | | | | | | | | |
| R | ? | ? | ? | ? | t″ | ? | + | ? | ? | ? | Transverse palmar crease | Clinodactyly of 5th finger | | | | | |
| L | ? | ? | ? | ? | t″ | ? | + | ? | ? | ? | Transverse palmar crease | Clinodactyly of 5th finger | | | 8 W | | |

R = Right; L = Left; UL = Ulnar loop; RL = Radial loop; W = Whorl; A = Arch; t = Present; – = Missing.

tion of isozyme distribution. In the white cells and fibroblasts, where subunit A is predominant, LDH activity was normal in the patient examined.

This observation confirmed localization of *LDH.B* on the short arm of chromosome 12, as well as an excess of enzyme activity produced by trisomy of that chromosomal segment.

## *4. Growth and Psychomotor Development*

Despite the deformities which have been cited and, with the exception of the patient observed by Fryns *et al.* (1974), who died from cardiac insufficiency during a lung illness, the patients were large, chubby infants who were extremely hypotonic (Fig. 21).

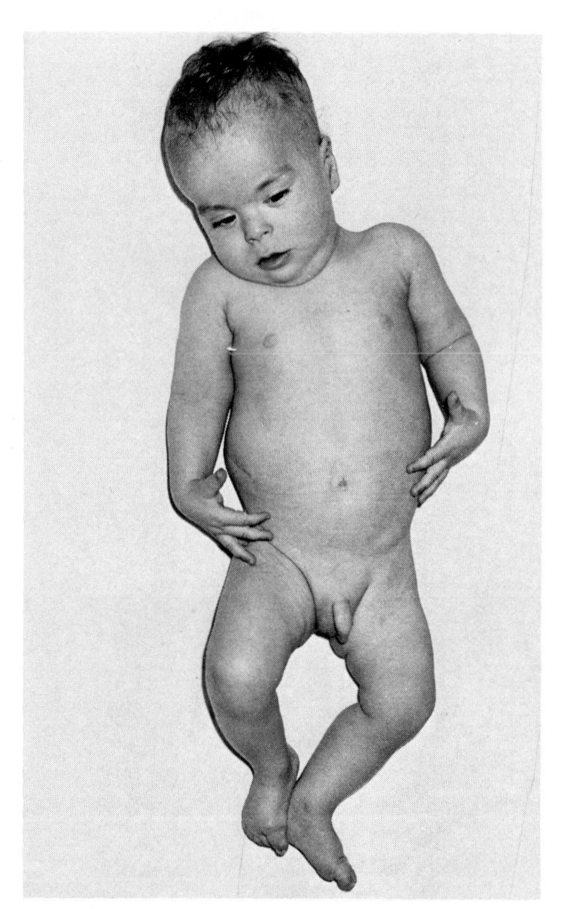

**Fig. 21.** Patient with 12p trisomy, 3-months-old (Rethoré *et al.,* 1975). Notice the excellent nutritional condition, with hypotonia.

Mental retardation was severe. The intelligence quotient at 4 months was 50 (Rethoré *et al.,* 1975) and at 2 years and 10 months, 29 (Armendares *et al.,* case 1, 1975).

### *5. Family Data*

All of the cases resulted from a balanced parental translocation described in Table XIX.

Family investigations are summarized as follows:

(a) Paternal translocation occurred in the patient's sister (Uchida and Lin, 1973).

(b) Chromosome examinations could not be performed for the family of the patient's father. The paternal grandmother had had five children with a normal phenotype: two daughters and three sons, one of whom was the patient's father. She had had three miscarriages, two of them having been early and one during the sixth month. (Fryns *et al.,* 1974).

(c) Chromosome examinations could not be performed for the family of the patient's mother. The karyotype of the patient's elder sister was normal (Rethoré *et al.,* 1975).

(d) Balanced maternal translocation was encountered in the two sisters of the patient. The older brother of the patient, who was not examined, had died when he was 3 days old: "He was a large child whose ears were malformed." One of the mother's sisters had a balanced translocation. Her two children had normal phenotypes, but they were not examined. The mother's other sister was case No. 2. The maternal grandmother's karyotype was normal, but that of the grandfather was not examined (Armendares *et al.,* 1975).

### *6. Cytogenetic Data (Figs. 22 and 23)*

Among the five patients, trisomy resulted from improper segregation of a balanced parental translocation.

Although chromosome analysis could only be performed for one of the two cases reported by Armendares *et al.* (1975), phenotype similarities between the two patients, an aunt and her niece, permits one to conclude that for both subjects chromosome defects had been the same. Both cases corresponded to a complete 12p trisomy which was a pure trisomy since, for these patients, the entire short arm of chromosome 12 existed in triplicate. Monosomy of the short arm of chromosome 21, resulting from parental translocation, did not produce any change in the phenotype as is demonstrated by the normal phenotype for subjects who have a balanced t(DqGq) or t(DqDq) translocation.

In three other cases, 12p trisomy was accompanied by another change of proportions: monosomy of the telomere portion of the short arm of chromosome 8 (Uchida and Lin, 1973); monosomy of the telomere portion of the long arm of

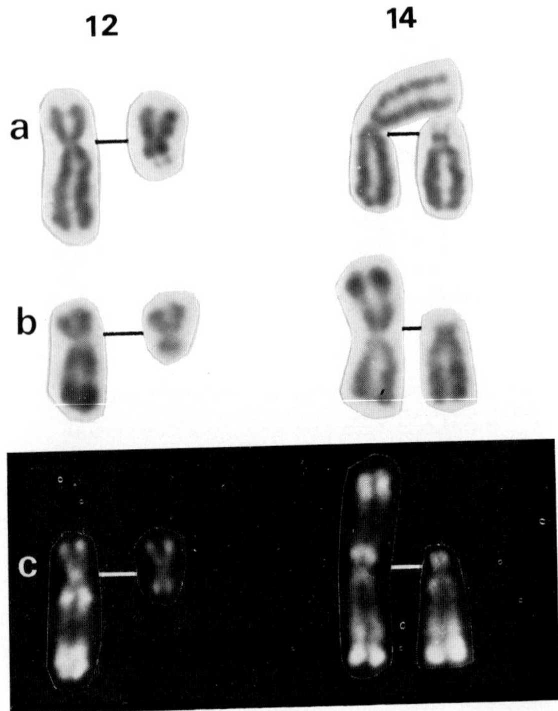


Fig. 22. View of maternal translocation t(12;14)(q12;p11) (Rethoré *et al.*, 1975). (a) Standard Giemsa staining; (b) R-bands; (c) staining with acridine orange.

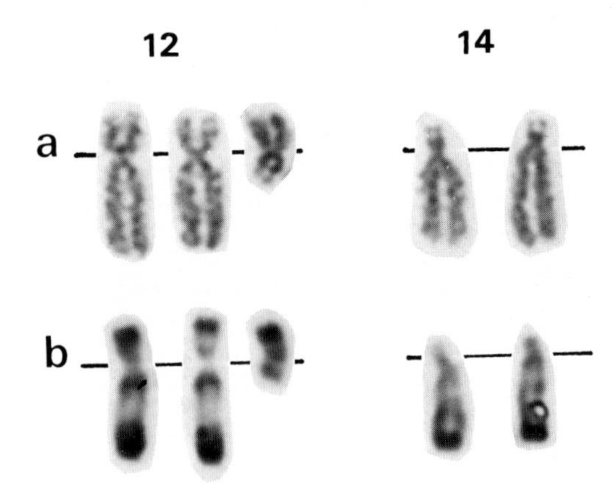


Fig. 23. View of 12p trisomy (Rethoré *et al.*, 1975). (a) Standard Giemsa staining; (b) R-bands.

chromosome 6 (Fryns *et al.,* 1974); trisomy of the proximal portions of the long arm of chromosome 12 and the short arm of chromosome 14 (Rethoré *et al.,* 1975) (Figs. 22 and 23).

## D. Conclusion

Analysis of defects of the short arm of chromosome 12 demonstrates that it may be possible to regard 12p trisomy as a recognizable entity although the same would not be true for partial 12p monosomy for which additional cases are needed to establish a defined syndrome. Partial monosomies, however, have allowed for the localization of the lactate dehydrogenase B gene within the midportion of band 12p12.

## REFERENCES


Alfi, O. S., Donnell, G. N., Crandall, B. F., Derencsenyi, A., and Menon, R. (1973). *Ann. Genet.* **16,** 17–22.

Alfi, O. S., Donnell, G. N., Allerdice, P. W., and Derencsenyi, A. )1976). *Ann. Genet.* **19,** 11–16.

André, M. J., Aurias, A., de Berranger, P., Gillot, F., Lefranc, G., and Lejeune, J. (1976). *Ann. Genet.* (in press).

Armendares, S., Salamanca, F., Nava, S., Ramirez, S., and Cantu, J. M. (1975). *Ann. Genet.* **18,** 89–94.

Aula, P., Leisti, J., and von Koskull, H. (1973). *Clin. Genet.* **4,** 241–251.

Baccichetti, C., and Tenconi, R. (1973). *J. Med. Genet.* **10,** 296–299.

Baccichetti, C., Tenconi, R., Anglani, F., and Zacchello, F. (1975). *J. Med. Genet.* **12,** 425–427.

Balicek, P., Zizka, J., and Lichy, J. (1975). *Humangenetik* **27,** 353–358.

Blank, C. E., Colver, D. C. B., Potter, A. M., McHugh, J., and Lorber, J. (1975). *Clin. Genet.* **7,** 261–273.

Bobrow, M., and Pearson, P. L. (1971). *J. Med. Genet.* **8,** 240–243.

Bobrow, M., Jones, L. F., and Clarke, G. (1971). *J. Med. Genet.* **8,** 235–239.

Boué, J., Boué, A., Deluchat, C., Perraudin, N., and Yvert, F. (1975). *J. Med. Genet.* **12,** 265–268.

Bowen, M. D., Ying, K. L., and Chung, G. S. H. (1974). *J. Pediatr.* **85,** 95–97.

Butler, L. J., Eades, S. M., and France, N. E. (1969). *Ann. Genet.* **12,** 15–27.

Canki, N., Rethoré, M. O., Ferrand, J., and Lejeune, J. (1975). *Lijec. Vjesn.* **97,** 103–105.

Cantu, J. M., Buentello, L., and Armendares, S. (1971). *Ann. Genet.* **14,** 177–186.

Carnevale, A., and De los Cobos, L. (1973). *J. Med. Genet.* **10,** 376–379.

Carter, R., Baker, E., and Hayman, D. (1969). *Med. Genet.* **6,** 224–227.

Caspersson, T., Zech, L., and Johansson, C. (1970). *Exp. Cell Res.* **62,** 490–492.

Centerwall, W. R., Mayeski, C. A., and Cha, C. C. (1975). *Humangenetik,* **29,** 91–98.

Chen, T. R., McMorris, F. A., Creagan, R., Ricciuti, F., Tischfield, J., and Ruddle, F. H. (1973). *Am. J. Hum. Genet.* **25,** 200–207.

Dallaire, L. (1969). *Birth Defects, Orig. Artic. Ser.* **5,** 114–116.

Dallapiccola, B. (1975). *Proc. Annu. Meet. Cytogenet., 8th, 1975.*

Dallapiccola, B., Capra, L., Preto, G., Covic, M., and Dutrillaux, B. (1974). *Ann. Genet.* **17,** 115–118.
Dallapiccola, B., Giovannelli, G., and Forabosco, A. (1975). *Pediatr. Radiol.* **3,** 34–40.
Darmady, J. M., and Seabright, M. (1975). *J. Med. Genet.* **12,** 408–411.
de la Chapelle, A., Koivisto, M., and Schroder, X. X. (1973). *J. Med. Genet.* **10,** 384–389.
Drets, M. E., and Shaw, M. W. (1971). *Proc. Natl. Acad. Sci. U.S.A.* **68,** 2073.
Dutrillaux, B., (1973). *Chromosoma* **41,** 395–402.
Dutrillaux, B., and Lejeune, J. (1971). *C.R. Hebd. Seances Acad. Sci.* **272,** 2638–2640.
Dutrillaux, B., Laurent, C., Couturier, J., and Lejeune, J. (1973). *C.R. Hebd. Seances Acad. Sci.* **276,** 3179.
Dutrillaux, B., Laurent, C., Forabosco, A., Noel, B., Suérinc, E., Biémont, M. C., and Cotton, J. B. (1975). *Ann. Genet.* **18,** 21–27.
Ebbin, A. J., Wilson, M. G., Towner, J. W., and Slaughter, J. P. (1973). *J. Med. Genet.* **10,** 65–69.
Edwards, J. H., Fraccaro, M., Davies, P., and Young, R. B. (1962). *Ann. Hum. Genet.* **26,** 163–178.
Faed, M., Stewart, A., and Keay, A. J. (1969). *J. Med. Genet.* **6,** 342–346.
Feingold, M., and Atkins, L. (1973). *J. Med. Genet.* **10,** 184–187.
Fonatsch, C., Flatz, S. D., and Hurter, P. (1975). *Humangenetik* **25,** 227–233.
Fraisse, J., Lauras, B., Ooghe, M. J., Freycon, F., and Rethoré, M. O. (1974). *Ann. Genet.* **17,** 175–180.
Francke, U. (1972). *Am. J. Hum. Genet.* **24,** 189–213.
Francke, U., Benirschke, K., and Jones, O. W. (1975). *Humangenetik* **29,** 243–250.
Fryns, J. P., Uchida, I. A., Hotz, H., Miller, R. C., and Cassiman, J. J. (1974). *Humangenetik* **24,** 247–252.
Fujita, H., Abe, T., Yamamoto, K., and Furuyama, J. (1974). *Humangenetik* **25,** 83–92.
Furbetta, M., Rosi, G., Cossu, P., and Cao, A. (1975). *Humangenetik* **25,** 87–91.
Ghymers, D., Hermann, B., Disteche, C., and Frederic, J. (1973). *Humangenetik* **20,** 273–282.
Giovannelli, G., Rossi, L., and Forabosco, A. (1973). *Helv. Paediatr. Acta* **28,** 543–552.
Giovannelli, G., Forabosco, A., and Dutrillaux, B. (1974). *Ann. Genet.* **17,** 119–124.
Giraud, F., Mattei, J. F., Mattei, M. G., Ayme, S., and Bernard, R. (1975). *Humangenetik* **30,** 99–108.
Golbus, M. S., Conte, F. A., and Daentl, D. L. (1973). *J. Med. Genet.* **10,** 83–85.
Gouw, W. L., Anders, G. J. P. A., and Kate, L. P. (1972). *Humangenetik* **16,** 251–259.
Gustavson, K. H., Finley, S. C., Finley, W. H., and Jallings, B. (1964). *Acta Paediatr. (Stockholm)* **53,** 172–181.
Haslam, R. H. A., Broske, S. P., Moore, C. M., Thomas, G. H., and Neill, C. A. (1973). *J. Med. Genet.* **10,** 180–183.
Hecht, F. (1969). *Birth Defects, Orig. Artic. Ser.* **5,** 106–113.
Hoehn, H., Sander, C., and Sander. L. Z. (1971a). *Ann. Genet.* **14,** 187–192.
Hoehn, H., Engel, W., and Reinwein, H. (1971b). *Humangenetik* **12,** 175–181.
Hustinx, T. W. J., Gabreels, F. J. M., Kirkels, V. G. H. J., Korten, J. J., Scheres, J. M. J. C., Joosten, E. M. G., and Rutten, F. J. (1975). *Ann. Genet.* **18,** 13–19.
Jacobsen, P., Hobolth, N., and Mikkelsen, M. (1975). *Clin. Genet.* **7,** 317–324.
Kajii, T., Ohama, K., Niikawa, N., Ferrier, A., and Avirachan, S. (1973). *Am. J. Hum. Genet.* **25,** 539–547.
Kistenmacher, M. L., and Punnett, H. H. (1970). *Am. J. Hum. Genet.* **22,** 304–318.
Kistenmacher, M. L., Punnett, H. H., Aronson, M., Miller, R. C., Greene, A. E., and Coriell, L. L. (1975). *Cytogenet. Cell Genet.* **15,** 122–123.

Knörr-Gärtner, H., Knörr, K., Haas, B., Vogel, W., and Siebers, J. W. (1974). *Humangenetik* **21**, 315–321.

Kurnick, J., Atkins, L., Feingold, M., Hills, J., and Dvorak, A. (1974). *Hum. Pathol.* **5**, 223–232.

Laurent, C., Biemont, M. C., and Dutrillaux, B. (1975). *Humangenetik* **26**, 35–46.

Leisti, J. T., Kaback, M. M., and Rimoin, D. L. (1975). *Am. J. Hum. Genet.* **27**, 441–453.

Lejeune, J. (1965). *Ann. Genet.* **8**, 9–10.

Lejeune, J. (1966). *In* "Journées parisiennes de Pédiatrie," p. 75. Flammarion, Paris.

Lejeune, J. (1968). *Ann. Genet.* **11**, 71–77.

Lejeune, J., Berger, R., Rethoré, M. O., Salmon, C., and Kaplan, M. (1966). *Ann. Genet.* **9**, 12–18.

Lewandowski, R. C., Yunis, J. J., Lehrke, R., O'Leary, J., Swaiman, K. F., and Sanchez, O. (1976). *Am. J. Dis. Child.* **130**, 663–667.

Lin, C. C., Holman, G., and Sewell, L. (1974). *Am. J. Hum. Genet.* **26**, 54 A.

Lin, C. C., Uchida, U. A., Hotz, H., Miller, R. C., Greene, A. E., and Coriell, L. L. (1975). *Cytogenetics* **14**, 78–79.

London Conference. (1963). *Cytogenetics* **2**, 264–268.

Lord, P. M., Casey, M. D., and Laurence, B. M. (1967). *J. Med. Genet.* **4**, 169–176.

Magnelli, N. C., and Therman, E. (1975). *J. Med. Genet.* **12**, 105–108.

Malpuech, G., Kaplan, J. C., Rethoré, M. O., Junien, C., and Geneix, A. (1975). *Lyon Med.* **233**, 275–279.

Mason, M. K., Spencer, D. A., and Rutter, A. (1975). *J. Med. Genet.* **12**, 310–314.

Mayeda, K., Weiss, L., Lindahl, R., and Dully, M. (1974). *Am. J. Hum. Genet.* **26**, 59–64.

Metz, F., Bier, L., and Pfeiffer, R. A. (1973). *Humangenetik* **18**, 207–211.

Mulcahy, M. T., and Jenkyn, J. (1975). *Clin. Genet.* **8**, 199–204.

Newton, M. S., Cunningham, C., Jacobs, P. A., Price, W. H., and Fraser, I. A. (1972). *Clin. Genet.* **3**, 226–248.

Niebuhr, E. (1974). *Humangenetik* **21**, 99–101.

Ockey, C. H., Feldman, G. V., Macaulay, M. E., and Delaney, M. J. (1967). *Arch. Dis. Child.* **42**, 428–434.

Orye, E., and Craen, M. (1975). *Humangenetik* **28**, 335–342.

Orye, E., Verhaaren, H., Van Egmond, H., and Devloo-Blancquaert, A. (1975). *Clin. Genet.* **7**, 134–143.

Owen, L., Martin, B., Blank, C. E., and Harris, F. (1974). *J. Med. Genet.* **11**, 291–295.

Penchaszadeh, V. B., and Coco, R. (1975). *J. Med. Genet.* **12**, 301–305.

Prieur, M., Rethoré, M. O., Boeswillwald, M., Lafourcade, J., Loewe-Lyon, S., and Lejeune, J. (1977a). *Ann. Genet.* (in press).

Prieur, M., Couturier, J., Herrault, A., Lepintre, J., and Lejeune, J. (1977b). *Ann. Genet.* (in press).

Rethoré, M. O. (1976). *In* "Handbook of Clinical Neurology" (P. J. Vinken and G. W. Bruyn, eds.), Vols. XVI–XXVII. North-Holland Publ., Amsterdam (in press).

Rethoré, M. O., and Lafourcade, J. (1974). *In* "Journées parisiennes de Pediatrie, 1974," pp. 379–390. Flammarion, Paris.

Rethoré, M. O., Larget-Petit, L., Abonyi, D., Boeswillwald, M., Berger, R., Carpentier, S., Cruveiller, J., Dutrillaux, B., Lafourcade, J., Penneau, M., and Lejeune, J. (1970). *Ann. Genet.* **13**, 217–232.

Rethoré, M. O., Hoehn, H., Rott, H. D., Couturier, H., Dutrillaux, B., and Lejeune, J. (1973). *Humangenetik* **18**, 129–138.

Rethoré, M. O., Dutrillaux, B., Job, J. C., and Lejeune, J. (1974a). *Ann. Genet.* **17**, 109–114.

Rethoré, M. O., Dutrillaux, B., Giovannelli, G., Forabosco, A., Dallapiccola, B., and Lejeune, J. (1974b). *Ann. Genet.* **17,** 125–128.

Rethoré, M. O., Ferrand, J., Dutrillaux, B., and Lejeune, J. (1974c). *Ann. Genet.* **17,** 157–161.

Rethoré, M. O., Kaplan, J. C., Junien, C., Cruveiller, J. Dutrillaux, B., Aurias, A., Carpentier, S., Lafourcade, J., and Lejeune, J. (1975). *Ann. Genet.* **18,** 81–87.

Rethoré, M. O., Debray Ritzen, P., Vivier, R., and Lejeune, J. (1977). *Ann. Genet.* (in press).

Rott, H. D., Schwanitz, G., and Grosse, K. P. (1971). *Z. Kinderheilkd.* **109,** 293–299.

Rutten, F. J., Scheres, J. M. J. C., Hustinx, T. W. J., and Haar, B. G. A. (1974). *Humangenetik* **25,** 163–170.

Rybak, M., Kozlowski, K., Kleczkowska, A., Lewandowski, J., Sckolowski, J., and Soltysik-Wilk, E. (1971). *Am. J. Dis. Child.* **121,** 490–495.

Sartori, A., Tenconi, R., Baccichetti, C., and Pujatti, G. (1974). *Acta Paediatr. Scand.* **63,** 631–635.

Schinzel, A., and Schmid, W. (1972). *Humangenetik* **15,** 163–171.

Schinzel, A., Hayashi, K., and Schmid, W. (1974). *Humangenetik* **25,** 171–177.

Schröcksnadel, H., Feichtinger, C., and Scheminzky, C. (1975). *Humangenetik* **29,** 329–335.

Schrott, H. G., Sakaguchi, S., Francke, U., Luzzatti, L., and Fialkow, P. J. (1974). *J. Med. Genet.* **11,** 201–205.

Schwanitz, G., and Grosse, K. P. (1973). *Ann. Genet.* **16,** 263–266.

Schwanitz, G., Schamberger, U., Rott, H. D., and Wieczorek, V. (1974). *Ann. Genet.* **17,** 163–166.

Schwingshakl, A., and Ganner, E. (1973). *Pediatr. Pado.* **8,** 362–371.

Shaw, M. W., Cohen, M. M., and Hildebrandt, H. M. (1965). *Am. J. Hum. Genet.* **17,** 54–70.

Sinet, P. J., Couturier, J., Dutrillaux, B., Poissonnier, M., Raoul, O., Rethoré, M. O., Allard, D., Lejeune, J., and Jerome, H. (1976). *Exp. Cell Res.* **97,** 47–55.

Sparkes, R. S., and Francke, U. (1973). *Am. J. Hum. Genet.* **25,** 73A.

Stoll, C., Levy, J. M., and Gardea, A. (1976). *Humangenetik* **27,** 269–274.

Sumner, A. T., Evans, H. J., and Buckland, R. A. (1971). *Nature (London), New Biol.* **232,** 31–32.

Surana, R. B., and Conen, P. E. (1972). *Ann. Genet.* **15,** 191–194.

Surana, R. B., Bailey, J. D., and Conen, P. E. (1971). *J. Med. Genet.* **8,** 517–521.

Tenconi, R., Baccichetti, C., Anglani, F., Pellegrino, P. A., Kaplan, J. C., and Junien, C. (1975). *Ann. Genet.* **18,** 95–98.

Therkelsen, A. J., Grunet, N., Hjort, T., Myhrejensen, O., Jonasson, J., Lauritsen, J. G., Lindsten, J., and Petersen, B. (1973). *In* "Chromosomal Errors in Relation to Reproductive Failure" (A. Boué and C. Thibault, eds.), pp. 81–93. INSERM, Paris.

Turleau, C., de Grouchy, J., Chavin-Collin, F., Roubin, M., and Langmaid, H. (1974). *Ann. Genet.* **17,** 167–174.

Turleau, C., de Grouchy, J., Roubin, M., Chavin-Colin, F., and Cachin, O. (1975a). *Ann. Genet.* **18,** 125–129.

Turleau, C., de Grouchy, J., Chavin-Colin, F., Roubin, M., Brissaud, P. E., Repesse, G., Safar, A., and Borniche, P. (1975b). *Humangenetik* **29,** 233–241.

Uchida, I. A., and Lin, C. C. (1973). *J. Pediatr.* **82,** 269–272.

Van Kempen, C. (1975). *J. Med. Genet.* **12,** 204–207.

Vesel, E. S., and Bearn, A. G. L. (1961). *J. Clin. Invest.* **40,** 586–592.

Vogel, W., Siebers, J. W., Gunkel, J., Haas, B., Knörr-Gartner, H., Niethammer, D. G., and Noel, B. (1975). *Humangenetik* **28,** 103–112.

Wahrman, J., Goitein, R., Richler, C., Goldman, B., Akstein, E., and Chaki, R. (1974). *Leiden Chromosome Conf., 1974* Abstracts, p. 87.

Williams, J. D., Summitt, R. L., Martens, P. R., and Kimbrell, R. A. (1975). *Am. J. Hum. Genet.* **27,** 478–485.

Wilson, M. G., Towner, J. W., Coffin, G. S., and Forsman, I. (1970). *Am. J. Hum. Genet.* **22,** 679–690.

Wolf, U., Reinwein, H., Porsch, R., Schroter, R., and Baitsch, H. (1965). *Humangenetik* **1,** 397–413.

Zaremba, J., Zdzienicka, E., Glogowska, I., Abramowicz, T., and Taracha, B. (1974). *J. Ment. Defic. Res.* **18,** 153–190.

# 5

# Partial Duplication 7q

WALTHER VOGEL



## I. CLINICAL FINDINGS IN PARTIAL DUPLICATION 7q

The possibility of delineating a clinical entity based on a few cases with partial trisomy 7q could only be conjectured by Vogel *et al.* in 1973. In the meantime, additional cases have been published which allow for a more detailed analysis of the clinical symptoms present in partial trisomy 7q.

A well-defined clinical syndrome is evident in 4 cases (Cases 1–4, Tables I and II). These patients carry the same duplicated segment (q31→qter) without a detectable deficiency. The main characteristics of this syndrome consist of low birth weight, growth and mental retardation, cleft palate, microretrognathia, small nose, small palpebral fissures, and hypertelorism (Fig. 1). Less frequent symptoms include coloboma of the iris, transverse palmar creases, and skeletal anomalies such as rib aplasia and syndactyly.

Two cases (5 and 6) with smaller duplicated segments (q32→qter) have several of the features observed in Cases 1–4 (Table II, Fig. 2). The main difference in both cases is the absence of cleft palate and hypertelorism. These two cases, however, show additional symptoms, such as kyphoscoliosis and dislocation of the hip, which became evident at the older age reached by these two patients (Table II).

**TABLE I**
**Patients with Partial Trisomy 7q**

| Case | Duplication | Balanced translocation | Deficiency | Reference |
|---|---|---|---|---|
| 1 | q22→qter | 46,XX,t(7;12)(q22;q24) | 12qter uncertain | Lejeune *et al.* (1968) |
| | | | | Carpenteier *et al.* (1972) |
| 2 | q31→qter | 46,XX,t(7;21)(q31;p13) | None | Vogel *et al.* (1973) |
| 3 | q31→qter | 46,XY,t(7;14)(q31;qter) | 14q ter uncertain | Alfi *et al.* (1973) |
| 4 | q31→qter | 46,XY,t(5;7)(q35;q31) | 5q ter uncertain | Saadi and Moghadam (1976) |
| 5 | q32→qter | 46,XX,t(7;19)(q32;q13) | 19q13 uncertain | Newton *et al.* (1972) |
| 6 | q32→qter | 46,XY,t(7;21)(q32;q22) | 21q ter uncertain | Bass *et al.* (1973) |
| 7 | q22→q31 | 46,XY,del(7q22;q31)ins(13q31) | None | Serville *et al.* (1975) |
| 8 | q22→q31 | 46,XX,del(7q22;q31)ins(3q27) | None | Grace *et al.* (1972) |
| | | | | Grace *et al.* (1973) |
| 9 | q22→q31 | 46,XX,del(7q22;q31)t(5;7;17)(q31;q22;q31;p13) | None | Berger *et al.* (1974) |

**TABLE II**
**Clinical Symptoms of Patients with Partial Trisomy 7q**

| | Case number | | | | | | | Case number | | | |
|---|---|---|---|---|---|---|---|---|---|---|---|
| | 1 | 2 | 3 | 4 | 5 | 6 | | 7 | 8 | 9 | |
| Age of patient: | few hours | 2 months | 5 months | 20 hours | 12 years | 18 months | Frequency of symp- | 7 months | 10 months | 13 months | Frequency of symp- |
| Duplication 7q: | q22→qter | q31→qter | q31→qter | q31→qter | q32→qter | q32→qter | toms | q22→q31 | q22→q32 | q22→q31 | toms |
| Low birth weight | + | + | + | + | + | + | 6/6 | + | – | – | 1/3 |
| Retardation of development | ? | + | + | ? | + | + | 4/4 | + | + | + | 3/3 |
| Cleft palate | + | + | + | + | – | – | 4/5 | – | – | – | 0/3 |
| Microretro-gnathia | + | + | + | + | – | + | 5/6 | – | – | – | 0/3 |
| Small nose | + | + | + | ? | ? | – | 3/4 | | | + | |
| Small palpebral fissures | + | + | + | ? | – | + | 4/5 | – | + | + | 2/3 |
| Large tongue | + | + | – | + | + | ? | 4/5 | – | – | – | 0/3 |
| Epicanthus | – | – | – | – | – | – | 0/6 | – | – | + | 1/3 |
| Hypertelorism | + | + | + | + | – | ? | 4/6 | ? | + | + | 2/3 |
| Low set ears | + | + | + | + | – | + | 5/6 | + | + | – | 2/3 |
| Ear anomalies | + | – | + | + | – | – | 3/6 | – | + | – | 1/3 |
| Wide open fontanel | + | + | + | ? | – | ? | 3/5 | – | – | – | 0/3 |
| Coloboma of iris | – | + | – | – | – | – | 1/6 | – | – | – | 0/3 |

*continued*

**TABLE II (*continued*)**

| | Case number | | | | | | | Case number | | | |
|---|---|---|---|---|---|---|---|---|---|---|---|
| | 1 | 2 | 3 | 4 | 5 | 6 | Frequency of symp- toms | 7 | 8 | 9 | Frequency of symp- toms |
| Age of patient: | few hours | 2 months | 5 months | 20 hours | 12 years | 18 months | | 7 months | 10 months | 13 months | |
| Duplication 7q: | q22→qter | q31→qter | q31→qter | q31→qter | q32→qter | q32→qter | | q22→q31 | q22→q32 | q22→q31 | |
| Strabismus | – | – | – | – | – | – | 0/6 | + | + | + | 3/3 |
| Fuzzy hair | + | + | – | ? | – | ? | 2/6 | + | + | – | 2/3 |
| Transverse palmar crease | + | + | – | – | – | – | 2/6 | – | – | – | 0/3 |
| Heart murmur | – | – | + | – | – | + | 2/5 | – | – | – | 0/3 |
| Skeletal anomalies | – | + | + | + | + | + | 5/6 | – | – | – | 0/3 |
| Aplasia of the 12th rib | – | + | – | – | – | – | 1/6 | – | – | – | 0/3 |
| Kyphoskoliosis | – | – | – | – | + | + | 2/6 | – | – | – | 0/3 |
| Dislocation of hip | – | – | – | – | + | + | 2/6 | – | – | – | 0/3 |
| Pes cavus | – | – | – | – | + | – | 1/6 | – | – | – | 0/3 |
| Syndactyly | – | – | + | – | – | – | 1/6 | – | – | – | 0/3 |
| Rocker bottom feet | – | – | – | + | – | – | 1/6 | – | – | – | 0/3 |

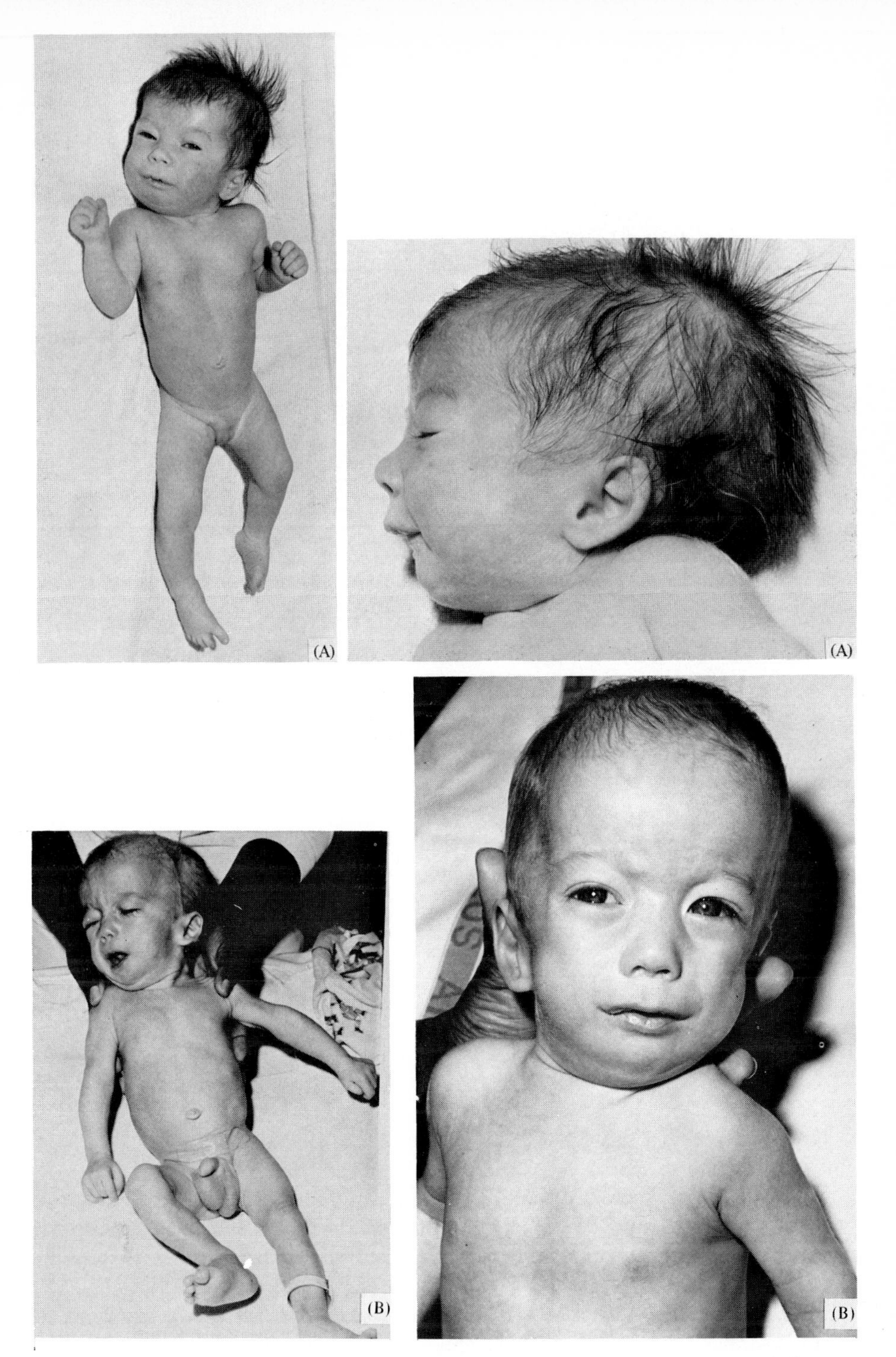

**Fig. 1.** Examples of patients carrying a duplication 7q22 or q31→qter. (A) Case 2 (from Vogel *et al.*, 1973). (B) Case 3 (with kind permission of Dr. O. S. Alfi).

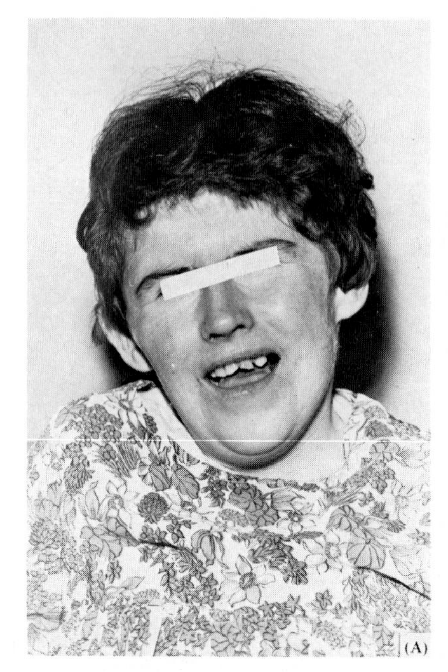

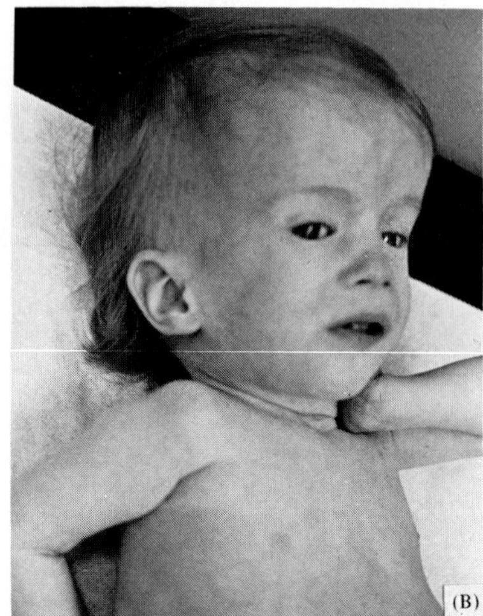

**Fig. 2.** Examples of patients carrying a duplication 7q32→qter. (A) Case 5 (with kind permission of Dr. M. S. Newton). (B) Case 6 (with kind permission of Dr. B. F. Crandall).

The remaining three cases (7, 8, and 9) (Fig. 3) show a different clinical picture without most of the conspicuous symptoms mentioned in Cases 1–4 (cleft palate, micrognathia, and skeletal anomalies). They have a duplication 7(q22→q31 or q32). Because of the clinical symptoms (Table II) and cytogenetic findings, these cases cannot be included as part of the duplication 7(q31→qter) syndrome, since they show a different clinical picture which includes retarded development, hypertelorism, small palpebral fissures, strabismus, and low set ears.

## II. CYTOGENETIC FINDINGS

According to the type of chromosomal rearrangement, there are two groups, one being the product of reciprocal translocations (Cases 1–6), and the other, the product of insertion translocations (Cases 7–9). In the group of reciprocal translocations, the patient of Case 2 had a pure duplication 7(q31→qter). If there is any deficiency at all, it would only involve a part of the satellite of chromosome 21 (Fig. 4), a material whose absence is known not to cause clinical symptoms.

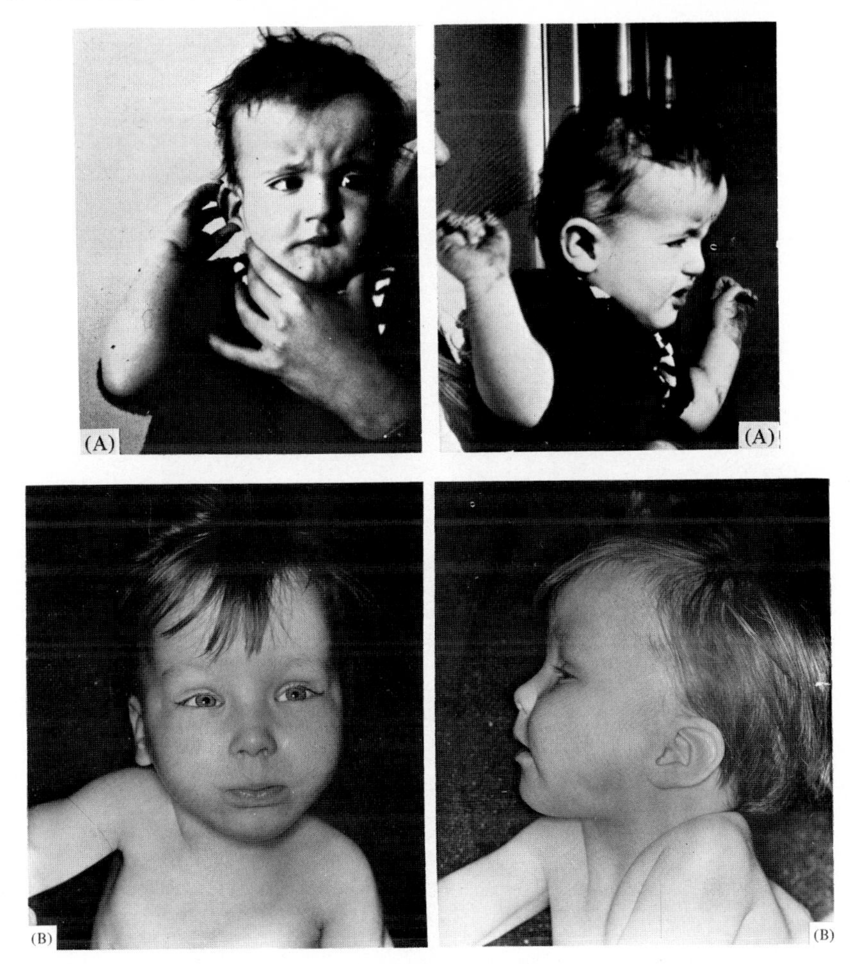


**Fig. 3.** Examples of patients carrying an interstitial duplication. (A) 7q22→q31 (Case 7) (with kind permission of Dr. F. Serville). (B) 7q22→q32 (Case 8) (with kind permission of Dr. E. Grace).

In the other cases of this group (1, 3–6), the distal part of 7q is translocated to a nonheterochromatic telomeric region of another chromosome. In this situation, it cannot be decided with certainty whether there is, in addition to the duplication, a deficiency which may affect the phenotype of the patients. It should be noted that a translocation onto a telomere may represent a phenomenon similar to telomeric fusion suggested by Lejeune (1973). A similar hypothesis has been discussed by Francke (1972).

It cannot be determined whether the translocation in Case 5 is also telomeric, since the band of chromosome 19 involved (q13) is large. The same argument

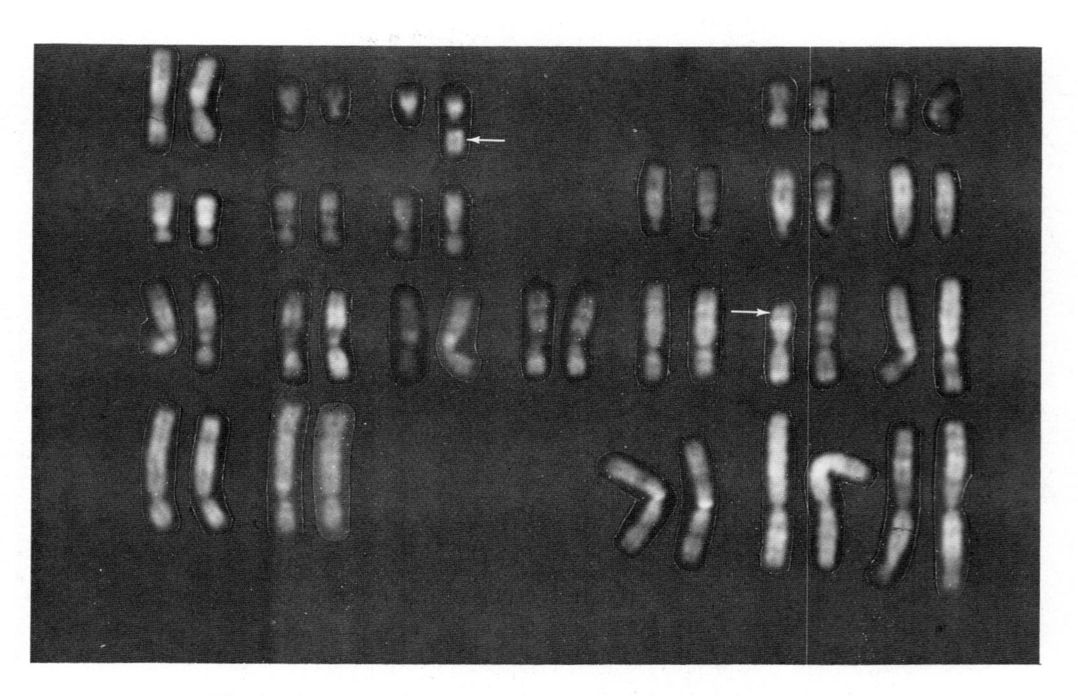

**Fig. 4.** Karyotype of mother of Case 2 with 46,XX,t(7;21)(q31;p13).

applies to Case 6 where the distal segment of 7q was translocated to 21q22. The clinical features of both patients could best be explained without the assumption of a deletion which causes additional clinical symptoms. A comparison of the clinical picture of all cases in this group is in accordance with this view and does not reveal symptoms which indicate the influence of a deficiency (Table II).

In the three cases of the second group (Cases 7–9) (q22→q31 or q32), an interstitial segment of 7q was inserted into another chromosome. In contrast to the other cases, these patients have no duplication of 7(q32→qter), and cytogenetic evidence excludes any assumption of a deficiency.

The observed break points in chromosome 7 are different in most cases. Although three break points are localized in band 7q31, the number of reported cases is too small to exclude or state the existence of a point predisposing to breakage. The segregation patterns in the different families concerned are presented in Table III. On the whole, the patterns are in accordance with the data obtained in large series (Hamerton, 1971). The heterogeneity of the different translocations does not permit a closer analysis or comparison except of Cases 3 and 6 and the family reported by Giraud *et al.* (1974). In all of these translocations, a distal segment of chromosome 7 was transferred onto the telomere of an acrocentric chromosome. In Case 6 and in that of Giraud *et al.* (1974), the acrocentric chromosome is No. 21, but in the latter case the translocated segment is considerably longer (7q22→qter) than chromosome 21, while it is smaller than chromosome 21 in Case 6. This difference may provide a sufficient explanation for the strikingly different segregation patterns in both translocations (see Table III). In Case 3, only one pregnancy is known and this resulted in

**TABLE III**
**Segregation Patterns in the Different Families with Partial Trisomy 7q**

| Family of case | Normal | Balanced | Unbalanced | Abortion | Pregnancies |
|---|---|---|---|---|---|
| 1 | 7 | 7 | 1 | 5 | 20 |
| 2 | – | – | 1 | – | 1 |
| 3 | – | – | 1 | – | 1 |
| 4 | 2 | 3 | 1 | 2 | 8 |
| 5 | (1)[a] | (1)[a] | 1 | – | 3 |
| 6 | – | – | 1 | 2 (1 deficiency) | 3 |
| 7 | 1 | – | 1 | 1 | 3 |
| 8 | – | 1 | 1 | – | 2 |
| 9 | – | – | 1 | 8 | 9 |
| Giraud *et al.* (1974) | – | – | 3 children + 21 | 4 | 7 |
| | 11 | 12 | 12 | 22 | 57 |

[a]Uncertain case.

the patient carrying the duplication similar to Case 6. On the other hand, a 3:1 segregation seems not to be unusual if a chromosome segment is translocated on the long arm of a chromosome 21.

## III. PARTIAL TRISOMIES AND CHROMOSOME NOMENCLATURE

Since partial trisomies in the majority of cases originate by segregation of parental reciprocal translocation, unrelated patients with partial trisomy usually do not have an identical duplication, and in many instances they also carry a deficiency of a different chromosome. If banding patterns are shown, the Paris Nomenclature allows for a most accurate description of these aberrant karyotypes. In particular, the origin and meiotic segregation of the aberrant chromosomes can be indicated using this nomenclature. This applies, however, to individual cases only, while the situation changes completely when similar cases are grouped together to be described as a uniform clinical picture. In partial trisomies, different break points and the varying second chromosome involved prevent the use of the Paris Nomenclature for common designations. All kinds of deletions as well as duplications of whole chromosome arms can be denoted and summarized unequivocally using the Chicago Nomenclature which was integrated into the Paris nomenclature. However, this does not apply to the partial trisomies because the translocated segments are attached to different chromosomes. In addition, difficulties concerning the nomenclature of partial trisomies may also arise from true genetic differences between cases. This aspect has to be considered if the clinical findings in patients with a partial trisomy are compared, and if the symptoms due to the duplication are to be defined. In the meantime, it is possible to distinguish two groups of patients with reciprocal translocations with duplication 7q31 or 7q32 to qter, and insertional translocations with dup 7q22→q31 or 7q22→q32. In this manner, it should become possible to define duplication syndromes as in partial duplication 7q with a brief and accurate nomenclature though they may include different deficiencies.

## REFERENCES

Alfi, O. S., Donnel, G. N., and Kramer, S. L. (1973). Partial trisomy of the long arm of chromosome No. 7. *J. Med. Genet.* **10,** 187–189.

Bass, N. H., Crandall, B. F., and Marcy, S. M. (1973). Two different chromosome abnormalities resulting from a translocation carrier father. *J. Pediat.* **83,** 1034–1038.

Berger, R., Derre, J., and Ortiz, M. S. (1974). Les trisomies partielles du bras long du chromosome 7. *Nouv. Presse Med.* **3,** 1801–1804.

Carpentier, S., Rethoré, M.-O., and Lejeune, J. (1972). Trisomie partielle 7q par translocation familiale t(7;12)(q22;q24). *Ann. Genet.* **15**, 283–286.

Chicago Conference (1966). Standardization in human cytogenetics. *Birth defects: Orig. Artic. Ser.*, **2**, No. 2. The National Foundation, New York.

Francke, U. (1972). Quinacrine mustard fluorescence of human chromosomes: Characterization of unusual translocations. *Am. J. Hum. Genet.* **24**, 189–213.

Giraud, F., Hartung, M., Mattei, J.-F., and Mattei, M.-G. (1974). t(7q–;21q+) et trisomie 21 familiale. *Annal. Genet.* **17**, 49–53.

Grace, E., Sutherland, G. R., and Bain, A. D. (1972). Familiar insertional translocation. *Lancet* II, 231.

Grace, E., Sutherland, G. R., Stark, G. D., and Bain, A. D. (1973). Partial trisomie of 7q resulting from a familial translocations. *Ann. Genet.* **16**, 51–54.

Hamerton, J. L. (1971). "Human Cytogenetics." Vol. I "General Cytogenetics." Academic Press, New York.

Lejeune, J. (1973). New autosomal syndromes. *Bull. Eur. Soc. Hum. Gen.* **Okt. pp. 57**–60.

Lejeune, J., Rethoré, M.-O., Berger, R., Abonyi, D., and Dutrillaux, B., See, G. (1968). Trisomie C partielle par translocation familiale t(Cq–;Cp+). *Ann. Genet.* **11**, 171–175.

Newton, M. S., Cunningham, C., Jacobs, P. A., Price, W. H., and Fraser, J. A. (1972). Chromosome survey of a hospital for the mentally subnormal Part 2: Autosome abnormalities. *Clin. Genet.* **3**, 226–248.

Paris Conference (1972). Standardization in human cytogenetics. *Birth defects: Orig. Artic. Ser.*, **8**, No.7. The National Foundation, New York.

Saadi, A., and Moghadam, H. (1976). Partial trisomy of the long arm of chromosome 7. *Clin. Genet.* **9**, 250–254.

Serville, F., Broustet, A., Sandler, B. Bourdeau., M.-J., and Leloup, M. (1975). Trisomie 7q partielle. *Ann. Genet.* **18**, 67–70.

Vogel, W., Siebers, J.-W., and Reinwein, H. (1973). Partial trisomy 7q. *Ann. Genet.* **16**, 227–280.

# 6

# Trisomy 8

R. A. PFEIFFER



## I. HISTORICAL BACKGROUND

In 1962, Pfeiffer *et al.* described a male infant exhibiting unusual, multiple dysplasias, later known to be the first reported case of trisomy 8. In 1969, based on three observations, Lejeune *et al.* introduced the concept of a "Trisomy C Syndrome." Since 1972, it has been shown (Bijlsma *et al.*, 1972; Caspersson *et al.*, 1972) that the supernumerary chromosome in the C-group was a No. 8.

## II. THE PHENOTYPE OF TRISOMY 8 MOSAICISM

By the end of 1975, to the best of our knowledge, thirty five cases in which the supernumerary autosome was identified were published. Most of these were based on lymphocyte cultures. In a few cases other tissues were studied. There seems to be no one case in which full trisomy was provided in more than one tissue. If in blood cells trisomy was noted in all analyzed cells, somatic mosaicism was not formally excluded. A few observations failed to show the trisomic cell line, at least in lymphocytes with increasing age (Schinzel *et al.*, 1974; G. Kosztolanyi, E. M. Bühler, P. Elmiger, and G. R. Stalder, unpublished manuscript).

## A. Growth Pattern

Birth weight of 19 males and 7 females was found to average 3570 gm and 3180 gm, respectively. Length at birth for both sexes was 52.2 cm ($n$ = 10). Growth in infancy and childhood has been considered to be within normal limits in 58%; growth was accelerated at least during the first decade in 8% and retarded in 35%. Only a few patients have been followed for a period of several years (see, for example, Cassidy *et al.*, 1975; Schinzel *et al.*, 1974; Stalder *et al.*, 1964; G. Z. Kostolanyi, E. M. Bühler, P. Elmiger and G. R. Stalder, unpublished manuscript).

From these data, it seems that a given growth pattern will usually be maintained. Height of three adult patients was 150 cm, 165 cm, and 176 cm (Laurent *et al.*, 1971; Crandall *et al.*, 1974). Because of progressive kyphoscoliosis and contractures of the hips and knees, the actual length would be slightly increased. In any case, the height achieved is remarkable since it exceeds that of all other autosomal hyperdiploidies. Weight parallels height but is some 25 percentiles lower. Adult weight may be below average (35, 57, 78 kg).

The body proportions are characterized by a long and slender trunk and long extremities (see Figs. 3, 6, 8). In three cases the body measurements are above the ninetieth percentile (Bijlsma *et al.*, 1972 [Case a]; Gustavson *et al.*, 1967; Schinzel *et al.*, 1974). In seven cases they are below the tenth percentile (Bijlsma *et al.*, 1972 [Case b]; Crandall *et al.*, 1974 [Case b]; Hustinx *et al.*, 1975; Riccardi *et al.*, 1970 [Case a; b]; Rützler *et al.*, 1974; Walravens *et al.*, 1974).

## B. Craniofacial Dysmorphia

The skull is large and square; the forehead is prominent and bulging (Figs. 1–3). Scaphocephaly is mentioned by Bijlsma *et al.* (1972) [Case a] and G. Kosztolanyi, E. M. Bühler, P. Elmiger, and G. R. Stalder (unpublished manuscript). Since the circumference of the head at birth was noted in but few cases, it may be expected to be within normal limits as it is in childhood. In three patients (Bijlsma *et al.*, 1972 [Case a]; Crandall *et al.*, 1974 [Case b]; Walravens *et al.*, 1944) all body measurements were below the third percentile.

The eyes are widely spaced. Hypertelorism is mentioned in six cases, but

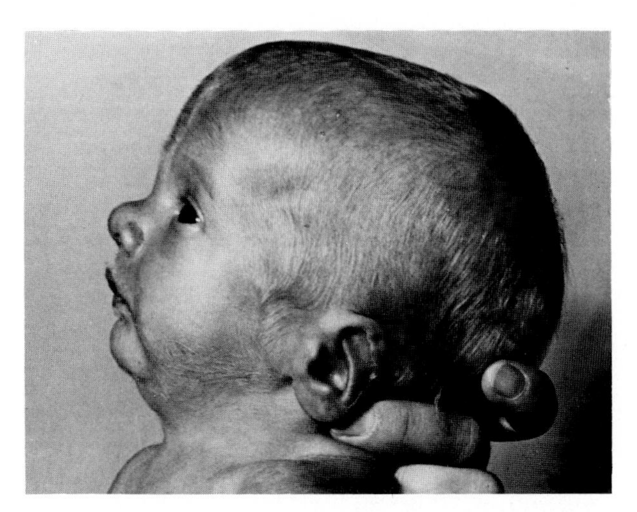

**Fig. 1.** Patient at the age of 6 weeks.

Laurent *et al.* (1971) [Case a] found hypotelorism. Abnormal slant of the palpebral fissures and epicanthus however are infrequent. The nose is broad and shallow; the large nostrils are upturned. The lips may be thick and everted. Hence the facies looks coarse and dull (Figs. 1–3). However, some photographs

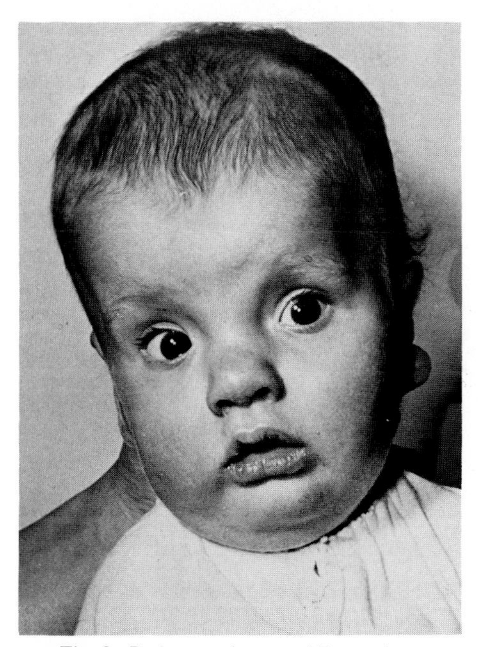

**Fig. 2.** Patient at the age of 7 months.

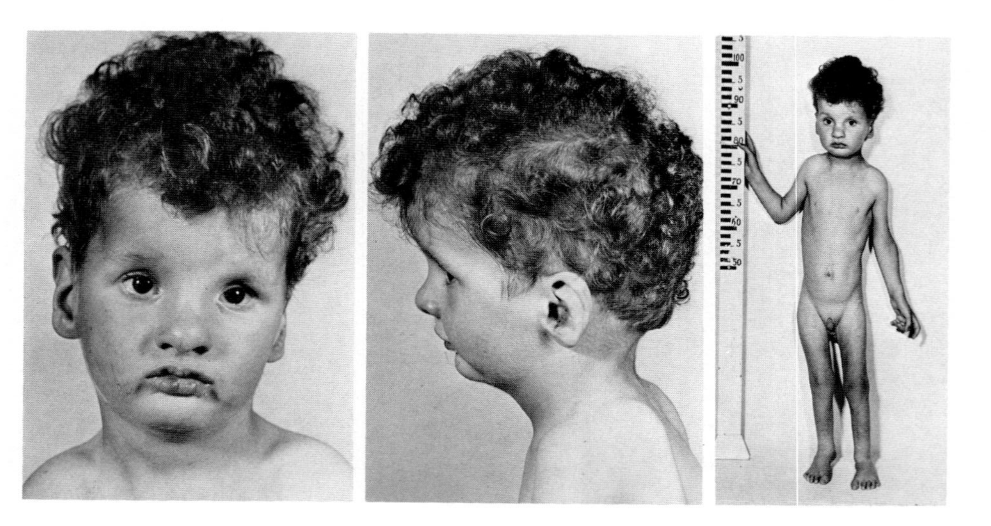

**Fig. 3.** Patient at the age of 3 years.

show handsome facies (see, for example, Giraud *et al.*, 1975; Hustinx *et al.*, 1975). Micrognathia is mentioned frequently, but prognathia also is noted. The ears are deep set and uncommonly shaped. They may be triangular. The lobule is large and rotated. The antihelix is prominent. The scapha is flat and broad. The helix may be irregularly folded in the upper parts only (Fig. 4). The neck is short and broad; the hairline low. The hair may be thick and tightly curled.

Ptosis of the upper eyelid was noted in two cases (Aller *et al.*, 1975; Crandall *et al.*, 1974 [Case b]); strabismus, generally convergent, in six cases. In three cases a unilateral corneal opacity was noted (Cassidy *et al.*, 1975; Fineman *et al.*, 1975; Riccardi *et al.*, 1970 [Case b]). Other anomalies of the eyes are left microphthalmia (Walravens *et al.*, 1974), left coloboma and cataract (Rützler *et al.*, 1974), congenital cataract and retrolental fibroplasia (Riccardi *et al.*, 1970 [Case b]), and heterochromia of the iris in two cases [Debray-Ritzen *et al.*, 1974; Higurashi *et al.*, 1969 [Case 1]).

The palate is highly arched. Cleft of the hard palate was noted in two cases (Caspersson *et al.*, 1972 [Case d]; Emberger *et al.*, 1970), and splitting of the soft palate in one case (Kakati *et al.*, 1973). Cleft uvula was mentioned in one case (Gustavson *et al.*, 1967).

## C. Body and Limbs

The body usually is slender and poorly shaped. The shoulders and the pelvic girdle are narrow (see Figs. 3 and 8–10). The movements of the joints are limited even if there is marked muscular hypotonia. However, hypertonia predominates, and the stiffness and contractures seem to get progressively worse. The fingers which are long and slender are flexed (Fig. 5). This condition is generally

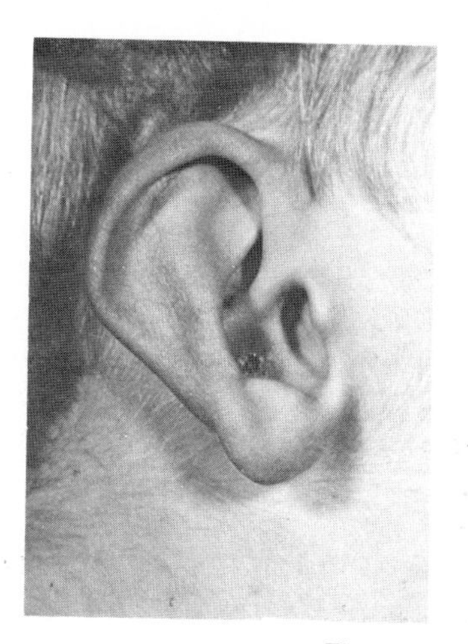
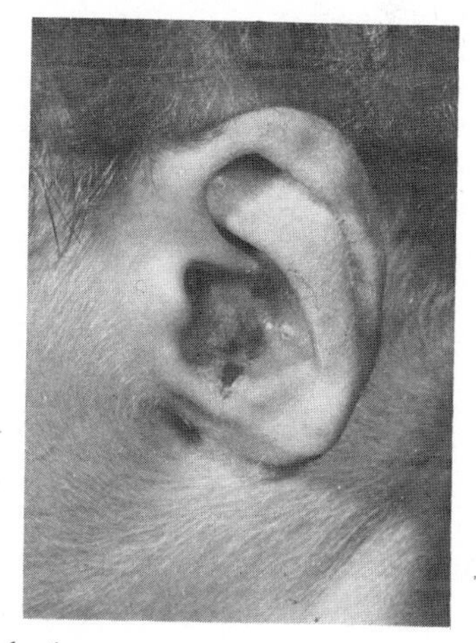

**Fig. 4.** Dysplastic ears.

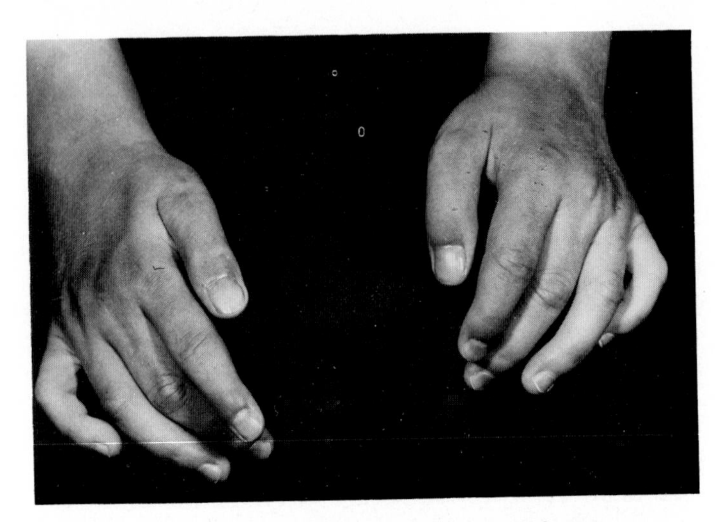

**Fig. 5.** Slender flexed fingers at the age of 3 years.

described as camptodactyly. In one case the flexion deformity of the fingers suggested trisomy 18 (Fineman *et al.*, 1975). The fifth fingers are short and crooked (brachyclinodactyly). The toes are also long and slender, may be irregularly inserted and closely moved together. The soles and palms exhibit deep skin furrows (plis capitonnés) (Fig. 6). However, this important finding does not persist beyond infancy. The nails, particularly the toenails, may be hypoplastic (Fig. 7).

## D. Skeletal Anomalies

Skeletal anomalies, although not specific, are more frequent than in any other chromosomal syndrome, and involve primarily bone dysplasias. Most frequently, accessory ribs have been noted. However, it remains to be confirmed as to whether the thirteenth rib belongs to the cervical or lumbar spine, or whether there are additional thoracic vertebrae, as noted in the very first observation (Fig. 8). Riccardi *et al.* (1970) and Walravens *et al.* (1974) emphasize that their patients had 6 lumbar vertebrae. The ribs may be narrow; fusion of ribs was noted by Caspersson *et al.* (1972) [Case d].

Scoliosis was present in at least four cases; spondylolysthesis was noted by G. Kosztolanyi, E. M. Bühler, P. Elmiger, and G. R. Stalder (unpublished manuscript). Abnormal ossification and shape of the sternum may give rise to pectus excavatum. Incomplete closure of the posterior arches (spina bifida) is seen in various segments of the spine, particularly in the lower cervical and upper

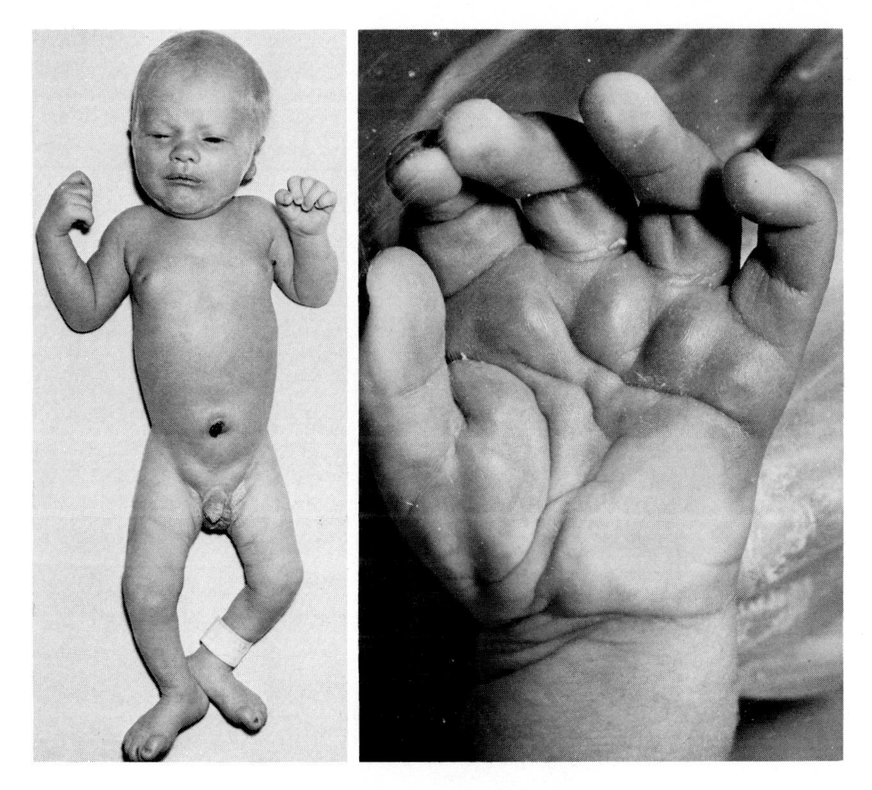

**Fig. 6.** Patient with trisomy 8 showing characteristic phenotype (left) and deep furrows of hand (right). From J. J. Yunis (personal communication, 1976).

thoracic segments and in the lumbar spine (Fig. 9). Abnormally shaped vertebrae (hemivertebrae, compression and fusion of vertebrae) are frequently noted. Anomalies of the upper cervical spine of the scapula (Gorlin *et al.*, 1975; Kakati *et al.*, 1973) and of the clavicula may cause Sprengel's deformity.

The iliac wings may be high and narrow due to abnormal rotation, as well as to abnormal shape (Fig. 10). From a lateral X-ray it was described as a "flat iliacal horn" (Rützler *et al.*, 1974) but this anomaly differs clearly from the deformity seen in osteo-onycho-dysplasia (Riccardi *et al.*, 1970). This very unusual configuration of the pelvis (G. Kosztolanyi, E. M. Bühler, P. Elmiger, and G. R. Stalder, unpublished manuscript; Schinzel *et al.*, 1974) was not noted in all patients. In several cases the pelvis was said to be normal. Coxa valga has been frequently noted but there is no mention of dysplasia of the femoral head or acetabulum.

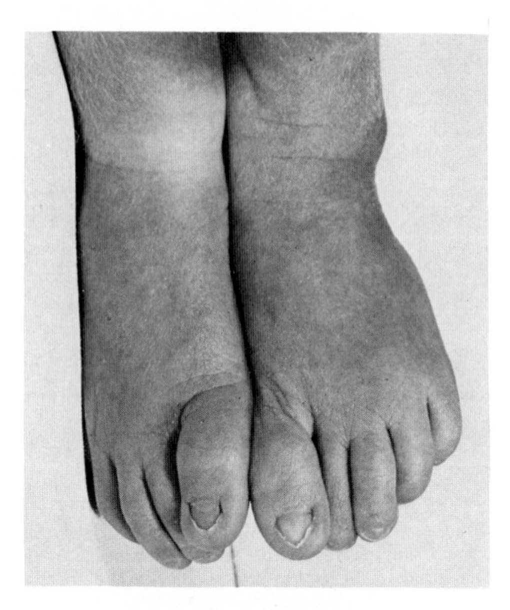

Fig. 7. Hypoplastic nails and irregularly inserted toes in infancy.

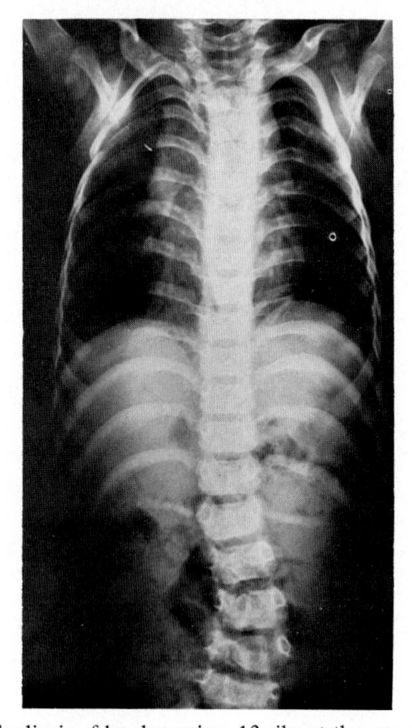

Fig. 8. Scoliosis of lumbar spine, 13 ribs at the age of 2 years.

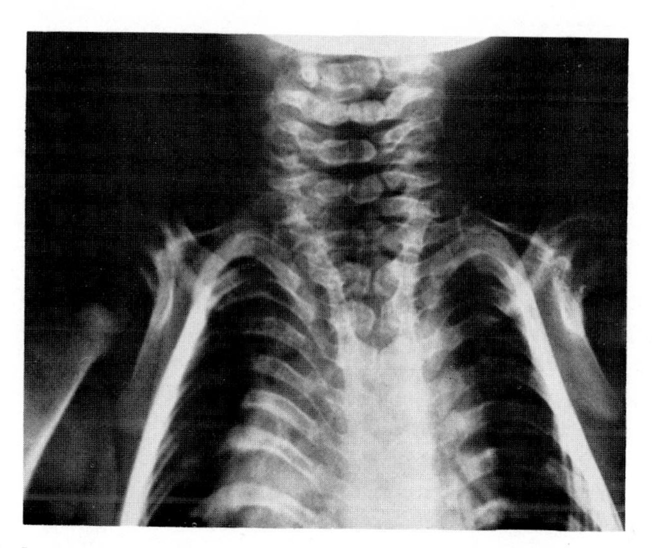

**Fig. 9.** Incomplete closure of posterior arches of the cervical and upper thoracic spine at the age of 7 months.

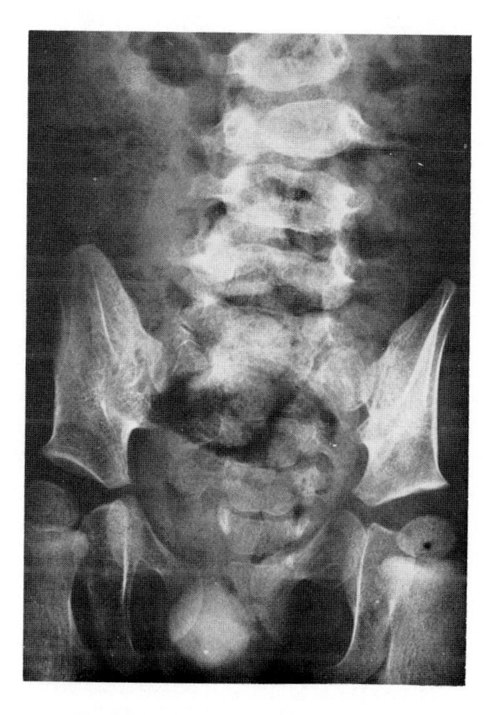

**Fig. 10.** Pelvis at the age of 2 years.

Flaring of the metaphysis and proximal diaphysis of the femur and humerus and also of the distal radius have been particularly noted (Fig. 11). The clavicles may be tortuous, irregularly thickened and sclerotic. It has been compared with a handle-bar (Fineman *et al.*, 1975). Metacarpal and metatarsal bones may be irregularly shortened and clumsy, but there is no pattern of any type of brachydactyly. Short fourth metacarpal and first metatarsal were noted by Cassidy *et al.* (1975) and Schinzel *et al.* (1974). Carpals may be abnormally shaped (Jacobsen *et al.*, 1974 [Case a]; Schinzel *et al.*, 1974) and even partially fused (Gorlin *et al.*, 1975). Short fifth fingers are due to brachymesophalangy. The middle phalanges of the toes may be lacking. Dislocation of the radial head was noted by Gorlin *et al.* (1975) and Rützler *et al.* (1974). Unusual findings have been talipes equinovarus (Schinzel *et al.*, 1974) and incomplete syndactyly between the third and fourth fingers (Emberger *et al.*, 1970).

Hypoplasia or aplasia of one or both patellae was observed in at least eight cases, even asymmetrically (Hustinx *et al.*, 1975). A hyperconvex configuration of the patella was described by Cassidy *et al.* (1975) and a patella bipartita mentioned by Laurent *et al.* (1974) [Case b].

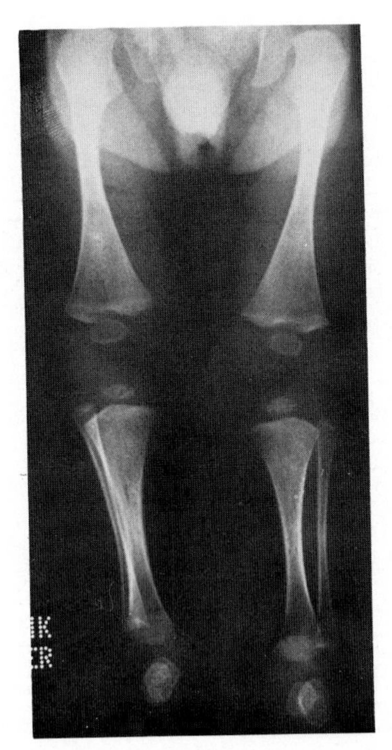

**Fig. 11.** Flaring of metaphyses of the long bones at the age of 2 months.

The bone age is considered to be mostly within normal range. It was said to be accelerated in the case of G. Kosztolanyi, E. M. Bühler, P. Elmiger, and G. R. Stalder (unpublished manuscript) and retarded in others. Asynchronous ossification leading to dissociated bone age (Fig. 12) can be expected.

The sella turcica was found to be large and deep (G. Kosztolanyi, E. M. Bühler, P. Elmiger, and G. R. Stalder, unpublished manuscript; Rützler *et al.*, 1974; Schinzel *et al.*, 1974).

## E. Dermatoglyphics

From the peculiar padding of the palms and soles which exhibit deep furrows (Bijlsma *et al.*, 1972), complex dermatoglyphic patterns should be expected. Indeed, whorls in the hypothenar, hallucal, and especially in the thenar areas are frequent. Although there is a tendency for an increased number of arches and ulnar loops and low total ridge count on fingers, whorls are irregularly dispersed among the digital patterns.

The distribution of digital patterns in twenty four patients with trisomy 8 shows that arches are present in 34%, which is 7 times more frequent than in the general population. Loops are found in 48%, versus 70%, and whorls in 18% versus 25%.

Arches are at least 10 times more frequent on the first, fourth and fifth digits than in the general population. On all digits, loops are less frequent and on the thumb and on the second fingers only, a whorl is found 2 and 3 times more

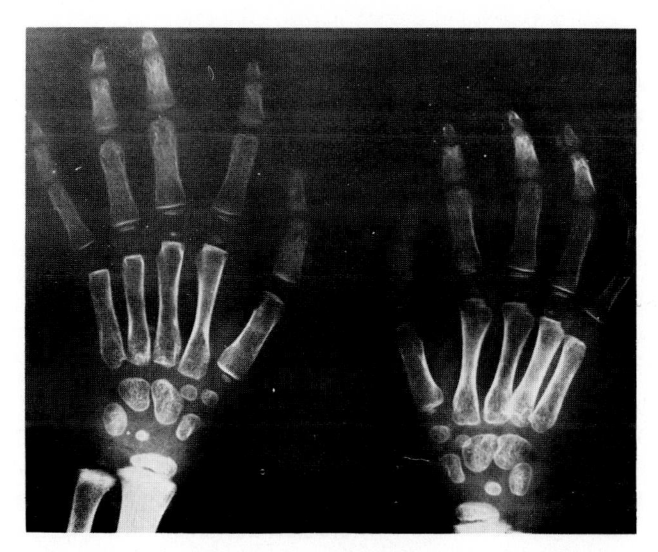

**Fig. 12.** Premature ossification of carpal bones (+3 years) at the age of 3 years.

frequently. These findings were first recognized by Penrose (1972) and confirmed by Schaumann *et al.* (1974).

## F. Occasional Malformations and Dysfunctions

### *1. Heart*

A ventricular septum defect (VSD) was present in two patients (Crandall *et al.*, 1974 [Case b]; Kakati *et al.*, 1973), an atrial septum defect (ASD) with pulmonary stenosis was suggested by Cassidy *et al.* (1975) and persistent ductus Botalli was noted in two more cases (Bijlsma *et al.*, 1972 [Case a]; Walravens *et al.*, 1974). Two types of anomalies of the great vessels were mentioned: a right-sided aortic arch "with a ductus ligament completing a vascular ring around the oesophagus" (Caspersson *et al.*, 1972 [Case a]), and a left vena cava superior and hypoplasia of the aorta (Caspersson *et al.*, 1972 [Case d]).

### *2. Kidneys*

Hydronephrosis was noted in at least five patients (Aller *et al.*, 1975; Caspersson *et al.*, 1972 [Case b]; Kakati *et al.*, 1973; Kondo *et al.*, 1975; Malpuech *et al.*, 1972). Cassidy *et al.* (1975) mentioned nonvisualization of the right kidney. Bifid pelvis was noted by Riccardi *et al.* (1970) and horseshoe kidney by Gorlin *et al.* (1975).

### *3. Genitalia*

Mild hypospadias was noted in two patients (Gorlin *et al.*, 1975; Hustinx *et al.*, 1975). The genitalia were hypoplastic in one child (Kakati *et al.*, 1973). Cryptorchidism seems to be very frequent, but normal puberty and secondary sexual characters have been noted in adolescent patients. Laurent *et al.* (1971) [Case b] mentioned testicular atrophy in a 26-year-old male.

### *4. Ear*

Atresia of the auditory canal was noted in one case (Hustinx *et al.*, 1975). In the patients examined by Gorlin *et al.* (1975) and Riccardi *et al.* (1970) conductive hearing loss was mentioned.

### *5. Deficiency of Clotting Factor VII (Hageman)*

de Grouchy *et al.* (1974) noted low concentration of Factor VII in three patients (ages 11, 15, 28 years) with trisomy 8 mosaicism, and suggested that a repressor gene might be presented on this autosome. In one more patient (3 months of age), the level of Factor VII was 42%, while other coagulation factors were normal. Stenbjerg *et al.* (1975) were not able to confirm these findings in three patients (ages 7, 10, 16 years).

### 6. *Intelligence and Brain Defects*

All patients but one (Caspersson *et al.*, 1972 [Case c]) who had been studied were retarded, most of them severely. Two previous observations of C-group mosaicism (Stolte and Blankenberg, 1964; Bishun, 1968) in normal females have not been reexamined.

IQ's range from 12 (Laurent *et al.*, 1970 [Case a]) to 94 (Hustinx *et al.*, 1975) with an average between 45 and 75. Most authors emphasize speech impairment and hence the importance of specific training (Riccardi *et al.*, 1970). In one patient (Debray-Ritzen *et al.*, 1974), this deficiency was analyzed more closely. At the age of 13, the IQ of the boy was 87 and equivalent to a mental age of 10 years. Speech comprehension was equivalent to the age of 6, vocabulary range of 3, phonetic level less than 3, syntax level less than 2, and verbal expression less than 1.5 years. His audiogram was normal. He had been suffering from seizures, but the EEG was not grossly altered.

In three patients, the diagnosis of agenesis of the corpus callosum was established by clinical examination (Caspersson *et al.*, 1972 [Case b]; Gustavson *et al.*, 1967; Malpuech *et al.*, 1972). In one infant (Walravens *et al.*, 1974), aqueductal stenosis was found. Higurashi *et al.* (1969) [Case 1] mentioned dilatation of the third ventricle in a Japanese child.

## III. CYTOGENETIC FINDINGS

### A. Trisomy 8

Parental age at birth of children with trisomy 8 was found to be slightly increased. Mean paternal age ($n$ = 26) was 31.8, and mean maternal age ($n$ = 28) 29.1 years. Since neither chromosomal nor genetic markers of this chromosome are known, the origin of trisomy 8 remains obscure. From trisomy 8 in abortuses (Boué *et al.*, 1975). and from the cases of Schinzel *et al.* (1974) and Crandall *et al.* (1974) [Case b] in which trisomy 8 was present in all fibroblasts from skin biopsies, postzygotic nondisjunction from an aneuploid zygote may be deduced.

In two cases (G. Kosztolanyi, E. M. Bühler, P. Elmiger, and G. R. Stalder, unpublished manuscript; Schinzel *et al.*, 1974), the aneuploid cell line in lymphocytes disappeared at age 13 and 10, respectively, while in skin fibroblasts it was still present.

Neu *et al.* (1969) were unable to find trisomy C at the age of $2\frac{1}{2}$ years, while 17% of the lymphocytes at the age of 7 months showed this anomaly. However, 7 out of 10 bone marrow cells contained the additional chromosome which was not late replicating. Therefore, aneuploidy cannot be excluded from normal findings in lymphocyte cultures alone if the clinical features suggest trisomy 8.

## B. Double Autosomal Trisomy

Double autosomal trisomy (47,XX,+21/48,XX,+8,+21) was found in a child with Down's syndrome of an 18-year-old mother (Wilson *et al.,* 1974). No morphological markers of the chromosomes 21 were described, and therefore the origin is not known. A zygote with trisomy 21 could have gained trisomy 8, but a zygote with both additional autosomes could have lost one chromosome 8 if negative selection were operating against this cell line. In this patient, no features of trisomy 8 were noted. In another patient studied by J. J. Yunis (personal communication, 1976) (47,XY,+8/47,XY,+21), the patient was found to have the typical phenotype of trisomy 8 except for the presence of mongoloid slants, hypotonia and prominent occiput (Fig. 13).

## C. Trisomy 8 in Abortuses

Trisomy 8 was identified in human abortuses. From the summary of findings published by Boué *et al.* (1975), trisomy 8 does not seem to be either favored or underrepresented. In one case of trisomy 8, the developmental arrest of the placenta was 19 days versus 25 days for the embryo. These authors emphasize the discrepancy between the very early lethal effect of trisomy 8 and the lack of lethal malformations in mosaics.

From the prevalence of males with trisomy 8 (25 males versus 9 females), one could expect that the early lethal effect is operating in females, but present data do not support this hypothesis.

## D. Clonal Trisomy 8

Trisomy 8 has been found in various disorders such as in the acute phase of chronic myeloid leukemia (CML), in acute leukemia (AL) and in polycythemia vera (PV) (de La Chapelle *et al.,* 1970, 1972, 1973; Hsu *et al.,* 1974). The anomaly was always noted in cells from bone marrow, but not in lymphocytes. Since the probability that random chromosomal gains involve the same chromosome is extremely small and translocations observed in leukemia also involve chromosome 8 preferentially, it is possible that duplications of this autosome may represent one step or mutation in a multiple step system of chromosomal gains (Rowley, 1975). Trisomy 8 has also been noted in different hematologic disorders such as sideroachrestic anemia which has been considered a preleukemic state (Jonasson *et al.,* 1974; Hellström *et al.,* 1971). Since the gene for gluthathione reductase was found to be located on this chromosome (Kucherlapati *et al.,* 1974), it has been suggested that an imbalance of enzyme activity could initiate or contribute to abnormal proliferation (Rowley, 1975).

If trisomy 8 were present in all tissues of patients with the trisomy 8 syndrome, hematologic disorders would be expected, provided that these pa-

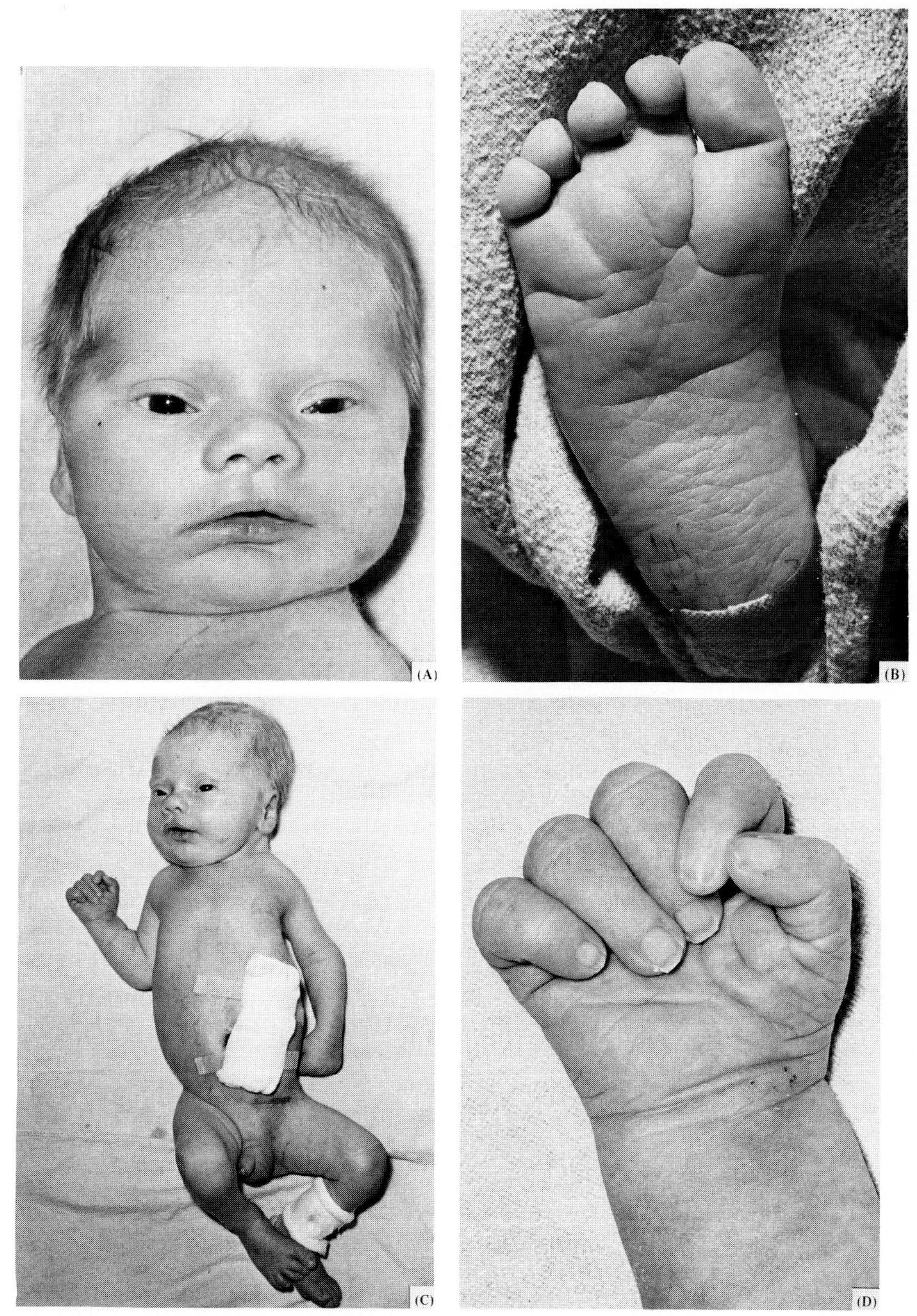


**Fig. 13.** Patient with trisomy 8/trisomy 21 mosaicism showing many of the typical features of trisomy 8 except for mongoloid slants, prominent occiput and hypotonia, commonly observed in trisomy 21. From J. J. Yunis (personal communication, 1976).

tients reach the critical age. The observation of Gafter *et al.* (1976) may be considered a clue to this important problem. A patient, a 40-year-old female who exhibited only minor features of trisomy 8 syndrome, first developed sideroplastic anemia and later on acute leukemia.

## E. Structural Aberrations of Chromosome 8

### 1. Partial Trisomy

***a. Trisomy of the Short Arm.*** This has been documented by the observations of Rosenthal *et al.* (1973), Yanagisawa and Hiraoka (1971), and Yanagisawa (1973). The patient of Rosenthal *et al.* (1973) suffered from seizures and exhibited asymmetry of the skull, micrognathia, small hands with a short fifth finger and dysplastic toenails. The patients of Yanagisawa (1973) had short stature, hypermobility of the joints, hypoplastic nails, brachymesophalangy of the second and fifth finger and an unusual facies.

In two more cases, trisomy 8p was associated with a deficiency of another chromosome, the significance of which is known. Chiyo *et al.* (1975) described a mentally defective boy with antimongoloid slanting, epicanthal folds, divergent strabism and deformed ears. The fingers were short, and the nails were dysplastic. There was only a flexion crease from the second to fourth fingers and a four-finger line. The third ventricle was dilated.

The patient described by Lejeune and Rethoré (1973) exhibited typical features of trisomy 8 such as 13 ribs, narrow iliac bones, limitation of the joints, and pli capitonné, and also a broad nose and hypertelorism. In this case, not only the short arm but also the proximal portion of the long arm but also the proximal portion of the long arm of No. 8 was trisomic.

***b. Trisomy of the Terminal Portion of the Long Arm.*** This is supported by several observations, particularly that of Sanchez and Yunis (1974). Both adult siblings examined by these authors had normal stature and exhibited scaphocephaly, triangular face, low hairline, prominent ears, webbing of the neck and a high palate. The chest was narrow. There was a pectus excavatum with a long and flat sternum. The pelvis was narrow, the mobility of the joints limited. Short fingers were due to brachymesophalangy; the middle phalanges of the toes were absent. The female had thoracic scoliosis. Recently, Laurent *et al.* (1974) made a remarkably similar observation in an adult male and female.

Several unusual anomalies exhibited by a patient of Fryns *et al.* (1974) may be due to the deficiency of chromosome 13 involved in interchange. There was a VSD, pulmonary stenosis, and an atopic perineal anus. The skeletal formation was said to be normal, but there was a cleft palate. In a patient of Fujimoto *et al.* (1975) a coloboma of the left eye and Fallot's tetralogy were noted. The child suffered from seizures since early infancy. She is carrier of a duplication following pericentric inversion which was transmitted through three generations of the family.

In a boy with mental retardation, absent patellae, cleft of the soft palate, mild hydronephrosis, contractures of the joints and other dysplastic features which were described earlier by Warkany *et al.* (1962) and Nakagome *et al.* (1968) 8q+ mosaicism was recently demonstrated although the mechanism of origin of this telocentric chromosome which represents the long arm of this chromosome remains unknown (Sh. W. Soukup, personal communication, 1975).

The full description is lacking in two brothers in whom reciprocal translocation between the terminal portions of the long arms of the chromosomes 8 and 22, equal in length, had conditioned a tertiary trisomy with mainly partial trisomy 8 (Lejeune *et al.*, 1972).

### *2. Deletion*

Cardiovascular malformations were noted in two of three children who were found to have a deletion of the short arm of chromosome no. 8. The patient of Lubs *et al.*, 1973, suffered from ASD, VSD, and a large left superior vena cava. Taillemite *et al.*, 1975, mention a VSD and infundibular pulmonary stenosis in a boy, aged 13, whose IQ was 30. Dysplastic features were lacking. Growth failure, epicanthus, short neck, and broad chest with wide-set nipples are the common features of the patients examined by Lubs *et al.* (1973) and Orye and Craen (1976) who, referring to their case, emphasized that there may be a craniofacial dysmorphia but that the phenotype of 8p– seems to be rather inconspicuous. Serological markers could not be linked to the missing segment (8p2) (Taillemite *et al.*, 1975).

## ACKNOWLEDGMENT

I am indebted to all my colleagues who helped me increase the number of patients with proved trisomy 8 mosaicism I have discussed: Dr. J. Antich, Barcelona, Spain; Dr. J. Battin, Bordeaux, France; Dr. J. M. Emberger, Montpellier, France; Dr. E. Engel, Nashville, Tennessee; Dr. K. H. Gustavson, Umea, Finland; Dr. M. Higurashi, Tokyo, Japan; Dr. S. L. Inhorn, Madison, Wisconsin; Dr. P. Jalbert, Grenoble, France; Dr. C. Laurent, Lyon, France; Dr. R. Neu, Syracuse, New York; Dr. M. Sasaki, Sapporo, Japan; Dr. F. R. Sergovich, London (Canada); Dr. Sh. Soukup, Cincinnati, Ohio; Dr. G. Stalder, Basel, Switzerland.

## REFERENCES

Aller, V., Abrisqueta, J. A., Perez, A., Martin, M. A., Goday, C., and Del Mazo, J. (1975). A case of trisomy 8 mosaicism. 47,XX,+8/46/XX. *Clin. Genet.* 7, 232–237.

Antich, J., and Sabater, J. (1973). C autosomal trisomy with mosaicism. *J. Ment. Defic. Res.* **17,** 33–45.

Atkins, L., Holmes, L. B., and Riccardi, V. M. (1974). Trisomy 8. *J. Pediatr.* **84,** 302–304.

Bargman, G. J., Neu, R. L., Kajii, T., Leao, J. C., and Gardner, L. I. (1967). Trisomy C mosaicism in a seven month old girl. *Humangenetik* **4,** 13–17.

Battin, J., Alberty, J., Azanda, X., and Joussein, M. (1968). Syndrome dysmorphique avec oligophrénie chez un garçon de 12 ans porteur d'une trisomie 6–12 en mosaique. *Arch. Franc. Ped.* **25,** 833–834.

Bijlsma, J. B., Wijfels, J. C. H. M., and Tegelaers, W. H. H. (1972). C8 trisomy mosaicism syndrome. *Helv. Paediatr. Acta* **27,** 281–298.

Bishun, N. P. (1968). Normal/trisomy C mosaicism in the mother of a "mongoloid" child. *Acta Paediatr. Scand.* **57,** 243–244.

Boué, J., Boué, A., Deluchat, C., Perraudin, N., and Yvert, F. (1975). Identification of C trisomies in human abortuses. *J. Med. Genet.* **12,** 265–268.

Caspersson, T., Lindsten, J., Zech, L., Buckton, K. E., and Price, W. H. (1972). Four patients with trisomy 8 identified by the fluorescence and Giemsa banding techniques. *J. Med. Genet.* **9,** 1–7.

Cassidy, S. B., McGee, B. J., van Eys, J., Nance, W. E., and Engel, E. (1975). Trisomy 8 syndrome. *Pediatrics* **56,** 826–831.

Chiyo, H.-A., Nakagome, Y., Matsui, I., Kuroki, Y., Kobayashi, H., and Ono, K. (1975). Two cases of 8p trisomy in one sibship. *Clin. Genet.* **7,** 328–333.

Crandall, B. F., Bass, H. N., Marcy, S. M., Glorsky, M., and Fish, C. H. (1974). The trisomy 8 syndrome: Two additional mosaic cases. *J. Med. Genet.* **2,** 393–398.

Debray-Ritzen, P., Bursztejn, C., Vivier, R., Rethoré, M. O., Prieur, M., and Lejeune, J. (1974). Troubles graves d'intégration du langage de l'enfant et aberrations chromosomiques à-propos de trois cas. *Rev. Neurol.* **130,** 357–365.

de Grouchy, J., Turleau, C., and Leonard, C. (1971). Etude en fluorescence d'une trisomie C mosaique probablement 8: 46,XY/47,XY,?8+. *Ann. Genet.* **14,** 69–72.

de Grouchy, J., Josso, F., Beguin, S., Turleau, C., Jalbert, P., and Laurent, C. (1974). Déficit en facteur VII de la coagulation chez trois sujets trisomiques 8. *Ann. Genet.* **17,** 105–108.

de la Chapelle, A., Wennström, J., Wasastjerna, C., Knutar, F., Stenman, U.-H., and Weber, T. H. (1970). Apparent C trisomy in bone marrow cells. *Scand. J. Haematol.* **7,** 112–122.

de la Chapelle, A., Schröder, J., and Vuopio, P. (1972). 8–trisomy in the bone marrow. Report of two cases. *Clin. Gent.* **3,** 470–476.

de la Chapelle, A., Schröder, J., and Vuopio, P. (1973). Cytogenetical and clinical aspects of trisomy-8 in bone marrow. *Chromosome Indent., Proc. Nobel Symp., 23rd, 1972* pp. 201–204.

Emberger, J.-M., Rey, J., Rieu, D., Dossa, D., Bonnet, H., and Jean, R. (1970). Trisomie du groupe C (47,XX,C+). *Arch. Fr. Pediatr.* **27,** 1081–1088.

Fineman, R. M., Ablow, R. C., Howard, R. O., Albright, J., and Breg, W. R. (1975). Trisomy 8 mosaicism syndrome. *Pediatrics* **56,** 762–767.

Frezza, M., Perona, G., Corrocher, R., Vettore, L., and de Sandre, G. (1967). Probabile mosaico di tipo "normale/trisomia 6–12/monosomia 6–12" in una bambina ipoevaluta con lievi note dismorfiche. *Fol. Hered. Path.* **17,** 47–56.

Fryns, J. P., Verresen, H., van den Berghe, H., van Kerckvoorde, J., and Cassiman, J. J. (1974). Partial trisomy 8: Trisomy of the distal part of the long arm of chromosome number 8+(8q2) in a severely retarded and malformed girl. *Humangenetik* **24,** 241–246.

Fujimoto, A., Wilson, M. G., and Towner, J. W. (1975). Familial inversion of chromosome No. 8. *Humangenetik* **27**, 67–73.

Gafter, U., Shabtal, F., Kahn, Y., Halbrecht, I., and Djaldetti, M. (1976). Aplastic anemia followed by leukemia in congenital trisomy 8 mosaicism. *Clin. Genet.* **9**, 134–142.

Giraud, F., Mattei, J.-F., Blanc-Pardigon, M., Mattei, M.-G., and Bernard, R. (1975). Trisomie 8 en mosaique. *Arch. Fr. Pediatr.* **32**, 177–183.

Gorlin, R. J., Cervenka, J., Moller, K., Horrobin, M., and Witkop, C. J. (1975). A selected miscellany. Trisomy 8 mosaicism syndrome. *Birth Defects, Orig. Artic. Ser.* **11**, No. 2, 48–50.

Gustavson, K. H., Hagberg, B., and Santesson, B. (1967). Mosaic trisomy of an autosome in the 6–12 group in a patient with multiple congenital anomalies. *Acta Paediatr. Scand.* **56**, 681–686.

Hellström, K., Hagenfeldt, L., Larsson, A., Lindsten, J., Sundelin, P., and Tiepolo, L. (1971). An extra C chromosome and various metabolic abnormalities in the bone marrow from a patient with refractory sideroblastic anemia. *Scand. J. Haematol.* **8**, 293–306.

Higurashi, M., Naganuma, M., Matsui, I., and Kamoshita, S. (1969). Two cases of trisomy C6–12 mosaicism with multiple congenital malformations. *J. Med. Genet.* **6**, 429–434.

Hsu, L. Y. F., Alter, A. V., and Hirschhorn, K. (1974). Trisomy 8 in bone marrow cells of patients with polycythemia vera and myelogenous leukemia. *Clin. Genet.* **6**, 258–264.

Hustinx, T. W. J., ter Haar, B. G. A., Rutten, F. J., Scheres, J. M. J. C., and Janssen, A. H. (1975). Trisomy 8 mosaicism. *Maandschr. Kindergeneeskd.* **43**, 138–150.

Jacobs, P. A., Harnden, D. G., Buckton, K. E., Court-Brown, W. M., King, M. J., McBride, J. A., McGregor, T. N., and McLean, N. (1961). Cytogenetic studies in primary amenorrhoea. *Lancet* **I**, 1183–1188.

Jacobsen, P., Mikkelsen, M., and Rosleff, F. (1974). The trisomy 8 syndrome: Report of two further cases. *Ann. Genet.* **17**, 87–94.

Jalbert, P., Jobert, J., Patet, J., Mouriquand, C., and Roget, J. (1966). Un nouveau cas de trisomie présumée 6–12. *Ann. Genet.* **9**, 109–112.

Jonasson, J., Gahrton, G., Lindsten, J., Simonsson-Lindemalm, C., and Zech, L. (1974). Trisomy 8 in acute myeloblastic leukemia and sideroachrestic anemia. *Blood* **43**, 557–563.

Kakati, S., Nihill, M., and Sinha, A. K. (1973). An attempt to establish trisomy 8 syndrome. *Humangenetik* **19**, 293–300.

Kondo, I., Tomisawa, T., Matsuura, A., Ibuki, Y., Yamashita, A., and Hara, Y. (1975). A case of trisomy 9 mosaicism. *Ann. Paediatr. Jpn.* **21**, 48–50.

Kucherlapati, R. S., Nichols, E. A., Creagan, R. P., Chen, S., Borgaonkar, D. S., and Ruddle, F. H. (1974). Assignment of the gene for glutathione reductase to human chromosome 8 by somatic cell hybridization. *Am. J. Hum. Genet.* **26**, 51a.

Laurent, C., Robert, J. M., Grambert, J., and Dutrillaux, B. (1971). Observations cliniques et cytogénétiques de deux adultes trisomiques C en mosaique. Individualisation du chromosome surnuméraire par la technique moderne de dénaturation. 47,XY, ?8+. *Lyon Med.* **226**, 827–833.

Laurent, C., Biemont, M. C., Midenet, M., Couturier, P., and Dutrillaux, B. (1974). Diagnostic chromosomique d'un Dp+ par l'association de plusieurs techniques de marquage. *Lyon Med.* **232**, 609–615.

Lejeune, J., and Rethoré, M. O. (1973). Trisomies of chromosome no. 8. *Chromosome Ident., Proc. Nobel Symp., 23rd, 1972* pp. 214–219.

Lejeune, J., Dutrillaux, B., Rethoré, M. O., Berger, R., Debray, H., Veron, P., Gorce, F., and Grossiord, A. (1969). Sur trois cas de trisomy C. *Ann. Genet.* **12,** 28–35.

Lejeune, J., Rethoré, M. O., Dutrillaux, B., and Martin, G., (1972). Translocation 8–22 sans changement de longueur et trisomie partielle 8q. *Exp. Cell Res.* **74,** 293–295.

Lubs, H. A., and Lubs, M.-L. (1973). New cytogenetic technics applied to a series of children with mental retardation. *Chromosome Ident., Proc. Nobel Symp. 23rd, 1972* p. 241.

Malpuech, G., Dutrillaux, B., Fonck, Y., Gaulme, J., and Bouche, B. (1972). Trisomie 8 en mosaique. *Arch. Fr. Pediatr.* **29,** 853–859.

Matsaniotis, N., Metaxotou-Stavridaki, A., Economou-Mavrou, C., and Tsenchi, C. (1968). Trisomy C-Report of a case studied by triatiated Thymidine. *Bull. Univ. Children's Hospital S. Sophia.* **15,** 450–457.

Moulies, J. P. (1969). Une nouvelle aberration chromosomique. La trisomie 6–12. Thesis, Bordeaux.

Monnet, P., Willemin-Clog, L., Gauthier, J., Peytel, J., Mme Laurent, C., Gay, Y., and Poncet, J. (1967). La trisomie 6–12. *Arch. Franc. Ped.* **24,** 869–879.

Nakagome, Y., Warkany, J., and Rubinstein, J. H. (1968). Mental retardation, absence of patellae and other malformations with chromosomal mosaicism. A follow-up report. *J. Pediatr.* **72,** 695–697.

Neu, R. L., Bargman, G. J., and Gardner, L. I. (1969). Disappearance of a 47,XX,C+ leucocyte cell line in an infant who had previously exhibited 46,XX/47,XX,C+ mosaicism. *Pediatrics* **43,** 624–626.

Oikawa, K., Kajii, T., Shimba, H., and Sasaki, M. (1969). 46,XY/47,XY,C+ mosaicism in a male infant with multiple anomalies. *Ann. Genet.* **12,** 102–106.

Orye, E., and Craen, M. (1976). A new chromosome deletion syndrome. Report of a patient with a 46,XY,8p– chromosome constitution. *Clin. Genet.* **9,** 289–301.

Penrose, L. S. (1972). Dematoglyphic patterns in a case of trisomy 8. *Lancet* **I,** 957.

Pfeiffer, R. A., and Lenard, H. G. (1973). Ringchromosom 8 (46,XY,8r) bei einem debilen Jungen. *Klin. Paediatr.* **185,** 187–191.

Pfeiffer, R. A., Schellong, G., and Kosenow, W. (1962). Chromosomenanomalien in den Blutzellen eines Kindes mit multiplen Abartungen. *Klin. Wochenschr.* **40,** 1058–1067.

Riccardi, V. M., Atkins, L., and Holmes, L. B. (1970). Absent patellae, mild mental retardation, skeletal and genitourinary anomalies, and C group autosomal mosaicism. *J. Pediatr.* **77,** 664–672.

Rowley, J. D. (1975). Nonrandom chromosomal abnormalities in hematologic disorders of man. *Proc. Natl. Acad. Sci. U. S. A.* **72,** 152–156.

Rützler, L., Briner, J., Saur, F., and Schmid, W. (1974). Mosaik-Trisomie-8. *Helv. Paediatr. Acta* **29,** 541–553.

Sanchez, O., and Yunis, J. J. (1974). Partial trisomy 8(8q24) and the trisomy 8-syndrome. *Humangenetik* **23,** 297–303.

Schaumann, B., Cervenka, J., and Gorlin, R. J. (1974). Dermatoglyphics in trisomy 8 mosaicism. *Humangenetik* **24,** 201–205.

Schinzel, A., Biro, Z., Schmid, W., and Hayashi, K. (1974). Trisomy 8 mosaicism syndrome. *Helv. Paediatr. Acta* **29,** 531–540.

Sergovich, F. R. (1967). Cytogenetic practice in a mental retardation clinique. *Canad. Psychiat. Assoc. J.* **12,** 35–52.

Stalder, G. R., Bühler, E. M., and Weber, J. R. (1963). Possible trisomy in chromosome group 6–12. *Lancet* **I,** 139. (See G. Kosztolanyi, E. M. Bühler, P. Elmiger and G. R. Stalder, unpublished manuscript.)

Stalder, G., Bühler, E. M., Brehme, H., Bühler, U., and Weber, J. R. (1964). Mosaik

Normal/Trisomie C bei einem schwachsinigen Kind aus einer $G^{e-}\rightarrow D^{k}\rightarrow$Translokationsfamilie. *Arch. Julius Klaus-Stift. Vererbungsfersch. Sozialonthropol. Rassenhyg.* **39,** 92–105. (see G. Kosztolanyi, E. M. Bühler, P. Elmiger, and G. R. Stalder, unpublished manuscript).

Stenbjerg, S., Husted, S., Bernsen, A., Jacobsen, P., Nielsen, J., and Rasmussen, K. (1975). Coagulation studies in patients with trisomy 8 syndrome. *Ann. Genet.* **18,** 241–242.

Stolte, L., and Blankenberg, G. E. (1964). Possible trisomy in chromosome group 6–12 in a normal woman. *Lancet* **2,** 1379.

Taillemite, J.-L., Channarond, J., Tinel, H., Mulliez, N., and Roux, C. (1975). Délétion partielle du bras court du chromosome 8. *Ann. Genet.* **18,** 251–255.

Tuncbilek, E., Atasu, M., and Say, B. (1972). Dermatoglyphics in trisomy 8. *Lancet,* **2,** 821.

Tuncbilek, E., Halicioglu, C., and Say, B. (1974). Trisomy-8 syndrome. *Humangenetik* **23,** 23–29.

Van Eys, J., Nance, W. E., and Engel, E. (1970). C autosomal trisomy with mosaicism. A new syndrome? *Pediatrics* **45,** 665–676.

Walravens, P. A., Greensher, A., Sparks, J. W., and Wesenberg, R. L. (1974). Trisomy 8 mosaicism. *Am. J. Dis. Child.* **128,** 564–566.

Warkany, J., Rubinstein, J. H., Soukup, S. W., and Curless, M. C. (1962). Mental retardation, absence of patellae, other malformation with chromosomal mosaicism. *J. Pediatr.* **61,** 803–812.

Wilson, M. G., Fujimoto, A., and Alfi, O. S. (1974). Double autosomal trisomy and mosaicism for chromosomes no. 8 and no. 21. *J. Med. Genet.* **11,** 96–101.

Yanagisawa, S. (1973). Partial trisomy 8: further observation of a familial C/G translocation. Chromosome identified by the Q-staining methods. *J. Ment. Defic. Res.* **17,** 28–32.

Yanagisawa, S., and Hiraoka, K. (1971). Familial C/G translocation in three relatives associated with severe mental retardation, short stature, unusual dermatogplyhics, and other malformations. *J. Ment. Defic. Res.* **15,** 136–146.

# 7

# Partial Duplication 10q and Duplication 10p Syndromes

JORGE J. YUNIS and RAYMOND C. LEWANDOWSKI, JR.



## I. INTRODUCTION

There are presently more than thirty case reports involving abnormalities of chromosome 10. Trisomy for the distal segment of the long arm represents a readily identifiable syndrome (Yunis and Sanchez, 1974). Trisomy for the short arm is also clinically definable, but further case reports are needed for the detailed delineation of the syndrome. Deletion for the short arm has only been reported in two patients without apparent common characteristic features (Shokeir *et al.,* 1975; Francke *et al.,* 1975). Finally, complete trisomy 10 has been found in four abortuses (Kajii *et al.,* 1973; T. Kajii, unpublished data, 1976), all of which showed considerable developmental delay (Table I).

**TABLE I**
**10-Trisomic Abortuses[a]**

| Code No. | Sex chromosomes | Maternal age (yr) | Previous pregnancies (term) | Ovulation age (days) | Developmental age (days) | Specimens |
|---|---|---|---|---|---|---|
| G72E042 | XY | 28 | 0 | 56 | 29 | A 2 mm nodular embryonic mass in intact amniotic sac |
| G72E116[b] | XX | 16 | 0 | 67 | – | Broken pieces of placenta. No amnion, no cord or embryo |
| G73E010 | XX | 25 | 0 | 86 | 34 | Ruptured amniotic sac; no trace of cord or embryo |
| G74E120 | XX | 27 | 1 | 88 | 24 | A 3 mm embryo arising from an intact amniotic sac |

[a]From Kajii, unpublished data, Geneva, Jan. 1976.
[b]G72E116 was previously reported by Kajii *et al.*, *Am. J. Hum. Genet.* **25**, 539–547 (1973).

## II. PARTIAL DUPLICATION 10q SYNDROME

In 1974, Yunis and Sanchez described a diagnostic phenotype in four cases with partial trisomy 10q. The first reported case was that of Bühler *et al.* (1967), originally described as partial trisomy 12 and most recently found to have partial trisomy 10q (Tsuchimoto and Bühler, 1974).

In this review, the detailed clinical and cytogenetic analysis of seventeen reported cases is made to establish the detailed characteristics of this distinctive chromosome disorder.

### A. Phenotype

Patients with duplication of the distal segment of the long arm of chromosome 10 represent a clinical syndrome with unique facial features and unusual somatic anomalies, particularly of hands and feet (Table II, Figs. 1–7); seventeen cases in fifteen families have been reported. Most patients reveal microcephaly, and all but one (Forabosco *et al.,* 1975) showed marked mental and somatic retardation within the first year of life.

The facial features are pathognomonic and found in most cases (Table II; Figs. 1, 3a, 3b, 4, 6, 7). They include microcephaly, spacious forehead, an oval and flat-appearing face, arched and widespread eyebrows, slight antimongoloid slants, small palpebral fissures, microphthalmia, small nose with anteverted nostrils and flat nasal bridge, prominent malar areas, bow-shaped mouth with prominent upper lip, micrognathia and malformed and/or low-set ears. From the side, the face is not truly flat. The spacious and full forehead, the small nose with depressed nasal bridge, prominent malar areas, and micrognathia give the impression of flatness from the front. Other facial features observed in approximately one-half of the cases include hypertelorism, ptosis of eyelids, cleft soft or bony palate, and long philtrum (Table II). Three patients were reported with prominent occiput (Laurent *et al.,* 1973; Yunis and Sanchez, 1974, Patient 2; J. J. Yunis, unpublished data, 1976), three with large ears (Laurent *et al.,* 1973; Roux *et al.,* 1974; Prieur *et al.,* 1975), and two with narrow auditory canals (Roux *et al.,* 1974; Krøyer and Niebuhr, 1975). The presence of true hypertelorism should be investigated by measurement, since patients may give the impression of wide-set eyes secondary to blepharophimosis, epicanthus, broad and flat nasal bridge, and high arched and wide-set eyebrows.

Most patients have defects of hands and feet (Figs. 2, 3b and c, 5). The most commonly found anomalies include camptodactyly, proximally implanted thumbs and/or great toes, a wide space between first and second toes, and bilateral syndactyly between second and third toes. Approximately one-third of all patients also show overlapping and/or fusiform fingers, deep plantar furrows, and rocker bottom feet (Table II). The deep plantar furrows, although not a

**TABLE II**
**Partial Duplication 10q Syndrome**

| | Bühler *et al.*, 1967; Tsuchimoto and Bühler, 1974, Patient 1 | Kersey *et al.*, 1971; Yunis and Sanchez, 1974, Case 2 | Yunis and Sanchez, 1974, Case 1 | de Grouchy *et al.*, 1972[a] | Francke, 1972[a] | Talvik *et al.*, 1973 | Laurent *et al.*, 1973 | Dutrillaux *et al.*, 1973 | Roux *et al.*, 1974 | Mulcahy *et al.*, 1974 | Krøyer and Niebuhr, 1975 | Forabosco *et al.*, 1975 | Moreno-Fuenmayor *et al.*, 1975 Patient 1 | Moreno-Fuenmayor *et al.*, 1975, Patient 2 | Prieur *et al.*, 1975, Patient 1 | Tsuchimoto and Bühler, 1974, Patient 2 | J. J. Yunis, unpublished data, 1976 |
|---|---|---|---|---|---|---|---|---|---|---|---|---|---|---|---|---|---|
| Sex: | M | F | M | F | F | M | F | M | M | M | F | M | M | M | M | M | M |
| Severe mental retardation | + | | + | + | + | + | + | + | + | | + | – | + | | + | | |
| Low birth weight | + | | + | + | | – | + | – | | + | + | – | – | + | – | + | + |
| Severe hypotrophy | | + | + | + | + | + | + | + | + | + | + | – | + | | + | | |
| Microcephaly | + | | + | | | + | + | – | + | + | + | – | + | + | + | | |
| Tall and spacious forehead | + | + | + | + | + | +[b] | + | + | + | +? | + | + | + | | + | | + |
| "Flat" face | + | + | + | | | +[b] | + | + | + | | | + | +[b] | +[b] | + | | + |
| Oval face | + | + | + | +[b] | | +[b] | | + | +[b] | | + | + | +[b] | +[b] | + | | + |
| Slight antimongoloid slants | + | + | + | | + | + | + | + | | | + | + | | + | + | | + |
| Ptsosis eyelid | + | | + | | + | | + | + | | | + | + | | | + | | |
| Arched and/or spaced eyebrows | + | + | + | | | +[b] | + | + | | | +[b] | + | + | + | + | | + |
| Epicanthus | + | | | | | | | | | | + | + | | | + | | |
| Blepharophimosis | + | + | + | | + | + | + | + | + | + | + | + | + | + | + | | – |

|  |  |  |  |  |  |  |  |  |  |  |  |  |  |  |  |  |  |
|---|---|---|---|---|---|---|---|---|---|---|---|---|---|---|---|---|---|
| Microphthalmia | – | + | + |  | + | + | + | + | – |  | + | + | + | – | + | + | + |
| Ocular anomaly |  |  | + |  |  | + | + |  |  |  |  |  |  | + |  |  |  |
| Hypertelorism[c] | – | – | – | + | + | + | + | + | + |  | + | + | – | – | + | – | – |
| Broad flat nasal bridge | + | + | – | + | + | + | + | + | +? | ? | + | + | + |  |  |  | + |
| Small nose with depressed nasal bridge | + | + | + |  | + | + | + | + | + | + |  | + |  |  |  |  |  |
| Anteverted nostrils | + | + | + |  |  | + |  | + | + |  | + | + |  |  | + |  | + |
| Bow-shaped mouth and/or prominent upper lip | + | + | + | +[b] | + | + | + | + | + |  | + | + | + |  | + |  | + |
| Cleft palate | + | – | – | + | + |  | + | + | – | +[e] |  | – |  | +[e] |  | + | – |
| Long philtrum | + | – | + |  |  | +[b] | + |  | + |  | –[b] | + |  | + | + |  | + |
| Micrognathia | + | + | + |  | + | + | + | + | + | + |  | + |  |  | + | + | + |
| Prominent malar areas | + | + | + |  |  | +[b] | +[b] | +[b] | + |  | + | + | +[b] | +[b] | + |  | + |
| Low-set ears | + | + | + | + | + | + | + | + | + | + | + | + | + |  | + | + | + |
| Short neck |  | + | + |  |  | + | + | + |  |  |  | + | + |  | + | + | + |
| Thin ribs |  |  | + |  |  |  | +[d] |  |  |  |  |  | + | + |  |  | + |
| Scoliosis |  |  | + |  |  | + | + | – | – |  | + | + |  |  | –? |  |  |
| Skeletal anomalies |  |  | + |  |  | + | + | + |  |  | + |  |  |  |  |  |  |
| Delayed bone age |  |  | + | – |  | + | + |  | – |  |  | – |  |  |  |  |  |
| Heart defect | +? | – | + |  |  | + | + | – | – | + |  | – | + | + | – | + | + |
| Cryptorchidism | + |  | + |  |  | + |  | – | – | + |  | + | + |  | + | + |  |
| Camptodactyly | +? | + | + |  |  | +? |  | + |  | + |  |  | + | + |  | + | + |
| Proximally implanted thumbs or toes |  |  |  |  |  | + | + | + | + | + | + |  | + | + |  | + | + |
| Overlapping fingers |  | + | + |  |  |  |  |  |  |  |  |  | + |  |  |  | + |
| Tapering fingers |  |  | + |  |  |  | + |  |  |  | + | +? |  | + |  |  |  |
| Wide space between 1st and 2nd toes |  | + | + |  |  | + |  | + | + |  |  | + | + | + | + |  | + |
| Syndactyly 2nd and 3rd toes |  | + | + |  |  |  | + | + | + | + |  |  |  | + |  |  | + |

*continued*

**TABLE II (*continued*)**

| | Bühler *et al.*, 1967; Tsuchimoto and Bühler, 1974, Patient 1 | Kersey *et al.*, 1971; Yunis and Sanchez, 1974, Case 2 | Yunis and Sanchez, 1974, Case 1 | de Grouchy *et al.*, 1972[a] | Francke, 1972[a] | Talvik *et al.*, 1973 | Laurent *et al.*, 1973 | Dutrillaux *et al.*, 1973 | Roux *et al.*, 1974 | Mulcahy *et al.*, 1974 | Krφyer and Niebuhr, 1975 | Forabosco *et al.*, 1975 | Moreno-Fuenmayor *et al.*, 1975 Patient 1 | Moreno-Fuenmayor *et al.*, 1975, Patient 2 | Prieur *et al.*, 1975, Patient 1 | Tsuchimoto and Bühler, 1974, Patient 2 | J. J. Yunis, unpublished data, 1976 |
|---|---|---|---|---|---|---|---|---|---|---|---|---|---|---|---|---|---|
| Sex: | M | F | M | F | F | M | F | M | M | M | F | M | M | M | M | M | M |
| Deep plantar furrows | | + | + | | | | | | + | | | | | + | + | | – |
| Rocker bottom feet | | + | + | +? | | | | | | + | | | | | | | +? |
| Simian crease | | + | + | | | + | | – | + | | – | – | | | – | | + |
| Abnormal muscle tone | + | | + | | | + | | + | – | | + | | + | + | | | + |
| Decreased renal function or hypoplastic kidneys | + | + | + | | | + | | | – | + | | – | | + | | + | – |

[a]Briefly reported.
[b]Identified in photographs, not reported.
[c]See text for details.
[d]Absent 12th ribs.
[e]Soft cleft palate.

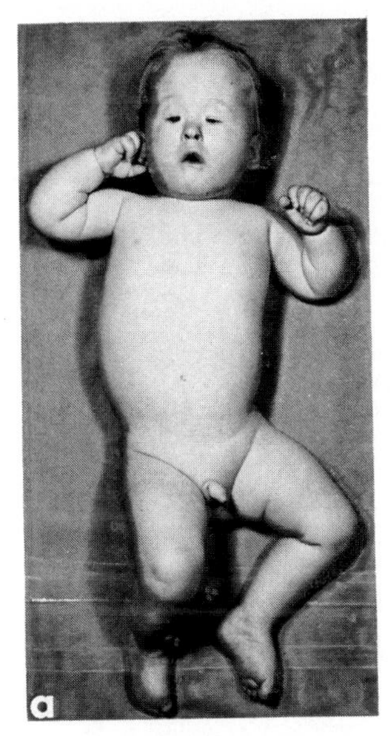


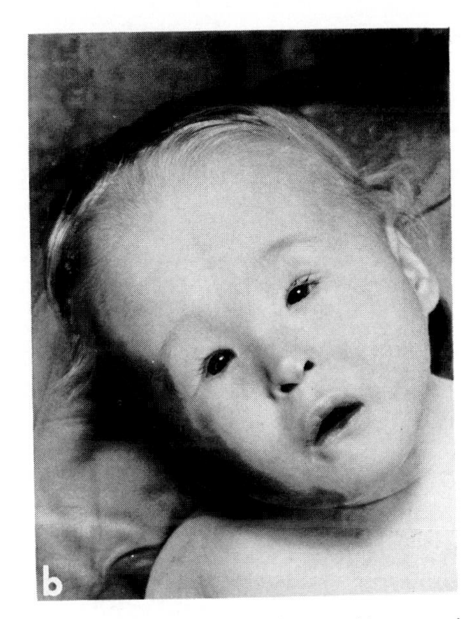


**Fig. 1** Five-month-old child with partial 10q duplication syndrome. Note spacious forehead, oval face, arched and wide-spread eyebrows, blepharophimosis, microphthalmia, small nose with depressed nasal bridge and anteverted nostrils, bow-shaped mouth with prominent upper lip, low-set ears, short neck, and undescended testicles (from Yunis and Sanchez, 1974).

constant feature (30%), represent a discriminating defect previously found in trisomy 8 syndrome. In the duplication 10q syndrome, the plantar furrows run posteriorly from between the first and second toes, turning laterally to follow the distal border of the hypothenar region. Sometimes they also branch into the distal border of the thenar area (Fig. 2c).

Common trunk anomalies include short neck, heart defects, and cryptorchidism (7/12 males). The heart defects vary, although a ventricular septal defect is frequently found (5/9). Yunis and Sanchez (1974, Patient 1) reported ventricular septal defect with an A-V canal and pulmonary hypertension. Talvik *et al.*'s patient (1973) had a patent foramen ovale, and Mulcahy *et al.*'s (1974) a ventricular septal defect. Moreno-Fuenmayor *et al.* (1975, Case 2) described a ventricular septal defect, pulmonary atresia, patent foramen ovale, and intramural atherosclerosis. Tsuchimoto and Bühler (1974, Patient 2) reported an atrial septal defect and patent ductus arteriosus. The patient of Bühler *et al.* (1967) was felt to have a ventricular septal defect, and that of Moreno-Fuenmayor *et al.* (1975, Case 2) a tetralogy, but definitive studies were not performed. Laurent *et al.* (1973) reported dextrocardia with ventricular septal

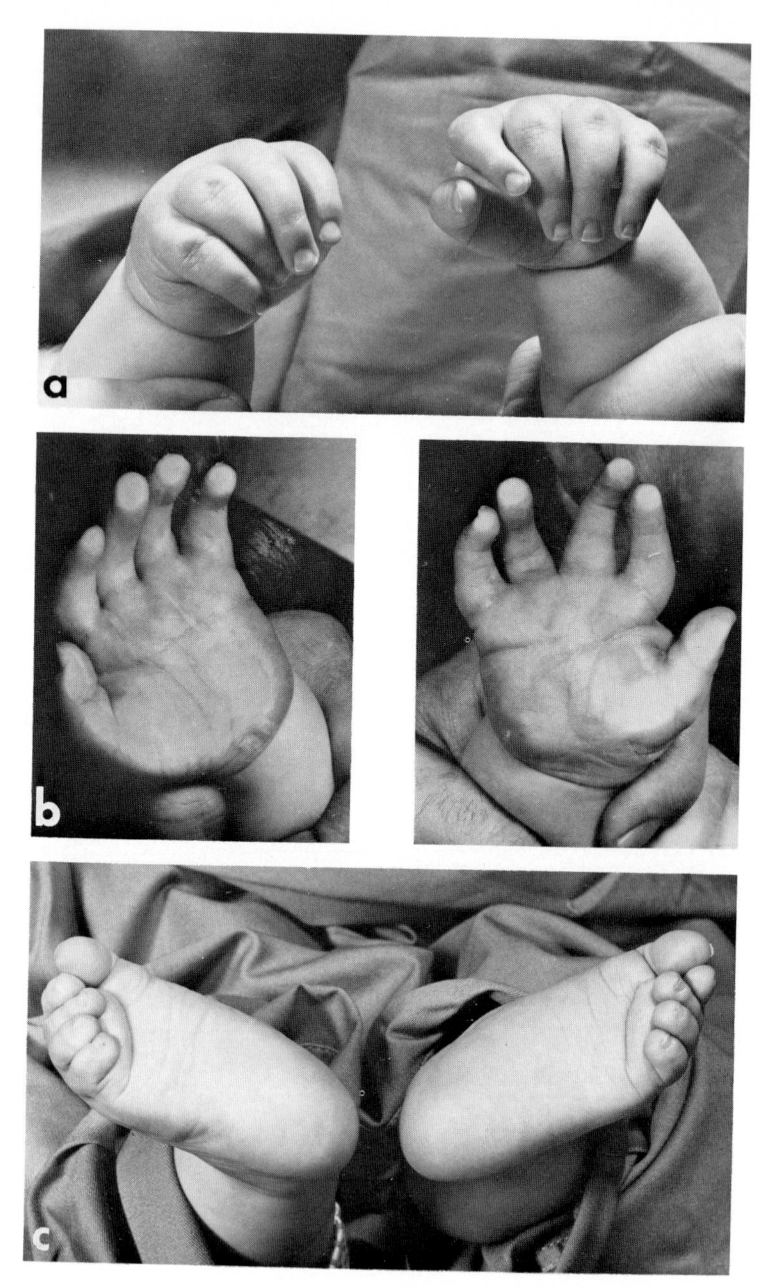


**Fig. 2.** Hands and feet of patient in Fig. 1. Note overlapping fingers and dimples (a), camptodactyly, ulnar and radial deviation of fingers and simian crease (b), and deep plantar furrows (c).

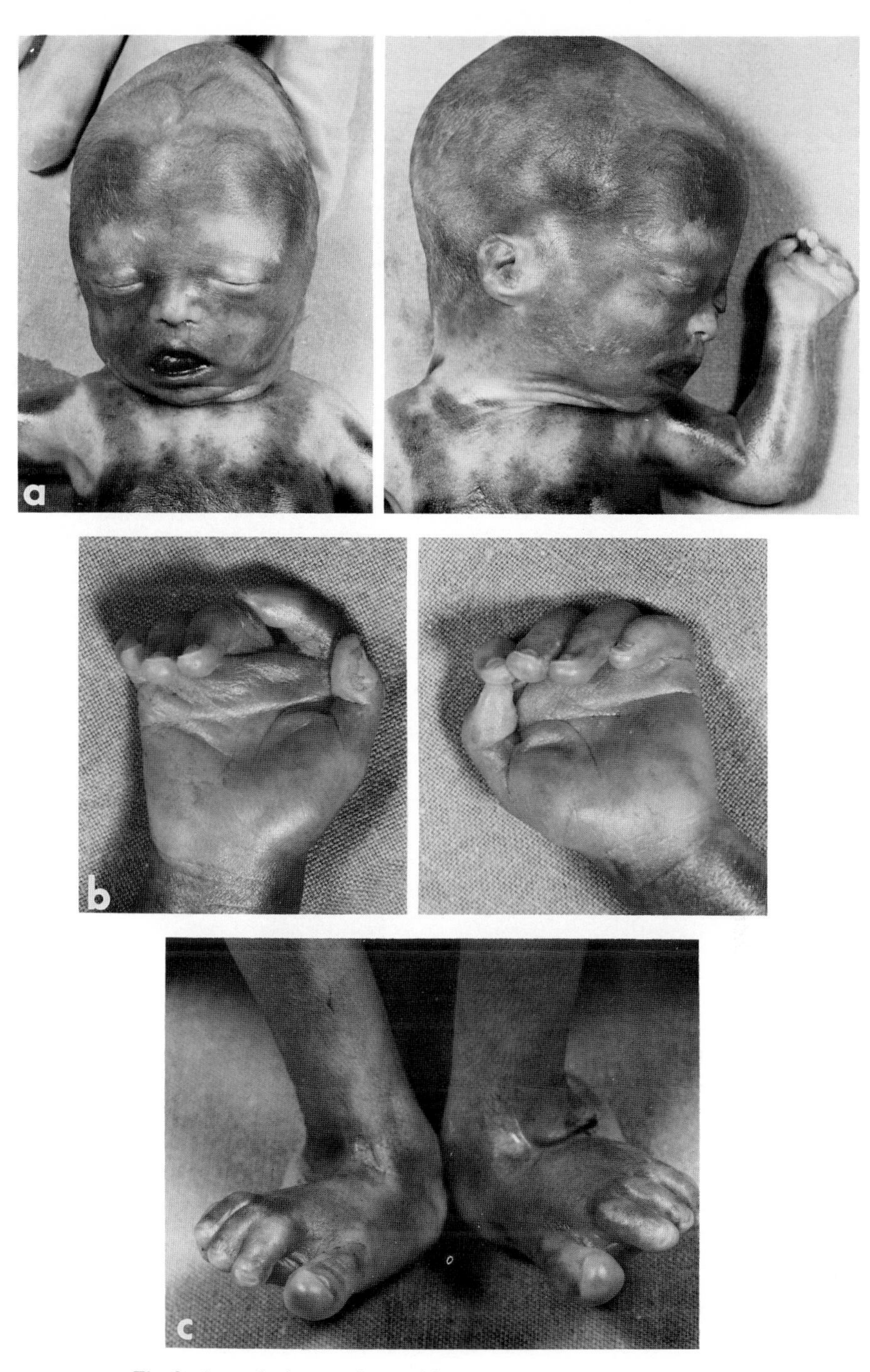


**Fig. 3.** Second trimester fetus with partial 10q duplication syndrome.

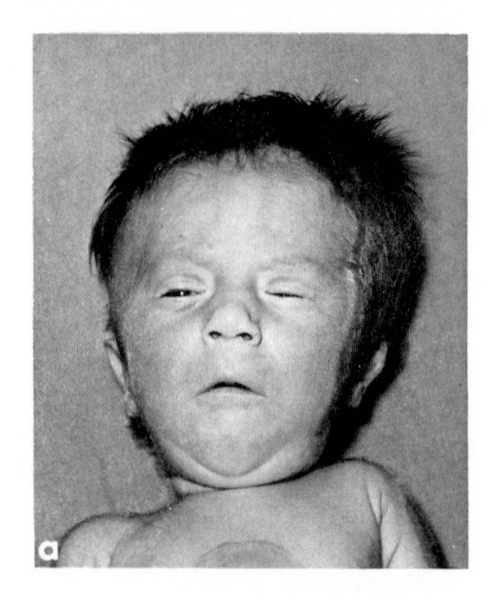

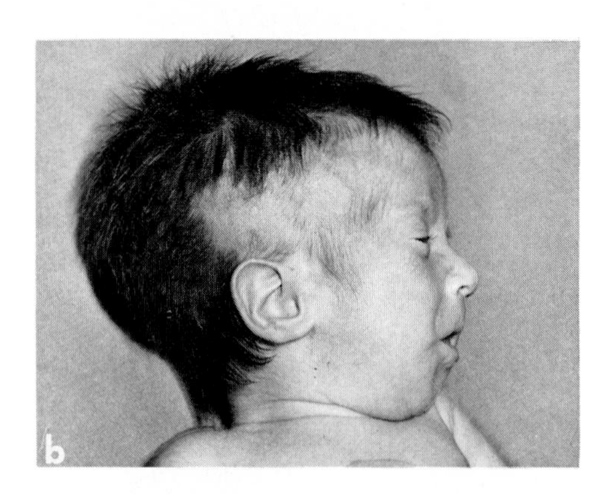

Fig. 4.

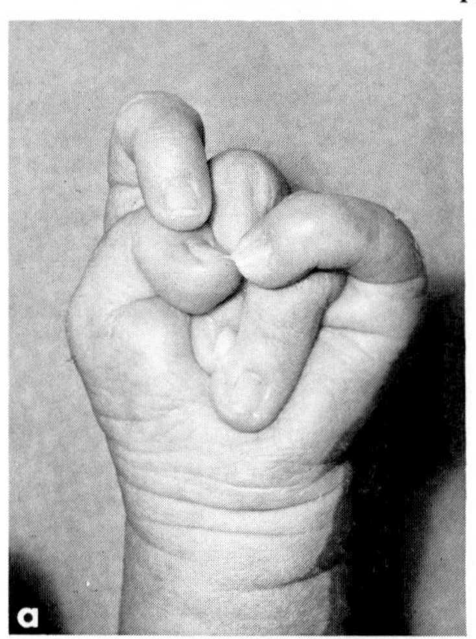

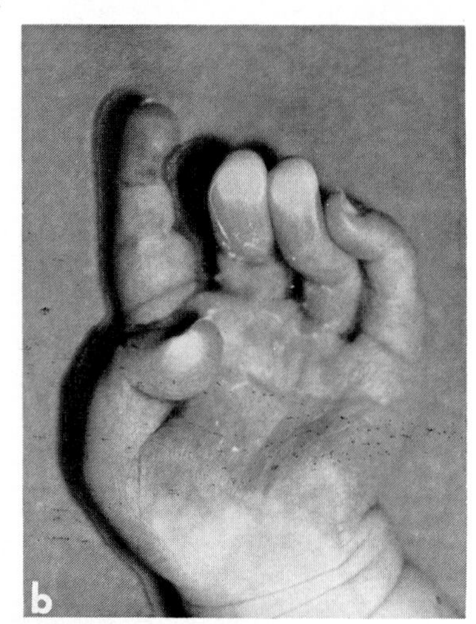

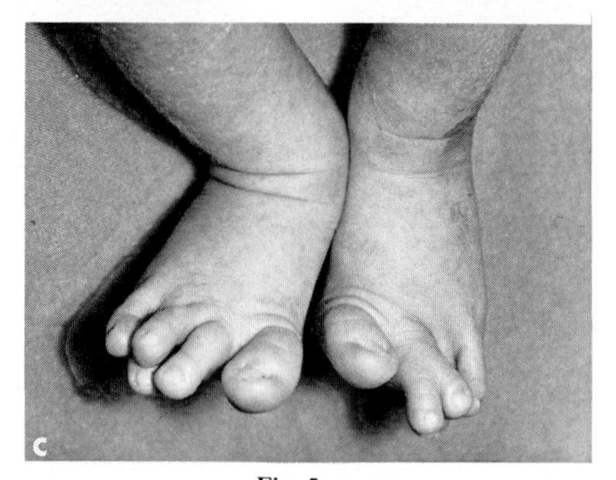

Fig. 5.

**Fig. 4a and b and 5a–c.** One-year-old boy with *de novo* partial trisomy 10q. Note similarity with patient in Figs. 1 and 2.

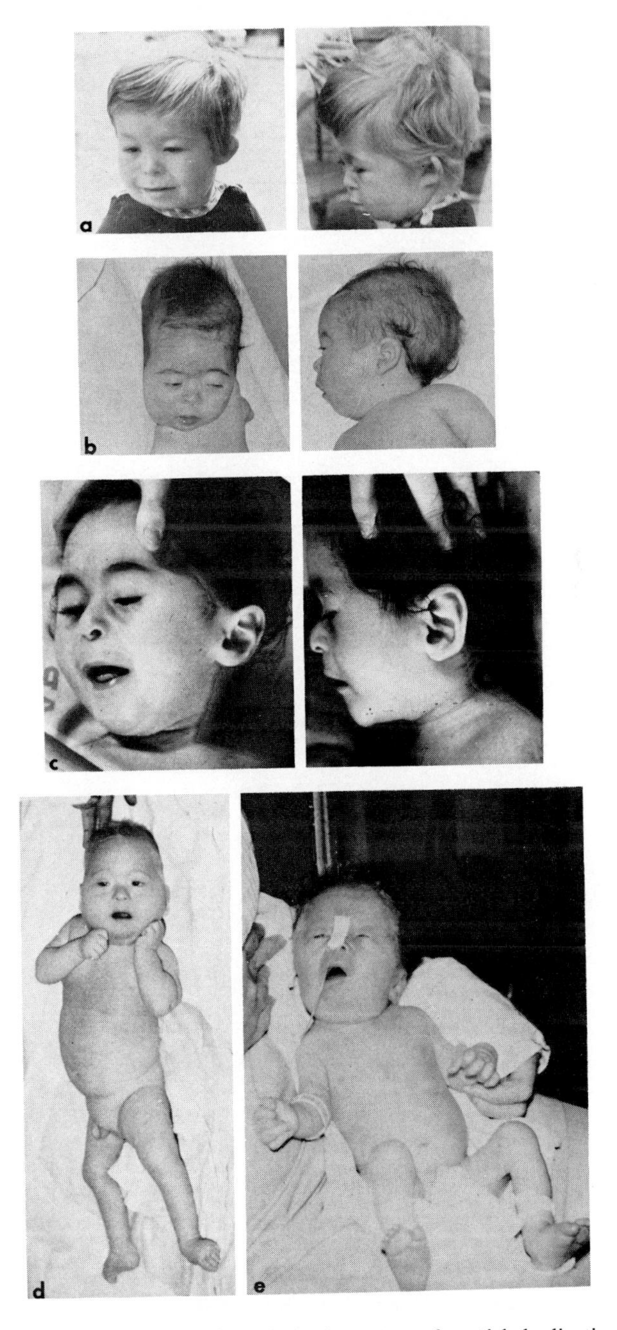

**Fig. 6.** Examples of patients with typical phenotype of partial duplication 10q syndrome reported in the literature: (a) from Forabosco *et al.* (1975); (b) from Dutrillaux *et al.* (1973); (c) from Prieur *et al.* (1975), Patient 1; (d) from Talvik *et al.* (1973); (e) from Moreno-Fuenmayor *et al.* (1975), Patient 2.

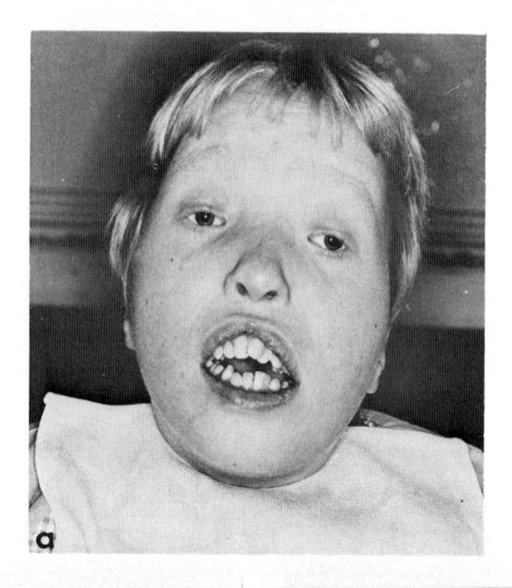

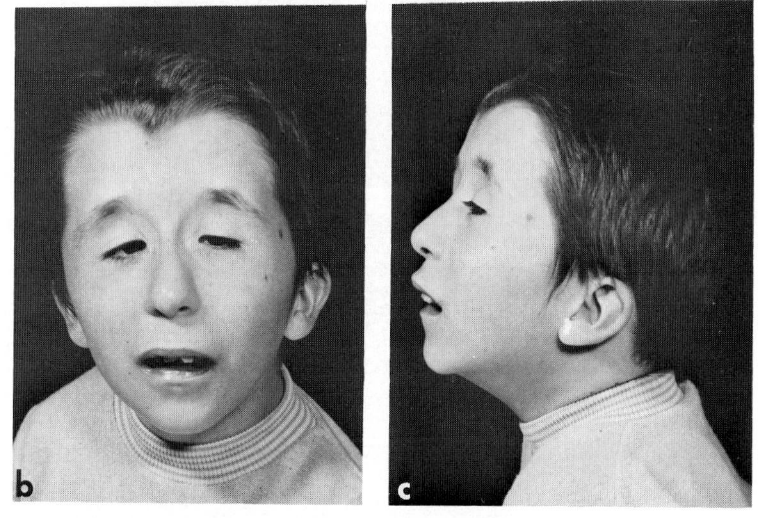

**Fig. 7.** Eighteen-year-old (a) and ten-year-old (b and c) patients with the unique facial features of the partial duplication 10q syndrome. From Krøyer and Niebuhr (1975) and Bühler *et al.* (1967), respectively.

defect. Other defects seen in one-fourth to one-third of all patients include delayed bone age, scoliosis, thin ribs, and abnormal renal function. The two patients of Yunis and Sanchez (1974) and the patient of Tsuchimoto and Bühler (1974, Patient 2) had hypoplasia of kidneys. Moreno-Fuenmayor *et al.* (1975, Case 2) reported absence of fetal lobulation and cortical microcysts. Talvik *et al.*'s patient (1973) had cystic kidneys, and Mulcahy *et al.* (1974) described hydronephrosis, hydroureter, glomerular and interstitial fibrosis, small tubules, and tubular cysts. Bühler *et al.* (1967) reported decreased renal function and blunted pelvis on IVP. Webbed neck, pectus excavatum, osteoporosis, and skin dimples were each noted in two to three patients.

In duplication 10q syndrome, there appears to be a preponderance of affected males (12/17), and all but one case (Table II, J. J. Yunis, unpublished data, 1976; Figs. 4, 5, and 10) resulted from a familial balanced translocation. The true incidence of familial versus sporadic cases is not known, since the characteristic phenotype for the duplication 10q syndromes was only recently described and the common use of metaphase banding studies makes it difficult to identify the origin of a *de novo* chromosomal rearrangement.

The duplication of 10q syndrome is a severe entity. Approximately 50% of the patients are known to have died of respiratory infection or cardiovascular insufficiency before 1 year of age (Table III). Once the patients reach 12 months of age, however, they may survive for several years but are usually severely retarded. One patient, trisomic for the distal fourth of the long arm, is alive at 18 years of age (Fig. 7a; Krøyer and Niebuhr, 1975), and two patients trisomic for the distal third are alive at 7 and 10 years of age, respectively (Figs. 1, 7b, 7c) (Yunis and Sanchez, 1974; Bühler *et al.*, 1967). The three older patients show marked hypotrophy, are bedridden, and are unable to communicate.

Prenatal diagnosis has been performed in one case (Kersey *et al.*, 1971; Yunis and Sanchez, 1974). The fetus had most of the features of the duplication 10q syndrome (Fig. 3), demonstrating that the phenotype appears early in development. Based on this and the analysis of older patients (Fig. 7), the phenotype appears to be constant and recognizable at any age. Also, the disorder has a universal occurrence, with cases reported in Japanese, Mexican, French, Russian, and Scandinavian stock.

## B. Dermatoglyphics

Frequent dermatoglyphic findings include transverse palmar creases, high axial triradius, and an increased number of ulnar loops. At least two patients have shown absence of the c–d triradius (Yunis and Sanchez, 1974; Krøyer and Niebuhr, 1975), and one a low a–b ridge count (Talvik *et al.*, 1973). These findings should be investigated more closely in all patients. The significance of excess ulnar loops in eight patients was not clear, since in three families one of the parents showed a similar finding (Talvik *et al.*, 1973; Yunis and Sanchez, 1974; Krøyer and Niebuhr, 1975), and parental study in three other cases was not reported (Roux *et al.*, 1974; Moreno-Fuenmayor *et al.*, 1975; J. J. Yunis, unpublished data, 1976).

**TABLE III**
**Patients with Partial Duplication 10q**

| Age | Death | Balanced translocation carrier | Patient | | Reference |
|---|---|---|---|---|---|
| | | | Duplication | Deficiency | |
| 10 years | | 46,XX,t(10;18)(q23;q23) | q23[a]→qter | 18qter, uncertain | Bühler *et al.*, 1967; Tsuchimoto and Bühler, 1974, Patient 1 |
| | Abortion | 46,XY,t(10;15)(q24;q26) | q24→qter | 15qter, uncertain | Kersey *et al.*, 1971; Yunis and Sanchez, 1974, Patient 2 |
| 7 years | | 46,XY,t(10;15)(q24;q26) | q24→qter | 15qter, uncertain | Yunis and Sanchez, 1974, Patient 1 |
| | Stillborn | 46,XX,t(10;12)(q22 or 23;p13) | q22?→qter | 12pter, uncertain | de Grouchy *et al.*, 1972 |
| | 4 years | 46,XX,t(10q–;15p+) | q22?→qter | 15pter, uncertain | Francke, 1972 |
| | 11 months | 46,XY,t(10;14)(q24;qter) | q24?→qter | 14qter, uncertain | Talvik *et al.*, 1973 |
| | 9 months | 46,XY,t(1;10)(q44;q22) | q23→qter | 1qter, uncertain | Laurent *et al.*, 1973 |
| | 4 months | 46,XX,inv(10)(p15;q24) | q24→qter | 10pter, uncertain | Dutrillaux *et al.*, 1973 |
| | 3 months | 46,XX,t(10;22)(q24;p12) | q24→qter | 22p13 | Roux *et al.*, 1974 |
| | 7 or 17 days | 46,XX,t(10;13)(q24;q34) | q24→qter | 13qter, uncertain | Mulcahy *et al.*, 1974 |
| 18 years | | 46,XY,t(10;18)(q25;q23) | q24→qter | 18qter, uncertain | Krøyer and Niebuhr, 1975 |
| 3 years | | 46,XX,t(10;18)(q24;p11) | q24→qter | 18pter, uncertain | Forabosco *et al.*, 1975 |
| 10 months | | 46,XY,t(10;17)(q24;p13) | q24→qter | 17pter, uncertain | Moreno-Fuenmayor *et al.*, 1975, Patient 1 |
| | 3 months | 46,XY,t(10;17)(q24;p13) | q24→qter | 17pter, uncertain | Moreno-Fuenmayor *et al.*, 1975, Patient 2 |
| 2 2/3 years | | 46,XY,t(10;17)(q24;q25) | q24→qter | 17qter, uncertain | Prieur *et al.*, 1975, Patient 1 |
| | 2 days | 46,XX,t(10;15)(q23;q26) | q23→qter | 15q26, uncertain | Tsuchimoto and Bühler, 1974, Patient 2 |
| 1 year | | – | q22→qter | – | Yunis, unpublished data 1976 |

[a]Distal portion of band q23.

## C. Cytogenetics

As mentioned earlier, seventeen patients with duplication 10q have been studied. Of these, fifteen were the product of a parental balanced translocation (Figs. 8–9), one from a paracentric inversion (Dutrillaux *et al.*, 1973), and another the result of a *de novo* translocation (J. J. Yunis, unpublished data, 1976) (Fig. 10).

Fourteen of the fifteen cases with balanced translocation resulted from a rearrangement involving the distal segment of the long arm of chromosome 10 and the telomere of another chromosome. Thus far, it appears that the preferential sites of translocation are the telomeres of chromosomes 15, 17, and 18 (Table III). It is believed that the translocation occurs either by telomeric fusion (Francke, 1972; Lejeune, 1973) or with a corresponding small deletion of the telomeric region of the recipient chromosome which could not be detected with the techniques used. If a small deletion occurred, it had no major clinical significance, since the phenotype of the patients was remarkably similar. In the patient observed by Tsuchimoto and Bühler (1974), it is not yet clear if the patient had a deletion of the distal light band of chromosome 15 (15q26). The only patient reported with a defined small deletion is that of Dutrillaux *et al.* (1973) in which the patient was the product of a paracentric inversion of chromosome 10. The general tendency of 10q to translocate to a telomeric region of another chromosome with either minimal or no reciprocal deletion

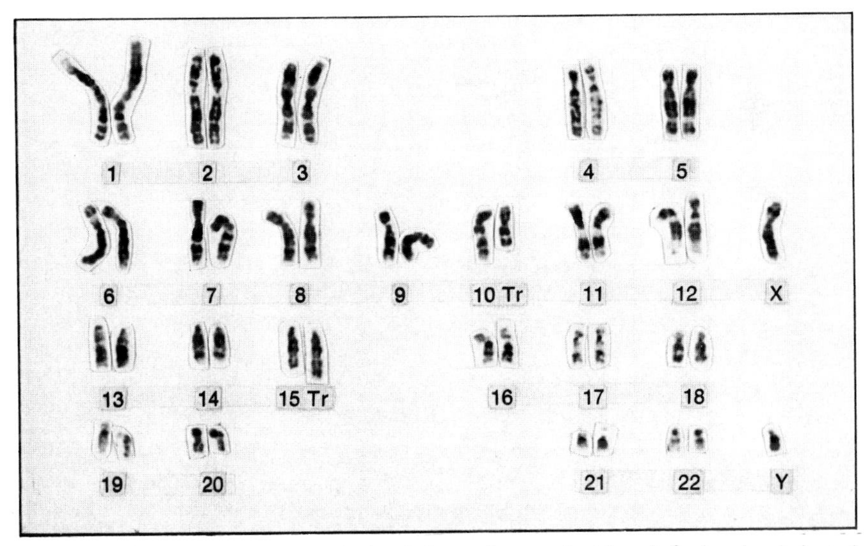


**Fig. 8.** G-banded karyotype of the father of cases in Figs. 1–3 showing balanced translocation 10/15 (from Yunis and Sanchez, 1974).

explains the remarkable similarity among patients and makes this conspicuous partial duplication syndrome one of the new syndromes easiest to diagnose on the basis of clinical findings.

Most of the seventeen patients reviewed are trisomic for the distal half (q23→qter) or distal third (q24→qter) of the long arm of chromosome 10 (Table III, Figs. 9 and 10). One patient, however, was found to be trisomic for only the distal fourth (q25→qter) (Krøyer and Niebuhr, 1975), while four others were trisomic for most of the long arm (q22→qter) (Francke, 1972; de Grouchy *et al.*, 1972; Laurent *et al.*, 1973; J. J. Yunis, unpublished data, 1976). Despite the difference in chromosome segment involved, all patients showed the characteristic phenotype, suggesting that trisomy for the distal fourth of the long arm of chromosome 10 (q25→qter) is critical for the production of the duplication 10q syndrome, and that the other segment (q22→q24) may not contribute significantly to the phenotype.

## D. Summary

Patients with trisomy for the distal segment of the long arm of chromosome 10 (q22→qter) represent a distinct clinical syndrome with unique facial features coupled with unusual anomalies of extremities. Indeed, the head features appear to be pathognomonic and include microcephaly, spacious forehead, oval face, arched and wide-set eyebrows, blepharophimosis, microphthalmia, small nose,

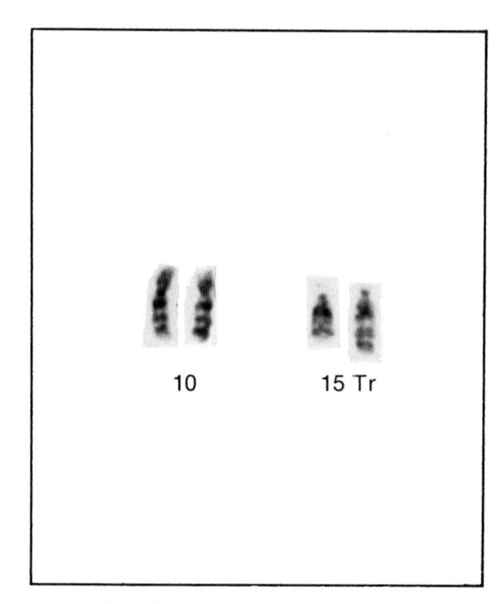


**Fig. 9.** Partial karyotype of patient in Figs. 1 and 2, demonstrating abnormal chromosome 15, making patient partially trisomic for segment q24→qter of the long arm of chromosome 10 (from Yunis and Sanchez, 1974).

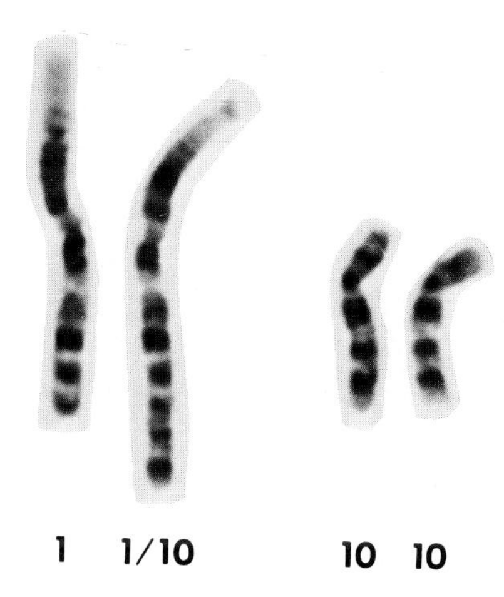


**Fig. 10.** Partial karyotype of patient in Figs. 4 and 5. The distal end of the long arm of one of the chromosomes demonstrates several extra bands arising from a *de novo* translocation. As can be seen, the extra bands correspond to the distal three-fourths of the long arm of chromosome 10 (q22→qter) making the patient partially trisomic for this segment.

prominent malar areas, bow-shaped mouth with prominent upper lip, micrognathia, low-set and/or malformed ears, and short neck. In the hands and feet, common findings include camptodactyly, proximally implanted thumbs and/or great toes, wide space between first and second toes, and bilateral syndactyly between second and third toes. The patients commonly demonstrate marked psychomotor retardation and hypotrophy. The main phenotype-determining chromosome segment appears to be localized to 10(q25→qter). The prognosis is poor since about half of the patients die before the first year of life.

## III. DUPLICATION 10p SYNDROME

In 1968, Insley *et al.* described the first patient that has been confirmed by banding to have trisomy for the short arm and the most proximal portion of the long arm of chromosome 10. Subsequently, an additional nine cases have appeared in the literature with 10p trisomy (Table IV). Although not previously described, the characteristic phenotype of these patients makes possible the delineation of a new clinical syndrome consisting of a peculiar facies coupled with flexion deformities of the extremities.

### A. Phenotype

Patients with duplication 10p demonstrate significant psychomotor and growth retardation and hypotonia. They have a high and full forehead, promi-

**TABLE IV**
**Duplication 10p Syndrome**

| | Insley *et al.*, 1968; Hirschhorn *et al.*, 1973 | Yanagisawa and Adachi, 1970; Nakagome and Kobayashi, 1975 | Hustinx *et al.*, 1974 | Schleiermacher *et al.*, 1974, Patient 1 | Schleiermacher *et al.*, 1974, Patient 2 | Cantu *et al.*, 1975, Patient 1 | Cantu *et al.*, 1975, Patient 2 | Grosse *et al.*, 1975 | Nakagome and Kobayashi, 1975 | Yunis *et al.*, 1976 |
|---|---|---|---|---|---|---|---|---|---|---|
| Sex: | F | M | F | M | F | F | F | M | F | M |
| Low birth weight | – | – | + | – | – | + | + | – | + | + |
| Psychomotor retardation | | | | + | + | + | + | + | | |
| Microsomatia | | + | + | + | + | + | + | + | | + |
| Hypotonia | + | | + | + | + | | | + | | |
| Dolicocephaly | + | | +?[a] | + | + | + | + | | | |
| Wide sagittal suture, fontanel | | | | + | | + | + | | | |
| High prominent forehead | +[a] | | +[a] | + | + | + | + | +[a] | | |
| Long face | +[a] | | –[a] | +[a] | + | + | | | | |
| Microphthalmia | | | | | | + | | | | + |
| Coloboma | | | | | | + | + | | | |
| Hypertelorism | | + | + | | | | | | + | |

| | | | | | | | | | | |
|---|---|---|---|---|---|---|---|---|---|---|
| Mongoloid slants | | | | + | + | | | + | | |
| Arched eyebrows | | | +[a] | + | + | | | | | |
| Wide or prominent nasal root | +[a] | | + | + | +[a] | +[a] | +[a] | + | | + |
| Anteverted nostrils | +[a] | | +[a] | +[a] | +[a] | +[a] | | +[a] | | |
| Turtle-like mouth | +[a] | | +[a] | +[a] | +[a] | +[a] | | | | |
| Cleft lip/cleft palate | | + | + | + | | | + | | + | + |
| Long philtrum | + | | + | | | + | | | | |
| Low-set, posteriorly rotated ears | + | + | + | + | + | + | + | | + | |
| Large or dysmorphic ears | +[a] | + | + | + | +[a] | + | + | + | | |
| Small jaw | | | + | | | | | | + | |
| Congenital heart disease | + | | + | | | + | + | | | |
| Pulmonary hypoplasia | | | + | | | | | | | + |
| Abnormal genitalia | + | + | + | + | | + | | | + | + |
| Flexion deformities | + | + | + | + | | | + | | | |
| Equinovarus deformity of feet | | + | + | + | + | + | + | | | + |
| Shortened carpals, phalanges | | | | + | + | + | + | + | | |
| Clinodactyly | | | | | | + | + | + | | |
| Abnormal dermatoglyphics | + | + | + | + | + | + | + | + | | |
| Renal abnormalities | | | + | | | | + | | | + |

[a]Identified in photographs, not reported.

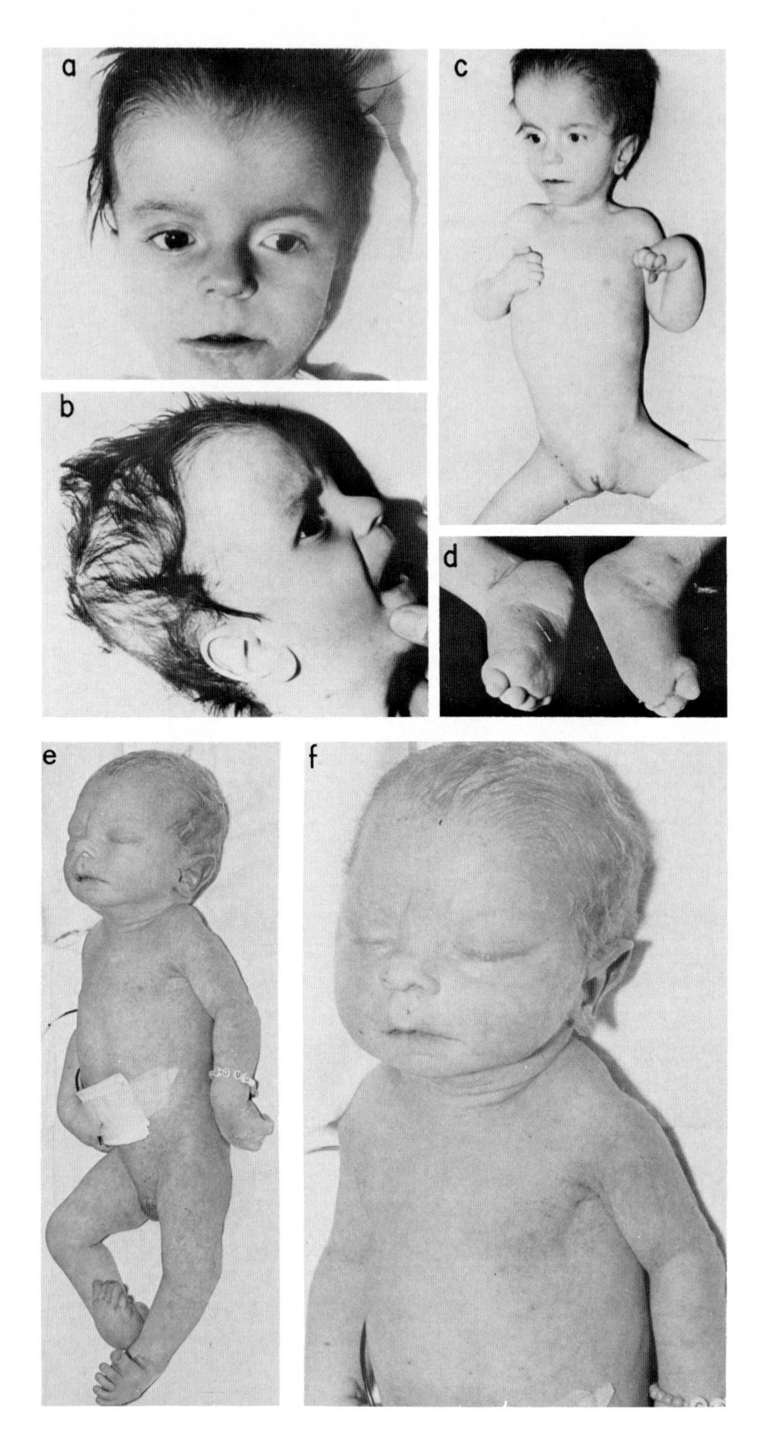
a
b
c
d
e
f

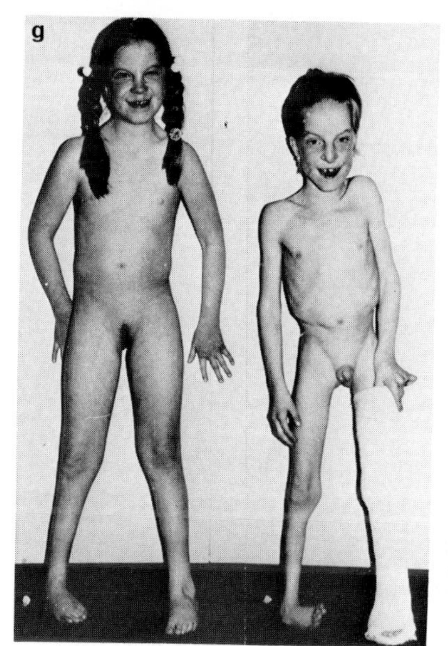


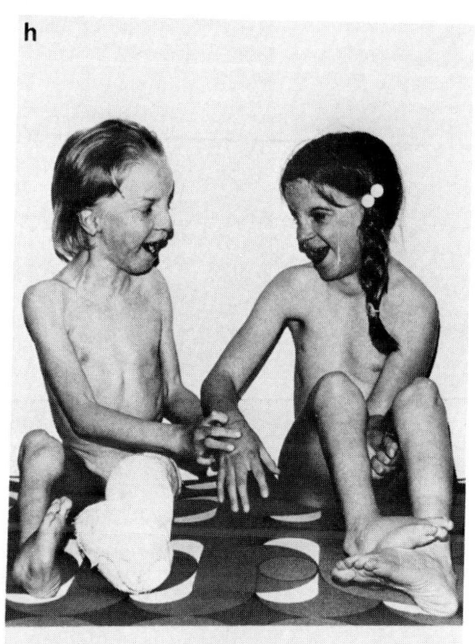


Fig. 11. (*continued*)

nent glabella or wide and prominent nasal root, anteverted nostrils, cleft lip/cleft palate and low-set posteriorly rotated and large/dysmorphic ears (Figs. 11 and 12; Table IV). Those without clefts have a very long philtrum. The mouths of these patients are unusual and appear turtle-like with a slightly arched, inverted, thin and protruding upper lip. The joint deformities are unusual and were seen in eight of the ten patients (Fig. 11). They consist of flexion abnormalities of elbows, wrists, fingers, hips, knees, ankles and toes. One patient showed all joint findings (Cantu *et al.*, 1975, Patient 1), while the rest had different combinations (Table IV). In addition, some patients showed shortened metacarpals or phalanges (Schleiermacher *et al.*, 1974, Cases 1 and 2; Cantu *et al.*, 1975, Case 2).

An interesting feature among patients with duplication 10p is genital anomaly. Two of the four males had a small penis (Schleiermacher *et al.*, 1974, Case 1; Yunis *et al.*, 1976). The first patient also had cryptorchidism, and a third male is reported as having abnormal genitalia without a specific description (Yanagi-

---

**Fig. 11.** Patients with trisomy for the short arm of chromosome 10: (a–d) from Cantu *et al.* (1975), Case 2; (e,f) from Hustinx *et al.* (1974); (g,h) from Schleiermacher *et al.* (1974). Note high prominent forehead, long philtrum, turtle-like mouth, large, low-set, posteriorly rotated and malformed ears, flexion deformities of elbows, wrists, fingers and feet, and enlarged clitoris.

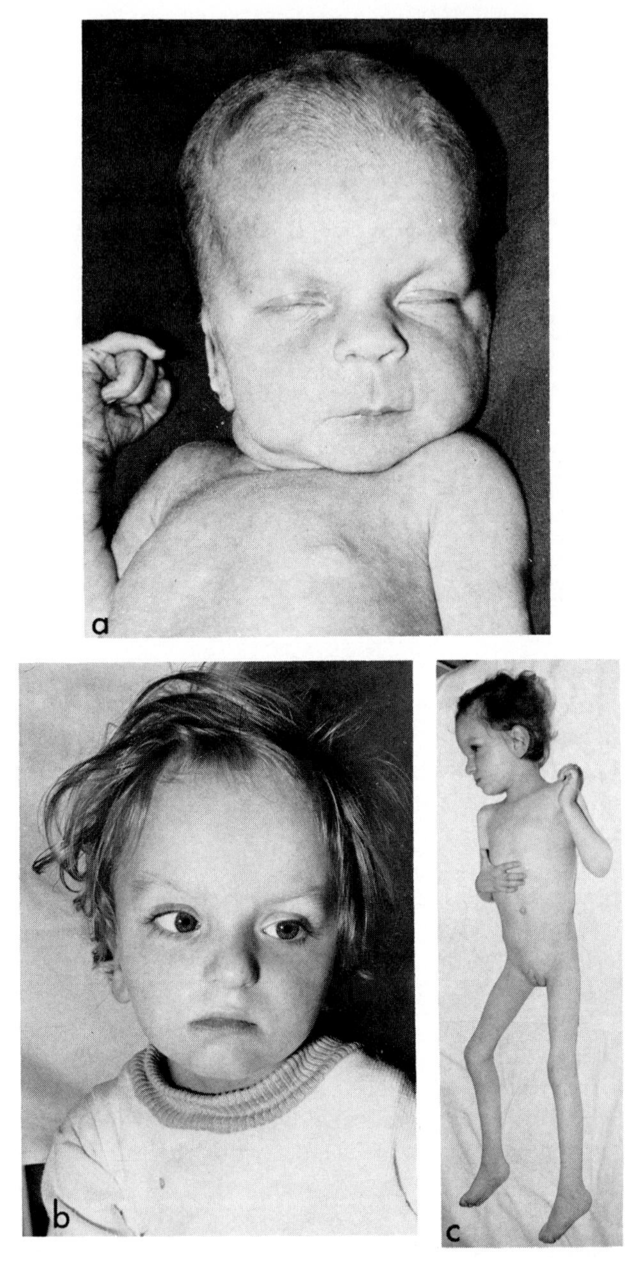

**Fig. 12.** Patients with duplication 10p: (a) The short arm and most proximal portion of the long arm is duplicated (Insley *et al.*, 1968). Note similarities to patient in Fig. 11; (b,c) The distal portion of the short arm is duplicated (p14→pter) (Grosse *et al.*, 1975). Note differences in phenotype compared to other patients.

sawa and Adachi, 1970). One female was intersexual with enlarged labia minora (Insley *et al.*, 1968), another had an enlarged clitoris (Cantu *et al.*, 1975, Case 1), and a third was reported to have abnormal genitalia without a specific description (Nakagome and Kobayashi, 1975).

Another unusual feature is the renal pathology. One of the patients had cystic disease of the kidneys (Hustinx *et al.*, 1974), one an absent kidney (Cantu *et al.*, 1975, Case 2), and a third showed renal dysplasia with cystic or atretic ducts and foci of cartilage (Yunis *et al.*, 1976).

Approximately one-half of the patients had congenital heart disease. Insley *et al.*'s (1968) patient had a patent foramen ovale, as did Yanagisawa and Adachi's (1970) in addition to a patent ductus arteriosus. The patient of Hustinx *et al.* (1974) had coarctation of the aorta. Cantu *et al.*'s patients (1975) were both felt to have congenital heart disease, but no definitive studies were performed.

The patient reported by Grosse *et al.* (1975) was not as severely affected and has a smaller segment duplicated (p14→pter) (Figs. 12b and c). This patient did not have dolicocephaly, a turtle-like mouth, low-set posteriorly rotated ears, congenital heart disease, joint deformities, or abnormal genitalia. Since this patient was not trisomic for the proximal portion of the short arm, that portion may be related to several of these anomalies.

## B. Dermatoglyphics

Most of the patients were reported as having abnormal dermatoglyphics but no consistent pattern has been found.

## C. Cytogenetics

Ten patients from seven unrelated families were studied. All resulted from familial balanced translocations (Table V). The break point in chromosome 10 in eight of the ten cases was at p11, making the patient trisomic for the entire short arm. Insley's patient had the break point at q11. The case of Grosse *et al.* (1975) had the break point at p14. In four of the seven families studied, the segment was translocated to the heterochromatic short arm of an acrocentric chromosome (Hustinx *et al.*, 1974; Nakagome and Kobayashi, 1975; Cantu *et al.*, 1975; Yunis *et al.*, 1976). One patient suffered a deficiency for 7q36 (Grosse *et al.*, 1975). The cases of Schleiermacher *et al.* (1974) involved a translocation to the telomere of the long arm of chromosome 7 which occurred either by telomeric fusion (Francke, 1972; Lejeune, 1973), or with a subsequent small but undectable telomeric deletion. Prenatal detection has been reported in one case (Nakagome and Kobayashi, 1975).

**TABLE V**
**Patients with Duplication 10p**

| Age | Death | Balanced translocation carrier | Patient | | Reference |
|---|---|---|---|---|---|
| | | | Duplication | Deletion | |
| | 8 months | 46,XX,t(10;15)(q11;q24) | q11→pter | None | Insley *et al*., 1968; Hirschhorn, *et al*., 1973 |
| | 101 days | 46,XX,t(10;22)(p11;p11) | p11→pter | 22(p12→pter) | Yanagisawa and Adachi, 1970; Nakagome and Kobayashi, 1975 |
| | 2 days | 46,XX,t(10;14)(p11;p11?) | p11→pter | 14(p12→pter) | Hustinx *et al*., 1974 |
| 10 years | | 46,XX,t(7;10)(p22;p11) | p11→pter | 7pter uncertain | Schleiermacher *et al*., 1975 Patient 1 |
| 12 years | | 46,XX,t(7;10)(p22;p11) | p11→pter | 7pter uncertain | Schleiermacher *et al*., 1975 Patient 2 |
| 8 years | | 46,XX,t(10;21)(p11;p11) | p11→pter | 21(p12→pter) | Cantu *et al*., 1975 Patient 1 |
| | 14 months | 46,XX,t(10;21)(p11;p11) | p11→pter | 21(p12→pter) | Cantu *et al*., 1975 Patient 2 |
| 4 9/12 years | | 46,XY,t(7;10)(q35;p14) | p14→pter | 7q36 | Grosse *et al*., 1975 |
| | Abortion | 46,XX,t(10;22)(p11;p11) | p11→pter | 22(p12→pter) | Nakagome and Kobayashi, 1975 |
| | Stillborn | 46,XY,t(10;21)(p11;p11) | p11→pter | 21(p11→pter) | Yunis *et al*., 1976 |

## D. Summary

The patients reported with duplication 10p appear to represent a syndrome with distinctive facial features, unusual flexion anomalies of upper and lower extremities, and abnormal genitalia. The most striking facial features include a prominent glabella and wide nasal root, a turtle-like mouth, long philtrum, and large, dysmorphic ears. The joint anomalies often include flexion deformity of the wrists and severe talipes equinovarus. Most patients also have either a small penis or cryptorchidism in males, or enlarged labia or clitoris in females. Of the known patients, all were severely affected, and half died early in life.

## ACKNOWLEDGMENT

This work was supported in part by NIH Grants No. HD01962 and GM05625.

## REFERENCES

Bühler, U. K., Bühler, E. M., Sartorius, J., and Stalder, G. R. (1967). *Helv. Paediatr. Acta* **1,** 41–53.

Cantu, J.-M., Salamanca, F., Buentello, L., Carnevale, A., and Armendares, S. (1975). *Ann. Genet.* **18,** 5–11.

de Grouchy, J., Finaz, C., Roubin, M., and Roy, J. (1972). *Ann. Genet.* **15,** 85–92.

Dutrillaux, B., Laurent, C., Robert, J. M., and Lejeune, J. (1973). *Cytogenet. Cell. Genet.* **12,** 245–253.

Forabosco, A., Bernasconi, S., Giovannelli, G., and Dutrillaux, B. (1975). *Helv. Paediatr. Acta* **30,** 289–295.

Francke, U. (1972). *Am. J. Hum. Genet.* **24,** 189–213.

Francke, U., Mahan, G. M., Dixson, B. K., and Jones, O. W. (1975). *Birth Defects, Orig. Artic. Ser.* **11,** No. 5.

Grosse, K.-P., Schwanitz, G., Singer, H., and Wieczorek, V. (1975). *Humangenetik* **29,** 141–144.

Hirschhorn, K., Lucas, M., and Wallace, I. (1973). *Ann. Hum. Genet.* **36,** 375–379.

Hustinx, T. W. J., ter Haar, B. G. A., Scheres, J. M. J. C., and Rutten, F. J. (1974). *Clin. Genet.* **6,** 408–415.

Insley, J., Rushton, D. I., and Everley Jones, H. W. (1968). *Ann. Genet.* **11,** 88–94.

Kajii, T., Ohama, K., Niikawa, N., Ferrier, A., and Avirachan, S. (1973). *Am. J. Hum. Genet.* **25,** 539–547.

Kersey, J. H., Yunis, J. J., Lee, J. C., and Bendel, R. P. (1971). *Lancet* **1,** 702.

Krøyer, S., and Niebuhr, E. (1975). *Ann. Genet.* **18,** 50–55.

Laurent, C., Bovier-Lapierre, M., and Dutrillaux, B. (1973). *Humangenetik* **18,** 321–327.

Lejeune, J. (1973). *Bull. Eur. Soc. Hum. Genet.* pp. 57–60.

Moreno-Fuenmayor, H., Zackai, E. H., Mellman, W. J., and Aronson, M. (1975). *Pediatrics* **56,** 756–761.

Mulcahy, M. T., Jenkyn, J., and Masters, P. L. (1974). *Clin. Genet.* **6,** 335–340.

Nakagome, Y., and Kobayashi, H. (1975). *J. Med. Genet.* **12,** 412–427.

Prieur, M., Forabosco, A., Dutrillaux, B., Laurent, C., Bernasconi, S., and Lejeune, J. (1975). *Ann. Genet.* **18,** 217–222.

Roux, C., Taillemite, J.-L., and Baheux-Morlier, G. (1974). *Ann. Genet.* **17,** 59–62.

Schleiermacher, E., Schliebitz, U., and Steffens, C. (1974). *Humangenetik* **23,** 163–172.

Shokeir, M. H. K., Ray, M., Hamerton, J. L., Bauder, F., and O'Brien, H. (1975). *J. Med. Genet.* **12,** 99–113.

Talvik, T., Mikelsaar, A.-V., Mikelsaar, R., Käosaar, M., and Tüür, S. (1973). *Humangenetik* **19,** 215–226.

Tsuchimoto, T., and Bühler, E. M. (1974). *4th Meet. Sect. Cytogenet., Soc. Anthropol. Hum. Genet., 1974.*

Yanagisawa, S., and Adachi, K. (1970). *Jpn. J. Hum. Genet.* **14,** 309–315.

Yunis, J. J., and Sanchez, O. (1974). *J. Pediatr.* **84,** 567–570.

Yunis, E. J., Silva, R., and Giraldo, A. (1976). *Ann. Genet.* **19,** 57–60.

# 8

# Abnormalities of Chromosomes 11 and 20

UTA FRANCKE



The rapidly accumulating knowledge about specific gene loci on the duplicated or deleted chromosomal segments has not yet helped our understanding of the complex pathogenetic mechanisms involved in the production of abnormal phenotypes by unbalanced karyotypes. Nevertheless, presentation of the current gene map of each chromosome, before discussion of the chromosomal syndromes, appears appropriate in view of such applications as the study of gene-dose relationships, deletion or duplication mapping, or mapping by cell hybridization using chromosomally abnormal human cells.

## I. CHROMOSOME 11

### A. Genetic Constitution

Assignments of gene loci to chromosome 11 were accomplished by studies of somatic cell hybrids. The gene for lactate dehydrogenase A(LDH A;EC 1.1.1.27) was the first one to be localized on chromosome 11 (Boone *et al.,* 1972). The genes for esterase $A_4$ ($EsA_4$;EC 3.1.1.1–6) (Shows, 1972) and lysosomal acid phosphatase ($AcP_2$; EC 3.1.3.2) (Bruns and Gerald, 1974) were shown to be syntenic with LDH A in man-mouse and man-Chinese hamster hybrids and were thus assigned to chromosome 11 indirectly.

The existence of a gene coding for an antigen which reacts with anti-human species antisera was initially suggested by Nabholz *et al.* (1969), and has been confirmed by Puck *et al.* (1971), using rabbit antihuman sera in a cytotoxicity assay. The so-called "lethal antigen" (AL) segregated concordantly with LDH A in man-hamster hybrids. Buck and Bodmer (1975) devised a more sensitive assay for detection of this antigen by raising an antiserum in a mouse strain against a man-mouse hybrid derived from the same mouse strain which contained a few human chromosomes including No. 11. The authors called the antigen human species antigen 1 (SA-1) and showed it to be syntenic with LDH A and chromosome 11. Recent evidence indicates that this antigen has several components which may be controlled by a cluster of tightly linked genes on chromosome 11 (Jones *et al.,* 1975).

Regional localization of these genes on chromosome 11 was achieved by detection of a spontaneous chromosome rearrangement in hybrids between thymidine kinase-deficient mouse ($3T3,TK^-$) cells and human t(11;17) translocation-containing fibroblasts. The rearranged marker chromosome consisted of the 11 short arm (11cen→11p15) and the TK-bearing portion of 17q, and could therefore be selected for, or against, by the appropriate selective media. In the absence of a normal chromosome 11, the concordant segregation of this marker chromosome with LDH A (Francke and Busby, 1975), $AcP_2$ (Busby *et al.,* 1976) and SA-1 (Buck *et al.,* 1976) places these gene loci on 11p. $EsA_4$ was not expressed in clones containing the marker chromosome, which may indicate the location of this gene on 11q (Busby *et al.,* 1976). Kucherlapati *et al.* (1974) have suggested that $EsA_4$ is not distal to 11q22.

Grzeschik (1975) reported cosegregation of LDH A with a translocation chromosome containing 11pter→11q13 which is consistent with our results. The provisional gene map of chromosome 11 is shown in Fig. 1.

## B. Partial Duplication 11p Syndrome

Three unrelated patients with duplication of nonidentical segments of the chromosome 11 short arm have been reported (Case I: Francke, 1972; Falk *et al.,* 1973; Case II: Sanchez *et al.,* 1974; Case III: Palmer *et al.,* 1976). All of them were unbalanced segregation products resulting from a balanced translocation in one parent. In two cases, the rearrangements involved three different chromosomes. Fig. 1 illustrates the extent of the duplicated segments. Cytogenetic and major clinical findings are summarized in Table 1.

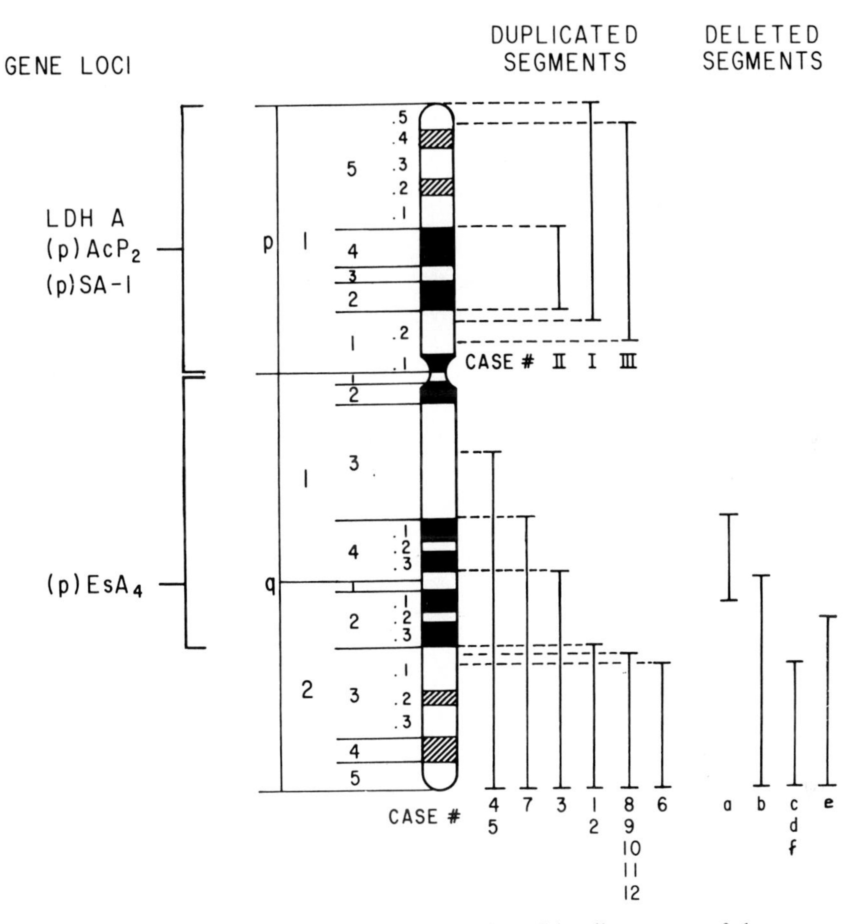


**Fig. 1.** Schematic representation of early metaphase G-banding pattern of chromosome 11. The numbering of bands is consistent with and expands the Paris Nomenclature. Duplicated and deleted segments in reported cases are indicated on the right. Gene assignments are listed on the left. For explanations of gene abbreviations, see text. (p) signifies the provisional nature of regional assignments.

The most obvious clinical manifestations upon comparison of the three cases are due to disturbance of craniofacial development (Fig. 2). Consistent findings are high prominent forehead with frontal upsweep of hair, flat supraorbital ridges and down-slanting palpebral fissures. Midline defects include delayed closure of the anterior fontanelle, wide glabella, broad flat nasal bridge, telecan-

**TABLE I**
**Cytogenetic and Clinical Data on 3 Reported Cases of Partial Duplication 11p Syndrome[a]**

| Case Number | I | II | III |
|---|---|---|---|
| References | Francke (Case 7) (1972); Falk *et al.* (1973) | Sanchez *et al.* (1974) | Palmer *et al.* (1976) |
| Chromosome banding method | Q | G | G,Q,R,C |
| Parental translocation | t(2;11)(q37;p11.2)pat | t(11;12;13)(p12 p15q14.1q23;p11 q24.1;q34)mat | t(3;11;20) (p13;p11;q13) mat |
| Duplicated segment | 11p11.2→11pter | 11p12→11p14 | 11p11→11p15 |
| Deleted segment | 2q37→2qter | – | – |
| Sex | M | F | F |
| Delivery | term | term | premature |
| Birth weight (gm) | 5,075 | . . . | 2,440 |
| Age at examination | 10 years | 3 years | 3 months |
| Growth retardation | + | + | + |
| Psychomotor retardation | profound | severe | . . . |
| Delayed closure of anterior fontanelle | . . . | + | + |
| Prominent forehead | + | + | + |
| Flat supraorbital ridges | – | + | + |
| Downslanting palpebral fissures | + | + | + |
| Epicanthal folds | + | – | + |
| Wide flat nasal bridge | + | + | + |
| Hypertelorism | – | + | + |
| Strabismus | + | + | + |
| Nystagmus | + | + | – |
| Palate | | | |
| High with bifid uvula | + | – | – |
| Cleft with cleft lip | – | + | + |
| Cryptorchidism | + | . . . | . . . |
| Broad fingers or toes | + | + | – |
| Hypotonic musculature | + | + | + |
| Spasticity of lower limbs | . . . | + | + |
| *Fig. 2:* | d,e | a,b | c |

[a]+, present; –, absent; . . ., not known or not applicable.

thus and/or hypertelorism, cleft palate, cleft lip and bifid uvula. The facies is further characterized by a broad short nose and round full cheeks.

Severe psychomotor retardation, strabismus, nystagmus, and generalized hypotonia with spasticity of the lower limbs indicate the presence of nervous system defects. Autopsy reports are not available as all three patients were alive

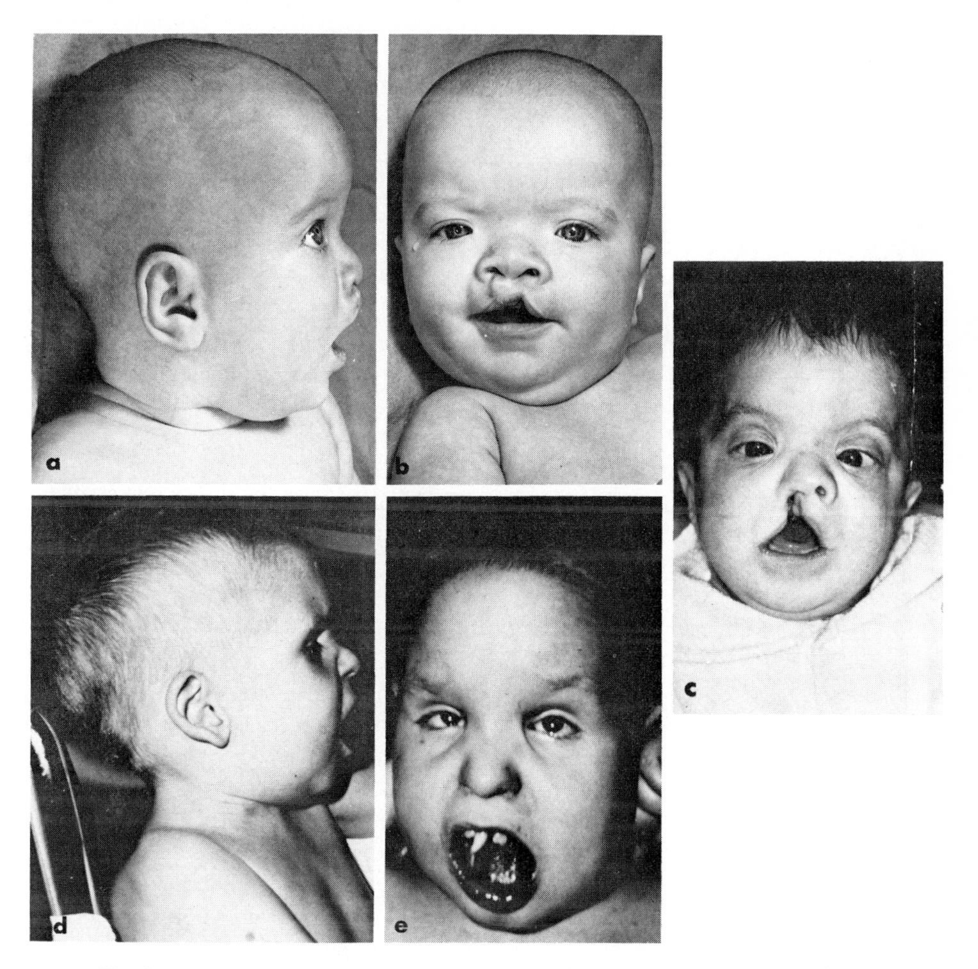


**Fig. 2.** Partial Dup11p syndrome. (a,b) Case II at 9 months, from Sanchez *et al.* (1974), (c) Case III at 3 months, from Palmer *et al.* (1976), (d,e) Case I at 9 years.

at the time of their report. It is possible that as yet unknown malformations of internal organs are part of the syndrome. Broad fingers or toes were mentioned in two cases. Dermatoglyphic findings were not consistent. The syndrome might be clinically recognizable based on the facies alone and should be considered in the differential diagnosis of patients with cleft lip and/or palate associated with developmental delay.

## C. Partial Duplication 11q Syndrome

Unbalanced karyotypes containing a partial duplication of the chromosome 11 long arm have been identified in twelve individuals from eleven presumably unrelated families in which balanced translocations involving a break in 11q were segregating. (Francke, 1972; Mann and Rafferty, 1972; Rott *et al.*, 1972; Tusques *et al.*, 1972; Jacobsen *et al.*, 1973; Laurent *et al.*, 1975; Aurias *et al.*, 1975; Giraud *et al.*, 1975; Francke *et al.*, 1976b). As depicted in Fig. 1, the extent of the duplicated segment varied in the different cases. Furthermore, in eight of them, the 11q duplication was associated with a terminal deletion of the recipient chromosome and of these, seven involved different chromosomal regions (Table II). The size of these deleted segments may also be variable, with only a small part of the telomeric region being involved in most cases. The translocations giving rise to Cases 1 and 3 are shown in Fig. 3. In Case 3, the small terminal 5p deletion in addition to the 11q duplication could account for the "cat cry" which was a presenting symptom in infancy (Mann and Rafferty, 1972).

In each of the four cases (Nos. 8, 10, 11 and 12) of pure 11q duplication without associated deletion, an apparently identical reciprocal translocation t(11;22) (q23.1;q11.1) was found in one of the parents. Because of the break point on 22q being located close to the centromere, this translocation can be expected to lead to unstable meiotic configurations. Nondisjunction involving the der(22) chromosome, interpreted as a 22 centromere with region 11q23.1→11qter attached to it, gave rise to the "47,+ small acrocentric" karyotypes found in the probands. The four families, all reported from France, were not known to be related. In two instances, the translocation was traced back over three and four generations (Aurias *et al.*, 1975); however, *de novo* occurrence of this t(11;22) translocation was not demonstrated in either one of the kindreds. Therefore, the number of original rearrangement events involving apparently identical regions of chromosomes 11 and 22 cannot be determined. Thus, these reports should not be taken as evidence for the existence of preferred sites of breakage and exchange between chromosomes 11 and 22, although it is intriguing to speculate on possible homologous regions in different chromosomes, which may give rise to the repeated occurrence of an identical translocation. As has been recently demonstrated by Allderdice and colleagues (1975) in the case of a pericentric inversion of chromosome 3, apparently unrelated families carrying identical chromosomal rearrangements may have a distant common ancestor.

The most consistent clinical findings of the twelve cases of chromosomally proven partial duplication 11q syndrome are listed in Table III. In addition, several kindreds contained one or more similarly affected members who were not karyotyped but are suspected to have had the same syndrome based on clinical history. These individuals are described in the following paragraphs since they are not included in Table III.

A maternal uncle of Case 1 was profoundly retarded and died at 9 years of age in an institution. At birth, he had micropenis and short tongue frenulum

**TABLE II**

Cytogenetic Findings in 12 Patients with Partial Duplication 11q syndrome

| Case Number | 1 | 2 | 3 | 4 | 5 | 6 |
|---|---|---|---|---|---|---|
| References | Francke *et al.* (1976b) | Francke (1972) (Case #9); Francke *et al.* (1976) | Mann and Rafferty (1972); Francke *et al.* (1976) | Rott *et al.* (1972) (Case IV-25) | Rott *et al.* (1972) (Case IV-26) | Jacobsen *et al.* (1973) (Case IV-4) |
| Banding method | Q,G | Q,G | Q,G | Q | Q | Q,G |
| Parental translocation | t(6;11) (q27;q23.1)mat | t(4;11) (q35;q23.1)mat | t(5;11) (p15;q21)mat | t(11;13)(q13;q32 to 34) pat | | t(11;21) q(23;q22)mat |
| Duplicated segment | 11q23.1→11qter | 11q23.1→11qter | 11q21→11qter | 11q13→11qter | | 11q23→11qter |
| Deleted segment | 6q27→6qter | 4q35→4qter | 5p15→5pter | 13q32 to 34→13qter | | 21q22→21qter |

| Case number | 7 | 8 | 9 | 10 | 11 | 12 |
|---|---|---|---|---|---|---|
| References | Tusques *et al.* (1972) | Laurent *et al.* (1975) (Case #1) | Laurent *et al.* (1975) (Case #2) | Aurias *et al.* (1975) (Case #1) | Aurias *et al.* (1975) (Case #2) | Giraud *et al.* (1975) |
| Banding method | R | R,T | R | R,T | R,G | R |
| Parental translocation | t(10;11) (q26;q13.3)pat | t(11;22) (q23.1;q11.1) mat | t(11;17) (q23.1;p13)mat | t(11;22) (q23.1;q11.1) mat | t(11;22) (q23.1;q11.1) mat | t(11;22) (q23;q11)mat |
| Duplicated segment | 11q13.3→11qter | 11q23.1→11qter 22pter→22q11.1 | 11q23.1→11qter | 11q23.1→11qter 22pter→22q11.1 | 11q23.1→11qter 22pter→22q11.1 | 11q23→11qter 22pter→22q11 |
| Deleted segment | 10q26→10qter | – | 17p13→17pter | – | – | – |

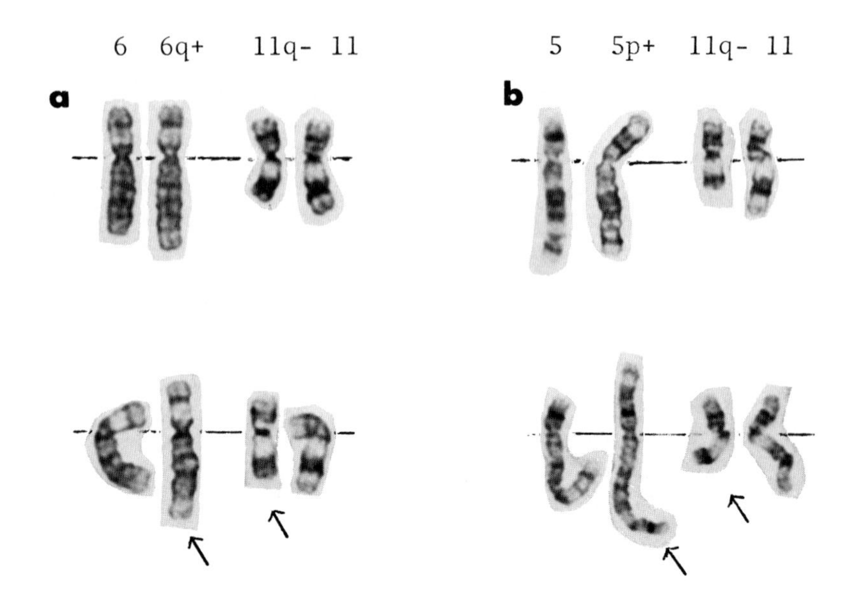


**Fig. 3.** Two sets each of G-banded partial karyotypes from two carriers of balanced reciprocal translocations involving 11q. (a) t(6;11)(q27;q23.1) in the mother of Case 1, (b) t(5;11)(p15;q21) in the mother of Case 3. Arrows indicate rearranged chromosomes.

exactly like the propositus. Similar facies, as is evident from family photographs, include small head, hypotelorism, short nose and elongated philtrum, low-set posteriorly rotated ears, retrognathia and retracted lower lip. At autopsy, slightly decreased brain size with thickened adherent areas of the dura and decreased numbers of pituitary cells were found.

An older brother of Case 3 was an underweight term baby with small low-set ears, a dimple below the lower lip, severe congenital heart defect and bilateral talipes equinovarus. He resembled Case 3 on photographs and had the same catlike cry. He failed to thrive and died at 1 month of age in congestive heart failure.

Cases 4 and 5 were siblings. An older sister (Case IV-23 in the report by Rott *et al.,* 1972) had died at 16 days from cyanotic congenital heart disease. She had dysplastic clavicles and meningomyelocele. Three additional children of translocation carriers from the same kindred died in infancy with malformations such as meningomyelocele, spina bifida, transposition of the great vessels, aplastic kidney or biliary atresia. Other significant findings were uterus unicornis, inguinal hernia and talipes.

Also not included in Tables II and III was a 980 gm stillborn male, product of a 33-week gestation, who had craniorachischisis (Wright *et al.,* 1974). At autopsy, the pituitary gland was hypoplastic and lacked eosinophilic cells. The published karyotype, interpreted in terms of the Paris Nomenclature, would most likely be 46,XY,der(6),t(6;11)(p25;q23)pat, the abnormality consisting of duplication of region 11q23→qter and deletion of the 6p telomere.

The major clinical findings in twelve live-born patients with partial 11q duplication are summarized in Table III and illustrated in Fig. 4. The following summary of the partial duplication 11q syndrome is based on the twelve cases in Table III, as well as on the probably likewise affected relatives described above.

Growth deficiency of prenatal onset, moderate to profound mental retardation and microcephaly were consistent manifestations. In the neonate, hypotonia of the trunk and hypertonia of the extremities, with abducted arms, flexed forearms and clenched fists, were associated with feeding difficulties due to inability to suck or swallow (Aurias and Laurent, 1975).

The facies were characterized by short nose with prominent tip, long philtrum and small mandible with retracted lower lip (Fig. 4a–n). Hypotelorism was present in Case 1 and in his deceased maternal uncle. Cases 4 and 5 had hypertelorism, while the interocular distances were apparently normal in the other patients. The external ears were large and often low-set or rotated backwards. They were characterized by poorly developed helices and prominent antihelices. Preauricular dimples or skin tags were common.

The palate was either cleft or described as high arched. Several cases presented with the complete Pierre Robin anomalad, consisting of microretrognathia, cleft posterior palate and glossoptosis.

Congenital defects of cardiac development, most often a septal defect and/or patent ductus arteriosus, were found in the majority of cases. The impression of wide-spaced nipples was reported in newborns with this syndrome (Cases 8–10). When internipple distances were measured in Cases 1 and 3, they were 20% and 22% of the chest circumferences respectively, which was in the normal range for age.

Inguinal or umbilical hernias, urinary tract abnormalities, and developmental defects of female internal genitalia were occasionally present, while common mesentery, aplastic hemidiaphragm and hypoplastic gallbladder were reported in single cases only.

Some skeletal defects were often seen such as dysplastic acetabulum with or without dislocated hip and talipes equinovarus, whereas kyphoscoliosis, radioulnar synostosis and absent fibulae represented isolated findings. A developmental defect of one or both clavicles was demonstrated in four patients. Cases 1 and 3 lacked fusion of the medial and lateral portions of the right clavicle, and two sibs reported by Rott *et al.* (1972) had dysplastic clavicles similar to the findings in cleidocranial dysostosis.

Micropenis without hypospadias (Fig. 4c) was present in six of seven males,

**TABLE III**
**Clinical Findings in 12 Cases of Partial Duplication 11q Syndrome**[a,b]

| Case Number | 1 | 2 | 3 | 4 | 5 | 6 | 7 | 8 | 9 | 10 | 11 | 12 |
|---|---|---|---|---|---|---|---|---|---|---|---|---|
| Sex | M | F | F | M | F | M | F | M | M | M | F | F |
| Gestation | term | term | term | 38 weeks | term | 37 weeks | 42 weeks | 42 weeks | term | term | term | 8 months |
| Birth weight (gm) | 2,300 | 2,600 | 2,250 | 1,800 | 2,380 | 2,250 | 2,380 | 2,500 | 3,040 | 2,330 | 2,350 | 2,350 |
| Birth length (cm) | 48 | 49 | . . . | 46 | 48 | 50 | 46 | 48 | 49 | 46 | 48 | 47 |
| Age at death | alive at 3 years | 5 weeks | alive at 6 years | 10 months | 1 week | alive at 11 years | 4 days | alive at 2 months | alive at 2 years | alive at 1 month | at birth | 24 hours |
| Growth retardation | + | . . . | + | + | . . . | + | . . . | + | – | . . . | . . . | . . . |
| Psychomotor retardation | severe | . . . | profound | + | . . . | IQ 65 | . . . | + | IQ 50 | . . . | . . . | . . . |
| Microcephalus | + | + | + | . . . | + | – | + | + | + | + | . . . | – |
| Short nose/long philtrum | + | . . . | + | + | + | + | + | + | + | + | . . . | + |
| Ears | | | | | | | | | | | | |
| Low-set | – | + | + | + | + | – | + | . . . | + | + | + | + |
| Posteriorly rotated | + | . . . | – | . . . | . . . | + | – | . . . | . . . | . . . | . . . | . . . |
| Prominent antihelix | + | . . . | + | . . . | . . . | – | + | + | + | + | . . . | + |

| | | | | | | | | | | | | |
|---|---|---|---|---|---|---|---|---|---|---|---|---|
| Palate | | | | | | | | | | | | |
| High arched | + | – | + | . . . | . . . | + | – | – | + | – | – | – |
| Cleft | – | + | – | – | – | – | – | + | – | + | + | + |
| Microretrognathia | + | + | + | + | + | + | + | + | + | + | + | + |
| Retracted lower lip | + | . . . | – | + | + | – | + | + | + | + | + | + |
| Clavicular defect | + | . . . | + | . . . | + | . . . | . . . | . . . | . . . | . . . | . . . | . . . |
| Cardiac defect | + | + | + | + | + | . . . | + | + | . . . | + | + | + |
| Urinary tract malformation | – | + | . . . | + | – | . . . | – | . . . | + | . . . | + | + |
| Micropenis | + | . . . | . . . | + | . . . | – | . . . | + | + | + | . . . | . . . |
| Hip dislocation/dysplastic acetabulum | + | + | + | . . . | . . . | – | + | . . . | . . . | . . . | . . . | . . . |
| Talipes equinovarus | + | . . . | – | . . . | . . . | – | . . . | . . . | + | + | + | . . . |
| Muscle tone | | | | | | | | | | | | |
| Hypotonia (central) | + | . . . | + | – | . . . | . . . | + | . . . | . . . | . . . | . . . | . . . |
| Hypertonia (peripheral) | + | . . . | – | + | . . . | . . . | + | + | . . . | + | . . . | . . . |
| Cutis laxa | – | – | + | + | . . . | – | + | + | . . . | + | . . . | + |
| Fig. 4 | a,b,c | – | d,e | h | i | f,g | k | m,n | – | j,l | – | – |

[a]For references, see Table II.
[b]+, present; –, absent; . . ., not applicable or not known.

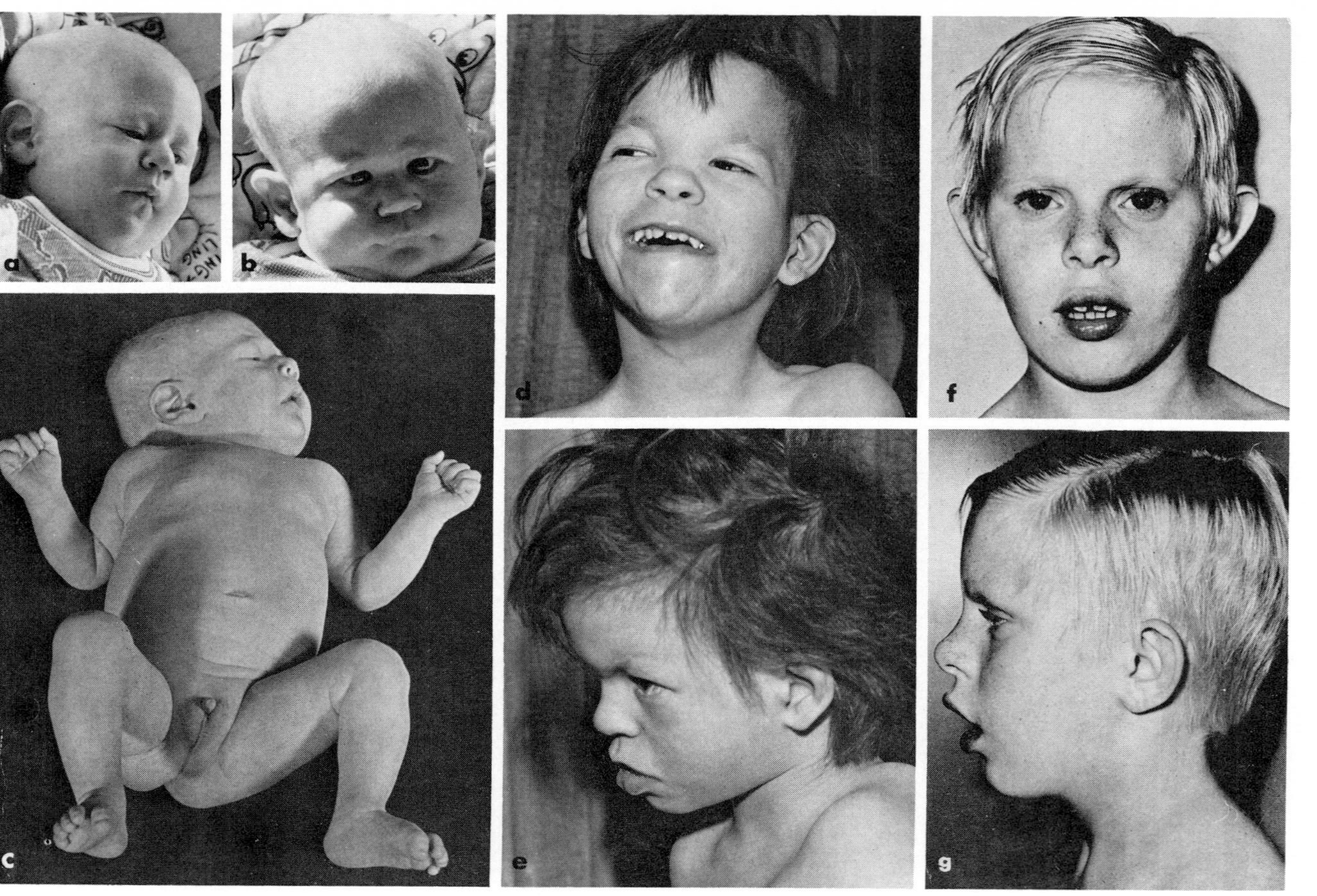


Fig. 4.

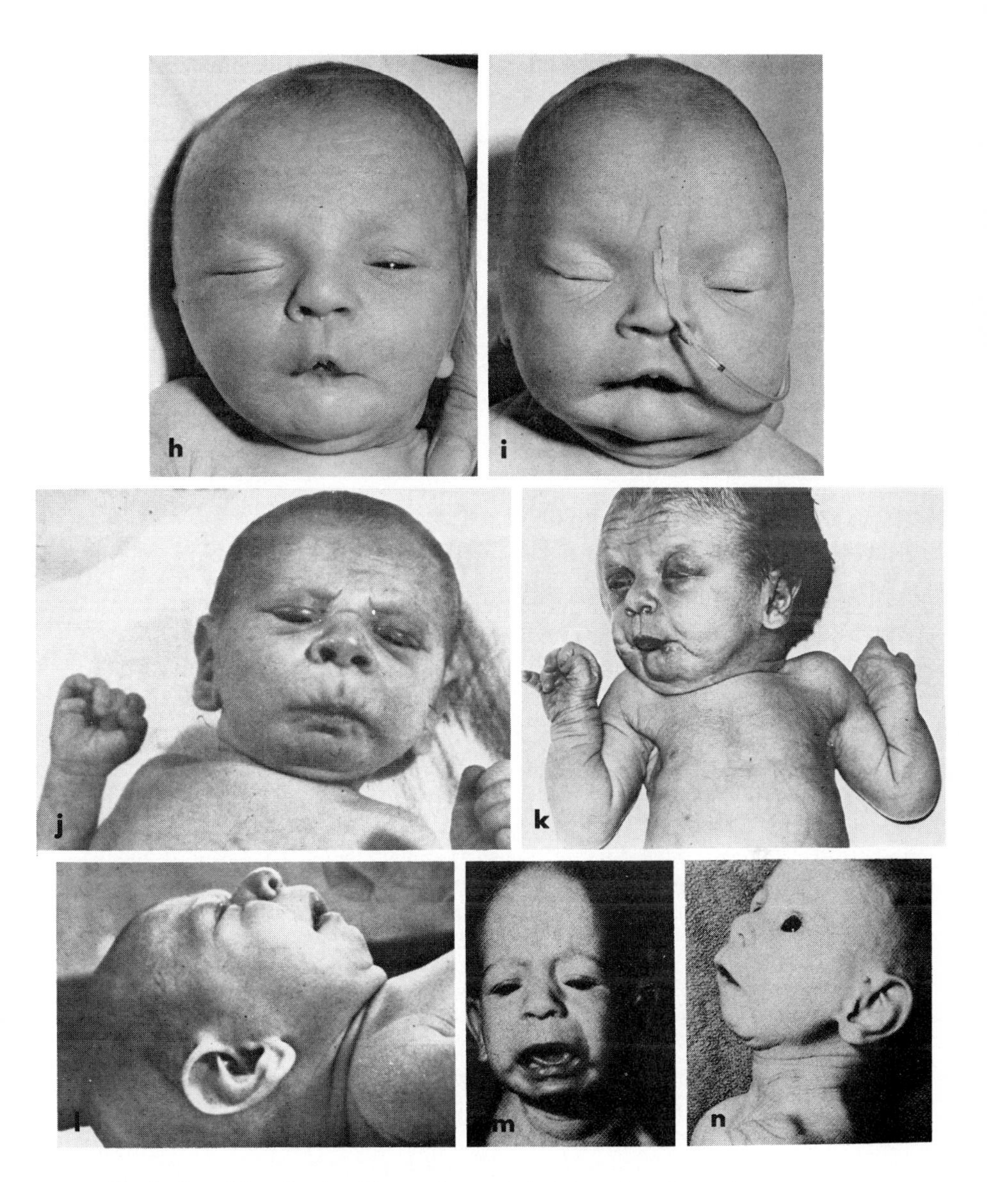


**Fig. 4.** Partial Duplication 11q syndrome. (a,b,c) Case 1 at 6 months, (d,e) Case 3 at 6 years, (f,g) case 6 at 11 years, from Jacobsen *et al.*, (1973), (h) Case 4 at 12 days, (i) Case 5 at 2 days, from Rott *et al.* (1972), (j,l) Case 10 at 1 month, from Aurias *et al.* (1975), (k) Case 7 at birth, from Tusques *et al.* (1972), (m,n) Case 8, from Laurent *et al.* (1975).

including the maternal uncle of Case 1. Normal response to testosterone, administered orally in Cases 8 and 9 and locally in Case 1, was documented by accelerated penile growth. Case 6 was reported to have normal male genitalia at age 11 years. No information was given about the size of his penis at birth or any previous hormone treatment.

Central nervous system malformations were variable. Neural tube closure defects, such as craniorachischisis and meningomyelocele, were reported once each. Agenesis of the corpus callosum was found in two autopsied cases, and decreased cellular elements in the pituitary gland in two others.

Dermatoglyphics were not unusual except for an increased distance between the palmar triradii *a* and *b,* at the bases of the second and third digits, due to radial displacement of *a.* This has been documented in Cases 1, 3, 7 and 10.

The combination of retracted lower lip, dysplastic clavicle with nonfused medial and lateral segments, and micropenis in the male is most unusual and may become the hallmark of the dup(11)q syndrome.

## D. Partial Deletion 11q

Five individuals with a partial deletion of 11q have been reported (Jacobsen *et al.,* 1973; Faust *et al.,* 1974; Taillemite, *et al.,* 1975; Linarelli *et al.,* 1975; and one was recently identified by us; Table IV). Cases a, b, e and f were sporadic, and Cases c and d were children of balanced t(11;21) translocation carriers from the same kindred. Having inherited the der(11),t(11;21) derivative chromosome, they presumably had a duplication of the telomeric region of 21q as well. Despite this difference and the variation in size and location of the deleted 11q segment, similarities of clinical manifestations were evident.

Significant craniofacial features were prominent forehead (scaphocephaly), epicanthus, broad nasal bridge, short nose, short and poorly designed philtrum with thin upper lip and downturned angles of the mouth (Table IV, Fig. 5). A hypoplastic mandible was present in five of the six patients. Cases c, d, e and f had a prominent metopic suture giving rise to a keel-like forehead similar to the one described in the deletion 9p syndrome (Alfi *et al.,* 1974). Cases e and f had unilateral coloboma of the iris.

Cardiac defects were present in four cases and led to early death in two of them. Irregular shortness of digits was noted in three cases. The degree of psychomotor retardation was quite variable. Case b, who was apparently monosomic for the largest segment of 11q, was the least retarded. Speech was more significantly delayed than other areas of development.

The information from these cases does not yet permit the delineation of a clinically identifiable syndrome. However, the facies appear similar enough to raise suspicion in the clinician faced with such a patient. As in all these "new" chromosomal disorders, the diagnosis lastly depends on cytogenetic analysis using high resolution banding methods.

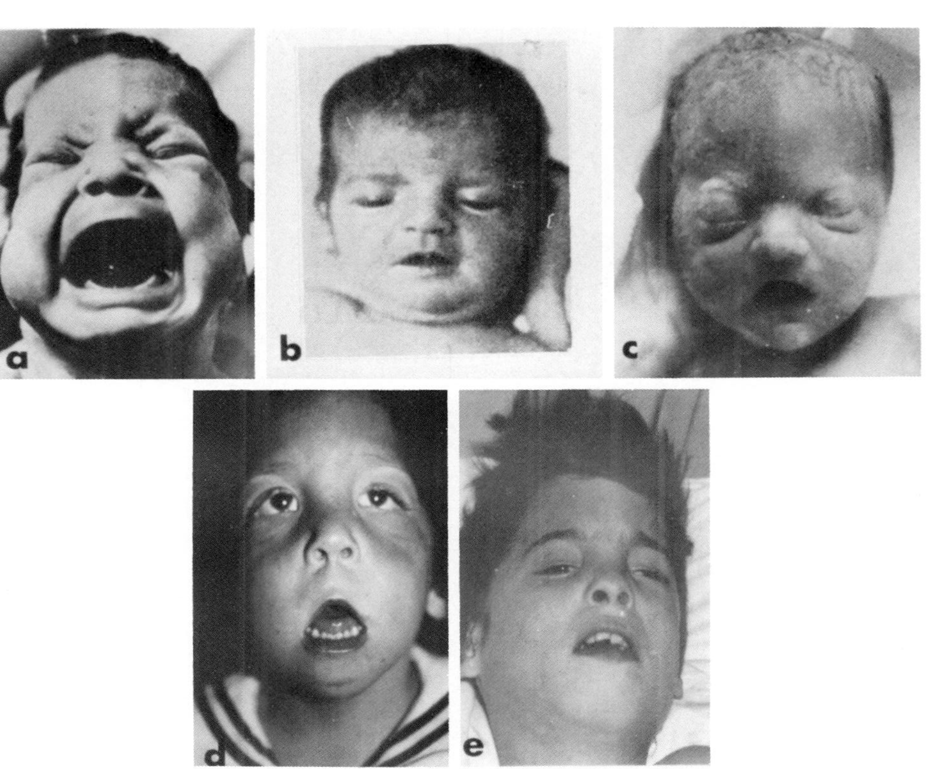


**Fig. 5.** Partial deletion 11q syndrome. (a) Case c, and (b) Case d, from Jacobsen *et al.* (1973), (c) Case a, from Taillemite *et al.* (1975), (d) Case f, (e), Case e.

**TABLE IV**
**Cytogenetic and Clinical Data on 6 Cases of Partial Deletion 11q[a]**

| Case Number | a | b | c | d | e | f |
|---|---|---|---|---|---|---|
| References | Taillemite *et al.* (1975) | Faust *et al.* (1974) | Jacobsen *et al.* (1973) (Case 1) | Jacobsen *et al.* (1973) (Case 2) | Linarelli *et al.* (1975) | Francke *et al.* (1976c) |
| Chromosome banding method | R,G | Q,G,R | G,Q | | G | G |
| Parental translocation | . . . | . . . | t(11;21)(q23;q22)pat | | . . . | . . . |
| Deleted segment | 11q14.1→ 11q22.1 | 11q21→ 11qter | 11q23→ 11qter | . . . | 11q22→11qter | 11q23→11qter |
| Sex | F | F | F | F | M | F |
| Gestation | term | term | 35 weeks | term | 38 weeks | 42 weeks |
| Birth weight (gm) | 3,420 | <2,000 | 2,000 | 1,700 | 2,353 | 3,840 |
| Birth length (cm) | 53 | . . . | 46 | 43 | . . . | . . . |
| Age at death | alive at 23 months | alive at 9 years | 19 months | 28 hours | alive at 12 years | alive at 3 years |
| Growth retardation | + | + | + | + | + | + |
| Psychomotor retardation | moderate | mild/moderate | severe | . . . | profound | moderate |

| | | | | | | |
|---|---|---|---|---|---|---|
| Scaphocephaly | – | + | + | + | + | + |
| Large square skull | + | – | – | – | – | – |
| Hypertelorism | + | + | + | + | – | – |
| Epicanthus | – | . . . | + | + | + | + |
| Palpebral fissures | | | | | | |
| Downslanting | + | – | – | – | + | + |
| Upslanting | – | – | + | + | – | – |
| Short nose | + | – | + | + | + | + |
| Short philtrum | + | + | + | + | + | + |
| Thin upper lip | + | + | . . . | . . . | . . . | + |
| Microretrognathia | + | – | + | + | + | + |
| External ears | | | | | | |
| Low-set | . . . | . . . | + | + | + | + |
| Malformed | + | . . . | + | + | + | + |
| Transverse palmar creases | + | . . . | + | + | + | + |
| Short fingers | . . . | . . . | + | . . . | + | + |
| Hypertonicity of limbs | + | . . . | + | – | . . . | – |
| Figure no. | 5b | – | 5a | 5c | 5e | 5d |

[a]+, present;–, absent; . . ., no information available.

## II. CHROMOSOME 20

### A. Genetic Constitution

Adenosine deaminase (ADA; EC 3.5.4.4) exists as multiple tissue-specific isoenzymes probably controlled by a single gene locus (Hirschhorn *et al.*, 1973), which has been assigned to chromosome 20 in man-mouse hybrids (Tischfield *et al.*, 1974). Complete ADA deficiency has been found in a number of children with combined immunodeficiency disease (Giblett *et al.*, 1972). The exact pathogenetic relationship between enzyme defect and clinical manifestations has not been elucidated, although there is evidence that excess adenosine, accumulating due to ADA deficiency, may have a detrimental effect on cells (Green and Chan, 1974).

The enzyme that converts desmosterol to cholesterol (D-C-E) has been provisionally assigned to chromosome 20 by Croce *et al.* (1974), because restoration of D-C-E activity in hybrids between mouse L cells, which lack this enzymatic activity, and human cells was dependent on the presence of human chromosome 20.

Inosine triphosphatase (ITP;EC 3.6.1.19) was found to cosegregate with ADA in man-hamster (Meera Khan *et al.*, 1976) and in man–mouse hybrids (Hopkinson *et al.*, 1976) indicating that the gene for ITP may be placed on chromosome 20.

No regional assignments on chromosome 20 have been accomplished as yet. We have studied levels of ADA activity in erythrocytes from patients with the duplication 20p syndrome from three different families (Cases 1–5 and 10–12 in Table V). The range of activity in controls was wide, some dup(20)p individuals had levels in the high normal range, and others had levels close to the mean (D. L. George and U. Francke, unpublished data). It is uncertain whether these results indicate location of the ADA gene on the long arm of chromosome 20, or whether levels of this enzyme are regulated independently from the gene-dose. ADA levels in individuals with trisomy 20 and dup(20)q need to be studied to distinguish between these possibilities.

Fig. 6 summarizes the gene loci on chromosome 20 and indicates the extent of the duplicated segments in 12 individuals with dup (20)p from six different families.

### B. Partial Duplication 20p

The clinical manifestations in 12 individuals identified as partially trisomic for 20p are compared in Table V. (Carrel *et al.*, 1971; Francke, 1972; Súbrt and Brýchnâc, 1974; Cohen *et al.*, 1975; Taylor *et al.*, 1976; Centerwall and Francke, 1976). It is important to note that most of them appeared to be normal at birth. They had normal prenatal and postnatal growth patterns which

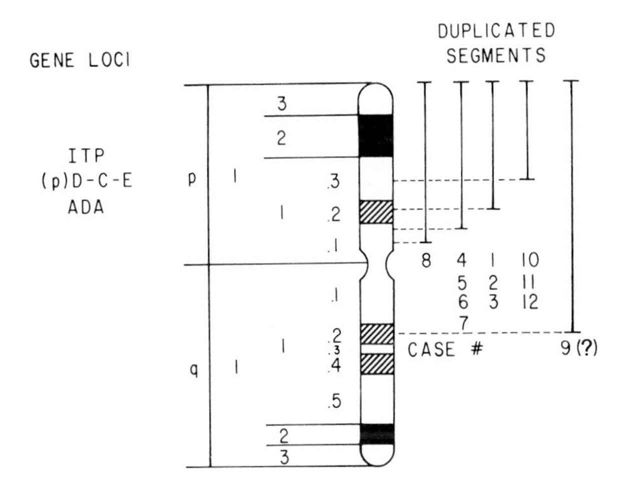


Fig. 6. Schematic representation of early metaphase G-banding pattern of chromosome 20. The numbering of bands is consistent with and expands the Paris Nomenclature. Gene loci are indicated on the left. For explanation of abbreviations, see text. (p) signifies provisional nature of gene assignment. Duplicated segments in reported cases are indicated on the right. In Cases 1 through 5, and 10 to 12, the break points were determined exactly by studying prophase cells, while in Cases 6 through 9 the break points were estimated from published karyotypes.

is unusual in chromosomal imbalance syndromes. For example, Case 4, an 18-year-old male, and Case 1, a 13-year-old female, were 171 cm and 166 cm tall, respectively, and had undergone normal pubertal development.

Psychomotor retardation was in the moderate range with IQ values around 50 in eight of eleven cases.

The phenotypic effects of this chromosomal imbalance appear to be rather variable. Close resemblance between affected members of the same sibship (e.g. Cases 1, 2, 3 and Cases 4 and 5) contrasts with great variability between patients from different families (Fig. 7). This could be due to slight differences in the extent of the duplicated segments of 20p as well as to the associated chromosomal deletions involving different regions such as 13q terminal, 18q terminal, 22q terminal (Fig. 8). On the other hand, the phenotypic effect of the dup(20)p abnormality might be modified by different genetic backgrounds. For example, Case 4 resembles in many aspects his father and unaffected brother more than his affected sister.

Nevertheless, the facies of nine of the 11 living patients, as compared in Fig. 7, contain similar features. The faces are round due to prominent cheeks and

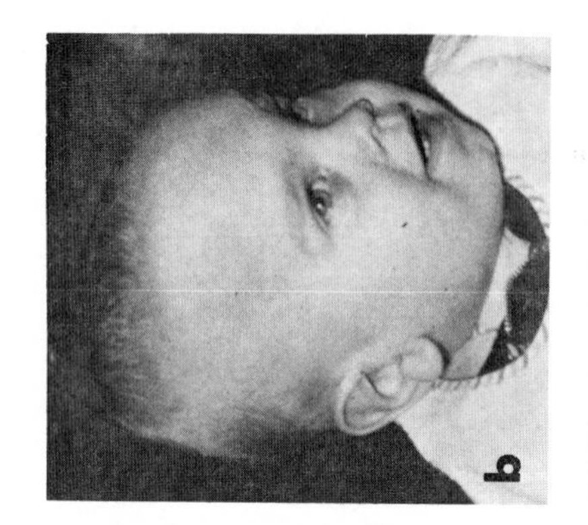
b

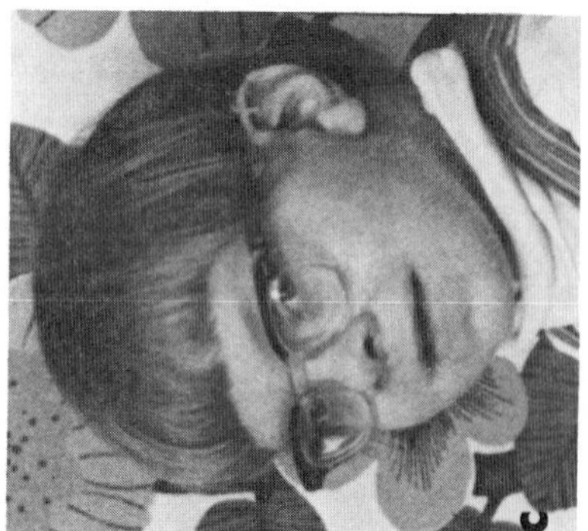
c

a

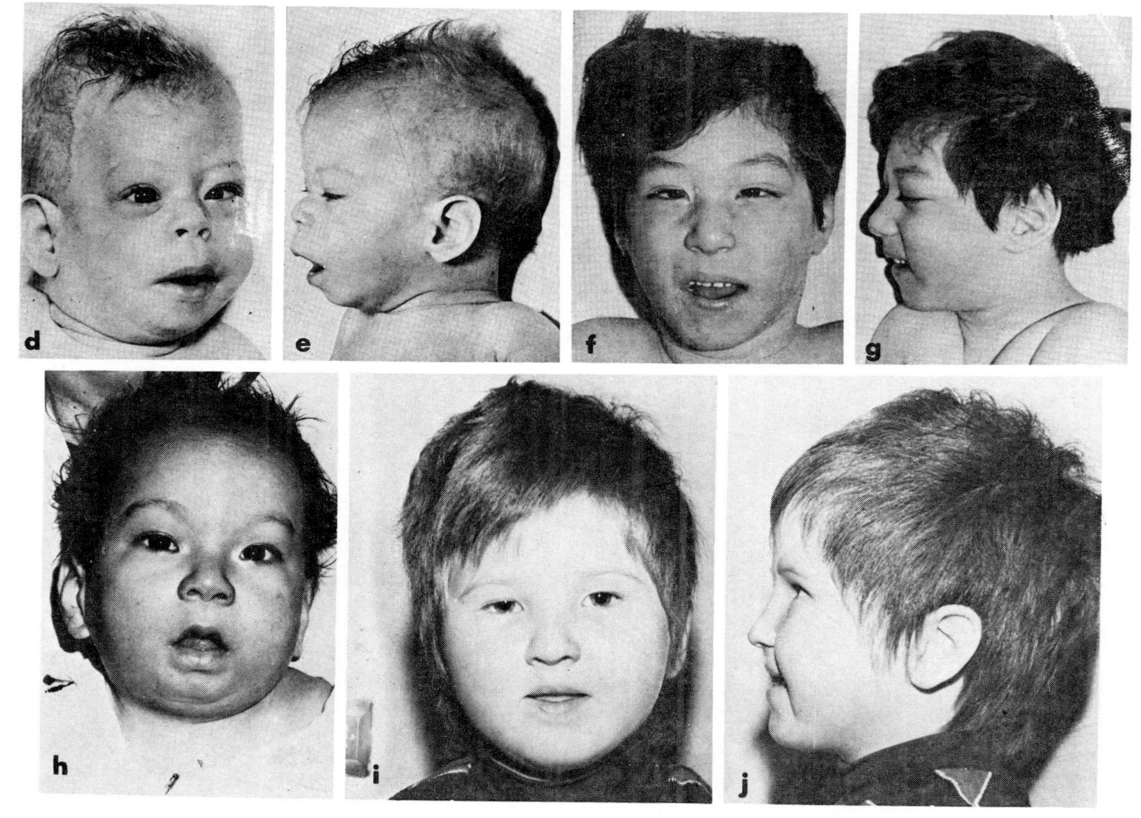


**Fig. 7.** Partial Duplication 20p. (a) Case 1 at 12 years (left), Case 2 at 11 years (right), Case 3 at 8 years (front), mother of Cases 1–3 (back), (b) Case 4 at 1 year, (c) Case 5 at 4 years, (d, ) Case 6 at 6 months and (f,g) Case 7 at 20 months, both from Cohen *et al.* (1975). (h) Case 9 at 7 months, from Krr potic *et al.* (1971), (i,j) Case 8 at 7 years, from Súbrt and Brýchnâc (1974).

**TABLE V**
**Cytogenetic and Clinical Findings in 10 Cases with Partial Duplication of the Short Arm of Chromosome 20 from 6 Unrelated Families**[a]

| Case | 1 | 2 | 3 | 4 | 5 |
|---|---|---|---|---|---|
| References | Carrel *et al.* (1971) Francke (case 11) (1972) Centerwall and Francke (1976) | | | Taylor *et al.* (1972) Taylor *et al.* (1976) (Case I) | (Case II) |
| Chromosome banding method | Q,G | | | G | |
| Parental translocation | t(13;20)(q34;p11.2) mat | | | t(18;20)(q23;p11.1) mat | |
| Duplicated segment | 20p11.2→20pter | | | 20p11.1→20pter | |
| Deleted segment | 13q34→13qter | | | 18q23→18qter | |
| Sex | F | F | M | M | F |
| Gestation | term | 33 weeks | term | term | term |
| Birth weight (gm) | 2,990 | 2,440 | 3,500 | 3,345 | 3,884 |
| Birth height (cm) | ... | ... | ... | 53.3 | 49.5 |
| Age at examination | 13 years | 12 years | 9 years | 18 years | 3 years |
| Psychomotor retardation | mod | mod | mod | mod | mod |
| Normal growth | + | + | + | + | + |
| Occipital flattening | + | + | + | – | – |
| Round face due to prominent cheeks and short chin | + | + | + | – | – |
| Oblique upward slanting palpebral fissures | + | + | + | – | – |
| Increased innercanthal distance | + | + | + | + | + |
| Epicanthal folds | – | – | – | – | – |
| Strabismus | – | – | – | + | + |
| Short nose with upturned tip and large nostrils | – | + | + | + | + |
| Dental abnormalities | + | + | + | + | + |
| Vertebral abnormalities | + | + | + | + | + |
| Cardiac abnormalities | – | + | + | – | – |
| Coarse hair | + | + | + | + | – |
| Poor coordination | + | + | + | + | + |
| Speech impediment | + | + | + | + | + |
| Fig. 7 | a | a | a | b | c |

*continued*

**TABLE V (*continued*)**

| Case | 6 | 7 | 8 | 9 |
|---|---|---|---|---|
| References | Cohen *et al.* (1975) (Case 1) | (Case 2) | Súbrt and Brýchnâc (1974) | Krmpotic *et al.* (1971) |
| Chromosome banding method | G | | G | Q |
| Parental translocation | t(20;22)(p11;q13) pat | | t(20;21) (p11;p11) mat | t(14q−;20q+) pat |
| Duplicated segment | 20p11→20pter | | 20p11→20pter | 20p?→20pter |
| Deleted segment | 22q13→22qter | | 21p11→21pter | – |
| Sex | F | M | F | M |
| Gestation | 7 months | term | term | term |
| Birth weight (gm) | 2,840 | 2,950 | 4,300 | 2,270 |
| Birth height (cm) | . . . | . . . | 54 | . . . |
| Age at examination | 8 months | 20 months | 5 years | 7 months |
| Psychomotor retardation | mod | severe | mild | mod |
| Normal growth | + | + | . . . | + |
| Occipital flattening | – | – | + | – |
| Round face due to prominent cheeks and short chin | + | + | + | + |
| Oblique upward slanting palpebral fissures | + | + | + | + |
| Increased innercanthal distance | . . . | . . . | + | . . . |
| Epicanthal folds | – | + | + | – |
| Strabismus | + | + | – | – |
| Short nose with upturned tip and large nostrils | + | – | – | + |
| Dental abnormalities | . . . | . . . | + | . . . |
| Vertebral abnormalities | . . . | . . . | + | . . . |
| Cardiac abnormalities | + | – | – | – |
| Coarse hair | . . . | . . . | + | . . . |
| Poor coordination | . . . | + | . . . | . . . |
| Speech impediment | . . . | . . . | – | . . . |
| Fig. 7 | d,e | f,g | i,j | h |

*continued*

TABLE V (*continued*)

| Case | 10 | 11 | 12 |
|---|---|---|---|
| References | Francke *et al.* (1976a) | Francke *et al.* (1976a) | Francke *et al.* (1976a) |
| Chromosome banding method | G | G | G,R |
| Parental translocation | t(20;21) (p11.3;p11) | t(20;21) (p11.3;p11) | t(20;21) (p11.3;p11) |
| Duplicated segment | mat | mat | mat |
| Deleted segment | 20p11.3→20pter 21p11→21pter | 20p11.3→20pter 21p11→21pter | 20p11.3→20pter 21p11→21pter |
| Sex | F | M | M |
| Gestation | 35 weeks | term | term |
| Birth weight (gm) | 2,100 | 3,375 | 3,500 |
| Birth height (cm) | 42 | . . . | . . . |
| Age at examination | died at 16 hours | 19 years | 45 years |
| Psychomotor retardation | – | mod | severe |
| Normal growth | – | + | + |
| Occipital flattening | – | + | – |
| Round face due to prominent cheeks and short chin | + | + | – |
| Oblique upward slanting palpebral fissures | + | – | + |
| Increased innercanthal distance | – | – | – |
| Epicanthal folds | + | – | – |
| Strabismus | – | + | – |
| Short nose with upturned tip and large nostrils | – | + | – |
| Dental abnormalities | . . . | – | + |
| Vertebral abnormalities | – | + | + |
| Cardiac abnormalities | + | – | – |
| Coarse hair | . . . | + | + |
| Poor coordination | . . . | + | . . . |
| Speech impediment | . . . | + | + |
| Fig. 7 | – | – | – |

[a]+, present; –, absent; . . ., not applicable or not known; mod, moderate.

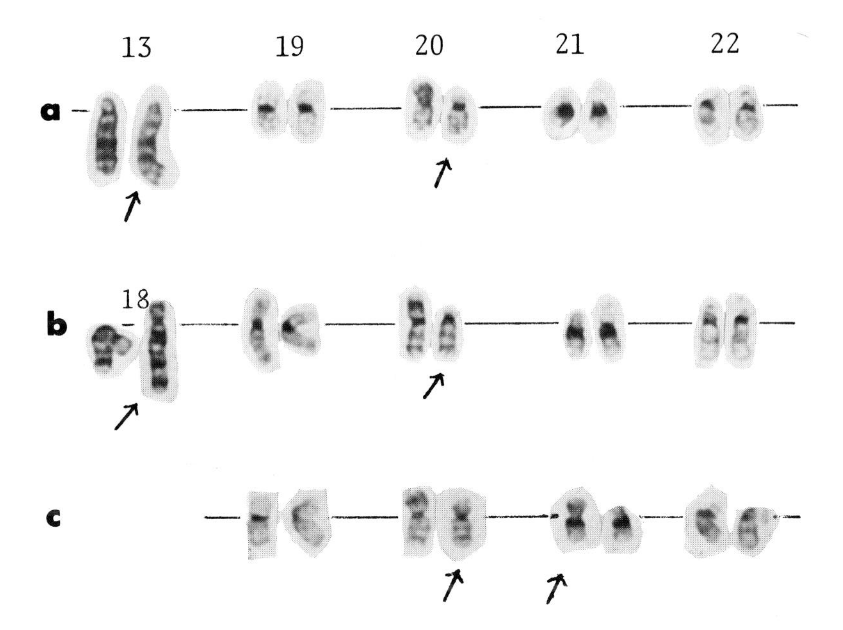


**Fig. 8.** Partial G-banded karyotypes of three carriers of balanced reciprocal translocations involving 20p. (a) t(13;20)(q34;p11.2) in the mother of Cases 1–3, (b) t(18; 20)(q23;p11.1) in the mother of Cases 4 and 5, (c) t(20;21)(p11.3;p11) in the mother of Case 10. Arrows indicate rearranged chromosomes.

short chin. The innercanthal distances are increased. The palpebral fissures are unusual with downward displacement of the medial corner and upward slant of the lateral part without giving a typical "mongoloid" appearance. The short noses have broad upturned tips and large nares. Dental abnormalities and coarse hair were rather consistent findings.

Vertebral abnormalities ranged from enlargement of cervical vertebrae in Case 5, irregularities in thoracic vertebrae in Cases 3 and 4, narrowed intervertebral disc in Case 11, to vertebral fusions in Cases 1, 2 and 4, and severe spondylar dysplasia in Cases 8 and 12. The cardiac abnormalities in Cases 2 and 3 were apparently mild without clinical significance. Coordination difficulties were paired with gross-motor clumsiness and occasional fine tremors. No evidence for a CNS malformation was found in Case 4 by EEG, PEG, brainscan, vertebral and carotid angiogram.

The speech defect in Cases 1 to 3 was due to specific articulation problems (Centerwall and Francke, 1976) with normal hearing, while in Cases 4 and 5, it may have been related to hearing deficit due to atresia of external ear canals.

Associated monosomy—although microscopically undetectable—for the distal 18q region in these patients could be responsible for the ear malformation.

Ocular abnormalities were inconsistent. Case 1 had bilateral coloboma of the iris. Cases 4 and 5 had myopia and exotropia. Cases 6 and 7 had strabismus.

Cases 1 to 3 were reported by Carrel *et al.* (1971) to have wide-spaced nipples, cubitus valgus, hypotonia and hypermobile joints. When reexamined 6 years later, the last three features were no longer present, and internipple distances expressed as percent of chest circumference were in the normal range.

Genito-urinary tract abnormalities were rare. Case 4 had unilateral hydronephrosis with duplicated collecting system and hypopadias. Case 10 presented with the "oligohydramnios tetrad" due to bilateral polycystic kidneys (Potter type II) and died shortly after birth (Francke *et al.,* 1976a). The renal malformation might have been coincidental and unrelated to the chromosomal abnormality. Cases 11 and 12, uncle and great-uncle of Case 10, who had the same duplication of the distal half of 20p, had no apparent renal abnormalities.

There were additional retarded individuals in most kindreds who phenotypically resembled the affected members listed in Table V, but whose chromosomes were not examined. The t(13;20) family contained two such females, one in the mother's, and one in the patient's sibship (Carrel *et al.,* 1971; Centerwall and Francke, 1976). Both died from congenital heart disease at the ages of 8 years and 4 months, respectively.

A maternal uncle of Cases 4 and 5 had rectal prolapse and umbilical hernia at birth. He is said to have had severe psychomotor retardation and hearing deficit of unknown origin and to have resembled Case 5. When he died from pneumonia at 13 years, no autopsy was performed (Taylor *et al.,* 1976).

In the family reported by Cohen *et al.* (1975), at least two additional individuals may have had partial duplication 20p.

In summary, the clinical manifestations associated with partial duplication 20p include round face, short upward slanting palpebral fissures, with downward displacement of the medial corner, vertebral irregularities or fusions, dental caries and coarse hair, moderate mental retardation, and normal physical growth and development. If significant cardiac or renal malformations are absent, lifespan may not be reduced.

The physical abnormalities are highly modified by the genetic background, with similar manifestations in affected sibs. Again, the diagnosis of this syndrome can only be suspected on clinical grounds alone and requires cytogenetic confirmation.

## ACKNOWLEDGMENTS

The competent technical assistance of M. G. Brown, and the typing of the manuscript by S. Benson are gratefully acknowledged.

This research was supported in part by the United States Public Health Service research grants GM-21110 and GM-17702, and a grant from the National Foundation, March of Dimes.

## REFERENCES

Alfi, O. S., Sanger, R. G., Sweeny, A. E., and Donnell, G. N. (1974). *Birth Defects, Orig. Artic. Ser.* **10,** No. 8, 27–34.

Allderdice, P. W., Browne, N., and Murphy, D. P. (1975). *Am. J. Hum. Genet.* **27,** 699–718.

Aurias, A., and Laurent, C. (1975). *Ann. Genet.* **18,** 189–191.

Aurias, A., Turc, C., Michiels, Y., Sinet, P.-M., Graveleau, D., and Lejeune, J. (1975). *Ann. Genet.* **18,** 185–188.

Boone C., Chen, T. R., and Ruddle, F. H. (1972). *Proc. Natl. Acad. Sci. U.S.A.* **69,** 510–514.

Bruns, G. A. P., and Gerald, P. S. (1974). *Science* **184,** 480–482.

Buck, D. W., and Bodmer, W. F. (1975). *Birth Defects, Orig. Artic. Ser.* **11,** No. 3, 87–89.

Buck, D. W., Bodmer, W. F., Bobrow, M., and Francke, U. *Birth Defects, Orig. Artic. Ser.* **12,** No. 7, 97–98.

Busby, N., Courval, J., and Francke, U. (1976). *Birth Defects, Orig. Artic. Ser.* **12,** No. 7, 105–107.

Carrel, R. E., Sparkes, R. S., and Wright, S. W. (1971). *J. Pediatr.* **78,** 664–672.

Centerwall, W., and Francke, U. *Ann. Genet.* submitted for publication.

Cohen, M. M., Davidson, R. G., and Brown, J. A. (1975). *Clin Genet.* **7,** 120–127.

Croce, C. M., Kieba, I., Koprowski, H., Molino, M., and Rothblat, G. H. (1974). *Proc. Nat. Acad. Sci. U.S.A.* **71,** 110–113.

Falk, R. E., Crandall, F., Carrel, E., and Valente, M. (1973). *Am. J. Ment. Defic.* **77,** 383–388.

Faust, T., Vogel, W., and Löning, B. (1974). *Clin. Genet.* **6,** 90–97.

Francke, U. (1972). *Am. J. Hum. Genet.* **24,** 189–213.

Francke, U., and Busby, N. (1975). *Birth Defects. Orig. Artic. Ser.* **11,** No. 3, 143–149.

Francke, U., Benirschke, K., Jones, K. L., Myhre, S., and Thuline, H. (1977a). In preparation.

Francke, U., Weber, F., Sparkes, R. S., Mattson, P. D., and Mann, J. (1977b). *Birth Defects: Orig. Artic. Ser.* In press.

Francke, U., Millet, V., Bay, C., and Jones, K. L. (1976c). In preparation.

Giblett, E. R., Anderson, J. E., Cohen, F., Pollara, B., and Meuwissen, H. J. (1972). *Lancet* **ii,** 1067.

Giraud, F., Mattei, J.-F., Mattei, M.-G., and Bernard, R. (1975). *Humangenetik* **28,** 343–347.

Green, H., and Chan, T.-S. (1974). *Science* **182,** 836–837.

Grzeschik, K. H. (1975). *Birth Defects. Orig. Artic. Ser.* **11,** No. 3, 172–175.

Hirschhorn, R., Levitska, V., Pollara, B., and Meuwissen, H. J. (1973). *Nature (London)* **246,** 200–202.

Hopkinson, D. A., Povey, S., Solomon, E., Bobrow, M., and Gormley, P. (1976). *Birth Defects, Orig. Artic. Ser.* **12,** No. 7, 159–160.

Jacobsen, P., Hauge, M., Henningsen, K., Hobolth, N., Mikkelsen, M., and Philip, J. (1973). *Hum. Hered.* **23,** 568–585.

Jones, C., Wuthier, P., and Puck, T. T. (1975). *Somatic Cell Genet.* **1,** 235–246.

Krmpotic, E., Rosenthal, I. M., Szego, K., and Bocian, M. (1971). *Ann. Genet.* **14,** 291–299.

Kucherlapati, R., McDougall, J. K., and Ruddle, F. H. (1974). *Birth Defects, Orig. Artic. Ser.* **10,** No. 3, 108–110.

Laurent, C., Biemont, M-C., Bethenod, M., Cret, L., and David, M. (1975), *Ann. Genet.* **18,** 179–184.

Linarelli, L. G., Pai, K. G., Pan, S. F., and Rubin, H. M. (1975). *J. Pediat.* **86,** 750–752.

Mann, J., and Rafferty, J. (1972). *J. Med. Genet.* **9,** 289–292.

Meera Khan, P., Pearson, P. L., Wijnen, L. L. L., Doppert, B. A., Westerveld, A., and Bootsma, D. (1976). *Birth Defects, Orig. Artic. Ser.* **12,** No. 7, 420–421.

Nabholz, M., Miggiano, V., and Bodmer, W. (1969). *Nature (London)* **223,** 358–363.

Palmer, C. G., Poland, C., Reed, T., and Kojetin, J. (1975). *Humangenetik* **31,** 219–225.

Puck, T. T., Wuthier, P., Jones, C., and Kao, F. T. (1971). *Proc. Nat. Acad. Sci. U.S.A.* **68,** 3102–3106.

Rott, H. D., Schwanitz, G., Grosse, K. P., and Alexandrow, G. (1972). *Humangenetik* **14,** 300–305.

Sanchez, O., Yunis, J. J., and Escobar, J. (1974). *Humangenetik* **22,** 59–65.

Shows, T. B. (1972). *Proc. Nat. Acad. Sci. U.S.A.* **69,** 348–352.

Súbrt, I., and Brýchnâc, V. (1974). *Humangenetik* **23,** 219–222.

Taillemite, J-L., Baheux-Morlier, G., and Roux, C. (1975). *Ann. Genet.* **18,** 61–63.

Taylor, K. M., Brown, M. G., and Wolfinger, H. L. (1972). *Am. J. Hum. Genet.* **24,** 67a.

Taylor, K. M., Wolfinger, H. L., Brown, M. G., Chadwick, D. L., and Francke, U. (1976). *Hum. Genet.* **34,** 155–162.

Tischfield, J. A., Creagan, R. P., Nichols, E. A., and Ruddle, F. H. (1974). *Hum. Hered.* **24,** 1–11.

Tusques, J., Grislain, J. R., Andre, M-J., Mainard, R., Rival, J. M., Caducal, J. L., Dutrillaux, B., and Lejeune, J. (1972). *Ann. Genet.* **15,** 167–172.

Wright, Y., Clark, E., and Breg, W. R. (1974). *J. Med. Genet.* **11,** 69–75.

# 9

# Partial Trisomies and Deletions of Chromosome 13

E. NIEBUHR



## I. INTRODUCTION

Chromosome 13 ($D_1$), present in triplicate in the well-known Patau's syndrome, represents the largest acrocentric chromosome of the human karyotype. Until recently, it could only be distinguished from other D-(13–15) group chromosomes by its pattern of DNA synthesis (Yunis *et al.*, 1964), a method not uniformly reliable in cases involving deletions (Wilson *et al.*, 1973). Introduction of differential staining techniques has made it possible to identify and describe each of the human chromosomes (Caspersson *et al.*, 1971) and consequently to arrive at a more precise delineation of chromosomal rearrangements. A diagram of chromosome 13 with band designations, as outlined at the Paris Conference, 1971 (1972), is shown in Fig. 1 together with normal, representative D-group chromosomes.

The clinical syndromes involving chromosome 13, other than the classical picture of Patau's syndrome, comprise a heterogeneous group, and result from mosaicism, unbalanced translocations, inversions, ring chromosomes, and terminal or interstitial deletions. The number of probands with such aberrations are few, and the clinical picture observed in each instance often shows little constancy, presumably because the amount of genetic material involved in any rearrangement differs from proband to proband. There are, however, some conspicuous somatic defects and hematologic changes in many individuals with aberrations of chromosome 13, making it promising to arrive at a more precise description of possible subphenotypes that may be correlated with the cytogenetic findings.

## II. TRISOMIES

### A. General Considerations

Characteristic features of the complete trisomy 13 syndrome include severe psychomotor retardation, arhinencephaly, sloping of the forehead, micro-

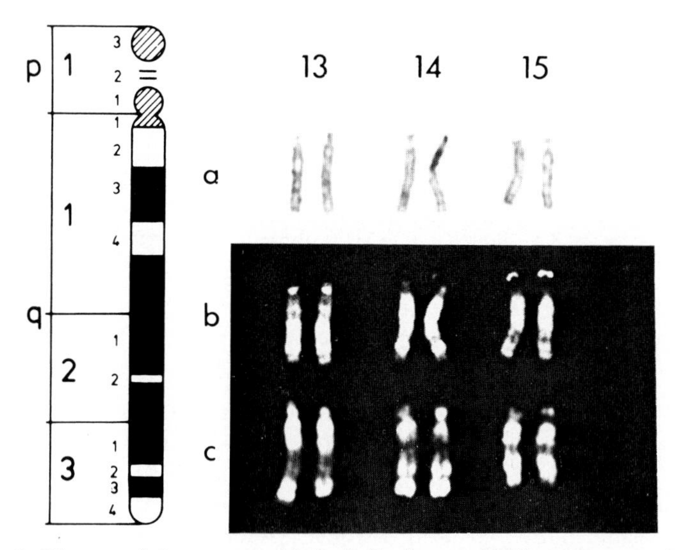


**Fig. 1.** Diagram of chromosome 13 (Paris Conference 1971) and D-group chromosomes stained with orcein (a), quinacrine mustard: Q-bands (b) and acridine orange: R-bands (c).

phthalmia, coloboma, cleft lip and palate, cardiac and renal defects, polydactyly and capillary hemangiomata (Table I, column A). In addition, hematological studies have revealed an increased number of nuclear projections in the neutrophils (Huehns *et al.*, 1964a) and changes in the type of hemoglobin with persistence of embryonic hemoglobin (HbGower 2) in newborns and elevated levels of fetal hemoglobin (HbF) in older children (Huehns *et al.*, 1964b).

Unlike the complete trisomy 13 syndrome, patients with mosaicism and partial trisomy 13 present a less severe clinical picture related to the relative number of trisomic cells and to the length of the chromosomal segment involved.

## B. Mosaicism

Of fifteen patients with mosaicism, eleven were found to have a normal and trisomic cell line (Conen and Erkman, 1963; Forteza *et al.*, 1964; Taylor and Polani, 1964; Bain *et al.*, 1965, two cases; Tzoneva-Maneva *et al.*, 1965, two cases; Stone *et al.*, 1966; Engel *et al.*, 1967; Moore and Engel, 1970; Schinzel *et al.*, 1974). In these cases, the frequency of cells trisomic for chromosome 13 varied between 2 and 66% (Taylor and Polani, 1964; Moore and Engel, 1970). Mosaicism in these cases result from postzygotic nondisjunction and it is essential to notice that the tentative cell line, monosomic of chromosome 13, has never been observed. One proband (Taylor, 1968) was found to have a tetrasomic cell line as well. The remaining three mosaic patients were found to have a trisomic cell line with a long arm 13 isochromosome (Therman *et al.*, 1963; Taylor *et al.*, 1970; Emberger *et al.*, 1972).

Positive identification of the abnormal chromosome as No. 13 was only presented in three cases (Moore and Engel, 1970; Emberger *et al.*, 1972; Schinzel *et al.*, 1974).

The clinical picture of these fifteen mosaic patients is summarized in Table I, column B. The frequency of congenital malformations varies. In general, patients with predominantly trisomic cells (Taylor *et al.*, 1970; Moore *et al.*, 1970) show an easily recognizable picture of Patau's syndrome, whereas cases with a minority of trisomic cells (Emberger *et al.*, 1972; Schinzel *et al.*, 1974) exhibit a less diagnostic clinical picture. When compared to cases with the complete trisomy (Table I, column A), mosaic patients have a higher mean birth weight and better chances of survival. Also, parental age does not appear to differ from that of the normal population. Dermatoglyphics were typical in a few cases with a distal axial triradius, a large tibial loop or a fibular arch on the hallucal area (e.g., Therman *et al.*, 1963) which may be of diagnostic value in clinical doubtful cases.

**TABLE I**
**Birth, Survival, and Clinical Findings in Patau's Syndrome and 38 Probands with a Partial Trisomy 13**

| | A[a] Complete trisomy | Partial trisomies | | | |
|---|---|---|---|---|---|
| | | | Translocations | | |
| | Patau's syndrome | B Mosaics | C[c] All cases | D[c] Trisomy prox.1/3–1/2 | E Trisomy distal 1/3–2/3 |
| Sex (F:M) | 35:29 | 8:7 | 14:9 | 4:2 | 10:7 |
| Maternal age (years) | 31.6 | 25.0 | 26.3 | 26.5 | 26.2 |
| | $n$=74 | $n$=10 | $n$=22 | $n$=5 | $n$=17 |
| Paternal age (years) | 31.9 | 25.5 | 29.8 | 29.6 | 29.9 |
| | $n$=74 | $n$=10 | $n$=19 | $n$=5 | $n$=14 |
| Gestation age (weeks) | 39.0 | 40.4 | 40.0 | 39.8 | 40.1 |
| | $n$=74 | $n$=8 | $n$=16 | $n$=4 | $n$=12 |
| Birth weight (gram) | 2610 | 3149 | 3183 | 2926 | 3277 |
| | $n$=74 | $n$=10 | $n$=15 | $n$=4 | $n$=11 |

| Clinical features | Patau's syndrome (%) | B Mosaics[b] (%) | C[c] All cases[b] (%) | D[c] Trisomy prox. 1/3–1/2[b] (%) | E Trisomy distal 1/3–2/3[b] (%) |
|---|---|---|---|---|---|
| Severe retardation | 100 | 14/15=93 | 22/24= 91 | 5/7=71 | 17/17=100 |
| Deafness | 53 | 2/7 =28 | 2/16= 12 | 1/6=16 | 1/10= 10 |

| | | | | | |
|---|---|---|---|---|---|
| Microcephaly | 59 | 3/7 =43 | 14/21= 66 | 3/5=60 | 11/16= 69 |
| Hypertelorism | 93 | 2/4 =50 | 10/19= 52 | 3/4=75 | 7/15= 46 |
| Hypotelorism | ? | 0/4 = 0 | 6/19 = 31 | 0/4= 0 | 6/15 = 40 |
| Epicanthus | 52 | 4/7 = 57 | 8/16 = 50 | 1/4=25 | 7/12 = 58 |
| Microphthalmos | 78 | 2/13=15 | 5/24 = 21 | 1/7=14 | 4/17 = 24 |
| Coloboma | 35 | 3/11=27 | 2/24= 8 | 0/7= 0 | 2/17= 12 |
| Harelip | 55 | 3/12=25 | 4/24= 16 | 0/7= 0 | 4/17= 24 |
| Cleft palate | 65 | 4/12=25 | 7/24= 29 | 2/7=29 | 5/17= 29 |
| Malformed ears | 81 | 4/6 =66 | 11/21= 52 | 2/6=33 | 9/15= 60 |
| Low-set ears | 11 | 7/8 =87 | 13/19= 68 | 3/5=60 | 10/14= 71 |
| Hemangioma | 73 | 2/12=16 | 13/23= 56 | 1/6=16 | 12/17= 71 |
| Polydactyly | 78 | 5/14=35 | 11/24= 46 | 0/7= 0 | 11/17= 65 |
| Heart disease | 76 | 5/10=50 | 4/20= 20 | 2/6=33 | 2/14= 14 |
| Renal anomaly | 52 | 3/8 =37 | 6/14= 43 | 0/4= 0 | 6/10= 60 |
| Biseptate uterus (♀) | 43 | 3/4 =75 | 0/1 = 0 | ?[f] | 0/1 = 0 |
| Absent olfactory bulbs | 71 | 2/5 =40 | 2/2 =100 | ?[f] | 2/2 =100 |
| Simian crease | 64 | 3/7 =43 | 8/16= 50 | 2/5=40 | 6/11= 55 |
| Distal t-triradius | 77 | 4/7 =57 | 13/19= 68 | 4/6=66 | 9/13= 69 |
| Elevated levels of HbF | >50[d] | 1/3 =33 | 8/15= 53 | 3/5=60 | 5/10= 50 |
| Abnormal nuclear projections | >50[d] | 2/3 =66 | 7/16= 44 | 5/6=83 | 2/10= 20 |
| Death prior to 6 months | 93[e] | 4/14=28 | 3/21= 14 | 0/5= 0 | 3/16= 19 |

[a] According to Taylor (1969).
[b] The numerator indicates the number of cases in which the sign is present and the denominator those in which the sign was recorded.
[c] Sex of one proband not known. (McIntyre *et al.*, 1964).
[d] According to Smith (1970).
[e] According to Crandall *et al.* (1974).
[f] ?, Finding not recorded.

## C. Partial Trisomies

In twenty-four reported cases, parental translocations or inversions and *de novo* translocations have resulted in partial trisomy 13. As was seen with mosaics, these patients as a group have a higher mean birth weight, a longer survival, and a reduced frequency of congenital malformations (Table I, column C). Of particular interest is the presence of hypotelorism in 25% of the probands; this sign was hardly ever observed in the series of complete trisomy published by Taylor (1969). However, orbital hypotelorism was a general feature in the craniofacial category 1 of $D_1$ trisomy described by Snodgrass *et al.* (1966). The relative low frequency of common trisomy 13 features (deafness, eye defects, harelip and cleft palate) is striking.

On the basis of the cytogenetic findings, the probands can be subdivided in two groups trisomic for, respectively, the proximal $\frac{1}{3}-\frac{1}{2}$ and the distal $\frac{1}{3}-\frac{2}{3}$ portion of the long arm of chromosome 13. Although there is an obvious cytogenetic overlap between the two groups, it is possible to postulate a connection between certain clinical and hematological features (Table I, columns D and E) and the length of the trisomic segment.

### *1. Trisomy, Proximal $\frac{1}{3}-\frac{1}{2}$ (Table I, column D)*

This group includes seven probands (MacIntyre *et al.*, 1964; Yunis and Hook, 1966, Case B; Bloom and Gerald, 1968, Case 1; Escobar and Yunis, 1974; Schinzel *et al.*, 1974, Case 3; Schwanitz *et al.*, 1974; Noel *et al.*, 1976, Case 1a). The clinical features are rather unspecific: psychomotor retardation, microcephaly, low-set ears, microstomia, micrognathia, elevated t-triradius and in-

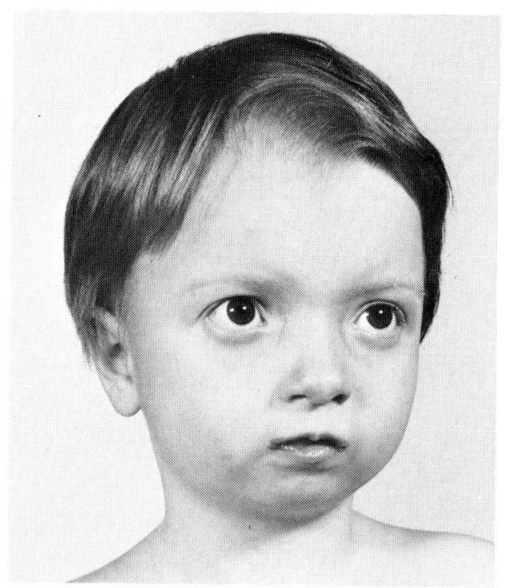
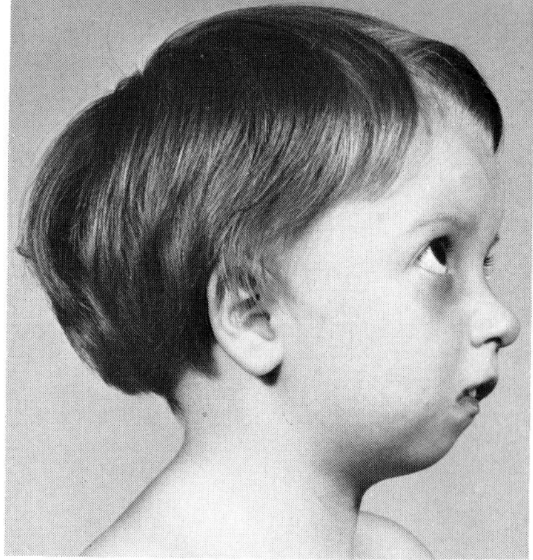

**Fig. 2.** A $5\frac{1}{2}$-year-old girl with partial trisomy 13, proximal $\frac{1}{3}$; note micrognathia and microstomia (from Escobar and Yunis, 1974; courtesy of American Journal of Diseases of Children, "copyright 1974, American Medical Association").

curved fifth fingers (Fig. 2). A few patients have also shown microphthalmia, hemangioma, cleft palate and epicanthic folds. Two probands were only slightly mentally retarded (Bloom and Gerald, 1968; Noel *et al.*, 1976). The clinical findings in a given patient are generally nondiagnostic. An increased number of nuclear projections in the neutrophils may be of value in the differential diagnosis.

Four probands studied with banding techniques were apparently trisomic for the following segments [(Paris Conference, 1971): 13pter→q12 (q13?) (Escobar and Yunis, 1974); 13pter→q13 (Schinzel *et al.*, 1974); 13pter→q14 (12?) (Schwanitz *et al.*, 1974); 13pter→q12 (Noel *et al.*, 1976)]. The correlation between break points, abnormal hemoglobin and neutrophils will be discussed in Section IV, B.

### *2. Trisomy, Distal $\frac{1}{3}-\frac{2}{3}$ (Table I, column E; Fig. 3)*

Seventeen patients are included in this group (Stalder *et al.*, 1964; Yunis and Hook, 1966, Case A; Wilson and Melnyk, 1967; Bloom and Gerald, 1968, Case 2; Hoehn *et al.*, 1971; Rosenkranz and Kaloud, 1972; Hauksdóttir *et al.*, 1972, Cases 1 and 2; Talvik *et al.*, 1973; Taysi *et al.*, 1973; Crandall *et al.*, 1974; Escobar *et al.*, 1974; Fryns *et al.*, 1974a; Schinzel *et al.*, 1974, Case 2; McDermott and Parrington, 1975; Noel *et al.*, 1976, Cases 1b and 2a). The main findings are not much different from the complete picture of Patau's syndrome, although deafness, eye malformations, cleft palate, harelip and cardiac defects are not so frequently observed in this group as in the complete trisomy 13 syndrome.

The presence of unusual features such as an open anterior fontanel, narrow temples and bossing of the forehead was noted by Escobar *et al.* (1974) and contrasted with the usual receding forehead observed in Patau's syndrome. Frontal bossing was also observed by Talvik *et al.* (1973) and Taysi *et al.* (1973). A normal, not receding, forehead was noticed in three further patients (Hauksdóttir *et al.*, 1972; Schinzel *et al.*, 1974). A sloping forehead together with a prominent metopic suture and narrow temples was observed by Crandall *et al.* (1974). Necropsy studies in two probands (Wilson and Melnyk, 1967; Fryns *et al.*, 1974a) both with a receding forehead, a small fontanel, harelip and cleft palate (Fig. 3c) showed the typical midbrain defects with agenesis of the olfactory bulbs, fusion of the frontal lobes and agenesis of the corpus callosum. Both patients were trisomic for at least $\frac{3}{4}$ of the long arm.

Essential features, besides severe mental retardation, polydactyly and hemangioma, are the combination of microcephaly, bossing of the foreband (Fig. 3a), narrow temples, open anterior fontanels, synophrys, curly eyelashes (Fig. 3b) and low-set, slightly malformed ears with small lobules.

In five patients, partial trisomy for the distal segment of the long arm of chromosome 13 was the result of crossing over within the meiotic loop of a parental pericentric inversion. These patients were trisomic for the following segments [(Paris Conference 1971): 13q31→qter (Hauksdóttir *et al.*, 1972, 2 patients); 13q14 (21?)→qter (Taysi *et al.*, 1973); 13q22→qter (Escobar *et al.*, 1974); 13q21→qter (McDermott and Parrington, 1975)]. The remaining thirteen reported patients had unbalanced translocations. The patient described by Wilson and Melnyk (1967) had an additional normal cell line in blood and bone marrow, but not in fibroblasts. The break points were localized in six cases, and these were apparently trisomic for the following segments: 13q22→qter (Talvik *et al.*, 1973); 13q13→qter (Crandall *et al.*, 1974); 13q14→qter (Fryns *et*

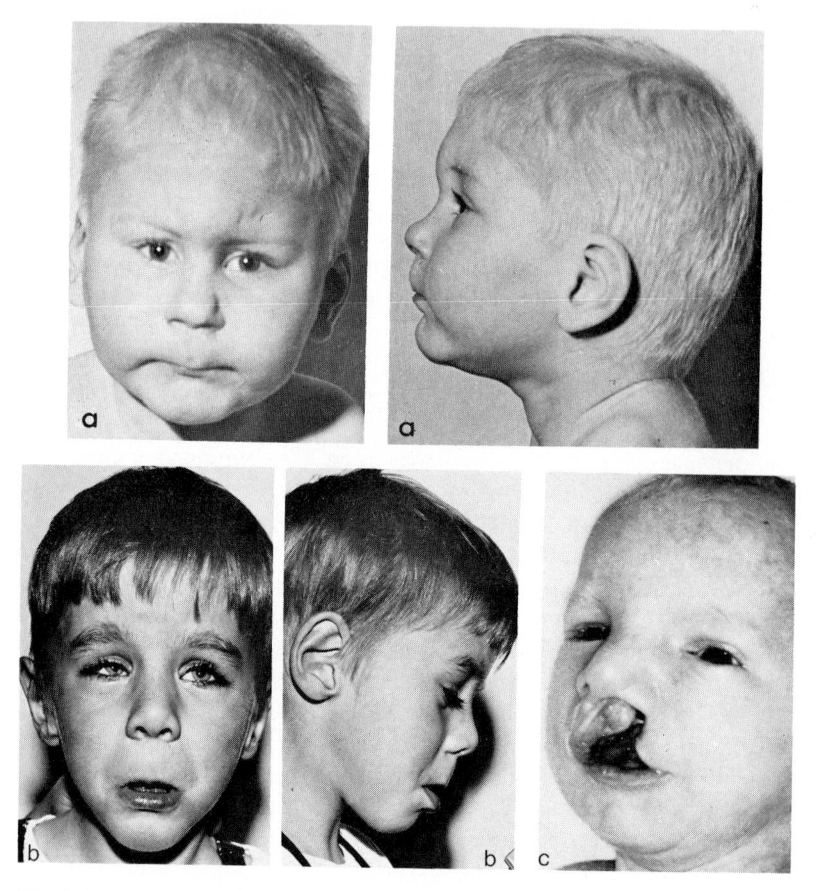


**Fig. 3.** Probands with partial trisomy 13, distal $\frac{1}{3}$ – $\frac{2}{3}$. (a) $1\frac{4}{12}$-year-old girl; note hypotelorism, frontal bossing and low set ears (from Talvik *et al.*, 1973; courtesy of *Humangenetik*). (b) $4\frac{1}{2}$-year-old boy with epicanthus, synophrys and curly eyelashes (from Hauksdóttir *et al.*, 1972; courtesy of *Journal of Medical Genetics*). (c) 1-month-old boy with harelip and cleft papate (from Fryns *et al.*, 1974a; courtesy of *Humangenetik*).

*al.*, 1974a); 13q14→qter (Schinzel *et al.*, 1974); 13q12→qter and 13q14→qter (Noel *et al.*, 1976, Cases 1b and 2a respectively). Duplication of the distal $\frac{1}{3}-\frac{1}{2}$ of the long arm of chromosome 13 seems to be critical for the production of polydactyly, hemangioma, frontal bossing and narrow temples (Escobar *et al.*, 1974). Development for features such as a receding forehead with midbrain defects, harelip and eye malformations probably requires a trisomy for most of the long arm. It is striking, that increased number of nuclear projections of neutrophils was only observed in two cases (Wilson and Melnyk, 1967; Noel *et al.*, 1976, Case 1b), both with trisomy for the greater part of the long arm. Fetal hemoglobin was normal in patients with trisomy for the distal $\frac{1}{3}$ of the long arm (Bloom and Gerald, 1968, Case 2; Talvik *et al.*, 1973; Escobar *et al.*, 1974; McDermott and Parrington, 1975), but elevated in the remaining examined cases trisomic for the distal $\frac{1}{2}-\frac{2}{3}$ with exception of Wilson and Melnyk (1967) and Schinzel *et al.* (1974, Case 2).

Dermatoglyphics of patients with trisomy for the proximal and distal portion of the long arm were abnormal in many patients with distally displaced triradii and simian creases. Two patients with trisomy for the proximal $\frac{1}{3}$ portion (Yunis and Hook, 1966; Case B; Schinzel *et al.*, 1974, Case 2) had a large tibial loop on the hallucal area. Two patients, trisomic for the distal portion (Yunis and Hook, 1966, Case A; Schinzel *et al.*, 1974, Case 2) had a fibular arch, but the data are too limited for further conclusions. Unusual findings in two patients (Wilson and Melnyk, 1967; Taysi *et al.*, 1973) are simple arches on all ten fingers, a finding more characteristic of trisomy 18.

## III. DELETIONS

### A. General Considerations

Seventy two patients with a presumptive deletion of chromosome 13 were studied for this review. Forty-four or 61% of these cases are associated with a ring chromosome formation (Wang *et al.*, 1962; Daniel, 1970; Picciano *et al.*, 1972; Salamanca *et al.*, 1972; Zink *et al.*, 1973; Fryns *et al.*, 1974b; Hoo *et al.*, 1974; Kuroki and Nagano, 1974; Fried *et al.*, 1975; Schmid *et al.*, 1975; Noel *et al.*, 1976; also 33 cases collected by Niebuhr and Ottosen, 1973). The ring chromosome has been identified as No. 13 in 57% of the reports. Twenty patients or 28% of all cases presented a Dq– karyotype with a terminal or interstitial deletion (Grosse and Schwanitz, 1973; Gilgenkrantz *et al.*, 1973; Howard *et al.*, 1974; Ikeuchi *et al.*, 1974; O'Grady *et al.*, 1974; Kučerová *et al.*, 1975; Noel *et al.*, 1976; also twelve cases summarized by Niebuhr and Ottosen, 1973). The deleted chromosome was identified as No. 13 in 70% of the reports. The remaining eight cases (Giraud *et al.*, 1967; O'Grady *et al.*, 1974; Gushina and Golovachev, 1975; also five probands reviewed by Niebuhr and Ottosen, 1973) had a normal cell line in addition to the Dq– cell line or a more complex aberration. The abnormal chromosomes were not identified in these eight cases.

## B. Phenotypic Picture

Birth, survival and selected clinical features of forty four patients with a D-ring chromosome and twenty eight patients with a Dq– karyotype or more complex aberrations are shown in Table II, columns A and B, respectively. Column C summarizes the data for both groups, i.e. all seventy two patients with a presumptive deletion of chromosome 13.

Four of the seventy two patients were stillborn (Bain and Gauld, 1963; Lazjuk *et al.*, 1973; Schmid *et al.*, 1975; Noel *et al.*, 1976). Of interest is the fact that the prosencephalic brain defect noted in trisomy 13 has also been observed in seven patients with a partial monosomy 13. Three patients had a holoprosencephalic cleavage defect (Juberg *et al.*, 1969; Opitz *et al.*, 1969; Leisti, 1971) associated with absence of the peripheral olfactory system. Arhinencephaly was found in four patients (Bain and Gauld, 1963; Biles *et al.*, 1969; Rethoré *et al.*, 1970; Orbeli *et al.*, 1971) and suspected in the patient of Sparkes *et al.* (1967) because of a sloping forehead. Two further brain defects, anencephaly and hydrocephaly were observed by Schmid *et al.* (1975) and Noel *et al.* (1976) respectively. Gross malformations of internal organs (other than brain defects) are common, but only a few cases have been subject to a necropsy study. Analysis of 12 necropsy reports (Bain and Gauld, 1963; Masterson *et al.*, 1968; Biles *et al.*, 1969; Opitz *et al.*, 1969; Varela and Sternberg, 1969; Cagianut and Theiler, 1970; Rethoré *et al.*, 1970; Taylor, 1970; Orbeli *et al.*, 1971; Sylvester *et al.*, 1971; Lazjuk *et al.*, 1973; Schmid *et al.*, 1975) showed that ventricular and atrial septal defects (each 33%), aplasia of the gallbladder (33%) and hypoplastic kidneys (60%) were leading malformations. Unilateral renal agenesis and hydronephrosis were reported in three cases. Malformations reported only once were aplasia of the pancreas, atresia oesophagi, and a polycystic kidney. X-Ray of two further probands (Sparkes *et al.*, 1967; Guschina and Golvachev, 1975) was compatible with diagnosis of tetralogy of Fallot.

The clinical features mentioned above are of little diagnostic value when considered alone, but together with more conspicuous features such as prominent nasofrontal bones, large ears with deep sulci helici and overdeveloped lobules, malformed, and malrotated ears, hypoplasia or agenesis of the thumbs, imperforate anus and retinoblastoma, it is indeed possible to individualize a specific 13 deletion syndrome (Lejeune *et al.*, 1968). Discriminating features are the broad prominent nose bridge, retinoblastoma, microphthalmia, coloboma and absent thumbs.

**TABLE II**
**Birth, Survival and Clinical Findings in 72 Probands with a Partial Deletion of Chromosome 13**

| | | | | Rings | | Dq– and complex aberrations | | |
|---|---|---|---|---|---|---|---|---|
| | A D rings | B[a] Dq– and complex aberrations | C[a] All cases | D Thumbs normal | E Thumb aplasia | F[a] Thumb aplasia | G Retinoblastoma | H Other cases |
| Sex (F:M) | 21:23 | 17:10 | 38:33 | 17:21 | 4:2 | 5:4 | 7:3 | 5:3 |
| Maternal age (years) | 26.1 | 26.3 | 26.1 | 26.4 | 24.3 | 25.5 | 25.5 | 28.3 |
| | *n*=35 | *n*=23 | *n*=58 | *n*=29 | *n*=6 | *n*=9 | *n*=8 | *n*=6 |
| Paternal age (years) | 28.4 | 29.6 | 28.9 | 28.8 | 26.0 | 28.7 | 28.6 | 32.8 |
| | *n*=28 | *n*=21 | *n*=49 | *n*=24 | *n*=4 | *n*=9 | *n*=7 | *n*=5 |
| Gestation age (weeks) | 39.6 | 39.5 | 39.6 | 39.3 | 40.8 | 40.6 | 39.2 | 40.5 |
| | *n*=33 | *n*=22 | *n*=55 | *n*=27 | *n*=6 | *n*=5 | *n*=9 | *n*=8 |
| Birth weight (gram) | 2117 | 2409 | 2229 | 2195 | 1711 | 2152 | 2722 | 2427 |
| | *n*=37 | *n*=23 | *n*=60 | *n*=31 | *n*=6 | *n*=9 | *n*=7 | *n*=7 |

| Clinical features | A D rings[b] (%) | B[a] Dq– and complex aberrations[b] (%) | C[a] All cases[b] (%) | D Thumbs normal (%) | E Thumb aplasia (%) | F[a] Thumb aplasia (%) | G Retinoblastoma (%) | H Other cases (%) |
|---|---|---|---|---|---|---|---|---|
| Severe retardation | 42/42=100 | 22/26= 85 | 64/68=94 | 100 | 100 | 100 | 70 | 87 |
| Microcephaly | 40/41= 97 | 15/18= 83 | 55/59=93 | 100 | 75 | 100 | 80 | 71 |
| Hypertelorism | 35/36= 97 | 17/19= 89 | 52/55=94 | 97 | 100 | 100 | 67 | 100 |

*continued*

**TABLE II (*continued*)**

| Clinical features | A D rings[b] (%) | B[a] Dq– and complex aberrations[b] (%) | C[a] All cases[b] (%) | Rings: D Thumbs normal (%) | Rings: E Thumb aplasia (%) | Dq– and complex aberrations: F[a] Thumb aplasia (%) | Dq– and complex aberrations: G Retinoblastoma (%) | H Other cases (%) |
|---|---|---|---|---|---|---|---|---|
| Prominent nasal bridge | 26/32= 81 | 6/16= 37 | 32/48=66 | 86 | 33 | 67 | 0 | 33 |
| Epicanthus | 26/31= 84 | 8/13= 61 | 34/44=77 | 85 | 75 | 66 | 80 | 40 |
| Microphthalmos | 8/41= 19 | 8/21= 38 | 16/62=25 | 16 | 40 | 100 | 22 | 25 |
| Coloboma | 7/39= 18 | 8/23= 35 | 15/62=24 | 17 | 25 | 80 | 20 | 25 |
| Retinoblastoma | 1/39= 2 | 10/23= 43 | 11/62=18 | 3 | 0 | 0 | 100 | 0 |
| Protruding maxilla | 14/18= 78 | 4/5 = 80 | 18/23=78 | 75 | 100 | 100 | 100 | 66 |
| Large ears | 31/36= 86 | 8/13= 61 | 39/49=79 | 93 | 25 | 100 | 40 | 71 |
| Low-set ears | 14/32= 44 | 11/15= 73 | 25/47=53 | 39 | 75 | 100 | 71 | 71 |
| Heart disease | 16/30= 53 | 7/12= 58 | 23/42=55 | 54 | 50 | 100 | 0 | 66 |
| Renal anomaly | 10/18= 55 | 4/6 = 66 | 14/24=58 | 57 | 50 | 100 | 0 | 66 |
| Imperforate anus | 8/41= 19 | 3/25= 12 | 11/66=16 | 19 | 20 | 43 | 0 | 0 |
| Hypospadias (♂) | 8/20= 40 | 3/9 = 33 | 11/29=38 | 33 | 100 | 66 | 33 | 0 |
| Bifid scrotum (♂) | 7/20= 35 | 2/8 = 25 | 9/28=32 | 28 | 100 | 66 | 0 | 0 |
| Biseptate uterus (♀) | 2/5 = 40 | 0/3 = 0 | 2/8 =25 | 33 | 50 | ?[c] | ?[c] | 0 |
| Thumb aplasia | 6/43= 14 | 13/28= 46 | 19/71=27 | 0 | 100 | 100 | 30 | 0 |
| Metacarpal fusion | 3/3 =100 | 4/5 = 80 | 7/8 =87 | 0 | 100 | 80 | ?[c] | ?[c] |
| Simian crease | 7/23= 30 | 7/15= 46 | 14/38=37 | 22 | 60 | 50 | 50 | 40 |
| Distal t-triradius | 6/27= 22 | 5/11= 45 | 11/38=29 | 21 | 33 | 50 | 0 | 66 |
| Trigonocephaly | 9/16= 56 | 4/4 =100 | 13/20=65 | 61 | 33 | 100 | 100 | 100 |
| Arhinencephaly | 4/6 = 66 | 3/4 = 75 | 7/10=70 | 66 | 66 | 100 | ?[c] | 50 |
| Death prior to 6 months | 7/43= 16 | 6/26= 23 | 13/69=19 | 10 | 50 | 50 | 0 | 25 |

[a]Sex of one proband not known (Case 2b, Noel *et al.*, 1976).
[b]The numerator indicates the number of cases in which the finding is present and the denominator those in which the sign was recorded.
[c]?, Finding not recorded.

## C. Phenotypic Subdivision

The seventy two patients with a deleted chromosome 13 are subdivided into 4 categories on the basis of specific clinical features. Patients with ring chromosomes and normal thumbs are classified into category 1, and those with abnormal thumbs into category 2a (Table II, columns D and E, respectively). Patients with a Dq− karyotype or complex aberrations are classified into category 2b, if their thumbs are abnormal, into category 3 if retinoblastoma is present, and into category 4 when the thumbs are normal and the patients show no sign of retinoblastoma (Table II, columns F, G and H, respectively). Each category of patients is discussed below.

### *1. Patients with Ring Chromosomes*

***a. Category 1: Patients with Normal Thumbs (Table II, column D; Fig. 4).*** The clinical phenotype was individualized by Lejeune *et al.* (1968). Essential features are microcephaly with true hypertelorism (Fig. 4c), epicanthic folds and hyperplasia of the median nasal process eliminating the bulge of the nose (Fig. 4f and g). In typical cases the cranium has a triangular configuration with a small pointed forehead and an increased biparietal diameter. Trigonocephaly with premature obliteration of the metopic suture and a small or closed anterior fontanel was observed by Rethoré et al. (1970), Tolksdorf *et al.* (1970), and Fried *et al.* (1975) (Fig. 4e). The ears are large with deep sulci helici and overdeveloped lobules (Fig. 4a, c and f). Protruding maxilla or upper incisors (Fig. 4f) are additional findings. Ocular abnormalities (microphthalmia and coloboma) and anogenital malformations (imperforate anus, perineal fistula, hypospadias, bifid scrotum) are occasional features. The constellation of malformations is so obvious that a clinical diagnosis is possible (Rethoré *et al.*, 1970). Hyperplasia of the nasal processes, the maxillary bones and the second branchial arch may be opposed to the hypoplasia observed in Patau's syndrome (Lejeune *et al.*, 1968). Dermatoglyphics in patients with category 1 features are not of diagnostic value.

The break points on the long arm of chromosome 13 were located to the bands 13q34 or q33 and at the short arm end to band 13p11 in five cases examined (Niebuhr, 1973; Fryns *et al.*, 1974b; Hoo *et al.*, 1974; Kuroki and Nagano, 1974; Noel *et al.*, 1976). Loss of the telomere and the most distal band 13q34 are thus essential for the development of category 1 features.

***b. Category 2a: Patients with Absent or Hypoplastic Thumbs (Table II, column E; Fig. 5).*** Craniofacial abnormalities are triangular shape of the calvarium, frontal bossing, hypertelorism, epicanthic folds and a protruding maxilla. The ears are slightly malformed, low-set and malrotated with absent or normal lobules (Fig. 5a). Microphthalmia, coloboma and genital malformations are probably more frequent than in category 1 patients. The general condition of

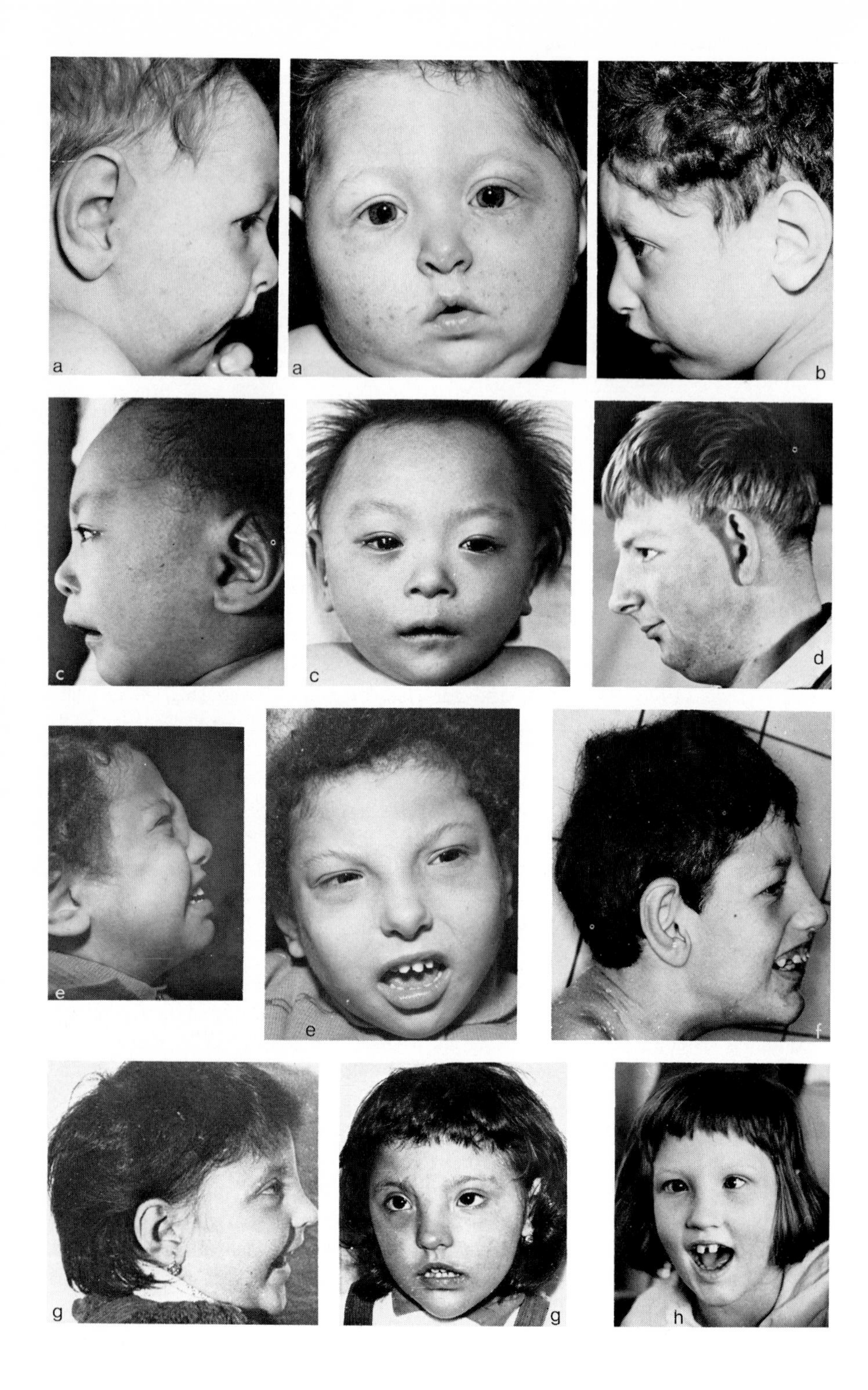
a
a
b
c
c
d
e
e
f
g
g
h

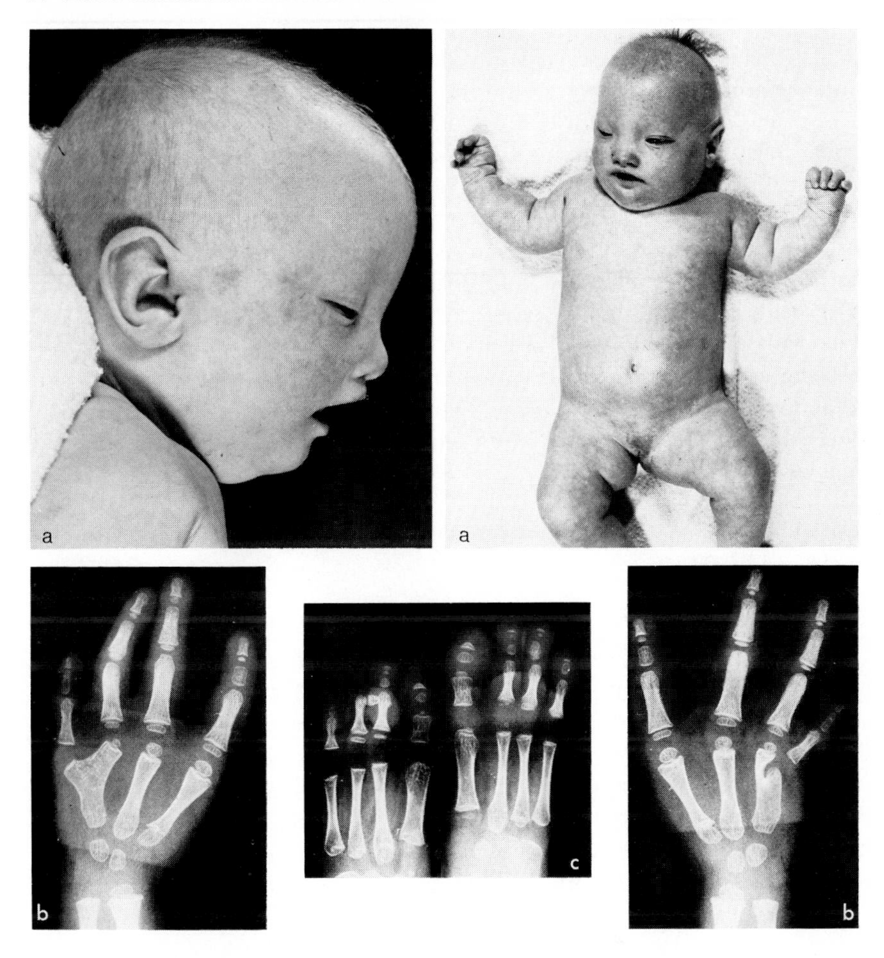

Fig. 5. (a) Three-month-old girl with a 13 ring. Note absent thumbs, malrotated ear and small mandible (from Juberg *et al.*, 1969; courtesy of *Journal of Medical Genetics*). X-ray of hands (b) and feet (c) of a 3-year-old 13 ring case (Niebuhr and Ottosen, 1973). Note absent metacarpals I, fusion of metacarpals IV and V, hypoplastic phalanx II dig.II, and absent metatarsals V.

---

Fig. 4. Probands with ring 13 and category 1 features. Note hypertelorism, saliant nasofrontal bones, protruding maxilla or upper incisors and large abnormal ears. (a) $\frac{1}{2}$ year; (b) $2\frac{1}{2}$ years; Case 1 from Grace *et al.*, 1971; courtesy of *Journal of Medical Genetics.* (c) 1-year-old boy from Kuroki and Nagano, 1974; (d) 17-year-old boy from Sylvester *et al.*, 1971; courtesy of *Journal of Mental Deficiency Research.* (e) $4\frac{1}{2}$-year-old girl from Fried *et al.*, 1975; courtesy of *Clinical Genetics;* (f) 9-year-old boy, personal observation; (g) $2\frac{4}{12}$-year-old girl from Moric-Petrovic *et al.*, 1970; courtesy of *Annales de Génétique*; (h) $6\frac{1}{6}$-year-old girl from Coffin and Wilson, 1970; courtesy of *American Journal of Diseases of Children*, "copyright 1970, American Medical Association."

the patients is poor with severely affected intrauterine growth, obvious psychomotor retardation and early death (50%). According to the selection criterion, all probands showed agenesis of the thumbs, ranging from an unilateral boneless, rudimentary fold (Masterson *et al.,* 1968) to bilateral absence of the thumbs without rudiments (Sparkes *et al.,* 1967; Juberg *et al.,* 1969; Niebuhr and Ottosen, 1973). The patient reported by Bain and Gauld (1963) showed an absent right thumb and a rudimentary left thumb. Accompanying features are absence of the first metacarpal bone, fusion of the fourth and fifth metacarpals (Y-shaped), and hypoplasia or aplasia of the middle phalanx of the fifth fingers. Hypoplasia of the middle phalanx of the second digits was obvious in one case (Niebuhr and Ottosen, 1973) (Fig. 5b), and according to X-ray photographs also in the two other patients with absent thumbs (Sparkes *et al.,* 1967; Juberg *et al.,* 1969). Bilateral fusion of the fourth and fifth toes was observed by Bain and Gauld (1963) and Sparkes *et al.* (1967); absent fifth toes with only four metatarsals was noticed in one patient (Niebuhr and Ottosen, 1973) (Fig. 5c). Break points in one case (Niebuhr and Ottosen, 1973) were determined to be 13p11 and 13q21.

### *2. Patients with a Dq– Karyotype and Complex Aberrations*

***a. Category 2b: Patients with Absent or Hypoplastic Thumbs (Table II, column F).*** Patients in this group do not differ clinically from those with absent thumbs and a ring chromosome and are therefore considered to belong to the same category. Most cases show only hypoplastic thumbs and first metacarpal bones, although absent thumbs were observed by Allderdice *et al.* (1969), Orbeli *et al.* (1971), Leisti (1971), and Noel *et al.* (1976). Hypoplasia of the second phalanx of the index fingers, absent middle phalanx of the fifth fingers was obvious in one case (Orbeli *et al.,* 1971). This patient showed only four metatarsals bilaterally and an absent left fifth toe. Hypoplastic or absent thumbs has been opposed to the tendency of polydactyly in Patau's syndrome (Lejeune *et al.,* 1968). However, it is more correct to oppose polydactyly with the tendency of ulnar (fibular) structures to fuse or to reduce, as extra fingers (toes) are always on the ulnar (fibular) site (Laurent *et al.,* 1967). One patient had a terminal deletion of the distal two thirds (13q14→qter) with an accompanying partial trisomy 5p (Noel *et al.,* 1976). The remaining patients were not studied by differential staining techniques; five had a simple deletion, two a normal cell line in addition (Orbeli *et al.,* 1971; Guschina and Golovachev, 1975) and two a complex aberration [(Giraud *et al.,* 1967): 46,t(DqDq)+/45,t(DqDq)tan; Lazjuk *et al.,* 1973): 46,XY/45,XY,–D/46,XY,Dq+].

Dermatoglyphics of patients with hypoplastic or absent thumbs do generally show an easily recognizable palmar configuration. When a hand has no thumb, there is no axial triradius, and the palmar ridges have a transversal course (Holt, 1972). These abnormalities were obvious in three probands (Giraud *et al.,* 1967;

Sparkes *et al.,* 1967; Juberg *et al.,* 1969) but not in a personally observed case with complex palmar patterns (Niebuhr and Ottosen, 1973). Additional findings are absent triradii (b and c or c and d) and fusion of the c– and d– triradii. These abnormalities may be considered secondary to the limb malformations.

***b. Category 3: Patients with Retinoblastoma (Table II, column G; Fig. 6).*** The occurrence of retinoblastoma in 10 patients with a Dq– karyotype

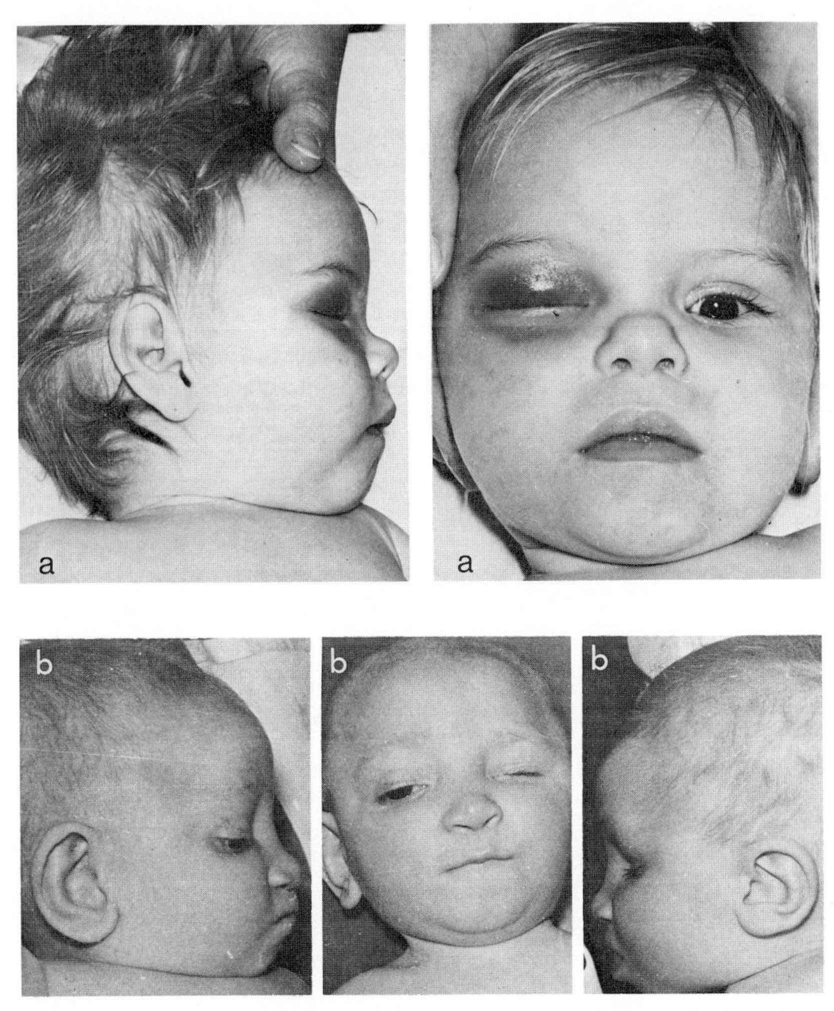


Fig. 6. Probands with retinoblastoma. (a) Girl at the age of 10 months (from Orye *et al.,* 1974); courtesy of *Clinical Genetics*); (b) 8-month-old girl, left eye enucleated (Case 2 from Taylor, 1970; courtesy of *Humangenetik*).

(Lele *et al.,* 1963; Thompson and Lyons, 1965; Van Kempen, 1966; Wilson *et al.,* 1969; Gey, 1970; Taylor, 1970, Case 2; O'Grady *et al.,* 1974, 2 cases; Orye *et al.,* 1974; Howard *et al.,* 1974) seems to be more than a coincidence. The tumor was bilateral in six patients. Another case of bilateral retinoblastoma was clinically diagnosed in a 13 ring case by Grace *et al.* (1971, Case 2). Mild psychomotor retardation associated with very few congenital malformations were observed by Lele *et al.* (1963), Wilson *et al.* (1969), and Orye *et al.* (1974) (Fig. 6a). Short, hypoplastic thumbs and other features of category 2 were observed by Thompson and Lyons (1965), Taylor (1970) (Fig. 6b), and O'Grady *et al.* (1974), Case 2. Hexadactyly was noticed in Case 1 published by O'Grady *et al.* (1974). A protruding maxilla was reported by Gey (1970). Dermatoglyphics of patients with category 3 features are usually normal. Six probands had a simple long arm deletion, one was a mosaic (Taylor, 1970) and three had more complex aberrations (Thompson and Lyons, 1965; Van Kempen, 1966; O'Grady *et al.,* 1974, Case 1). Break points were located in three patients. Howard *et al.* (1974) showed an interstitial deletion of the proximal $\frac{1}{3}$ of the long arm, probably involving the most proximal part of 13q21. Orye *et al.* (1974) demonstrated a deletion between 13q14 and 13q22. Wilson *et al.* (1973) suggested an interstitial deletion of the more distal band 13q31, but as argued by Orye *et al.* (1974) this third proband may also have a deletion of the 13q21 band.

***c. Category 4: Patients with Normal Thumbs and Absence of Eye Tumor (Table II, column H; Fig. 7).*** These eight patients do not necessarily represent a clinical entity as they were not selected on account of certain clinical features. They all have a verified deletion of chromosome 13 with one exception (Cagianut and Theiler, 1970). One patient described by Gilgenkrantz *et al.* (1973) had a terminal deletion of the very distal part of the long arm, and belongs, also according to clinical features (Fig. 7e), to category 1.

The phenotypic picture of the remaining seven patients (Opitz *et al.,* 1969; Cagianut and Theiler, 1970; Taylor, 1970, Case 1; Grosse and Schwanitz, 1973; Ikeuchi *et al.,* 1974; Kučerová *et al.,* 1975; Noel *et al.,* 1976, Case 4) includes psychomotor retardation, microcephaly hypertelorism, large but well-formed ears, cardiac defects, and abnormal dermatoglyphics with a distally displaced t-triradius and simian creases. Microstomia and mongoloid slant of the palpebral fissures were noticed in two patients (Grosse and Schwanitz, 1973; Kučerová *et al.,* 1975). The latter authors drew attention to the resemblance of their patient (Fig. 7d) with three others (Fig. 7a, b and c). Case 4 in the paper of Noel *et al.* (1976), a 10-year-old, slightly retarded girl, showed an interstitial deletion between 13q21 and 13q31. Examination of more cases is needed before conclusions can be drawn regarding the presence of a distinct phenotypic picture in category 4 patients.

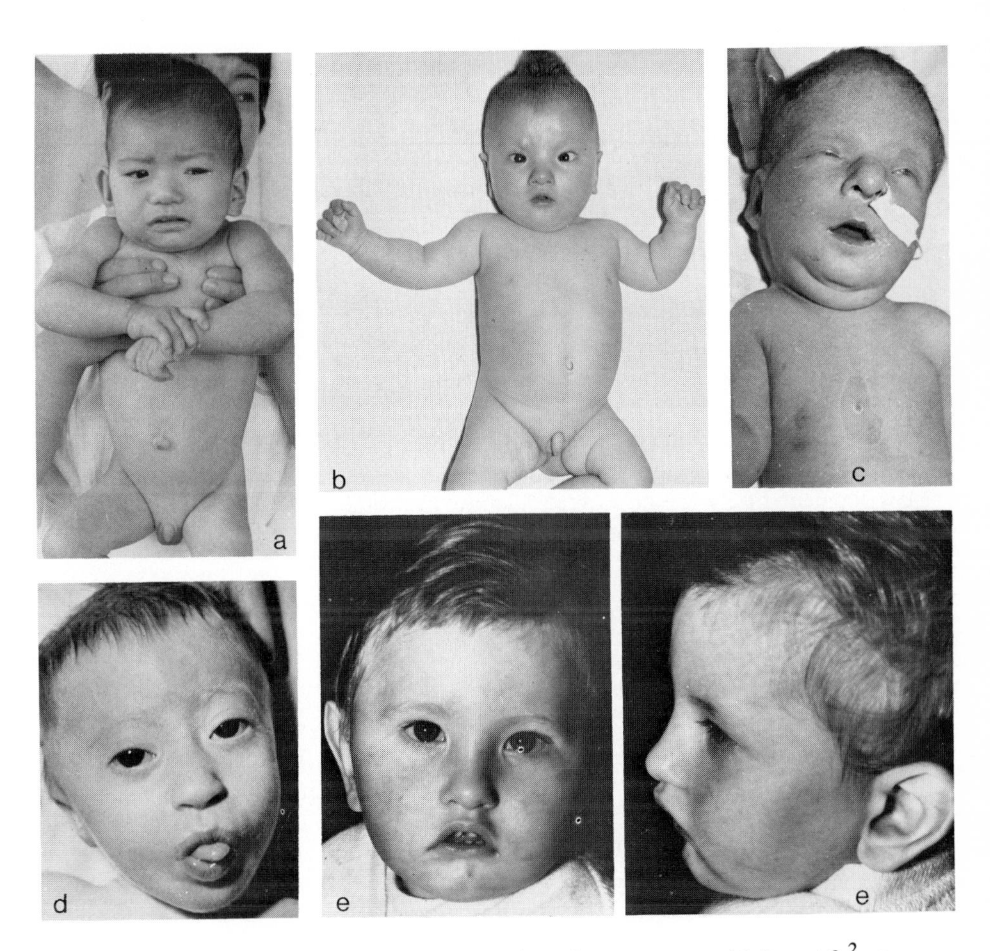


**Fig. 7.** Probands with a 13q–, normal thumbs and no eye tumor. (a) Boy at $2\frac{2}{12}$ years (from Ikeuchi *et al.*, 1974; courtesy of *Humangenetik*); (b) 4-month-old boy (from Grosse and Schwanitz, 1973; courtesy of *Klinische Pädiatrie*); (c) newborn girl (from Taylor, 1970, Case 1; courtesy of *Humangenetik*); (d) $1\frac{3}{12}$-year-old girl (from Kučerová *et al.*, 1975, courtesy of *Humangenetik*); (e) 1-year-old boy with facial features of category 1 (from Gilgenkrantz *et al.*, 1973; courtesy of *Annales de Génétique*).

## IV. PHENOTYPE–GENOTYPE

### A. General Considerations

Earlier attempts towards a clinical mapping of human chromosomes (de Grouchy, 1965) have been refuted as an oversimplification of little value and even misleading (Hamerton, 1971). From a genetic point of view, this rejection may be accepted, at least concerning general features as developmental retardation, microcephaly, and epicanthus. Whether the amount of genetic material is increased or decreased, the greater part of these unspecific clinical features are more or less the same, and also to a large extent independent of the specific autosomes involved (Vogel, 1973). However, from a clinical point of view, it is useful to assume, that a specific relationship exists between the combination of certain clinical features and the absence or triplication of some particular genetic elements. Nobody will probably repudiate that a certain phenotype makes it possible at least to suspect the cytogenetic diagnosis in most cases of Down's syndrome and Patau's syndrome, although most of the clinical features per se are unspecific. Consequently, it is likely that a particular genotype causes the development of a given phenotypic picture. Concerning chromosome 13, there are several clinical and hematological abnormalities justifying a phenotypic mapping attempt.

### B. Partial Trisomies

As early as 1966, Yunis and Hook suggested that genes involved in the control of multiple nuclear projections and neutrophils were located near the centromere on the long arm of chromosome 13, while genes involved in the control of fetal hemoglobin were located more distally on the proximal portion of the long arm. Two cases in the present survey may add additional information. Case 1b (Noel *et al.,* 1976) with trisomy 13q12→qter showed elevated HbF and abnormal neutrophils. Case 2 (Schinzel *et al.,* 1974) with trisomy 13q14→qter had normal values, but the brother (Case 3) with trisomy 13pter→q14 showed elevated HbF and abnormal neutrophils, indicating that genes regulating both parameters are close to each other between 13q12 and 13q14 with those regulating nuclear projections closest to the centromere (Fig. 8). This location is in agreement with normal HbF values and abnormal neutrophils [(Escobar and Yunis, 1974): 13pter→13q12 (13?)] and the reverse condition in the case of Crandall *et al.* (1974) with trisomy 13q13→qter. The remaining cases do not give contradictory information. The presence of polydactyly in three patients with trisomy 13q31→qter (Escobar *et al.,* 1974; Hauksdóttir *et al.,* 1972, Cases 1 and 2), but not in patients trisomic for the proximal $\frac{1}{3}$–$\frac{1}{2}$ strongly indicate a location distally to 13q22 (Fig. 8). Other features such as hemangioma and eye anomalies

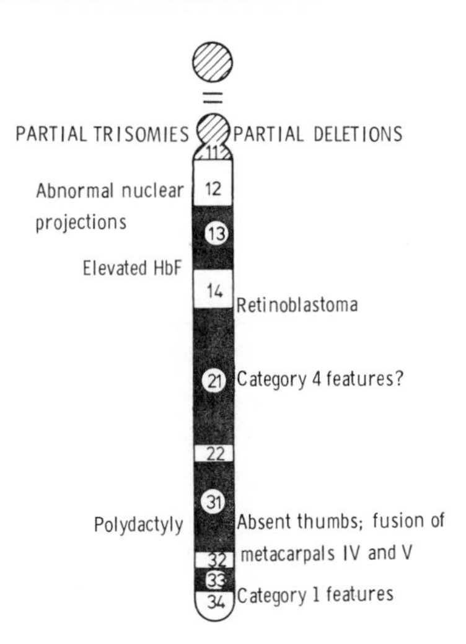


**Fig. 8.** Diagram of chromosome 13 with band designations (Paris Conference 1971) and tentative phenotypic mapping.

cannot be located with certainty, perhaps due to several regulatory sites on chromosome 13.

## C. Partial Deletions

Clinical features in category 1 are due mainly to loss of the most distal segment of the long arm (Fig. 8). It may be inaccurate to attempt mapping using the observations of ring chromosomes, due to instability of some rings with duplication/deletion effects or even loss of the ring. Typical category 1 features in one patient with a terminal deletion (Gilgenkrantz *et al.,* 1973) and the stability of many 13 rings make it *unlikely* to assume that a small percentage of cells with a duplicated/deleted ring should be important for the development of typical category 1 features. Isolated, unexpected clinical findings in category-1 patients as retinoblastoma (Grace *et al.,* 1971, Case 2) and hexadactyly (Biles *et al.,* 1969) may, however, be explained by a localized limited effect of an unstable ring. The clinical influence of a cell line without a ring may also be rejected, since a similar cell line never has been observed in patients with 46/47,+13 mosaicism, suggesting that the 45,−13 cell line is not viable *in vivo.* This argumentation may be supported by the fact that cells with loss of the ring were much more frequent in long-term fibroblast cultures than in blood cultures and absent in bone marrow (Lejeune *et al.,* 1968) indicating an *in vitro* phenomenon.

Absent or hypoplastic thumbs, absent I metacarpals, absent fifth toes and synostosis between metacarpals IV and V may be caused by a deletion distal to 13q22 (Fig. 8), which fits with the localization of polydactyly. The combination of polydactyly and absent I metacarpal bone in one patient with facial features of category 1 (Biles *et al.,* 1969) may reflect a peculiar behaviour of the ring in early embryonic life.

Retinoblastoma, the essential feature in category 3 patients, is a rare childhood cancer occurring in less than one in 20,000 live births. The idea that retinoblastoma is a direct effect of an autosomal dominant gene has been challenged (Zimmerman, 1970) and a two-mutation model (Knudson, 1971) fits better the data on retinoblastoma. Cytogenetic evidence (Howard *et al.,* 1974; Orye *et al.,* 1974) suggests that a locus for retinoblastoma is on the long arm of chromosome 13 in the proximal part of 13q21 or the adjacent 13q14 region (Fig. 8). A deletion of this specific region might affect normal retinal development, predisposing the retinal cells to the action of a second mutational event.

A more detailed location of break points in category 4 patients and examination of additional patients with clinical features different from the first three categories is needed to confirm the presence of a specific phenotypic category 4 picture. If this exists, it might be attributed to a deletion of the more distal part of 13q21 (Fig. 8).

## V. SUMMARY AND CONCLUDING REMARKS

A phenotypic subdivision of twenty four patients with a partial trisomy 13 and seventy two patients with a deletion of chromosome 13 has been attempted to facilitate the establishment of a correct diagnosis in cases with only minor features of a chromosome 13 abnormality. A tentative chromosome 13 map is constructed on the basis of an observed correlation between conspicuous clinical findings/hematologic changes and specific chromosome regions. Exact location of break points in previously published cases and a detailed description of clinical, hematological and cytogenetic data in further patients may give additional information regarding the phenotypic subdivision and the proposed chromosome 13 map. A more detailed evaluation will be possible when additional loci for structural genes are located. The assignment of esterase-D to chromosome 13 (van Heyningen *et al.,* 1975) may be the next step towards the evaluation of a more detailed chromosome 13 map.

## ACKNOWLEDGMENTS

I am grateful to my colleagues for their kind permissions to reproduce original material. I am also indebted to Mrs. Lone Quedens for typing the manuscript.

Research was supported in part by grants from the P. Carl Petersens Foundation (No. B. 885) and the Danish Medical Research Council 512-4276).

## REFERENCES


Allderdice, P. W., Davis, J. G., Miller, O. J., Klinger, H. P., Warburton, D., Miller, D. A., Allen, F. H., Abrams, C. A. L., and McGilvray, E. (1969). The 13— Deletion Syndrome. *Am. J. Hum. Genet.* **21**, 499–512.

Bain, A. D., and Gauld, I. K. (1963). Multiple congenital abnormalities associated with ring chromosome. *Lancet* **2**; 304–305.

Bain, A. D., Insley, J., Douglas, D. M., Gauld, I. K., and Scott, H. A. (1965). Normal/ trisomy 13–15 mosaicism in two infants. *Arch. Dis. Child.* **40**, 442–445.

Biles, A. R., Lüers, T., and Sperling, K. (1969). Multiple Fehlbildungen bei einem Neugeborenen mit einem $D_1$-ringchromosom. *Med. Welt* **20**, 1771–1775.

Bloom, G. E., and Gerald, P. S. (1968). Localization of genes on chromosome 13: Analysis of two kindreds. *Am. J. Hum. Genet.* **20**, 495–511.

Cagianut, B., and Theiler, K. (1970). Bilateral colobomas of iris and choroid. *Arch. Ophthalmol.* **83**, 141–144.

Caspersson, T., Lomakka, G., and Zech, L. (1971). The 24 fluorescence patterns of the human metaphase chromosomes–distinguishing characters and variability. *Heriditas* **67**, 89–102.

Coffin, G. S., and Wilson, M. G. (1970). Ring chromosome D(13). *Am. J. Dis. Child.* **119**, 370–373.

Conen, P. E., and Erkman, B. (1963). A mosaic normal–"D" trisomy boy with a radiation chimera father. *Am. J. Pathol.* **43**, 28a–29a.

Crandall, B. F., Carrel, R. E., Howard, J., Schroeder, W. A., and Müller, H. (1974). Trisomy 13 with a 13-X translocation. *Am. J. Hum. Genet.* **26**, 385–392.

Daniel, W. L. (1970). Aberrant serum protein inheritance in a patient with a ring D chromosome. *Pediatrics* **46**, 120–123.

de Grouchy, J. (1965). Chromosome 18: A topologic approach. *J. Pediatr.* **66**, 414–431.

Emberger, J. M., Nègre, C., and Lafon, R. (1972). Trisomie 13 en mosaique avec isochromosome: 46,XX/46,XX, 13–, 13qi. *Ann. Genet.* **15**, 111–114.

Engel, E., Haddow, J. E., Lewis, J. F., Tipton, R. E., Overall, J. C., McGhee, B. J., Levrat, O. J., and Engel-de Montmollin, M. (1967). Three unusual trisomic patterns in children. *Am. J. Dis. Child.* **113**, 322–328.

Escobar, J. I., and Yunis, J. J. (1974). Trisomy for the proximal segment of the long arm of chromosome 13. *Am. J. Dis. Child.* **128**, 221–222.

Escobar, J. I., Sanchez, O., and Yunis, J. J. (1974). Trisomy for the distal segment of chromosome 13. *Am. J. Dis. Child.* **128**, 217–220.

Forteza, G., Báguena, R., Amat, E., Barcia, B., and Juan, A. (1964). Mosaico trisomia $D_1$/normal en una niña de seis años con un sindrome de trisomia $D_1$ incompleto. *Med. Esp.* **51**, 83–93.

Fried, K., Rosenblatt, M., Mundel, G., and Krikler, R. (1975). Ring chromosome 13 syndrome. *Clin. Genet.* **7**, 203–208.

Fryns, J. P., and Eggermont, E., Verresen, H., and van den Berghe, H. (1974a). Partial trisomy 13: Karyotype 46,XY, –6,+t(13q,6q). *Humangenetik* **21**, 47–54.

Fryns, J. P., Deroover, J., van den Berghe, H., Cassiman, J. J., Goffaux, P., and Lebas, E. (1974b). Malformative syndrome with ring chromosome 13. *Humangenetik* **24**, 235–240.

Gey, W. (1970). Dq–, multiple Missbildungen und Retinoblastom. *Humangenetik* **10,** 362–365.

Gilgenkrantz, S., Pierson, M., and Manuuary, G. (1973). Chromosome 13q+ par translocation probable d'un Y surnuméraire. *Ann. Genet.* **16,** 167–172.

Giraud, F., Hartung, M., Brusquet, Y., Stahl, A., and Bernard, R. (1967). Mosäique chromosomique complexe: Trisomie D/Disomie partielle pour un grand télocentrique. *Pediatrie* **22,** 711–718.

Grace, E., Drennan, J., Colver, D., and Gordon, R. R. (1971). The 13q– deletion syndrome. *J. Med. Genet.* **8,** 351–357.

Grosse, K.-P., and Schwanitz, G. (1973). Ein neues Syndrom durch Chromosomenstückverlust (13q–). *Klin. Paediatr.* **185,** 468–473.

Guschina, L. A., and Golovachev, G. D. (1975). Dq– syndrome with mosaicism in a newborn boy. *Pediatriya (Moscow)* **4,** 71–73.

Hamerton, J. L. (1971). "Human Cytogenetics," Vol. 2. Academic Press, New York.

Hauksdóttir, H., Halldórsson, S., Jensson, Ó., Mikkelsen, M., and McDermott, A. (1972). Pericentric inversion of chromosome no. 13 in a large family leading to duplication deficiency causing congenital malformations in three individuals. *J. Med. Genet.* **9,** 413–421.

Hoehn, H., Wolf, U., Schumacher, H., and Wehinger, H. (1971). A chromosome 13q+ in a patient with characteristics of the trisomy 13 syndrome. *Humangenetik* **13,** 34–42.

Holt, S. B. (1972). The effect of absence of thumb on palmar dermatoglyphics. *J. Med. Genet.* **9,** 448–450.

Hoo, J. J., Obermann, U., and Cramer, H. (1974). The behavior of ring chromosome 13. *Humangenetik* **24,** 161–171.

Howard, R. O., Breg, W. R., Albert, D. M., and Lesser, R. L. (1974). Retinoblastoma and chromosome abnormality. *Arch. Ophthalmol.* **92,** 490–493.

Huehns, E. R., Lutzner, M., and Hecht, F. (1964a). Nuclear abnormalities of the neutrophils in $D_1$ (13–15)–trisomy syndrome. *Lancet* **1,** 589–590.

Huehns, E. R., Hecht, F., Keil, J. V., and Motulsky, A. G. (1964b). Developmental hemoglobin anomalies in a chromosomal triplication: $D_1$ trisomy syndrome. *Proc. Natl. Acad. Sci. U.S.A.* **51,** 89–97.

Ikeuchi, T., Sonta, S., Sasaki, M., Hujita, M., and Tsunematsu, K. (1974). Chromosome banding patterns in an infant with 13q– syndrome. *Humangenetik* **21,** 309–314.

Juberg, R. C., Adams, M. S., Venema, W. J., and Hart, M. G. (1969). Multiple congenital anomalies associated with a ring-D chromosome. *J. Med. Genet.* **6,** 314–321.

Knudson, A. G. (1971). Mutation and cancer: Statistical study of retinoblastoma. *Proc. Natl. Acad. Sci. U.S.A.* **68,** 820–823.

Kučerová, M., Polívková, Z., and Pokorná, M. (1975). Deletion of long arms of chromosome 13. *Humangenetik* **27,** 255–257.

Kuroki, Y., and Nagano, Y. (1974). On the ring 13 chromosome in a malformed infant with special regard to the break point. *Proc. Jpn. Acad.* **50,** 645–647.

Laurent, C., Cotton, J.-B., Nivelon, A., and Freycon, M.-T. (1967). Délétion partielle du bras long d'un chromosome du groupe D (13–15): Dq–. *Ann. Genet.* **10,** 25–31.

Lazjuk, G. I., Lurie, I. W., Kravtzova, G. I., and Usoev, S. S. (1973). New cytogenetic variant of Orbeli's syndrome (46,XY/45,XY,–D/46,XY,Dq+). *Humangenetik* **20,** 219–221.

Leisti, J. (1971). Structural variations in human mitotic chromosomes. *Ann. Acad. Sci. Fenn., Ser. A4* **179,** 1–69.

Lejeune, J., Lafourcade, J., Berger, R., Cruveiller, J., Rethoré, M-O., Dutrillaux, B., Abonyi, D., and Jérôme, H. (1968). Le phénotype (Dr). Etude de trois cas de chromosomes D en anneau. *Ann. Genet.* **11,** 79–87.

Lele, K. P., Penrose, L. S., and Stallard, H. B. (1963). Chromosome deletion in a case of retinoblastoma. *Ann. Hum. Genet.* **27**, 171–174.

McDermott, A., and Parrington, J. M. (1975). Elucidation of a pericentric inversion of a D-group chromosome in the mother of a child with Patau's syndrome. *Ann. Hum. Genet.* **38**, 305–307.

MacIntyre, M. N., Staples, W. I., and Lapolla, J. J. (1964). Partial $D_1$ trisomy in a child whose mother and maternal grandmother demonstrate a D/F translocation. *Am. Soc. Hum. Genet.* **(abstract) p. 21.**

Masterson, J. G., Law, E. M., Rashad, M. N., Cahalane, S. F., and Kavanagh, T. M. (1968). A malformation syndrome with ring D chromosome. *J. Ir. Med. Assoc.* **61**, 398–399.

Moore, M. K., and Engel, E. (1970). Clinical, cytogenetic and autoradiographic studies in 10 cases with rare chromosome disorders. III. Cases 6, 7 and 8. *Ann. Genet.* **13**, 207–212.

Moric-Petrovic, S., Garzicic, B., Despotovic, M., and Kalicanin, P. (1970). Une observation de chromosome D en anneau. *Ann. Genet.* **13**, 265–267.

Niebuhr, E. (1973). Reexamination of a family with a t(13q14q) and a ring D(13) child. *Ann. Genet.* **16**, 199–202.

Niebuhr, E., and Ottosen, J. (1973). Ring chromosome D(13) associated with multiple congenital malformations. *Ann. Genet.* **3**, 157–166.

Noel, B., Quack, B., and Rethoré, M.-O. (1976). Partial deletions and trisomies of chromosome 13. Mapping of bands associated with particular malformations. *Clin. Genet.* **9**, 593–602.

O'Grady, R. B., Rothstein, T. B., and Romano, P. E. (1974). D-group deletion syndromes and retinoblastoma. *Am. J. Ophthalmol.* 77, 40–45.

Opitz, J. M., Slungaard, R., Edwards, R. H., Inhorn, S. L., Muller, J., and de Venecia, G. (1969). Report of a patient with a presumed Dq– syndrome. *Birth Defects, Orig. Artic. Ser.* **5**, 93–99.

Orbeli, D. J., Lurie, I. W., and Goroshenko, J. L. (1971). The syndrome associated with the partial D-monosomy. *Humangenetik* **13**, 296–308.

Orye, E., Delbeke, M. J., and Vandenabeele, B. (1974). Retinoblastoma and long arm deletion of chromosome 13. Attempts to define the deleted segment. *Clin. Genet.* **5**, 458–465.

Paris Conference 1971 (1972). Standardization in human cytogenetics. *Birth Defects, Orig. Artic. Ser.* **8**, No. 7. The National Foundation, New York.

Picciano, D. J., Berlin, C. M., Davenport, S. L. H., and Jacobson, C. B. (1972). Human ring chromosomes: A report of five cases. *Ann. Genet.* **15**, 241–247.

Rethoré, M.-O., Praud, E., le Loc'h, J., Joly, C., Saraux, H., Aussannaire, M., and Lejeune, J. (1970). *Presse Med.* **78**, 955–958.

Rosenkranz, W., and Kaloud, H. (1972). Nicht balancierte D/E–Translokation. *Paediatr. Paedol.* 7, 377–379.

Salamanca, F., Buentello, L., and Armendares, S. (1972). Ring $D_1$ chromosome with remarkable morphological variation in a boy with mental retardation. *Ann. Genet.* **15**, 183–186.

Schinzel, A., Schmid, W., and Mürset, G. (1974). Different forms of incomplete trisomy 13. Mosaicism and partial trisomy for the proximal and distal long arm. *Humangenetik* **22**, 287–298.

Schmid, W., Mühlethaler, J. P., Briner, J., and Knechtli, H. (1975). Ring chromosome 13 in a polymalformed anencephalic. *Humangenetik* **27**, 63–66.

Schwanitz, G., Grosse, K.-P., Semmelmayer, U., and Mangold, H. (1974). Partielle freie Trisomie 13 in einer Familie mit balancierter Translokation (13q–;16q+). *Monatsschr. Kinderheilkd.* **122**, 337–342.

Smith, D. W. (1970). "Recognizable Patterns of Human Malformation." Saunders, Philadelphia, Pennsylvania.

Snodgrass, G. J. A. I., Butler, L. J., France, N. E., Crome, L., and Russell, A. (1966). The D (13–15) trisomy syndrome: An analysis of 7 examples. *Arch. Dis. Child.* **41**, 250–261.

Sparkes, R. S., Carrel, R. E., and Wright, S. W. (1967). Absent thumbs with a ring D2 chromosome: A new deletion syndrome. *Am. J. Hum. Genet.* **19**, 644–659.

Stalder, G. R., Buhler, E. M., Gadola, G., Widmer, R., and Freuler, F. (1964). A family with balanced $D_1 \rightarrow C^S$–translocation carriers and unbalanced offspring. *Humangenetik* **1**, 197–200.

Stone, D., Akad, A. S., Noyes, C., and Lamson, E. (1966). 13–15 trisomy mosaicism in a normal-looking 14-year-old retarded girl. *J. Med. Genet.* **3**, 142–144.

Sylvester, P. E., Richards, B. W., Rundle, A. T., and Stewart, A. (1971). Pathological observations on a male patient with D-ring chromosome. *J. Ment. Defic. Res.* **15**, 207–223.

Talvik, T., Mikelsaar, A-V., Mikelsaar, R., Käsosaar, M., and Tüür, S. (1973). Inherited translocations in two families (t(14q+;10q–) and t(13q–;21q+)). *Humangenetik* **19**, 215–226.

Taylor, A. I. (1968). Autosomal trisomy syndromes: A detailed study of 27 cases of Edwards' Syndrome and 27 cases of Patau's syndrome. *J. Med. Genet.* **5**, 227–252.

Taylor, A. I. (1969). Autosome abnormalities in man. *In* "Handbook of Molecular Cytology" (A. Lima-de-Faria, ed.), pp. 804–834. North-Holland Publ., Amsterdam.

Taylor, A. I. (1970). Dq–, Dr and retinoblastoma. *Humangenetik* **10**, 209–217.

Taylor, A. I., and Polani, P. E. (1964). Autosomal trisomy syndromes, excluding Down's. *Guy's Hosp. Rep.* **113**, 231–249.

Taylor, M. B., Juberg, R. C., Jones, B., and Johnson, W. A. (1970). Chromosomal variability in the $D_1$ trisomy syndrome. *Am. J. Dis. Child.* **120**, 374–381.

Taysi, K., Bobrow, M., Balci, S., Madan, K., Atasu, M., and Say, B. (1973). Duplication/deficiency product of a pericentric inversion in man: A cause of $D_1$ trisomy syndrome. *J. Pediatr.* **82**, 263–268.

Therman, E., Patau, K., DeMars, R. I., Smith, D. W., and Inhorn, S. L. (1963). Iso/Telo-$D_1$ mosaicism in a child with an incomplete $D_1$ trisomy syndrome. *Port. Acta Biol., Ser. A* **7**, 211–224.

Thompson, H., and Lyons, R. B. (1965). Retinoblastoma and multiple congenital anomalies associated with complex mosaicism with deletion of D chromosome and probable D/C translocation. *Hum. Chromosome Newslett.* No. 15, p. 21.

Tolksdorf, M., Goll, U., Wiedemann, H.-R., and Pfeiffer, R. A. (1970). Die Symptomatik von Ringchromosomen der D-gruppe. *Arch. Kinderheilkd.* **181**, 282–295.

Tzoneva-Maneva, M. T., Petrov, B., and Bosajieva, E. (1965). Cytogenetic studies in two brothers with Laurence-Moon-Biedl's syndrome. *Hum. Chromosome Newslett.* No. 15, p. 22.

van Heyningen, V., Bobrow, M., Bodmer, W. F., Gardiner, S. E., Povey, S., and Hopkinson, D. A. (1975). Chromosome assignment of some human enzyme loci: Mitochrondrial malate dehydrogenase to 7, mannosephosphate isomerase and pyruvate kinase to 15 and probably, esterase D to 13. *Ann. Hum. Genet.* **38**, 295–303.

Van Kempen, C. (1966). A case of retinoblastoma, combined with severe mental retardation and a few other congenital anomalies, associated with complex aberrations of the karyotype. *Maandschr. Kindergeneeskd.* **34**, 90–95.

Varela, M. A., and Sternberg, W. H. (1969). Ring chromosomes in two infants with congenital malformations. *J. Med. Genet.* **6**, 334–341.

Vogel, F. (1973). Genotype and phenotype in human chromosome aberrations and in the minute mutants of *Drosophila melanogaster. Humangenetik* **19**, 41–56.

Wang, H.-C., Melnyk, J., McDonald, L. T., and Uchida, I. A. (1962). Ring chromosomes in human beings. *Nature (London)* **195**, 733–734.

Wilson, M. G., and Melnyk, J. (1967). Translocation/normal mosaicism in $D_1$ trisomy. *Pediatrics* **40,** 842–846.
Wilson, M. G., Melnyk, J., and Towner, J. W. (1969). Retinoblastoma and deletion D(14) syndrome. *J. Med. Genet.* **6,** 322–327.
Wilson, M. G., Towner, J. W., and Fujimoto, A. (1973). Retinoblastoma and D-chromosome deletions. *Am. J. Hum. Genet.* **25,** 57–61.
Yunis, J. J., and Hook, E. B. (1966). Deoxyribonucleic acid replication and mapping of the $D_1$ chromosome. *Am. J. Dis. Child.* **111,** 83–89.
Yunis, J. J., Hook, E. B., and Mayer, M. (1964). DNA replication pattern of trisomy $D_1$. *Lancet* **2,** 935–937.
Zimmerman, L. E., (1970). Changing concepts concerning the pathogenesis of infectious disease. *Am. J. Ophthalmol.* **69,** 947–964.
Zink, U., Rix, R., Grosse, K.-P., and Schwanitz, G. (1973). Ringchromosom $D_{13}$. Kasuistik und Übersicht. *Klin. Paediatr.* **185,** 192–197.

# 10

# Abnormal Chromosomes 14 and 15 in Abortions, Syndromes, and Malignancy

HERMAN E. WYANDT, R. ELLEN MAGENIS,
and FREDERICK HECHT



## I. INTRODUCTION

Prior to 1970–1971, when banding revolutionized chromosome identification, the individual D (13–15) chromosomes were distinguishable only by late labeling and autoradiography (Yunis *et al.*, 1964; Schmid and Carnes, 1965; Gianelli and Howlett, 1966; Miller, 1970). Even so, small chromosomes or chromosomal rearrangements derived from Nos. 14 or 15 were difficult to identify. These problems have fortunately been solved by banding, and now chromosomes 13, 14, and 15 can be easily separated and small segments often identified.

## II. FULL TRISOMY OF CHROMOSOMES 14 AND 15

Complete trisomy 14 and 15 have been identified in spontaneous abortions. Whether they occur in live-borns is still conjectural. Evidence to date suggests that full trisomy 14 or 15 is not compatible with live birth.

### A. Live-borns

Since the advent of banding, one case with trisomy 14/normal mosaicism has been reported (Rethoré *et al.*, 1975). This case involves an infant girl who had about 10% of cells with an extra chromosome 14, the rest being normal. She had severe congenital anomalies and died at 3 months of age.

On the basis of autoradiographic identification, Murken *et al.* (1970) had previously described a patient with total trisomy 14. This was a $2\frac{1}{2}$ year-old female with surprisingly mild anomalies (as compared to individuals partially trisomic for chromosome 14); nevertheless, many of her clinical findings were similar to those seen in patients trisomic for part of 14q. It is possible that this girl may have had a normal cell line.

No case of trisomy 15 in live-borns has been reported thus far.

### B. Spontaneous Abortions

Three to 4% of spontaneous abortions have D trisomy (Carr, 1965; Boué *et al.*, 1967). A banded study of sixteen abortuses revealed two with trisomy 13, ten with trisomy 14, and four with trisomy 15 (Rethoré *et al.*, 1975). Another study of fifteen abortuses revealed one with trisomy 13, four with trisomy 14, and ten with trisomy 15 (Therkelson *et al.*, 1973). The inversion of the ratios of trisomy 14:15 in these two studies is unexplained, but may reflect the small sample sizes. The sum of both studies, however, indicates that the majority (28/31) of D trisomic abortuses have an extra No. 14 or 15 chromosome.

The frequencies of trisomy 13, 14 and 15 conceptuses appear roughly equal (Table 1). The deficit of trisomy 14 and 15 live-borns is apparently due to *in utero* loss.

**TABLE I**
**Frequencies of Trisomy 13, 14, and 15 in Live-borns, Abortuses, and Total Conceptions[a]**

| | Trisomy | | |
|---|---|---|---|
| | 13 | 14 | 15 |
| Live-borns | ~1/4000 | – | – |
| Abortuses | ~1/2000 | ~1/1000 | ~1/1000 |
| Total conceptions | ~1/1000 | ~1/1000 | ~1/1000 |

[a]The frequency of trisomy 13 in live-borns is from Carr (1965). The frequencies in abortuses were calculated by assuming that 10% had trisomy 13, 45% trisomy 14, and 45% trisomy 15, that 3–4% of abortuses have D trisomy, and that 15% of pregnancies spontaneously end in abortion. The above calculations are approximations.

Cultured cells with trisomy 14 have been characterized by low growth potential, variations in cell size and morphology, and metabolic disturbances when compared to normal diploid cells (Kuliev *et al.*, 1974).

### C. Robertsonian Rearrangements

Robertsonian ("centric-fusion") translocations constitute one of the major categories of chromosome rearrangements. Approximately 1/1100 live-borns are balanced Robertsonian translocation carriers (Hamerton *et al.*, 1975). Among Robertsonian rearrangements, those of the 13;14 and 14;21 type predominate (Hecht and Kimberling, 1971; Jacobs *et al.*, 1974). In spite of this, *no* case of trisomy for the long arm of chromosome 14 has been reported in the many families with 13;14 or 14;21 translocations. In one study of a large kindred involving a Robertsonian translocation between chromosomes 14 and 22, about one-third (13/37) of known conceptions resulted in abortions and there were no live-borns with trisomy for the long arm of 14 or 22 (Neu *et al.*, 1975).

Robertsonian rearrangements involving chromosomes 15 have similarly not resulted in trisomy for the long arm of 15. One woman with a history of 13 abortions was studied and found to have a Robertsonian rearrangement between two No. 15 chromosomes (Lucas, 1969). Examination of the last abortus showed a similar translocation plus 5 normal D group chromosomes. The abortus was thus trisomic for the long arm of chromosome 15.

These data are in keeping with the concept that total trisomy of the long arm of chromosome 14 or 15 is lethal *in utero*.

# III. PARTIAL ANEUPLOIDY FOR CHROMOSOMES 14 AND 15

## A. Partial Trisomies 14 and 15

### *1. Partial Trisomies of Chromosome 14*

*a. Proximal Long Arm.* Nine cases with trisomy of the proximal portion of the long arm of chromosome 14 are summarized in Table II. Note that five cases (cases 3–7) involve approximately identical segments (all of the short arm plus the proximal half of the long arm).

One case, not included in Table II, was reported by Muldal *et al.* (1973). In this case an extra small acrocentric chromosome, believed to be a partially deleted No. 14 was present in a 16-year-old female with surprisingly mild anomalies when compared with cases 3–7. However, many of the same anomalies were present (Fig. 4) and included: growth and mental retardation, broad glabella and broad nose, low-set ears, and anomalies of the joints and extremities.

*b. Distal Long Arm.* Three cases trisomic for the distal segment of the long arm of 14 have been reported. One case (not included in Table II) involved a fetus therapeutically aborted at 19 weeks (Wahlström, 1974). Chromosome studies revealed a translocation between Nos. 14 and 22 (Table III). The fetus was trisomic for the distal portion of chromosome 14 (qter→q22). The only clinical features mentioned were an umbilical hernia and possible rocker bottom feet. Autopsy, however, revealed anomalies of the lung and liver, and a small thymus.

One of the cases in Table II (case 10) is of particular interest because she was a first cousin of another child (case 9) with trisomy for the proximal portion of 14q. Thus cases 9 and 10 are trisomic for complementary segments of chromosome 14 (Figs. 1 and 2). A detailed description of case 9 has been published by Reiss *et al.* (1972).

Case 10 has not been previously reported. Information on this patient has been provided by L. Luzzatti (personal communication). In addition to the data in Table II on this case, it is pertinent that the patient was admitted to the hospital at 3 months of age and demonstrated failure to thrive (no gain over birth weight), widely open anterior and posterior fontanelles and sagittal suture, right orbit lower than left, whistling mouth, contractures of fourth fingers, overlapping of third and fifth fingers bilaterally, and valgus deformity of the right foot.*

---

*Case 10 was born to individual III 3 in the pedigree published by Reiss *et al.* (1972). Case 10 was a subsequently born sib of IV 5 and 6.

**TABLE II**
**Summary of Clinical Findings in Cases of Partial 14 Trisomy**

| Segment of chromosome 14 that is trisomic | pter→ q12 | pter→ q13 | pter→q22 or 23 | | | | | pter→q24 (mosaic) | | q24→ qter | q22→ qter | Total trisomy (mosaic) |
|---|---|---|---|---|---|---|---|---|---|---|---|---|
| Case number | 1 | 2[a] | 3 | 4 | 5 | 6 | 7 | 8 | 9 | 10 | 11 | 12 |
| Reference | Fryns *et al.* (1974) | Laurent *et al.* (1973) | Raoul *et al.* (1975) | Turleau *et al.* (1975) | Fawcett *et al.* (1975) | Francke *et al.* (unpublished) | Allderdice *et al.* (1971) | Short *et al.* (1972) | Reiss *et al.* (1972) | Luzzatti (personal communication) | Pfeiffer *et al.* (1973) | Rethoré *et al.* (1975) |
| Age studied (‡Decreased) | 7 years | 3 months | 3 years | 18 months | 6 months | 5 years | 4 years | 60 hours‡ | 10 months | 3 months‡ | 15 months | 3 months‡ |
| Sex | F | M | M | M | F | F | F | M | M | F | F | F |
| Length of gestation (weeks) | 40 | 39 | 28 | 39 | 38 | 40 | 40 | term | term | term | – | 43 |
| Birth weight (gm) | 2200 | 2100 | 1800 | 2650 | 2220 | 2495 | 2600 | 2190 | 2400 | 3070 | 3650 | 3140 |
| General | | | | | | | | | | | | |
| Motor retardation | + | | | + | | + | + | | + | | + | |
| Growth retardation | + | + | + | + | + | + | + | + | + | | | |
| Mental retardation | + | | + | + | | + | + | | + | | + | |
| Seizures | | | | | + | | + | + | | | | |
| Tone | | | | | Hyper | | | | Hyper | Hyper | Hypo | |

*continued*

**TABLE II (*continued*)**

| Segment of chromosome 14 that is trisomic: | pter→ q12 | pter→ q13 | pter→q22 or 23 | | | | | pter→q24 (mosaic) | | q24→ qter | q22→ qter | Total trisomy (mosaic) |
|---|---|---|---|---|---|---|---|---|---|---|---|---|
| Case number: | 1 | 2[a] | 3 | 4 | 5 | 6 | 7 | 8 | 9 | 10 | 11 | 12 |
| Reference: | Fryns *et al.* (1974) | Laurent *et al.* (1973) | Raoul *et al.* (1975) | Turleau *et al.* (1975) | Fawcett *et al.* (1975) | Francke *et al.* (unpublished observations) | Allderdice *et al.* (1971) | Short *et al.* (1972) | Reiss *et al.* (1972) | Luzzatti (personal communication) | Pfeiffer *et al.* (1973) | Rethoré *et al.* (1975) |
| Head | | | | | | | | | | | | |
| Microcephaly | Mild | + | + | | | + | | + | Mild | + | | + |
| Brachycephaly | | | | | + | | | | | | + | |
| Low hair line | + | | + | + | + | + | | | + | | | |
| High forehead | | | | | | | | | | + | | + |
| Eyes | | | | | | | | | | | | |
| Telorism | Hyper | | | Hypo | Hypo | | Hypo | Hyper | Hypo | | | Hyper |
| Epicanthal folds | | | | | | + | | | | + | | + |
| Small palpebral fissures | | | | | + | | + | | + | | | |
| Ptosis | | | + | | | | + | | | | | |
| Microphthalmia | | + | + | | | | | + | | | | |

| | | | | | | | | | | | | |
|---|---|---|---|---|---|---|---|---|---|---|---|---|
| Antimongoloid slant | | | | | | | | | | + | + | |
| Strabismus | | | | | | + | | | + | | | |
| Ears | | | | | | | | | | | | |
| Low set | | | | | + | + | + | + | + | + | | + |
| Malformed | | | | | + | + | | + | + | | | |
| Nose | | | | | | | | | | | | |
| Broad nasal bridge | + | + | | | | + | | | | | | + |
| Prominent tip | + | | + | + | + | + | | + | + | | | + |
| Palate | | | | | | | | | | | | |
| High arched | + | | | | + | | + | | | + | | + |
| Cleft | | | + | + | + | | | | + | | | |
| Mouth and chin | | | | | | | | | | | | |
| Unusual philtrum | | | | | + | + | | | + | | | |
| Unusual mobile mouth | + | + | + | + | + | | | | + | | | + |
| Protruding upper lip | | | | | | | | | | + | | |
| Buccal fat pads | | | | | | | | | | + | | |
| Micrognathia | | | | | | | + | + | + | + | | + |

*continued*

**TABLE II** (*continued*)

| Segment of chromosome 14 that is trisomic: | pter→q12 | pter→q13 | pter→q22 or 23 | | | | | pter→q24 (mosaic) | | q24→qter | q22→qter | Total trisomy (mosaic) |
|---|---|---|---|---|---|---|---|---|---|---|---|---|
| Case number: | 1 | 2[a] | 3 | 4 | 5 | 6 | 7 | 8 | 9 | 10 | 11 | 12 |
| Reference: | Fryns *et al.* (1974) | Laurent *et al.* (1973) | Raoul *et al.* (1975) | Turleau *et al.* (1975) | Fawcett *et al.* (1975) | Francke *et al.* (unpublished observations) | Allderdice *et al.* (1971) | Short *et al.* (1972) | Reiss *et al.* (1972) | Luzzatti (personal communication) | Pfeiffer *et al.* (1973) | Rethoré *et al.* (1975) |
| Short neck | + | + | + | + | + | + | | + | + | | | + |
| Hands | | | | | | | | | | | | |
| Tapered fingers | + | + | + | | | + | | | | | | |
| Clinodactyly | | | + | | + | | | | | | | |
| Camptodactyly | | + | | | + | | | + | + | + | | + |

| | | | | | | | | | |
|---|---|---|---|---|---|---|---|---|---|
| Skeletal anomalies | | | | | | | | | |
| Club feet | + | + | | | + | | + | | |
| Dislocated or other hip anomaly | + | + | | | | + | | | + |
| Missing left radius | + | | | | | | | | |
| Kyphotic spine | | | | | + | | | | |
| Hypoplastic twelfth rib | | | | | | + | | | |
| Cardiovascular | | | | | | | | | |
| Cyanosis | | | Acro | | + | + | + | + | + |
| Heart defects | | + | | | + | + | + | | + |
| Genital anomalies | | + | | + | + | + | | | |

[a]Monosomic for proximal 21q plus centromere and short arm (i.e. 21pter→q21)

**TABLE III**
**Summary of Chromosomal Findings in Cases of Partial 14 Trisomy**

| Paris nomenclature[a] | Parental origin | Mechanism | References |
|---|---|---|---|
| 47,XX,+der(14),t(14;19)(q11;?) | Mat | 3:1 Meiotic segregation error | Fryns *et al.* (1974) |
| 46,XY,–21,+der(14),t(14;21)(q13;q22) | Pat | Meiotic adjacent 2 segregation | Laurent *et al.* (1973) |
| 47,XY,+der(14),t(10;14)(p15;q22) | Mat | Segregation error | Raoul *et al.* (1975) |
| 47,XY,+der(14),t(12;14)(q24;q21) | Mat | 3:1 Meiotic segregation error | Turleau *et al.* (1975) |
| 47,XX,+der(14),t(14;20)(q22;q13) | Mat | 3:1 Meiotic segregation error | Fawcett *et al.* (1975) |
| 47,XX,+der(14),t(X;14)(p22;q21) | Mat | 3:1 Meiotic segregation error | U. Francke (personal communication) |

| | | | |
|---|---|---|---|
| 46,XX,–6,–14,t(6q;14q)t(20p;14q),t(6q; 20p)+der(14),t(6q;14q) | Mat | Meiotic adjacent 2 (complex) segregation | Allderdice *et al.* (1971) |
| 47,XY,+der(14),t(9p;14q)(p?;q24) | Mat | 3:1 Meiotic segregation error | Short *et al.* (1972) |
| 46,XY,t(2;14)(q37;q24)/ 47,XY,+der(14),t(2;14)(q37;q24) | Pat | Meiosis II or mitotic nondisjunction | Reiss *et al.* (1972) |
| 46,XX,der(21),t(14;21)(q22;q22) | Pat | Meiotic adjacent 1 segregation | Pfeiffer *et al.* (1973) |
| 46,XY,der(22),t(24;22)(q22;q22) | Pat | Meiotic adjacent 1 segregation | Wahlström (1974) |
| 46,XX,der(2),t(2;14)(q37;q24)? | Mat | Meiotic adjacent 1 segregation | L. Luzzatti (personal communication) |

[a]Descriptions of chromosomal findings are according to Paris Conference 1971: *Standardization in Human Cytogenetics.*

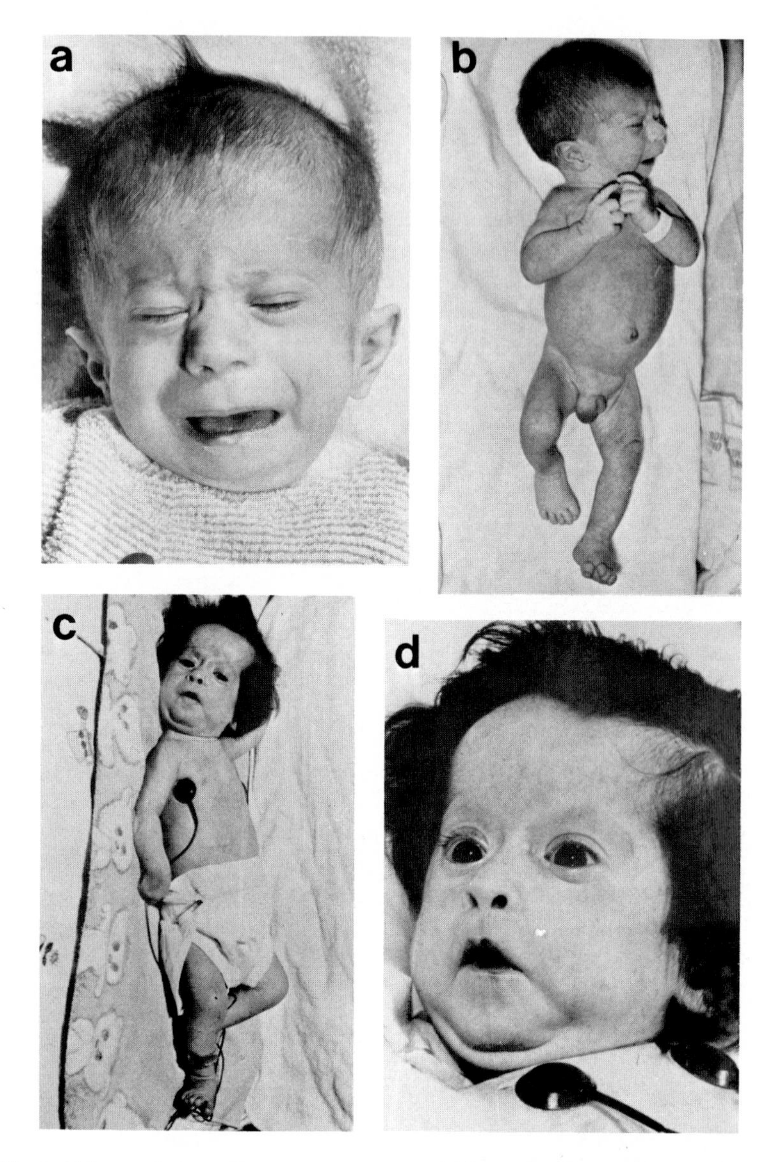


**Fig. 1.** Trisomy for complementary segments of 14q. (a,b) Clinical views of child (Case 9, Table II) with trisomy for proximal 14q in some of his cells. (c,d) First cousin (Case 10, Table II), who is trisomic for the complementary distal segment of 14q. (Photographs c and d provided by L. Luzzatti).

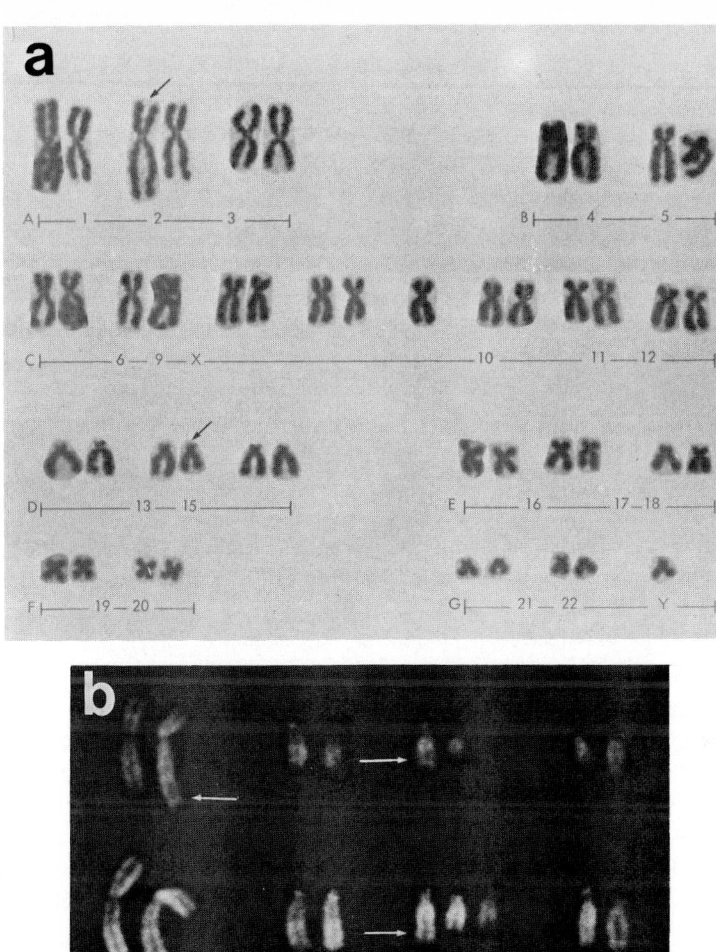


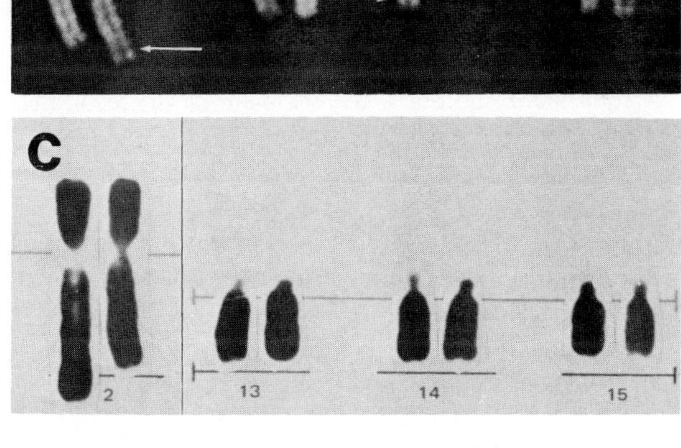


**Fig. 2.** (a) Unbanded karyotype showing a balanced 2;14 translocation present in father, grandfather and paternal aunt of Case 9 (Table II). (b) Partial karyotypes (by quinacrine staining) showing two cell lines present in child (Fig. 1a and b, Case 9): a balanced cell line (top) similar to his parent's and a second cell line (bottom) with two copies of partially deleted No. 14. Arrows point to the break points. (Reprinted with modification from Reiss *et al.*, 1972). (c) Partial unbanded karyotype from Case 10 showing long number 2. Child is trisomic for distal 14q. (Latter karyotype was provided by L. Luzzatti).

*c. Clinical Findings.* The general features of trisomy for the proximal portion of 14q (Table II) include low birth weight, mental retardation, generalized growth and psychomotor retardation, microcephaly, microphthalmia, hypotelorism, low hair line, unusually large or mobile mouth (large when crying) with typically thin lips down-turning at the corners, a prominent pointed or bulbous tipped nose, high arched or cleft palate, short neck and anomalies of the extremities (i.e., clinodactyly, camptodactyly, and tapered fingers, (Fig. 3). Additional features such as low-set or malformed ears, more severe skeletal anomalies, and anomalies of the extremities have been noted in several of the cases.

Trisomy for the distal portion of 14q appears to produce a different spectrum of anomalies. Features such as antimongoloid slant, cyanosis, anomalous blood vessels and heart defects are more typical of the latter cases. Birth weights are normal. The one case (case 10) presented above particularly shows facial features not seen in the cases involving the proximal segment. These include high forehead, unusual appearing mouth with a pursed or protruding upper lip, and prominent buccal fat pads.

As might be expected, the phenotypic consequences of partial trisomy 14 generally appear milder, the smaller the segment that is trisomic (Fig. 4). Trisomy for the proximal portion of 14q appears to produce milder clinical manifestations than trisomy for the distal portion of 14q. It is difficult, however, to make specific karyotype-phenotype comparisons on the basis of length of chromosomal segment involved since only six of these cases (3–8) are trisomic for approximately an identical segment (e.g., pter→q22 or 23), and the remaining cases involve segments of varying length.

Case 12, Table II (Rethoré *et al.*, 1975) was mosaic with some cells trisomic for an entire chromosome 14 (see discussion, Section II, A).

*d. Chromosomal Data.* The majority of reported cases occurred as the result of a translocation with another chromosome. The break points were typically in q22–23 of chromosome 14 and near the telomere of the second chromosome. Six of the nine cases (Table III) partially trisomic for the proximal portion of 14 are due to a 3:1 meiotic segregation error; all six are *maternal* in origin. The remaining three cases involve (1) *complex mosaicism,* (2) a *complex rearrangement* between three chromosomes and (3) *monosomy* for part of 21 as well as trisomy for proximal 14q. These three cases are less consistent with regard to parental origin and mechanisms of error.

Of the cases partially trisomic for the distal portion of 14q, all three are due to adjacent 1, meiotic segregation; one of these is of paternal origin.

One can probably explain the consistent 3:1 meiotic segregation error in cases of trisomy for the proximal portion of 14q on the basis of (1) the mechanics of meiotic segregation involving translocations, and (2) the relative effect that other possible abnormal products would have on phenotype. Trisomy for the distal portion of 14q can only arise from an adjacent 1 segregation. One of the outcomes of adjacent 2 segregation would be trisomy of proximal 14q plus

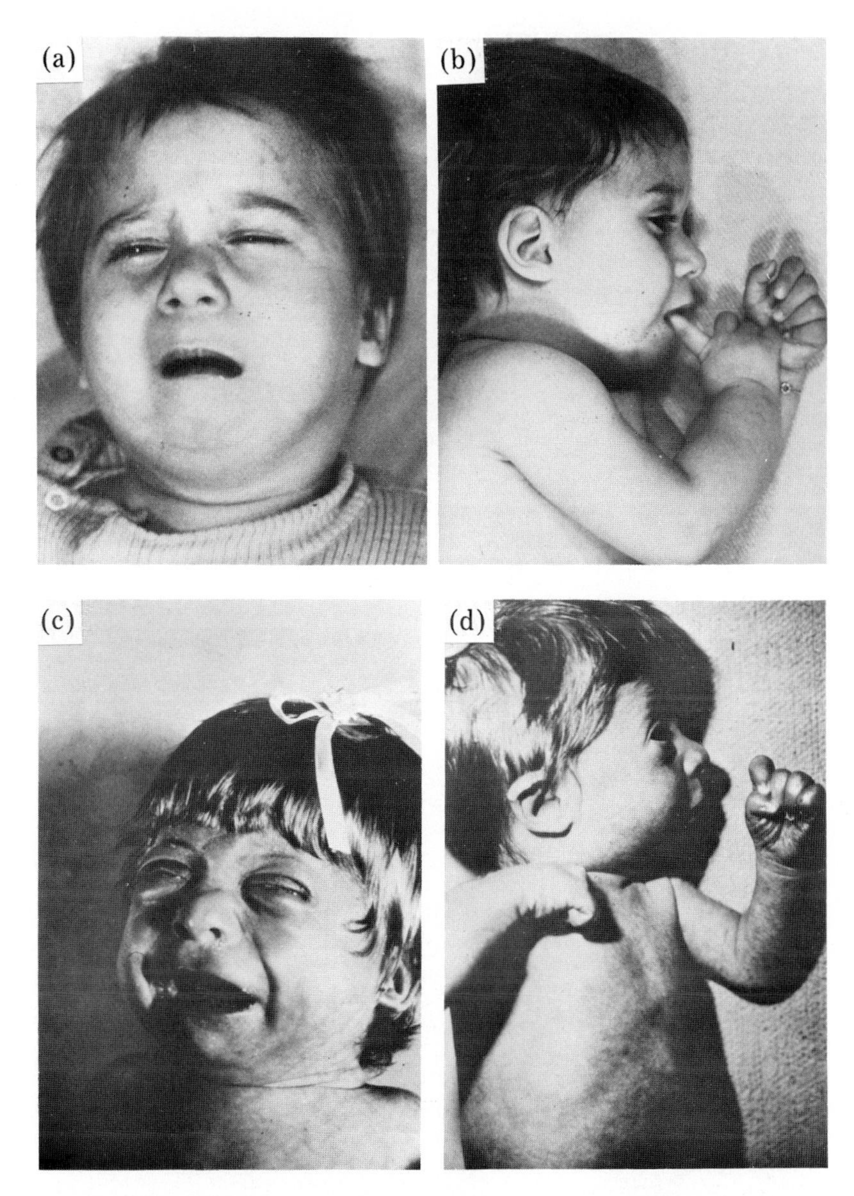


**Fig. 3.** Trisomy for proximal half of 14q. Clinical views of two different patients. (a,b) Case 4 (Table II). (Reprinted by permission from Turleau *et al.*, 1975). (c,d) Case 5 (Table II); photographs provided by U. Francke (case reported by Fawcett *et al.*, 1975).

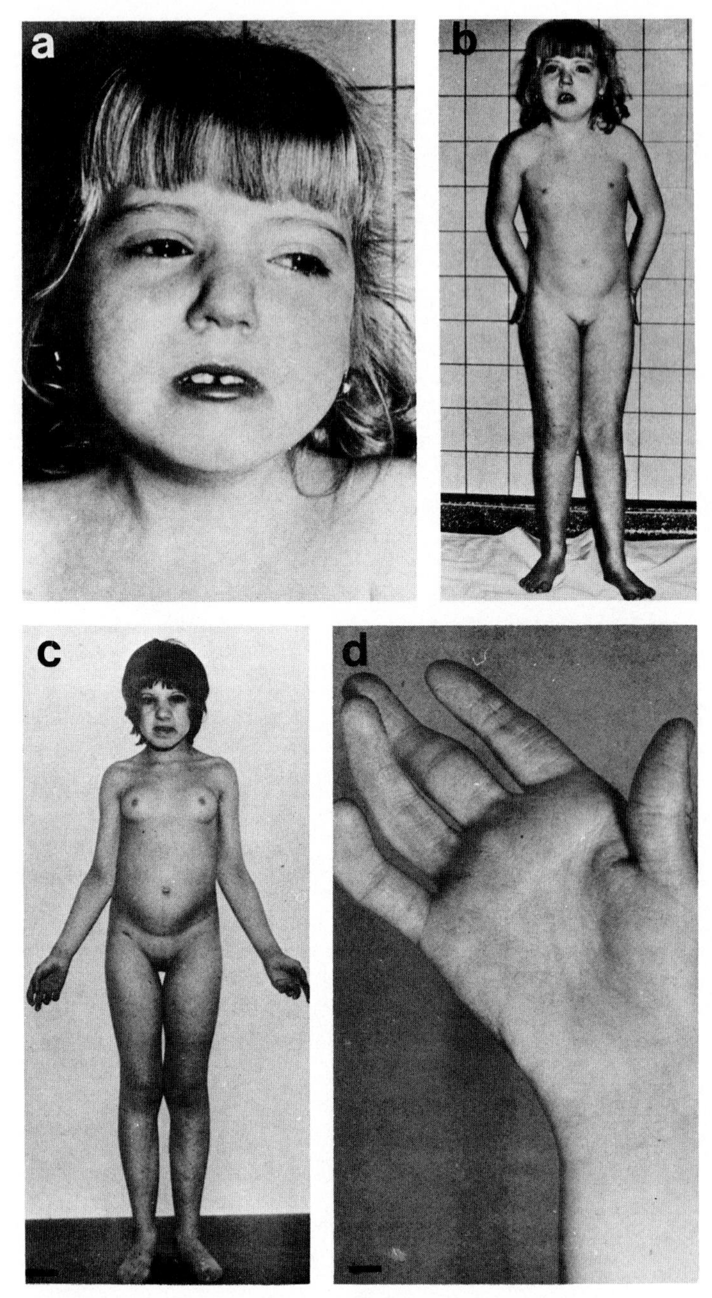


**Fig. 4.** Trisomy for proximal half of 14q. Clinical views of two mildly affected individuals. (a,b) Case 1 (Table II). (Reprinted by permission from Fryns *et al.,* 1974); (c,d) photographs reprinted with permission from Muldal *et al.,* 1973 (see text for discussion).

monosomy for most of the second chromosome (i.e., all except the telomeric portion). It is thus of interest to note that the one case (Case 2), also involving partial monosomy of 21, where the break point was not near the telomere, was able to survive and did involve an adjacent 2 rather than a 3:1 segregation error. Such data tend to support the concept that trisomies for distal 14q are less frequent and probably more often lethal.

Less easy to answer is why all cases involving a 3:1 meiotic segregation error, discussed above, are maternal in origin.

### 2. *Partial Trisomies of Chromosome 15*

*a. Proximal Long Arm.* There are seventeen cases of duplication of the proximal portion of the long arm of chromosome 15 (Tables IV and V). Clinically they fall into at least two distinct groups. One group of patients has few defects which include: moderate to severe mental retardation, hypotonia, strabismus, heavy eyebrows and seizures. Growth is normal (Figs. 5 and 6). The chromosomes of most of these mildly affected patients are satellited on both the short and long arms and/or label on the ends of both arms with the antibody, anti-5-methylcytosine (Breg *et al.,* 1974). This stain appears to label only the short arm satellite region of chromosome 15 and the heterochromatic regions of chromosome 1, 9, 16 and the Y (Lubit *et al.,* 1974). Thus, use of the stain should be of help in delineating *de novo* extra chromosome 15 material when the short arm is involved. The cases of Parker and Alfi (1972) and Bucher *et al.* (1973) (Table IV, Cases 10 and 11) though not shown to have bisatellited duplicated No. 15's, have the mild phenotype described above.

The cases of Bannister and Engel (1975), Cohen *et al.* (1975), and Magenis *et al.* (1972) (Table IV, Cases 3, 4, and 8) share many clinical features: growth retardation, profound mental retardation, low-set or posteriorly related ears, high arched palate, and by photograph, similar facies including almond shaped down-slanted eyes, small palpebral features, telecanthus and wide-spaced superiorly placed eyebrows lacking an arch. Cases 3 and 4 in addition exhibit by photograph anteriorly placed nostrils and increased nasolabial distance (Fig. 7). The partial trisomy in these two cases is the result of familial translocations; amniocentesis revealed a sibling to case 4 to be similarly affected, and the fetus was aborted. The third case is *de novo,* but when stained with anti-5-methylcytosine, only the short arms are definitely labeled.* The chromosome does not appear bisatellited, and the break point appears similar to the translocation cases.

The familial cases of Rethoré *et al.* (1973) are demonstrably monosomic for proximal chromosome 21 as well as trisomic for proximal 15q; this may explain the difference in phenotype. The familial translocation cases studied by Breg (Breg *et al.,* 1974) (Fig. 6; Table V), with features noted in the table he provided, (see Table V), also appear to be somewhat different; it may be that the

*Stained by O. J. Miller.

**TABLE IV**
**Summary of Clinical Findings in Cases of Partial 15q Trisomy**

| Type of chromosome 15 anomaly | Trisomy, proximal segment | | | | | | | | | | | Trisomy, distal segment |
|---|---|---|---|---|---|---|---|---|---|---|---|---|
| | Due to translocation | | | | *De novo* bisatellited | | | *De novo* | | | | |
| Case number | 1[a] | 2[a] | 3 | 4 | 5 | 6 | 7 | 8[b] | 9 | 10 | 11 | 12 |
| Reference | Rethoré *et al.* (1973) | Rethoré *et al.* (1973) | Bannister and Engel (1975) | Cohen *et al.* (1975) | Crandall *et al.* (1973) | Crandall *et al.* (1973) | Watson and Gordon (1974) | Magenis *et al.* (1972) | Howard *et al.* (1974), and Howard, personal communication | Parker and AIP; (1972) | Bucher *et al.* (1973) | Fujimoto *et al.* (1974) |
| Length of gestation | term | 8 months | term | 36 weeks | term | term | term | 39 weeks | 35 weeks | | | term |
| Paternal age (years) | | | 20 | 34 | | | 38 | 36 | 31 | | | |
| Maternal age (years) | | | 18 | 32 | 28 | 34 | 33 | 30 | 30 | | | 30 |
| Birth weight (grams) | 2600 | 2400 | 2410 | 2780 | 3232 | 3629 | 3400 | 2929 | 1956 | | | 2980 |
| Sex | F | M | M | F | M | M | F | F | F | F | | M |

| Age at report | 14 years | 5 years | 16 months | 4 years | 11 years | 11 years | 11 months | 16 years | 10 years | 10 years | | 3 years |
|---|---|---|---|---|---|---|---|---|---|---|---|---|
| General | | | | | | | | | | | | |
| Growth retardation | Mild | – | Mild | + | – | – | Mild | Mild | – | | | |
| Developmental delay | | | + | + | + | Mild | + | + | + | | | + |
| Mental retardation | + | + | + | + | + | + | + | + | + | + | + | + |
| Siezures | | | | + | + | + | + | | | | | + |
| Tone | Hyper | Hyper | | Hyper | Hypo | Hypo | Hypo | | | | | |
| Hyperactivity | | | | | ? | + | | + | + | + | + | |
| Head | | | | | | | | | | | | |
| High forehead | + | + | | | | | | | | | | |
| Prominent occiput | + | + | | | | | | | | | | + |
| Heavy brows | | | | | + | + | + | | | | | |
| Eyes | | | | | | | | | | | | |
| Antimongoloid slant | | | + | + | + | | | + | | | | + |

continued

TABLE IV (*continued*)

| Type of chromosome 15 anomaly | Trisomy, proximal segment: Due to translocation | | | | Trisomy, proximal segment: *De novo* bisatellited | | | Trisomy, proximal segment: *De novo* | | | | Trisomy, distal segment |
|---|---|---|---|---|---|---|---|---|---|---|---|---|
| Case number | 1[a] | 2[a] | 3 | 4 | 5 | 6 | 7 | 8[b] | 9 | 10 | 11 | 12 |
| Epicanthal folds | | | | | + | | | + | | | | |
| Strabismus | + | + | | | | + | + | + | + | + | + | + |
| Deep set eyes | + | + | | | | + | | | | | | |
| Cataracts | | | | + | | | | | | | | + |
| Nose | | | | | | | | | | | | |
| Beaked nose | + | + | + | | | | | | + | | | |
| Wide nasal bridge | | | | | | | | + | + | | | |
| Ears | | | | | | | | | | | | |
| Low or posteriorly set | + | + | + | + | | | | + | | | | |
| Malformed | | | + | + | | | | | | | | |

| | | | | | | | | |
|---|---|---|---|---|---|---|---|---|
| Palate | | | | | | | | |
| High arched palate | | | + | + | + | + | + | |
| Malpositioned teeth | | | | + | | + | | |
| Micrognathia | + | + | + | | | | | + |
| Short neck | + | + | | + | | | | |
| Extremities | | | | | | | | |
| Clinodactyly (hand) | | | + | | | | + | |
| Syndactyly (feet) | | | + | | + | | | |
| Club foot | | | | + | | | | |
| Dorsal kyphosis | + | + | | | | + | | |
| Genital anomalies | | | + | | | | | |

[a]Partially monosomic for proximal 21q.
[b]Restudied at age 16, with anti-methylcytosine stain by O. J. Miller (see text)

**TABLE V**
**Summary of Clinical Findings in 6 Cases of Proximal 15q Trisomy Studied by Staining with Anti-5-Methylcytosine[a]**

| | Chromosome anomaly | | | | | |
|---|---|---|---|---|---|---|
| | Translocation[b] | | *de novo*, bisatellited[c] | | | |
| | Case No. | | Case No. | | | |
| | 1 | 2 | 3 | 4 | 5 | 6 |
| Age when studied (years) | 5 | 13 | 32 | 34 | 23 | 11 |
| Sex | Male | Female | Male | Female | Male | Male |
| Growth failure | + | + | – | – | – | – |
| Mental retardation | + | + | + | + | + | + |
| Microcephaly | + | + | – | – | – | – |
| Ears: low-set | + | + | – | – | – | – |
| Nasal bridge flat | + | ± | – | – | – | – |
| Strabismus | + | + | + | + | – | + |
| Malocclusion | + | – | + | – | – | ± |
| Palate | Posterior sulcus | Cleft | High | – | Slightly high | – |
| Thumb: Abnormal insertion | + | + | – | – | – | – |
| Hypospadias | 1° | | – | | – | 1° |
| Hypotonia | + | + | – | – | – | – |
| Dislocated hips | – | + | – | – | – | |
| Club feet | – | + | – | – | – | + |
| Retarded bone age | + | ? | – | | | |
| Cubitus valgus | – | – | + | – | | – |
| Seizures | – | – | – | – | + | + (Infantile spasms) |

[a]Data compiled by W.R. Breg (1976)
[b]47,XY or XX +der(15),t(11;15)(q26;q2105) (first cousins)
[c]47,XY or XX, +15qs (both ends label with anti-5-methylcytosine)

composition of the extra chromosome is different. At this time it is difficult to point out specific diagnostic features peculiar to proximal 15q trisomy.

The nature of the process which results in a bisatellited chromosome, (with apparently identical short-arm, satellite material on both ends) is obscure. Crossing-over within an inversion loop can result in a duplication-deficiency product, but this rearranged chromosome is usually not seen as an additional chromosome. The latter would require a nondisjunctional error as well. Why the phenomenon appears to involve mostly chromosome 15 also is not clear.

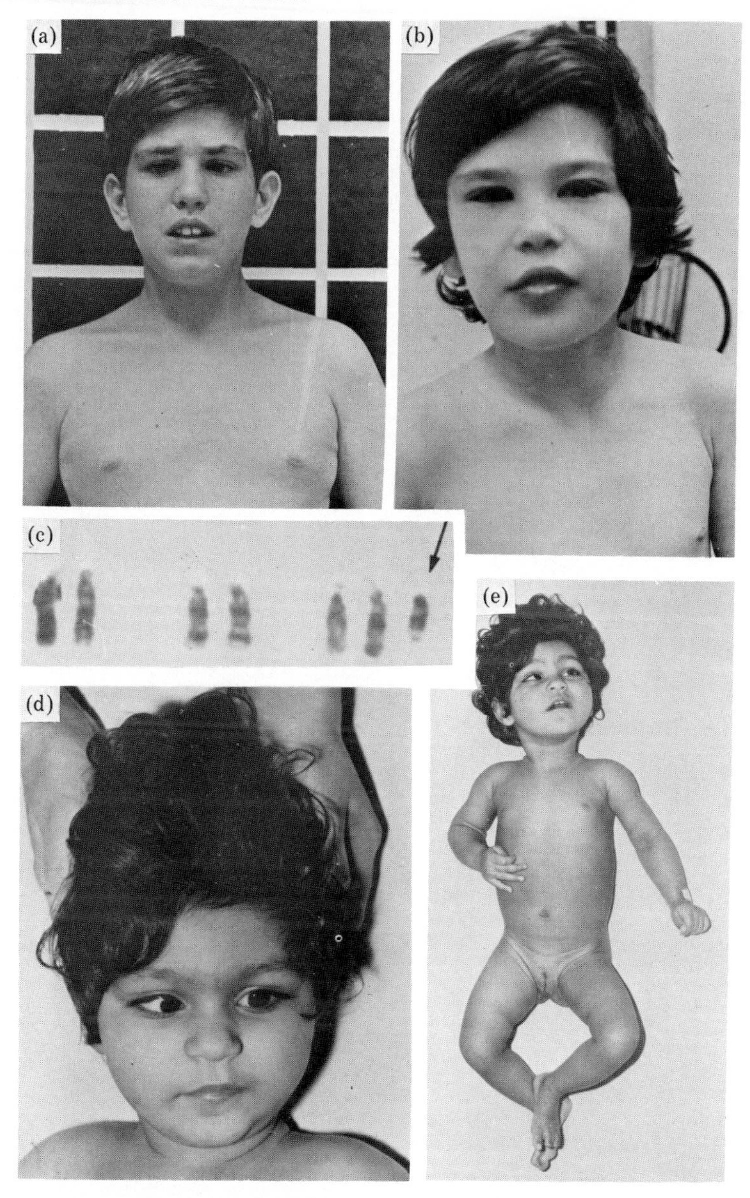


**Fig. 5.** Mildly affected individuals with *de novo* bisatellited proximal long arm 15 trisomy. (a,b) Eleven-year-old unrelated males (Cases 5, 6, Table IV); (c) G-banded partial karyotype of Case 6. Arrow points to extra bisatellited chromosome. (Reprinted with modification from Crandall *et al.*, 1973.) (d,e) Case 7 (Table II). (Reprinted with permission from Watson and Gordon, 1974).

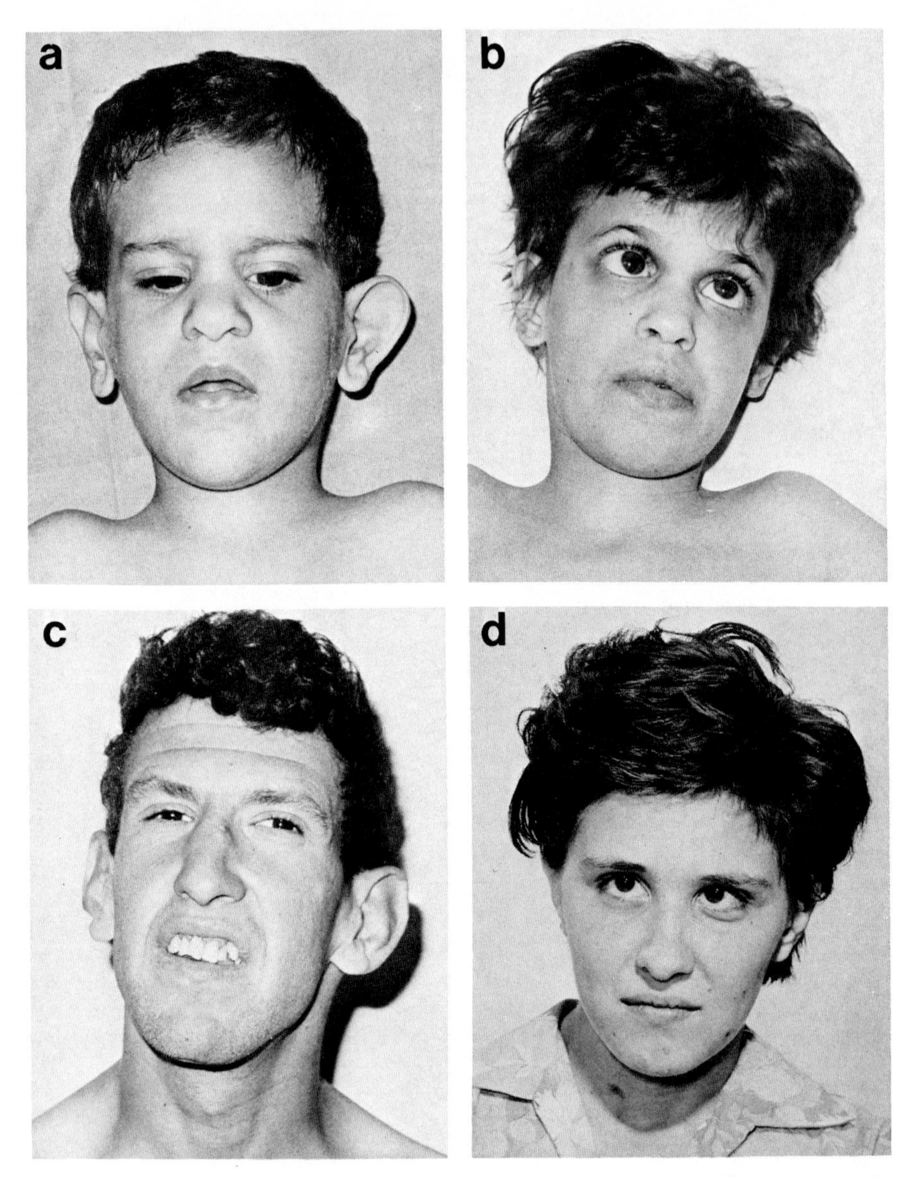


Fig. 6. Trisomy for proximal 15q. Four cases studied by R. Breg and colleagues (see text and Table V for details). (a,b) First cousins (Cases 1,2, Table V). (c,d) Cases 3,4, Table V.

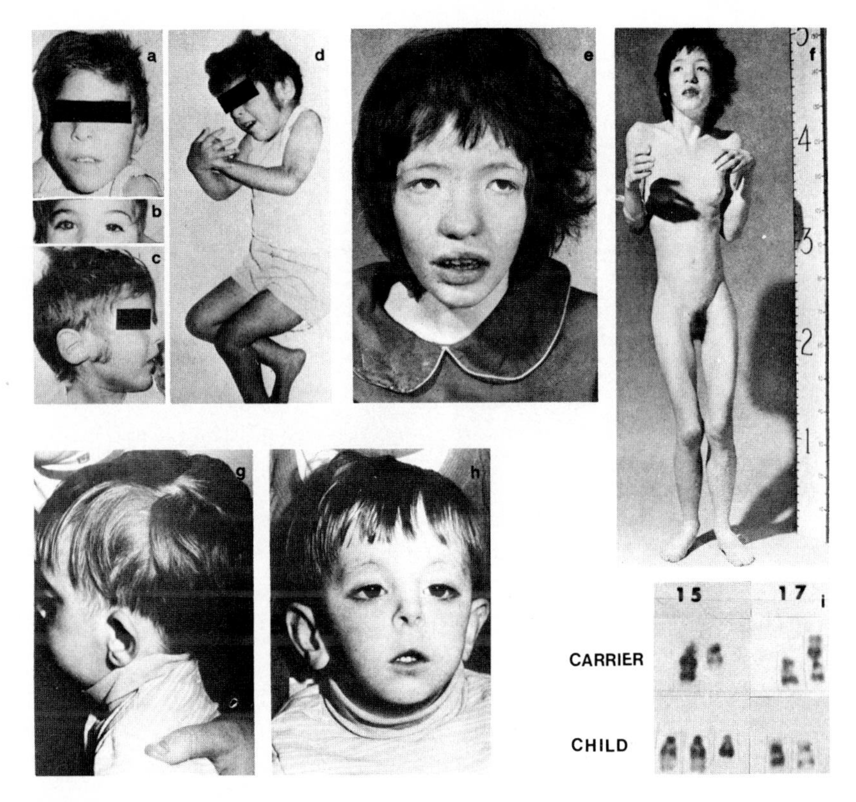


**Fig. 7.** Clinical views of patients trisomic for proximal 15q. (a–d) Case 4 (Table IV). (Reprinted with permission from Cohen *et al.*, 1975). (e,f) Case 8 (Table IV). [Unpublished photographs of case reported by Magenis *et al.* (1972).] (g–i) Case 3 (Table IV). Lower right, partial karyotypes (i) of chromosomes 15 and 17 from carrier parent with balanced rearrangement and from child with trisomy of proximal 15q. (Reprinted with modification from Bannister and Engel, 1975).

*b. Distal Long Arm.* There is only one reported case of trisomy for the distal segment of the long arm of chromosome 15 (Fig. 8); many of the clinical features appear distinct from those seen in trisomy for the proximal segment, as shown in Table IV (Fujimoto *et al.*, 1974).

## B. Partial Monosomy 15 Due to Translocation

Monosomy for part of chromosome 14 has not been reported to our knowledge.

We know of only one case of partial monosomy of 15q and studied this patient in detail. The patient, a female (Fig. 9), manifested severe motor delay; at 3 years of age, she had poor head control and could not sit, roll over, crawl or

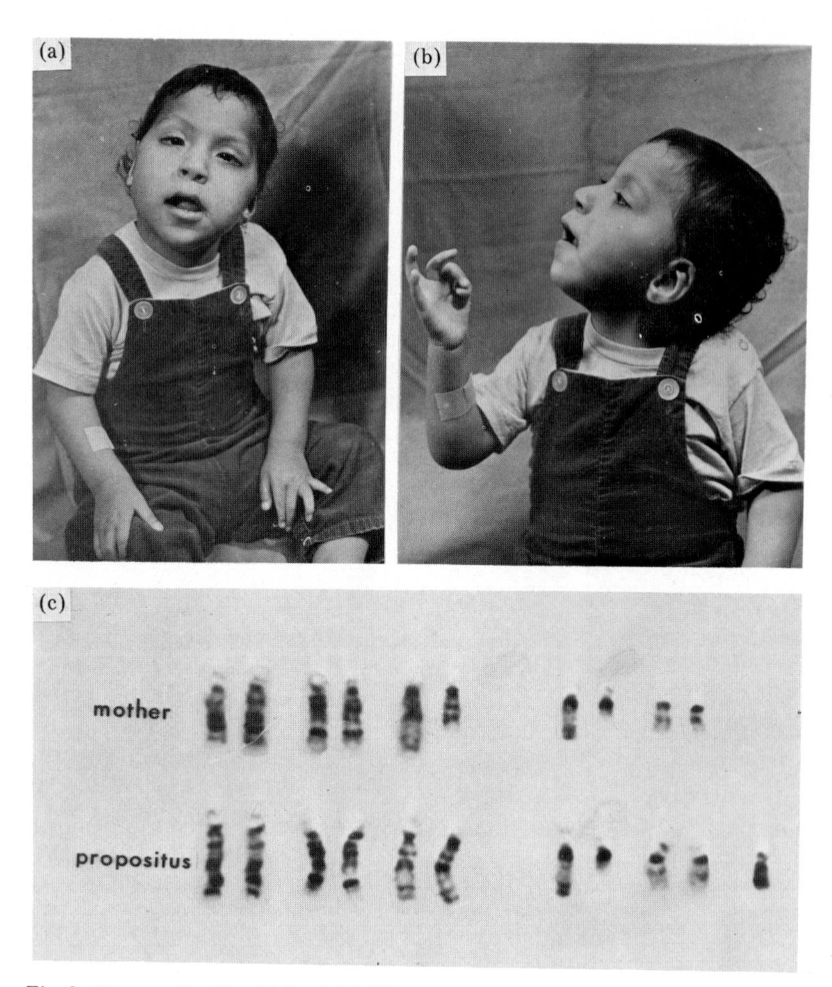


**Fig. 8.** Trisomy for distal 15q. (a,b) Clinical view of patient. (c) Partial karyotypes of D and G group chromosomes showing balanced 15;21 translocation in mother and extra 15 material on chromosome 21 in her male child. (Reprinted with permission from Fujimoto *et al.*, 1974).

grasp objects, and she had seizures many times daily. Growth was retarded; born 3 weeks past term, she weighed only 5 lbs. and has subsequently had slow weight gain. Other clinical findings included: head narrow and dolichocephalic, low hairline, low-set ears with large pinnae, nevus flammeus of the midforehead, small nose, narrow palate, short neck, sloping shoulders, increased AP diameter of the chest, marked dorsal kyphosis, and thoracolumbar scoliosis, an overlapping second toe, dimples over the elbows and medial malleoli, and stubby fifth fingers due to short proximal segments. She had limited hip abduction due to coxa valga configuration and obtuse cervicofemoral angles.

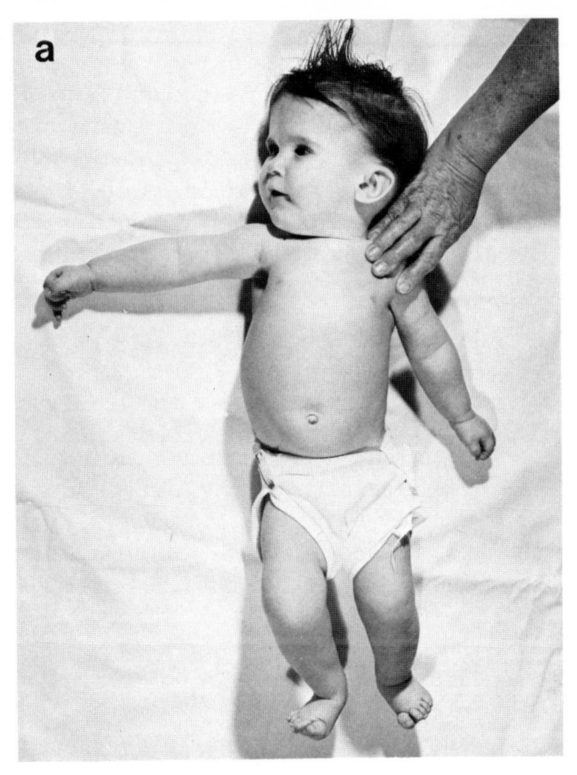


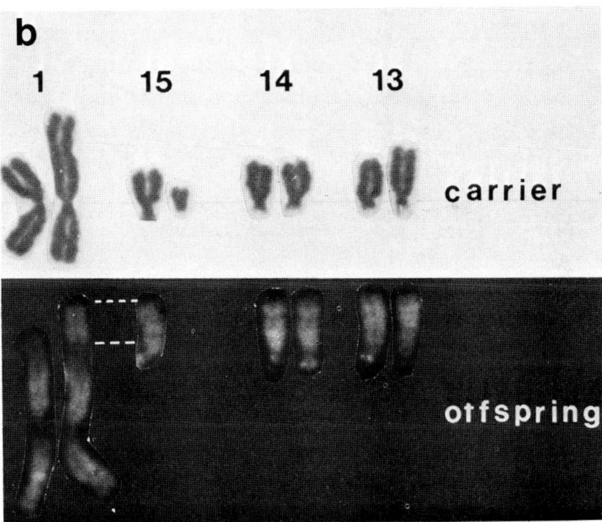


Fig. 9. Monosomy for proximal portion of 15q. (a) Child monosomic for the proximal portion of 15q. (b) Partial karyotypes of chromosome 1 and the D group from the mother who is a balanced carrier of the 1;15 translocation (by plain Giemsa) and from the child with monosomy for proximal 15q (by Q-banding). (Reprinted with permission from Hecht *et al.*, 1974.)

The patient's mother has a balanced translocation between chromosome 1 and 15 and had a history of several consecutive spontaneous abortions. The patient had 45 chromosomes with a missing chromosome 15 (Fig. 9). A portion of chromosome 15 (15q15→qter) was translocated to the long arm of chromosome 1. The patient was thus monosomic for the satellite, short arm, centromere and bands q11 to q14 of the long arm of chromosome 15 (Hecht *et al.*, 1974, 1975a; Prescott *et al.*, 1975).

### C. Partial Monosomy 15 Involving Rings

Gilgenkrantz *et al.* (1971) reviewed some twenty eight cases with D rings studied before banding including their own case which they believed, by autoradiography, involved a chromosome number 14. The authors attempted to compare the clinical features in these cases and suggested three distinctive ring syndromes involving 13, 14 and 15, respectively. The facts, however, that ring chromosomes are variable in size, that identification of ring chromosomes by autoradiography is unreliable, and that break points were not detectable by techniques available at the time make such distinctions questionable. There have been no cases reported, to our knowledge, with a ring chromosome 14 identified by banding.

We have reviewed three cases with a ring 15 chromosome. One case was identified by autoradiography (Emberger and Rossi, 1971). Two additional cases were studied by R-banding, but break points were not determined (Forabosco *et al.*, 1972; Stoll *et al.*, 1975). The clinical features common to the three cases were short stature, microcephaly, and mental retardation, but a lack of other striking features. Since these ring-15 patients manifest rather mild phenotypic anomalies, we presume that they are monosomic for only a small portion of the long arm of chromosome 15 (Fig. 10).

## IV. HETEROMORPHISMS OF CHROMOSOMES 14 AND 15

The short arms and satellites of chromosomes 14 and 15 are highly variable regions (Hoo *et al.*, 1974). For example, they may be absent or may appear as large bright structures with Q-banding of the size and brilliance of the Y long arm heterochromatic region. These variations have no apparent effect on the phenotype. A large and brilliant No. 14 or 15 satellite in interphase may be confused with the Y-body and incorrect sex chromosome assignment given (Hecht *et al.*, 1974). In addition, it may be difficult to distinguish between apparently normal variations and translocations occurring in this region, such as those described between the 15 short arm and the Y (Šubrt and Blehová, 1974) or between the short arms of 4 and 15 (Schröcksnadel *et al.*, 1975).

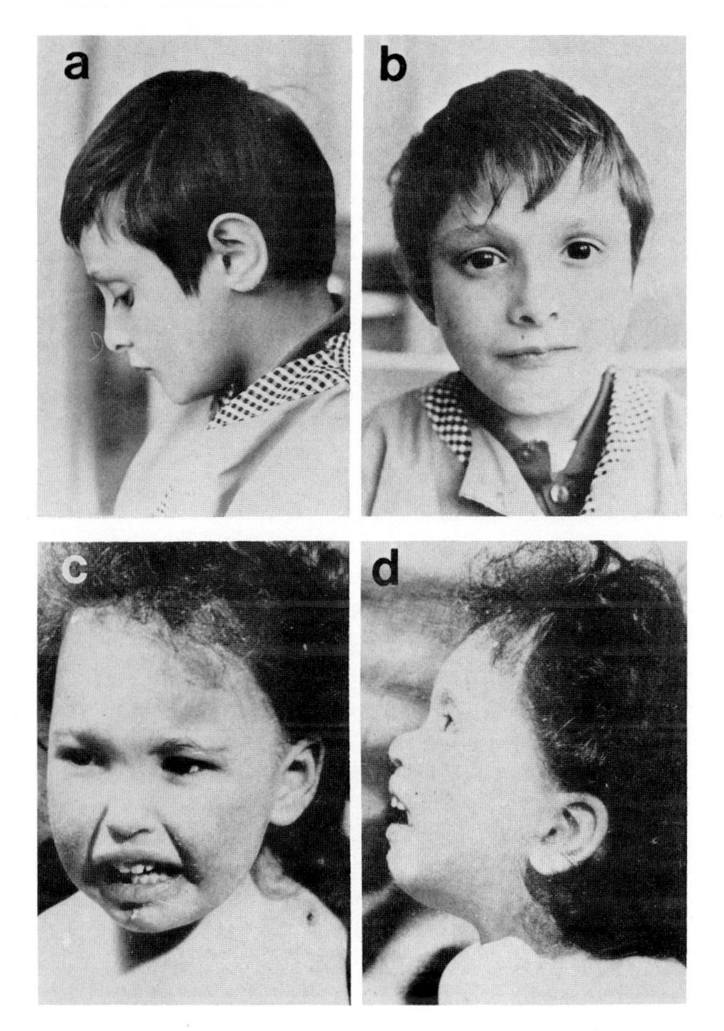


**Fig. 10.** Ring 15 chromosome. Two children with a ring 15 chromosome. a and b, reprinted with permission from Emberger and Rossi (1971). c and d, reprinted with permission from Stoll *et al.* (1975).

## V. CHROMOSOME 14 CHANGES IN BENIGN AND MALIGNANT LYMPHOID CLONES

The best known example of a specific chromosome change in malignancy is the Philadelphia chromosome (22q−) found in chronic myeloid leukemia. The Philadelphia chromosome is part of a rearrangement, usually with chromosome 9 (Rowley, 1973). However, it has recently been found that various neoplasias and premalignant conditions have specific chromosome defects such as those found in ataxia telangiectasia and lymphomas.

### A. 14q− Clones in Ataxia–Telangiectasia

#### *1. Clinical Picture*

Ataxia–telangiectasia (A–T) is an autosomal recessive disorder. It is characterized by progressive cerebellar ataxia, oculocutaneous telangiectasia and progressive sinopulmonary infections (Boder and Sedgwick, 1958). A–T commonly includes IgA deficiency and impaired cellular immunity. There may be hypoplasia of the thymus, a decrease in peripheral lympoid tissue, and a diminished number of circulating T lymphocytes. Lymphocytes are often sluggish in their response to phytohemagglutinin and other mitogens.

A–T patients are predisposed to lymphoreticular malignancies, principally lymphomas and lymphocytic leukemia.

#### *2. Cytogenetic Picture*

We first studied an 18-year-old male with A–T who had increased chromosome breakage in cultured lymphocytes, a frequent finding in A–T patients (Hecht *et al.,* 1966). In 60 lymphocytes there was one with D/D translocation. We undertook longitudinal studies of this patient over a period of 52 months until his death from pulmonary insufficiency. These studies revealed that (1) the D/D translocation involved both No. 14 chromosomes and was of the t(14q−; 14q+) type; (2) the lymphocytes with the translocation increased from 1–2% to the majority (78%) of cells sampled; and (3) chromosome breakage was less in the translocation cells. The data was correlated with information on other A–T patients and suggested that the 14q− translocation in A–T lymphocytes might be analogous to the Philadelphia chromosome translocation in chronic myeloid leukemia (Hecht *et al.,* 1973). Indeed, at present eight patients have been studied (McCaw *et al.,* 1975) and the clones in all the patients proved to have a structural rearrangement of 14 q. Seven showed a 14q− translocation with the break point being at band q12 (Fig. 11) and the remaining patient had a clone with a 14 ring. The key chromosome change in A–T clones appears to involve the *distal* portion of 14q. That portion is either lost (as in the ring 14) or presumably inactivated by position effect (through translocation).

One A–T patient had chromosome studies before *and* after the onset of chronic lymphocytic leukemia from which she subsequently died. Before the onset of leukemia, there was a minor lymphocyte clone with a translocation of the t(14q−;14q+) type. This clone constituted about 20% of cells sampled. After the onset of leukemia, 100% of cells had lost the 14q− chromosome and retained its 14q+ translocation partner (Fig. 12). Based on this one case, the odds appear

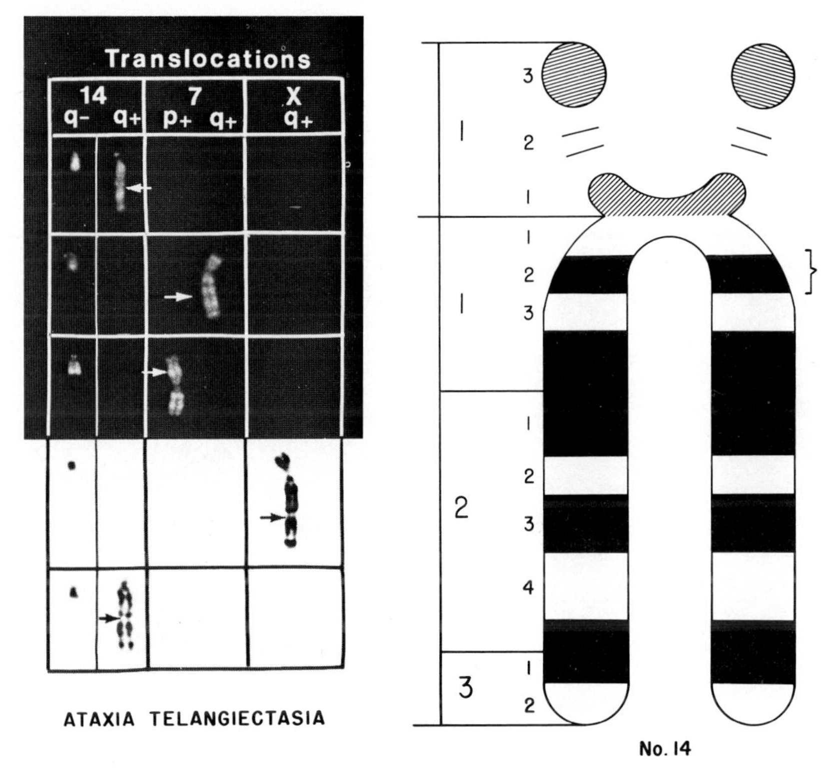


**Fig. 11.** Ataxia-telangiectasia 14q– clones. Translocation chromosomes marking pseudodiploid lymphocyte clones in 5 A-T patients. Note that the translocations in every case involve No. 14, resulting in a 14q– chromosome (left-most column). The break point in No. 14 is consistently in the q11–12 region indicated on the diagram by the bracket (right). (Reprinted with modification from McCaw *et al.*, 1975).

to be approximately 4:1 (79:21) that there may be a connection between the chromosome 14 change and malignancy.

## B. 14q+ Clones in Lymphoid Malignancies

Banded chromosome data are now accumulating on many malignant lymphoid tumors, one of the best studied being Burkitt's lymphoma.

### *1. Burkitt's Lymphoma*

***a. Clinical Picture.*** Burkitt's lymphoma is a disease of children. First described in Africa, it is known to occur in virtually epidemic proportions also in New Guinea. It usually presents with involvement of the peripheral lymph nodes (e.g., in the jaw) or with a lymphomatous mass in the lower abdomen. The tumor is characterized by small undifferentiated lymphoid cells in a rather

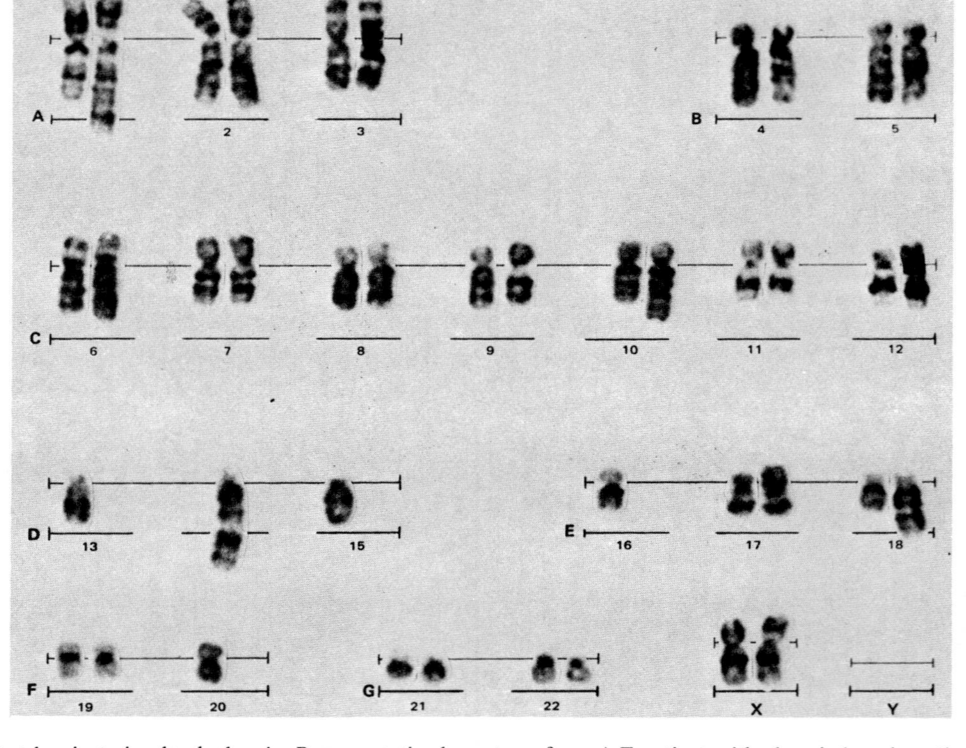


**Fig. 12.** Ataxia-telangiectasia plus leukemia. Representative karyotype from A-T patient with chronic lymphocytic leukemia. In addition to the original 14;14 translocation detected in lymphocytes in 1968 prior to the clinical onset of leukemia, the patient's lymphocytes now show a modal number of 41 chromosomes due to a series of other chromosome changes. These include a 1;13 and 15;18 translocation, extra material on a No. 10 chromosome, and absence of a No. 16 and a No. 20 chromosome. The karyotype is Giemsa-banded. (Reprinted with permission from McCaw *et al.*, 1975).

uniform histiocytic background, presented a "starry sky" appearance. The course of the disease is usually short, resulting in death within a few years.

Burkitt's lymphoma is now known to occur sporadically worldwide including America and Europe. In such areas the children tend to be a little older, to be more resistant to drug therapy, and to die even more quickly after diagnosis than in Africa or New Guinea. For the purposes of the following discussion, we will refer to African-type Burkitt's lymphoma (AfBL) and to American-type Burkitt's lymphoma (AmBL).

Perhaps the most interesting difference between AfBL and AmBL relates to the Epstein-Barr virus (EBV). The large majority of cases of AfBL are EBV-*positive* while the majority of AmBL patients are EBV-*negative.*

***b. Cytogenetic Picture. i. African-type Burkitt's lymphoma (AfBL).*** Biopsies and cell cultures from patients with AfBL demonstrate a chromosome change: elongation of the long arm of a No. 14 chromosome (Manolov and Manolova, 1972). We have confirmed this finding in 3 AfBL cell lines. All were EBV-positive. All three also contained the 14q+ marker chromosome (McCaw *et al.,* 1976) and in one AfBL line we were able to determine that the extra chromosome material was coming from 8q (McCaw *et al.,* 1977). In similar studies of AfBL cell lines and biopsies by workers in Sweden (Zech *et al.,* 1976) the findings appear the same; the 14q+ chromosome is one of two partners in a t(8q−;14q+) translocation.

***ii. American-type Burkitt's lymphoma (AmBL).*** We studied cells cultured from two children with American Burkitt's lymploma (Epstein *et al.,* 1976). One cell line contains no detectable EBV, but shows translocation of material onto the long arm (q) of chromosome 14. The other cell line contains EBV and also has a translocation onto 14q. In both cases the translocation appears to be of the t(8q−;14q+) type and is indistinguishable from that observed in African-type Burkitt's lymphoma (McCaw *et al.,* 1977).

***iii. Other Lymphomas and Lymphoid Leukemias.*** Chromosome 14 involvement is apparent beyond A–T and Burkitt's lymphoma. Changes in 14 have been reported in a number of different lymphoid tumors including multiple myeloma and lymphosarcoma (McCaw *et al.,* 1975, 1977).

## C. Role of Chromosome 14 in Lymphoid Cells

We have reviewed data on 14q− lymphocyte clones in ataxia-telangiectasia and on 14q+ malignant clones in Burkitt's lymphoma and other lymphoid neoplasms. We and others have also noted the nonrandom occurrence of chromosome 14 translocations in cultured lymphocytes (Hecht *et al.,* 1975b; Welch and Lee, 1975; Beatty-DeSana *et al.,* 1975) (Fig. 13). In a sample of 32,300 metaphases, the frequency of 7;14 translocations was $4 \times 10^{-4}$. These translocations closely resemble those observed in A–T clones. It is possible that there is a

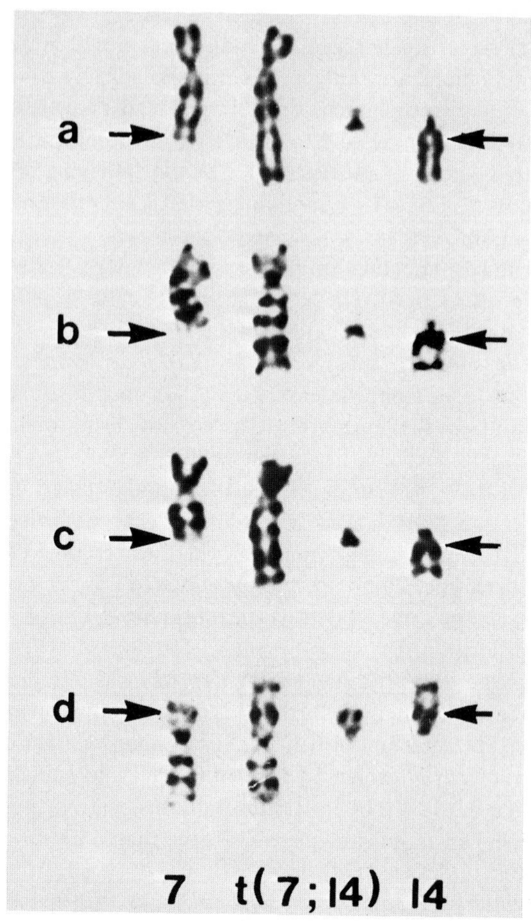


**Fig. 13.** Translocations in solitary lymphocytes cultured from four individuals (a–d). The arrows on the left indicate the point at which the distal part of chromosome 14 is translocated to the long arm (a–c) or to the short arm (d) of chromosome 7. The arrows on the right indicated the break points in chromosome 14. (Reprinted with permission from Hecht *et al.*, 1975b).

segment below band q12 on the long arm of chromosome 14 which affects lymphoid cell growth.

## VI. GENES ON CHROMOSOMES 14 AND 15

Genes have been assigned to chromosomes 14 and 15 by two techniques: *in situ* DNA–RNA annealing of ribosomal RNA, and somatic cell hybridization of rodent and human cells. Chromosomes 14 and 15 contain genes for ribosomal RNA. These genes are located on the short arms of chromosomes 14 and 15, as well as on the short arms of the other acrocentric autosomes.

The long arm of chromosome 14 is thought to contain genes for nucleoside phosphorylase and for tryptophanyl-tRNA synthetase. The gene for nucleoside phosphorylase is probably located between the centromere and band q21.

The long arm of chromosome 15 is now believed to bear genes for beta-2-microglobulin, manosephosphate isomerase, pyruvate kinase-3, hexosaminidase A, and possibly mitochondrial isocitrate dehydrogenase. These map assignments are summarized in the Workshops on Human Gene Mapping (1974, 1975, 1976).

The mapping of genes on chromosomes 14 and 15 is proceeding rapidly. For example, a recent issue of *Somatic Cell Genetics* contains back-to-back papers pertaining to chromosome 15. Soloman and colleagues (1976) used an X/autosome translocation, fused the human cells with a mouse cell line, selected for hybrid cells, and studied them. They found that (aside from several X-linked gene products) pyruvate kinase ($PK_{m2}$), mannose phosphate isomerase (MPI), N-acetyl hexosaminadase A ($HEX_A$), and $\beta_2$-microglobulin ($\beta_2$-m) all segregated in concordance with the X;15 translocation chromosome. These markers have all been assigned to chromosome 15. The breakpoint in chromosome 15 is in band q11, so all these markers are presumably located on the long arm below that band. The second article is by Faber and co-workers (1976). They demonstrate that the $\beta_2$-microglobulin locus is clearly on chromosome 15, a finding which Goodfellow *et al.* (1975) first reported. This is of special interest, because the major histocompatibility complex in man (HL–A) is carried on chromosome 6 and so is not linked to $\beta_2$-microglobulin.

## VII. SUMMARY

Trisomy 14 or 15 is known to cause spontaneous abortions. Ring 15 results in partial monosomy and is compatible with life. Partial trisomy 14 occurs and is responsible for the proximal 14q trisomy syndrome and probably for a separate distal 14q trisomy syndrome. Similarly, several different syndromes due to partial 15q trisomy appear to be emerging.

Heteromorphic variations involving the short arms of 14 and 15 are common. Banding and family studies are essential in distinguishing these normal heteromorphisms from true aberrations.

Clones of lymphocytes in ataxia-telangiectasia patients are marked by a 14q− chromosome arrangement. Lymphomas and possibly lymphocytic leukemic cells may also be marked by a 14q+ rearrangement, usually of the t(8q−;14q+) type.

## ACKNOWLEDGMENTS

This work was supported in part by NIH Grants HD 07997, HD 08236, CA 16747 and by Maternal and Child Health Services 970.

## REFERENCES

Allderdice, P. W., Miller, O. J., and Miller, D. A. (1971). *Humangenetik* **13**, 205.

Bannister, D. L., and Engel, E. (1975). *J. Pediatr.* **86**, 916.

Beatty-DeSana, J. W., Hoggard, M. J., and Cooledge, J. W. (1975). *Nature (London)* **255**, 242.

Boder, E., and Sedgwick, R. P. (1958). *Pediatrics* **21**, 526.

Boué, J. G., Boué, A., and Lazar, P. (1967). *Ann. Genet.* **10**, 179.

Breg, W. R., Schreck, R. R., Erlanger, B. F., and Miller, O. J. (1974). *Am. J. Hum. Genet.* **26**, 17A.

Bucher, W., Parker, C. E., Crandall, B., and Alfi, O. S. (1973). *Lancet* **2**, 1250.

Carr, D. H. (1965). *Obstet. Gynecol.* **26**, 308.

Cohen, M. M., Ornoy, A., Rosenmann, A., and Kohn, G. (1975). *Ann. Genet.* **18**, 99.

Crandall, B. F., Muller, H. M., and Bass, H. N. (1973). *Am. J. Ment. Defic.* **77**, 571.

Emberger, J. M., and Rossi, D. (1971). *Humangenetik* **11**, 295.

Epstein, A. L., Henle, W., Henle, G., Hewetson, J. F., and Kaplan, H. S. (1976). *Proc. Natl. Acad. Sci. U.S.A.* (in press).

Faber, H. E., Kucherlapati, R. S., Poulik, M. D., Ruddle, F. H., and Smithies, O. (1976). *Somatic Cell Genet.* **2**, 141–153.

Forabosco, A., Dutrillaux, B., Vazzolu, G., and Lejeune, J. (1972). *Ann. Genet.* **15**, 267.

Fawcett, W. A., McCord, W. K., and Francke, U. (1975). *Birth Defects, Orig. Artic. Ser.* **11**, 223.

Fryns, J. P., Cassiman, J. J., and van den Berghe, H. (1974). *Humangenetik* **24**, 71.

Fujimoto, A., Towner, J. W., Ebbën, A. J., Kahlström, E. J., and Wilson, M. G. (1974). *J. Med. Genet.* **11**, 287.

Gianelli, F., and Howlett, R. M. (1966). *Cytogenetics* **5**, 186.

Gilgenkrantz, S., Cabrol, C., Lausecker, C., Hartleyb, M. E., and Bohe, B. (1971). *Ann. Genet.* **14**, 23.

Goodfellow, P. N., Jones, E. A., van Heyningen, V., Solomon, E., Bobrow, M., Miggiano, J., and Bodmer, W. F. (1975). *Nature (London)* **254**, 267–269.

Hamerton, J. L., Canning, N., Ray, M., and Smith, S. (1975). *Clin. Genet.* **8**, 223.

Hecht, F., and Kimberling, W. J. (1971). *Am. J. Hum. Genet.* **23**, 361.

Hecht, F., Koler, R. D., Rigas, D. A. *et al.* (1966). *Lancet* **2**, 1193.

Hecht, F., McCaw, B. K., and Koler, R. D. (1973). *N. Engl. J. Med.* **289**, 286.

Hecht, F., Wyandt, H. E., and Magenis, R. E. (1974). *In* "The Cell Nucleus" (H. Busch, ed.), Vol. 2, p. 33. Academic Press, New York.

Hecht, F., Lovrien, E. W., Magenis-Heath, R. E., McCaw, B. K., and Wyandt, H. E. (1975a). *In* "Endocrine and Genetic Diseases of Childhood and Adolescence" (L. I. Gardner, ed.), 2nd ed., p. 692. Saunders, Philadelphia, Pennsylvania.

Hecht, F., McCaw, B. K., Peakman, D., and Robinson, A. (1975b). *Nature (London)* **255**, 243.

Hoo, J. J., Hillig, U., Cramer, H., Hansen, S., and Hermann, F. (1974). *Humangenetik* **21**, 283.

Howard, P. N., Stoddard, G. R., and Yarbrough, K. M. (1974). *Am. J. Hum. Genet.*, **26**, 41A.
Jacobs, P. A., Buckton, K. E., Cunningham, C., and Newton, M. (1974). *J. Med. Genet.* **11**, 50.
Kuliev, A. M., Kukharenko, V. I., Grinberg, K. N., Terskikh, V. V., Tamarkina, A. D., Bogomazav, E. A., Redkin, P. S., and Vasileysky, S. S. (1974). *Humangenetik* **21**, 1.
Laurent, C., Dutrillaux, B., Biemont, M.-Cl., Genoud, J., and Bethenod, M. (1973). *Ann. Genet.* **16**, 281.
Lubit, B. W., Schreck, R. R., Miller, O. J., and Erlanger, B. F. (1974). *Exp. Cell Res.* **89**, 426.
Lucus, M. (1969). *Ann. Hum. Genet.* **32**, 347.
McCaw, B. K., Hecht, F., Harnden, D. G., and Teplitz, R. L. (1975). *Proc. Natl. Acad. Sci. U.S.A.* **72**, 2071.
McCaw, B. K., Epstein, A. L., Kaplan, H. S., and Hecht, F. (1976). *Clin. Res.* **24**, 191A.
McCaw, B. K., Epstein, A. L., Kaplan, H. S., and Hecht, F. (1977). *Int. J. Cancer* (in press).
Magenis, R. E., Overton, K. M., Reiss, J. A., Macfarlane, J., and Hecht, F. (1972). *Lancet* **2**, 1365.
Manolov, G., and Manolova, Y. (1972). *Nature (London)* **237**, 33.
Miller, O. J. (1970). *Adv. Hum. Genet.* **1**, 35.
Muldal, S., Enoch, B. A., Ahmed, A., and Harris, R. (1973). *Clin. Genets.* **4**, 480.
Murken, J. D., Bauchinger, M., Paltizsch, D., Pfeifer, H., Suschke, J., and Haendle, H. (1970). *Humangenetik* **10**, 254.
Neu, R. L., Valentine, F. A., and Gardner, L. I. (1975). *Clin. Genet.* **8**, 30.
Paris Conference, 1971. (1972). *Birth Defects, Orig. Artic. Ser. 8,* No. 7.
Parker, C. E., and Alfi, O. S. (1972). *Lancet* **1**, 1073.
Pfeiffer, R. A., Büttinghaus, K., and Struck, H. (1973). *Humangenetik* **20**, 187.
Prescott, G. H., McCaw, B. K., Tolby, B. E., Hecht, F., Miller, R. D., Green, A. E., and Coriell, L. L. (1975). *Cytogenet. Cell Genet.* **14**, 84.
Raoul, O., Rethoré, M.-O., Dutrillaux, B., Michon, L., and Lejeune, J. (1975). *Ann. Genet.* **18**, 35.
Reiss, J. A., Wyandt, H. E., Magenis, R. E., Lovrien, E. W., and Hecht, F. (1972). *J. Med. Genet.* **9**, 280.
Rethoré, M.-O., Dutrillaux, B., and Lejeune, J. (1973). *Ann. Genet.* **16**, 271.
Rethoré, M.-O., Couturier, J., Carpentier, S., Ferrand, J., and Lejeune, J. (1975). *Ann. Genet.* **18**, 71.
Rowley, J. D. (1973). *Nature (London)* **243**, 290.
Schmid, W., and Carnes, J. D. (1965). *In* "Human Chromosome Methodology" (J. J. Yunis, ed.), p. 91. Academic Press: New York.
Schröcksnadel, H., Feichtinger, C., and Scheminzky, C. H. (1975). *Humangenetik* **29**, 329.
Short, E. M., Solitare, G. B., and Breg, W. R. (1972). *J. Med. Genet.* **9**, 367.
Solomon, E., Bobrow, M., Goodfellow, P. N., Bodmer, W. F., Swallow, D. M., Povey, S., and Noël, B. (1976). *Somantic Cell Genet.* **2**, 125–140.
Stoll, C., Juif, J. G., Luckel, J. C., and Lausecher, C. (1975). *Humangenetik* **27**, 259.
Šubrt, I., and Blehová, B. (1974). *Humangenetik* **23**, 305.
Therkelsen, A. J., Grunnet, N., Hjort, T., Myhre-Jensen, O. Johansson, J., Lauritsen, J. G., Lindsten, J., and Brunn-Petersen, G. (1973). *Colloq.* INSERM p. 81.
Turleau, C., de Grouchy, J., Bocquentin, F., Roubin, M., and Chavin-Colin, F. (1975). *Ann. Genet.* **18**, 41.
Wahlström, J. (1974). *Hereditas* **78**, 251.
Watson, J. E., and Gordon, R. R. (1974). *J. Med. Genet.* **11**, 400.

Welch, J. P., and Lee, C. L. Y. (1975). *Nature (London)* **255**, 241.
Workshop on Human Gene Mapping. (1974). *Cytogenet. Cell Genet.* **13**, 1–216.
Workshop on Human Gene Mapping. (1975). *Cytogenet. Cell Genet.* **14**, 162–480.
Workshop on Human Gene Mapping. (1976). *Cytogenet. Cell Genet.* **16**, 1–452.
Yunis, J. J., Hook, E. B., and Mayer, M. (1964). *Lancet* **2**, 935.
Zech, L., Haglund, U., Nilsson, K., and Klein, G. (1976). *Int. J. Cancer* **17**, 47.

# 11

# The Trisomy 22 Syndrome and the Cat Eye Syndrome

LILLIAN Y. F. HSU and KURT HIRSCHHORN



## I. TRISOMY 22

### A. Historical Review

Between 1960 and 1975, trisomy G in children without Down's syndrome and without XYY by conventional karyotyping was reported in at least twenty eight papers in thirty seven patients. Of these, trisomy 22 was cytogenetically identified in nineteen cases by either autoradiographic studies or the new banding techniques (Uchida *et al.*, 1968a; Hsu *et al.*, 1971; Gustavson *et al.*, 1972; Punnett *et al.*, 1973; Hirschhorn *et al.*, 1973; Bass *et al.*, 1973; Zackai *et al.*, 1973; Alfi *et al.*, 1975; Penchaszadeh and Coco, 1975; Goodman *et al.*, 1971; R. M. Goodman, personal communication, 1975; Perez-Castillo *et al.*, 1975; Vianello and Bonioli, 1975). Of the remaining eighteen cases, six had rather typical clinical features of trisomy 22 but were not studied with either banding or autoradiographic techniques [Ferguson and Pitt, 1963; Hall, 1963; Ishmael and Laurence, 1965; Giorgi *et al.*, 1967; Uchida *et al.*, 1968a (Case 1); Hsu *et al.*, 1971 (Case 1)]. The clinical features of eight other patients were not typical of trisomy 22 syndrome but showed some of its features. In seven of these eight cases, no special cytogenetic techniques for confirmation were employed (Hayward and Bower, 1960; Crawfurd, 1961; Zellweger *et al.*, 1962a; Koulischer and Perier, 1962; Kruger *et al.*, 1968; Walbaum *et al.*, 1970). One of these eight cases (Magenis *et al.*, 1971) was reported as a case of possible partial trisomy 15 by banding studies. The remaining four cytogenetically unconfirmed cases of trisomy G clinically appeared not to have trisomy 22 syndrome (Turner and Jennings, 1961; Dunn *et al.*, 1961; Biesele *et al.*, 1962).

In the eighteen cases of trisomy G in which no special identification techniques were used, trisomy 22 was suggested by the authors in six cases [Hayward and Bower, 1960; Turner and Jennings, 1961; Koulischer and Perier, 1962; Uchida *et al.*, 1968a (Case 1); Walbaum *et al.*, 1970; Hsu *et al.*, 1971 (Case 1)]. Hayward and Bower, in 1960, first suggested trisomy 22 in a boy with Sturge-Weber syndrome and an extra G chromosome. The clinical features of this patient were not typical for trisomy 22 syndrome but were compatible; the published karyotype showed no distinction between the depicted Y chromosome and the other 5 G group chromosomes. Thus, the possibility of 47,XYY as well as partial trisomy for other chromosomes remains to be considered in this case. This report raised the suspicion of an association between trisomy G and Sturge-Weber syndrome. This suspicion was subsequently determined to be unfounded (Lehmann and Forssman, 1960; Gustavson and Hook, 1961; Hall, 1961). Of the other five cases with initial suggestion of trisomy 22, two had clinically typical trisomy 22 syndrome [Uchida *et al.*, 1968a (Case 1); Hsu *et al.*, 1971 (Case 1)]; two were clinically compatible with trisomy 22 (Koulisher and Perier, 1962; Walbaum *et al.*, 1970); one probably did not have trisomy 22 syndrome (Turner and Jennings, 1961).

## B. Phenotypic Manifestations

In the nineteen cases of cytogenetically confirmed cases of trisomy 22, ten were females and nine were males. The age at the reporting date ranged from newborn to 10 years (twelve were under 2 years of age, and seven were from 4 to 10 years). Except for four young infants aged from newborn to 8 weeks who could not be evaluated for mental development, all had obvious mental retardation and growth retardation. The phenotypic manifestations of these nineteen patients are tabulated in Table I. Facial similarity was noted in fourteen patients out of eighteen patients with available photos. Seventy percent or more patients had the following congenital abnormalities: microcephaly (15/19), preauricular skin tags and/or sinus (15/19), congenital heart disease (16/19), micrognathia (14/19), long philtrum (14/19), low-set and/or malformed ears (14/19), long and/or beaked nose (14/19). Fifty percent or more had cleft palate (13/19), fingerlike, malopposed thumb and/or long slender fingers (12/19), congenital dislocated hip (9/19), hypotonia (10/19), and antimongoloid slant of eyes (10/19). Thirty percent or more had strabismus (8/19), and hypoplastic and/or low-set nipples (5/19). Cryptorchidism and/or small penis were noted in all males. It must be emphasized that the frequencies of fingerlike, malopposed thumb, strabismus, abnormal and low-set nipples, quoted here probably underestimate the true frequencies, since no comment was made in many of the reports in terms of these findings. The finding of cubitus valgus was noted in the two patients of Uchida *et al.* (1968a) and two of our three patients (Hsu *et al.*, 1971) and was not mentioned in the other reports. Photographs of two of our patients are shown in Figs. 1 and 2.

In six clinically and cytogenetically suggestive cases of trisomy 22, three were males and three were females. The clinical features of these six patients were tabulated in Table II. Low-set and/or malformed ears were present in all 6, typical facies by available photographs in 5, micrognathia in 5, microcephaly in 4, cleft palate in 4, evident mental and growth retardation in 4, hypotonia in 4, long and/or beaked nose in 3, preauricular skin tags or sinus in 3, congenital dislocated hip in 3, congenital heart disease in 2, and long philtrum in 2. In all three males, undescended testes and/or small penis were noted.

The mean maternal age in twenty-five patients (including nineteen confirmed cases and six suggestive cases with data available) was 30.5 years. The mean birth weight excluding premature birth in two was 2666.9 grams. Maternal mosaicism of 46,XX/47,XX,+G (presumably +22) was found in two (Uchida *et al.*, 1968a; Hsu *et al.*, 1971). Subfertility and/or history of fetal wastage were noted in three families (Hsu *et al.*, 1971; Bass *et al.*, 1973).

Eight other cases of trisomy G in children without Down's syndrome did not have typical clinical features of trisomy 22 syndrome but cannot be excluded from this diagnosis with certainty. The clinical features are tabulated in Table III.

By combining the nineteen cytogenetically confirmed cases of trisomy 22 and the six cases clinically and cytogenetically suggestive of trisomy 22.

The most common phenotypic manifestations (60% or more) of trisomy 22 syndrome are

| | | |
|---|---|---|
| 1. Mental and growth retardation (excluding the young infants) | (20/20) | 100% |
| 2. Small penis and/or undescended testes | (11/12) | 91.7% |
| 3. Typical facies | (18/22) | 81.8% |
| 4. Microcephaly | (20/25) | 80.0% |
| 5. Low-set and/or malformed ears | (20/25) | 80.0% |
| 6. Micrognathia | (19/25) | 76.0% |
| 7. Preauricular skin tag and/or sinus | (18/25) | 72.0% |
| 8. Congenital heart disease | (18/25) | 72.0% |

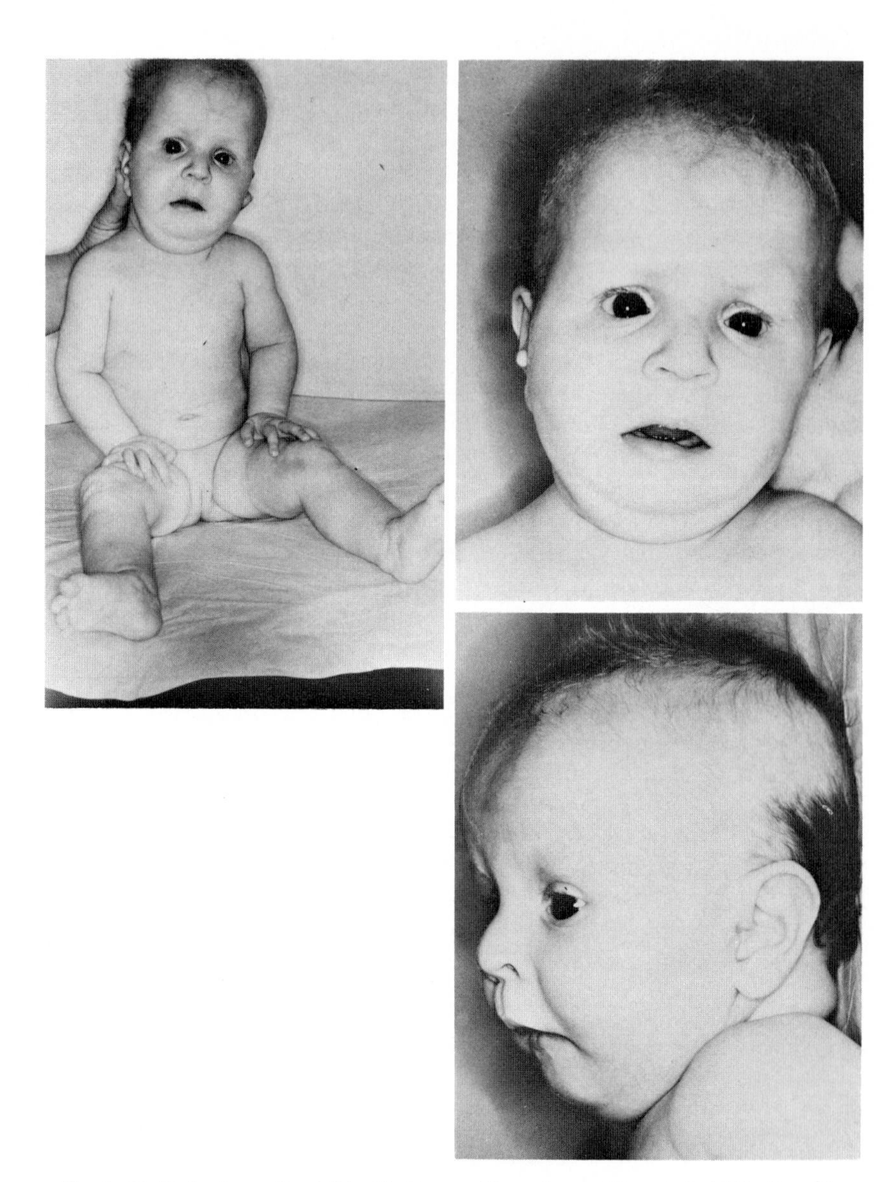

**Fig. 1.** Multiple views of a child with trisomy 22 syndrome (Hsu *et al.*, 1971, Case 2).

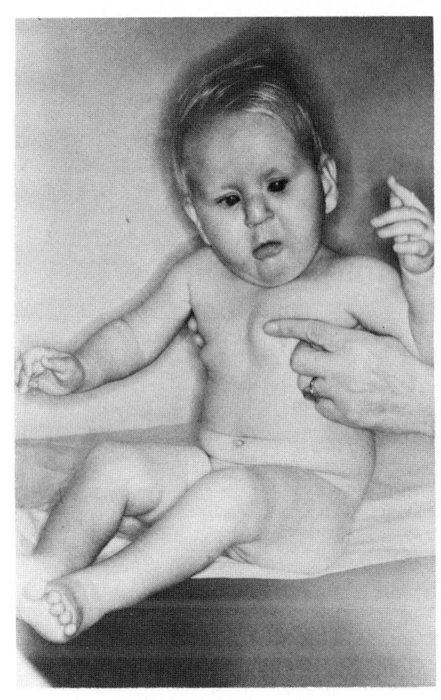
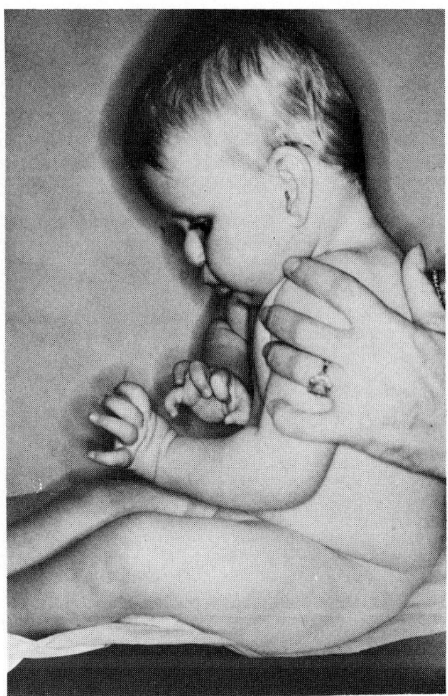

**Fig. 2.** Two views of a child with trisomy 22 syndrome (Hsu *et al.*, 1971, Case 3).

| | | |
|---|---|---|
| 9. Cleft palate and/or uvula | (17/25) | 68.0% |
| 10. Long philtrum | (16/25) | 64.0% |
| 11. Long and/or beaked nose | (16/25) | 64.0% |

The next frequent findings (about 50%) are

| | | |
|---|---|---|
| 12. Hypotonia | (14/25) | 56.0% |
| 13. Fingerlike thumb and/or long slender fingers | (13/25) | 52.0% |
| 14. Congenital dislocated hip | (11/25) | 44.0% |
| 15. Antimongoloid slant of eyes | (11/25) | 44.0% |

Twenty-five to forty percent were noted to have the following findings

| | | |
|---|---|---|
| 16. Strabismus | (9/25) | 36.0% |
| 17. Hypoplastic and/or low-set nipples | (6/25) | 24.0% |

## C. Etiology

Cytogenetic identification of trisomy 22 by autoradiographic studies was first made by Uchida and her co-workers in 1968 (1968a). They presented a family in

**TABLE I**
**Cytogenetically Confirmed Cases of Trisomy 22**

| Reference | Sex | Age at diagnosis | Parental age: Mother | Parental age: Father | Birth weight (gm) | Retardation: Mental | Retardation: Growth |
|---|---|---|---|---|---|---|---|
| Uchida *et al.* (1968a) | M[a] | 4 days died | 20 | 21 | 2660 | ? | ? |
| Hsu *et al.* (1971) | F | 7 months | 41 | 43 | 3080 | + | + |
| | F | 8½ months | 38 | 41 | 2410 | + | + |
| Gustavson *et al.* (1972) | F[b] | 7 years | 26 | 29 | 2750 | + | + |
| | F[b] | 5½ years | 29 | 31 | 3000 | + | + |
| | M | 4 years | 21 | 21 | 2470 | + | + |
| Punnett *et al.* (1973) | M | 5 years | 31 | 38 | 3312 | + | + |
| | F | 6 months | 32 | 35 | 2340 | + | + |
| | M[c] | 17 months | 29 | 28 | 3569 | + | + |
| Hirschhorn *et al.* (1973) | F | Newborn died | 36 | ? | 2260 | ? | + |
| Bass *et al.* (1973) | F | 8 years | 32 | ? | 1893 | + | + |
| Zackai *et al.* (1973) | F[b] | 6 weeks | 34 | 36 | 2818 | ? | ? |
| | M[b] | 5 years | 29 | 31 | 2820 | + | + |
| Penchaszadeh and Coco (1975) | F | 4 months | 40 | 53 | 3200 | + | + |
| | M | 18 months | 26 | 28 | 2500 | + | + |
| Alfi *et al.* (1975); Sparkes *et al.* (1966) | F | 10 years | 28 | – | 2730 | + | + |
| Goodman *et al.* (1971; R. M. Goodman, personal communication, 1975) | M | 10 months | 20 | 26 | 2150 | + | + |
| Perez-Castillo *et al.* (1975) | M | 2 months | 38 | 38 | 2000 | ? | + |
| Vianello and Bonioli (1975) | M | Infant | – | – | 2300 | + | |

[a]Sibling to one with trisomy 22 syndrome.
[b]Sibship to each other in the same report.
[c]Patient had imperforated anus.

| Microcephaly | Micrognathia | Low-set and/or malformed ears | Facial similarity (by photo) | Preauricular skin tag or sinus | Congenital heart disease | Cleft palate | Long or beaked nose | Long philtrum | Hypotonia | Congenital dislocated hip | Fingerlike thumbs or long slender fingers | Antimongoloid slant of eyes | Strabismus | Hypoplastic or low-set nipples | Cubitus valgus | Other skeletal abnormalities of lower extremities | Undescended testes and/or small penis |
|---|---|---|---|---|---|---|---|---|---|---|---|---|---|---|---|---|---|
| ? | + | + | + | – | + | – | + | + | ? | ? | – | ? | ? | – | + | – | ? |
| + | + | + | + | + | + | + | + | + | + | + | + | ± | – | + | + | – | |
| + | + | + | + | + | – | + | + | + | + | – | + | + | + | + | + | – | |
| + | + | + | + | + | – | + | + | + | + | + | + | ± | + | + | ? | ? | |
| + | + | + | + | + | + | + | + | + | + | – | + | ± | + | ? | ? | – | |
| + | + | + | ? | + | + | + | + | + | + | + | + | ± | ? | ? | ? | – | + |
| ? | + | + | + | + | + | – | + | + | ? | ? | + | + | ? | ? | ? | ? | + |
| + | ? | ? | + | + | + | – | ? | + | ? | ? | ? | + | ? | ? | ? | ? | |
| + | + | + | + | + | + | + | ± | ± | + | + | ? | + | ? | + | ? | ? | + |
| + | – | ? | ? | – | + | + | ? | ? | ? | – | + | ? | ? | ? | ? | + | |
| + | ? | – | ? | + | + | + | + | ? | + | ? | – | + | ? | ? | ? | ? | |
| – | + | + | + | + | + | – | – | – | ± | – | – | + | ? | + | ? | – | |
| + | ? | – | + | + | + | + | + | – | ± | + | – | ± | – | – | ? | – | + |
| + | + | + | + | + | + | + | + | + | + | + | + | + | + | ? | ? | – | |
| + | + | + | + | + | + | + | + | + | + | + | + | + | + | ? | ? | + | + |
| + | + | + | | ? | + | – | ? | ? | ? | + | + | ? | + | ? | ? | – | |
| + | + | + | + | + | – | – | + | + | – | + | + | – | + | ? | – | – | + |
| + | – | + | ± | + | + | + | + | ? | – | – | + | + | ? | ? | – | + | + |
| | + | + | + | – | + | + | + | + | + | – | | + | + | | | + | + |

**TABLE II**
**Clinically and Cytogenetically Suggestive Cases of Trisomy 22**

| Reference | Sex | Age at time of examination | Parental age: Mother | Parental age: Father | Birth weight (gm) | Retardation: Mental | Retardation: Growth | Microcephaly |
|---|---|---|---|---|---|---|---|---|
| Ferguson and Pitt (1963) | F | Infant | 30 | 34 | 2945 | + | ? | + |
| Hall (1963) | M | 7 years | 44 | 47 | 2220 | + | + | + |
| Ishmael and Laurence (1965) | M | Died 20 days | 22 | 30 | 3180 | ? | ? | – |
| Giorgi *et al.* (1967) | M | 3 years | 40 | ? | 2800 | + | + | ± |
| Uchida *et al.* (1968a) | F[a] | 18 months | 18 | 19 | 2350 | + | + | + |
| Hsu *et al.* (1971) | M | 2 months | 26 | 28 | 1300 | ? | + | + |

[a]Sibling to one with trisomy 22 syndrome.

which two sibs with multiple congenital anomalies had an extra G group chromosome. The autoradiographic identification was made in one of the affected sibs. The phenotypically normal mother was mosaic for trisomy G. Three additional cases of trisomy 22 were reported from our laboratory (Hsu *et al.*, 1971). Two were identified by autoradiographic studies. One mother was also mosaic for trisomy G, presumably trisomy 22. We reviewed twenty previously reported cases with an extra G group chromosome and found that ten of these had similar clinical features. On the basis of the phenotypic and cytogenetic findings in our cases and ten previous patients, we proposed that trisomy 22 may be a distinct clinical entity.

Cytogenetic confirmation of trisomy 22 by the new banding techniques was first made by Punnett and her co-workers in 1973. Using quinacrine fluorescence

| Micrognathia | Low-set and/or malformed ears | Facial similarity (by photo) | Preauricular skin tag or sinus | Congenital heart disease | Cleft palate or cleft uvula | Long or beaked nose | Long philtrum | Hypotonia | Congenital dislocated hip | Fingerlike thumbs or long slender fingers | Antimongoloid slant of eyes | Strabismus | Hypoplastic or low-set nipples | Cubitus valgus | Other skeletal abnormalities of lower extremities | Undescended testes and/or small penis |
|---|---|---|---|---|---|---|---|---|---|---|---|---|---|---|---|---|
| + | + | + | – | – | – | + | + | ? | + | ? | – | ? | ? | ? | + | |
| – | + | + | + | +? (murmur) | + | + | ? | + | – | ? | – | + | ? | ? | – | + |
| + | + | + | + | – | + | – | ? | ? | – | ? | + | ? | ? | ? | – | + |
| + | ? | + | – | ? | + | ? | ? | + | ? | ? | ± | – | ? | ? | ? | + |
| + | + | + | + | + | ? | + | + | + | – | + | ± | + | + | + | ? | |
| + | + | – | – | + | + | ± | ± | ? | + | – | ± | – | ? | – | + | |

staining (Q-banding) and Giemsa banding (G-banding) (Paris Conference, 1971), they identified three cases of trisomy 22 in children with similar clinical features as those described by Uchida *et al.* (1968a) and Hsu *et al.* (1971). Three additional cases of trisomy 22 may be those reported by Gustavson *et al.* (1972), as indeed from both autoradiographic and banding studies as well as the phenotypic manifestations, although the authors suggested partial D or E trisomy for these three cases. In 1973, four additional cases of trisomy 22 were identified by banding techniques [Hirschhorn *et al.*, 1973; Bass *et al.*, 1973; Zackai *et al.* (two cases), 1973]. A partial G-banding karyotype of trisomy 22 is shown in Fig. 3. Subsequently, to our knowledge, six more confirmed cases of trisomy 22 have been reported [Penchaszadeh and Coco (two cases), 1975; Alfi *et al.*, 1975 (initially reported by Sparkes *et al.*, in 1966); Goodman *et al.*,

**TABLE III**
**Cases of Trisomy G without Down Syndrome (and without Typical Features of Trisomy 22 Syndrome)**

| Reference | Sex | Age at time of examination | Parental age | | Birth weight (gm) | Retardation | |
|---|---|---|---|---|---|---|---|
| | | | Mother | Father | | Mental | Growth |
| Hayward and Bower (1960) | M | 3 years | 32 | – | ? | + | ? |
| Crawfurd (1961) | F | Infant | 40 | 49 | ? | + | |
| Zellweger *et al.* (1962a) | F | 2 years | 35 | 40 | 2700 | + | + |
| | F | 3 years | 33 | | 3500 | + | + |
| Koulischer and Perier (1962) | F | Infant | ? | – | ? | + | + |
| Kruger *et al.* (1968) | F[a] | 2 years | 29 | 32 | 3000 | + | + |
| Walbaum *et al.* (1970) | M | 3 years | ? | ? | 3490 | + | + |
| Magenis *et al.* (1972) | F | 12 years | 30 | – | 2920 | + | + |

[a]Patient had coloboma

initially reported in 1971 and confirmed in 1975 (R. M. Goodman, personal communication); Perez-Castillo *et al.*, 1975; Vianello and Bonioli, 1975]. Therefore, up to 1975, there have been a total of nineteen cases of trisomy 22 confirmed by special techniques (three by autoradiography and sixteen by banding methods).

The mean maternal age of the twenty four cases reviewed was 30.5 years as compared to the mean maternal age in the United States of 26.4 years (Vital Statistics of the United States, 1962). The comparatively advanced maternal age in the cases of trisomy 22 implies maternal meiotic nondisjunction of chromosome 22. Since only sixteen cases of trisomy 22 have been confirmed by the banding techniques, no attempt has yet been made to identify the origin of the

| Microcephaly | Micrognathia | Low-set and/or malformed ears | Facial similarity (by photo) | Preauricular skin tag or sinus | Congenital heart disease | Cleft palate | Long or beaked nose | Long philtrum | Hypotonia | Congenital dislocated hip | Fingerlike thumbs or long slender fingers | Antimongoloid slant of eyes | Strabismus | Hypoplastic or low-set nipples | Cubitus valgus | Other skeletal abnormalities of lower extremities | Undescended testes and/or small penis |
|---|---|---|---|---|---|---|---|---|---|---|---|---|---|---|---|---|---|
| + | – | – | + | – | – | – | ? | ? | ? | – | ? | – | – | | ? | – | ? |
| ? | + | + | | | + | ? | ? | ? | ? | + | ? | ? | ? | | ? | + | |
| + | – | ? | + | – | – | ? | ? | ? | + | ? | ? | + | ? | ? | ? | ? | |
| – | – | + | + | – | – | – | ? | ? | + | ? | ? | + | ? | ? | ? | ? | |
| + | ? | ? | ? | ? | + | ? | ? | ? | + | ? | ? | ? | ? | ? | ? | ? | |
| + | + | + | ? | – | + | – | – | ? | + | + | ? | ? | ? | ? | ? | – | |
| + | ? | ? | | | | | ± | ± | | | | – | | | | | ? |
| ? | – | + | ? | – | – | | ? | – | | – | – | + | + | | + | – | |

extra No. 22 chromosome. If one studies the patients with trisomy 22 and their parents for chromosome polymorphisms of the No. 22 chromosomes, one expects to be able to determine the origin of the extra No. 22 chromosome, namely from the mother or the father, and also the occurrence of nondisjunction at first or second meiotic division.

In addition to maternal age, other factors which have been implicated in chromosomal nondisjunction are irradiation (Uchida *et al.,* 1968b; Alberman *et al.,* 1972), viral infection (Stoller and Collmann, 1965; Kucera, 1970) and autoimmunity (Fialkow, 1966).

Parental structural chromosomal abnormalities such as translocation (Turpin and Lejeune, 1961) or inversion (Sparkes *et al.,* 1970) have also been implicated as predisposing factors for meiotic nondisjunction.

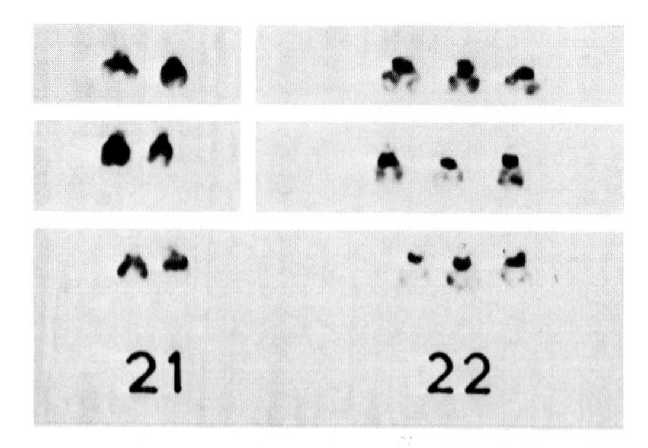


**Fig. 3.** Partial Giemsa banding karyotypes of G group chromosome showing trisomy 22.

It is interesting to note that trisomy G mosaicism (presumably trisomy 22 mosaicism) was found in two mothers of children affected with trisomy 22 (Uchida *et al.,* 1968a; Hsu *et al.,* 1971). Therefore parents of trisomy 22 children should be studied to rule out trisomy 22 mosaicism and also other chromosomal aberrations.

We have recently studied a case of partial trisomy 22 and 13 with an unbalanced 13/22 translocation (47,XX,+der(22),t(13;22) (q22;q12)mat (Kim *et al.,* 1975). From this patient, it appears that the presence of the extra proximal unbanded (G-band) euchromatic region of No. 22 (22q11) is responsible for the trisomy 22 syndrome. This was an infant clinically manifesting both trisomy 22 and 13 syndromes. The patient's phenotypically normal mother who had a history of fetal wastage was found to be a balanced translocation carrier 46,XX,t(13;22) (q22;q12).

The incidence of trisomy 22 is not yet known. In 43,558 consecutive live born studies by various groups, no case of trisomy 22 was detected (Jacobs *et al.,* 1974). Punnett *et al.* (1973) estimated that trisomy 22 occurs once in every 30,000 to 50,000 live births.

## II. CAT EYE SYNDROME

### A. Historical Review

Almost one century ago, the association of coloboma and anal atresia was described (Haab, 1878). In 1965, Schachenmann *et al.* found an extra acrocentric chromosome which was smaller than the G group in three patients with the combination of coloboma and anal atresia. The extra chromosome was also detected in one patient's mother who also had coloboma, while the patient's brother, a maternal uncle, and the maternal grandmother were found to have this extra chromosome in a small proportion of their cells cultured *in vitro.* These three relatives of the patient were phenotypically normal (Fig. 4). Since then, this syndrome had been coloquially called the cat eye syndrome because the vertical iridochoroidal coloboma of these patients gave the appearance of a cat's eye. Gerald *et al.* were the first to use this term in their paper presented in 1968, in which two patients with similar clinical and cytogenetic abnormalities were described. Until 1975, there have been at least fifteen reported cases of typical cat eye syndrome, that is with the combination of both coloboma and anal atresia (Table IV), and fifteen cases of incomplete cat eye syndrome (Table V), all associated with an extra small chromosome. Normal chromosomal constitution has been reported in at least four cases with cat eye syndrome (Table VI) (Zellweger *et al.*, 1962b; Neu *et al.*, 1970; Franklin and Parslow, 1972).

## B. Phenotypic Manifestations

Phenotypically, in addition to coloboma and anal atresia, the patients with cat eye syndrome may also have many other congenital abnormalities including

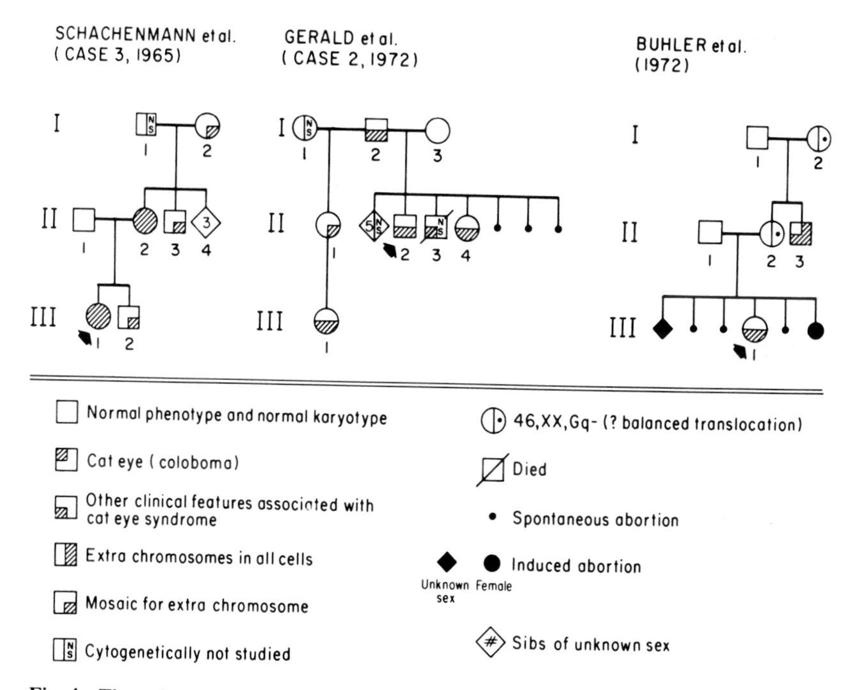


**Fig. 4.** Three families with cat eye syndrome: Pedigrees of Case 3 of Schachenmann *et al.* (1965), Case 2 of Gerald *et al.* (1972), and family of Bühler *et al.* (1972).

**TABLE IV**
**Patients with Complete Cat Eye Syndrome**

| Reference | Sex | Age | Parental age | | Birth weight (gm) | Retardation | | Coloboma | Anal atresia | Preauricular skin tag or sinus | Low-set and/or malformed ears | Downward slanting of eyes | Hypertelorism | Microphthalmia | Strabismus | Epicanthus | Depressed nasal bridge | Congenital heart disease | Renal anomalies | Hip dislocation | Long slender thumb |
|---|---|---|---|---|---|---|---|---|---|---|---|---|---|---|---|---|---|---|---|---|---|
| | | | Mother | Father | | Mental | Growth | | | | | | | | | | | | | | |
| Schachenmann *et al*. (1965) | F | 3 years | 27 | 40 | 2320 | – | ? | + | + | + | | | | + | | | | | | | |
| | F | 5 months | 34 | 40 | 3100 | + | + | + | + | + | | + | + | | | | | | | | |
| | F | 7 years | 30 | 29 | ? | + | + | + | + | | | | + | | + | | | | + | + | |
| Gerald *et al*. (1968, 1972) | M | 3 years | 37 | 40 | 4430 | ± | – | + | + | + | | | | – | + | | | + | + | – | |
| | F | 16 days | 23 | 22 | ? | | | + | + | + | + | | | – | | | | + | + | – | |
| Thomas *et al*. (1969) | M | 1 year | 31 | 30 | 3200 | + | + | + | + | + | + | | + | | | + | | | + | | |
| Pfeiffer *et al*. (1970) | F | Infant | 38 | 34 | 3100 | | | + | + | + | ? | + | + | | | | | | | | + |
| Noel and Quack (1970) | M | 21 years | | | | + | – | + | + | + | + | + | + | | | | + | + | + | | |
| Weber *et al*. (1970) | F | 2 years | 25 | 29 | 3555 | + | – | + | + | + | + | | | + | | | | | + | | |

| | | | | | | | | | | | | | | | | | | | |
|---|---|---|---|---|---|---|---|---|---|---|---|---|---|---|---|---|---|---|---|
| Darby and Hughes (1971) | F | 2 years | | | | | | + | + | + | ? | | | | | | | | |
| Fryns *et al*. (1972) | M | 8 weeks | 34 | | 3250 | ? | + | + | + | + | ? | + | + | | | | | | |
| Petit *et al*. (1973) | – | – | 32 | 36 | 3400 | + | | + | + | + | | + | | | | | + | | |
| Cory and Jamison (1974) | F | Infant | 24 | 30 | 3060 | ? | – | + | + | + | ? | – | + | | | | + | + | + |
| Kunze *et al*. (1975) | F | 5½ years | 32 | 31 | 3420 | – | + | + | + | + | + | – | + | + | | | – | + | |
| Pierson *et al.* (1975) | M[a] | 7 years | | | | – | + | + | + | + | + | + | ± | – | – | + | + | + | – |

[a]This patient also had hypopituitarism and other skeletal anomalies.

**TABLE V**
**Patients with Incomplete Cat Eye Syndrome**

| | | | Parental age | | | Retardation | | | | |
|---|---|---|---|---|---|---|---|---|---|---|
| Reference | Sex | Age | Mother | Father | Birth weight (gm) | Mental | Growth | Coloboma | Anal atresia | Preauricular skin tag or sinus |
| Zellweger *et al.* (1962b) | F | 6 months | 29 | – | 2400 | + | + | – | + | + |
| Schachenmann *et al.* (1965) | F | 37 years | – | – | – | + | + | + | – | |
| Taft *et al.* (1965) | M | 16 months | – | – | 3006 | + | + | + | – | – |
| Ishmael and Laurence (1965) | M | Died 20 days | 22 | 30 | 3180 | ? | – | – | + | + |
| Curcio (1967) | F | Newborn | 26 | 28 | 3150 | ? | ? | – | + | + |
| Ginsberg *et al.* (1968) | F | Died 14 weeks | 32 | 34 | 2665 | ? | + | + | – | – |
| Beyer *et al.* (1968) | M | 4 years | – | – | 3200 | ? | + | – | + | + |
| Gerald *et al.* (1968) | M[a] | 10 years | 32 | 40 | 3180 | + | – | – | + | |
| | M[a] | Died 6 months | 19 | 27 | | ? | ? | – | + | |
| | F[a] | 9 years | 34 | 42 | 2850 | ? | – | – | + | + |
| | F[a] | | 30 | ? | 2900 | ? | ? | – | + | + |
| Bühler *et al.* (1972) | F | 2 years (proband) | 25 | 29 | 2640 | + | – | – | – | + |
| | M | 17 years (uncle) | 28 | 28 | 2600 | + | | – | + | + |
| De Chieri *et al.* (1974) | M | Died 3 months | – | – | 2800 | | | + | – | + |
| Bofinger and Soukup (1977) | F | 20 months | 32 | 34 | 2450 | + | – | | + | + |

[a]One family (see fig. 4).

| Low-set and/or malformed ears | Downward slanting of eyes | Hypertelorism | Microphthalmia | Strabismus | Epicanthus | Depressed nasal bridge | Cleft palate | Congenital heart disease | Renal anomalies | Skeletal anomalies | Seizure | Other anomalies and remarks |
|---|---|---|---|---|---|---|---|---|---|---|---|---|
| + | + | + | – |  | – | – | + | + | + | + | + |  |
|  |  |  |  |  |  |  |  |  | + |  |  | Cataract |
| + |  |  |  |  | + |  |  |  |  |  |  |  |
| + | + | + | + |  | – | ± | + | – | + | – | – |  |
|  |  |  |  |  |  |  |  |  |  | + |  |  |
| + |  |  | + |  |  |  |  | + |  | + |  |  |
| + |  | + |  |  | + |  |  |  | + | + |  |  |
|  |  |  |  |  |  |  |  | + | + |  |  | Proband |
|  |  |  |  |  |  |  |  |  | + |  |  | Brother |
|  |  |  |  |  |  |  |  | – | – | – |  | Sister |
|  |  |  |  |  |  |  |  | – | – | – |  | Half niece |
| + | + | + |  | + |  |  |  |  |  | + |  |  |
|  | + | + |  | + |  |  |  |  |  | + |  |  |
| + | + | + |  |  |  | + |  | + |  | + |  |  |
| – |  |  | ± |  |  | – | – | +? |  | + |  | Redundant nuchal skin fold bulbous nose |

**TABLE VI**
**Patients with Cat Eye Syndrome but with Normal Chromosomal Constitution**

| Reference | Sex | Age | Parental age | | Birth weight (gm) | Retardation | | Coloboma | Anal atresia | Preauricular skin tag or sinus |
|---|---|---|---|---|---|---|---|---|---|---|
| | | | Mother | Father | | Mental | Growth | | | |
| Zellweger *et al.* (1962b) | F | 3 years | 25 | – | 2340 | + | | + | + | |
| Neu *et al.* (1970) | F | 3½ months | 43 | 45 | – | | | + | + | – |
| Franklin and Parslow (1972) | F[a] | 3 months | 27 | 28 | 2810 | + | + | – | + | + |
| | F[a] | Newborn | 30 | 31 | 2610 | + | + | + | + | + |

[a]Sibship to each other.

preauricular skin tags, dimples or sinuses, low-set and/or malformed ears, antimongoloid slant of the eyes, hypertelorism, microphthalmia, congenital heart disease, skeletal anomalies including hemivertebrae, rib anomalies, hip dislocation, long slender thumb or fingers, renal anomalies such as horseshoe kidneys or agenesis, depressed nasal bridge, epicanthus, strabismus, and cataracts. Mental retardation is frequently present. Children with cat eye syndrome are shown in Figs. 5 and 6. The coloboma is usually bilateral, iridal and/or choroidal. The anal atresia is usually associated with a rectovaginal or rectoperineal fistula.

In fifteen cases of incomplete cat eye syndrome, only one of the two major abnormalities, i.e., coloboma or anal atresia, was present (Table V). However, all of these patients had a few other congenital abnormalities seen in the cat eye syndrome, such as preauricular skin tags or sinuses, antimongoloid slant of eyes, microphthalmia, epicanthal folds, hypertelorism, low-set and/or malformed ears, congenital heart disease, renal anomalies and skeletal abnormalities (Table V). Four of these cases had coloboma (Schachenmann *et al.*, 1965; Taft *et al.*,

| Low set and/or malformed ears | Downward slanting of eyes | Hypertelorism | Microphthalmia | Strabismus | Epicanthus | Depressed nasal bridge | Cleft palate | Congenital heart disease | Renal anomalies | Skeletal anomalies | Other anomalies |
|---|---|---|---|---|---|---|---|---|---|---|---|
| + | | | | | | | + | + | | | Cleft lip, tracheo-esophageal fistula and simian lines |
| + | ± | + | − | + | + | + | + | + | + | + | Flat occiput, and simian lines |
| + | − | + | + | | + | + | − | + | − | + | Microcephaly, and simian lines. |
| + | − | + | + | | + | | − | + | − | + | Microcephaly, simian lines and hypoplastic nails |

1965; Ginsberg *et al.,* 1968; De Chieri *et al.,* 1974), and ten cases had anal atresia [Zellweger *et al.,* 1962b; Ishmael and Laurence, 1965; Curcio, 1967; Beyer *et al.,* 1968; Gerald *et al.*·(four cases), 1968, 1972; Bühler *et al.,* 1972; Bofinger and Soukup, 1977]. One patient (Bühler *et al.*, 1972) had no coloboma or anal atresia, but many other associated congenital anomalies such as preauricular skin tags, antimongoloid slant of eyes, and hypertelorism. This patient's uncle, however, had membranous anal atresia and other anomalies associated with the cat eye syndrome (Fig. 4).

It appears that the minimal diagnostic criteria for cat eye syndrome has gradually become less rigid. Initially, the combination of coloboma and anal atresia was the basic requirement for this diagnosis. As more and more other cases appear to fit into this diagnosis without the combination of coloboma and anal atresia, but with other associated congenital anomalies, it is suggested that perhaps the clinical diagnosis of cat eye syndrome need not be restricted to the combination of coloboma and anal atresia (see Table VII).

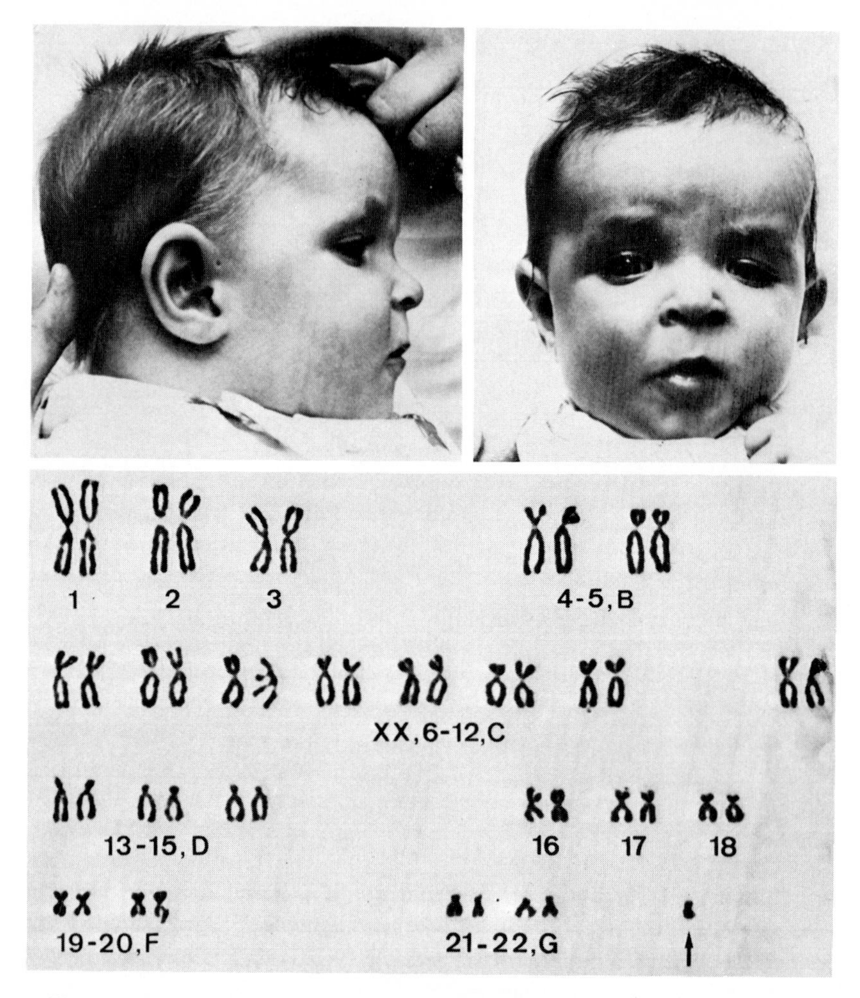


**Fig. 5.** Infant with cat eye syndrome, showing preauricular pit, coloboma of iris, and downward slanting of eyes. The karyotype of this patient shows an extra chromosome (arrow). (Reproduced by courtesy of Schmid, 1967.)

The average birth weight in twenty-seven cases (with recorded birth weight) of typical and incomplete cases of cat eye syndrome was 2881.32 grams. There was no intrauterine growth retardation. Postnatally, growth retardation was noted in fourteen of the twenty four cases with available information. Mental retardation was recorded in sixteen of twenty cases; two of the twenty were minimally retarded (Noel and Quack, 1970; Weber *et al.*, 1970). The ages of the thirty-four patients ranged from newborn to 37 years. Twenty-nine patients

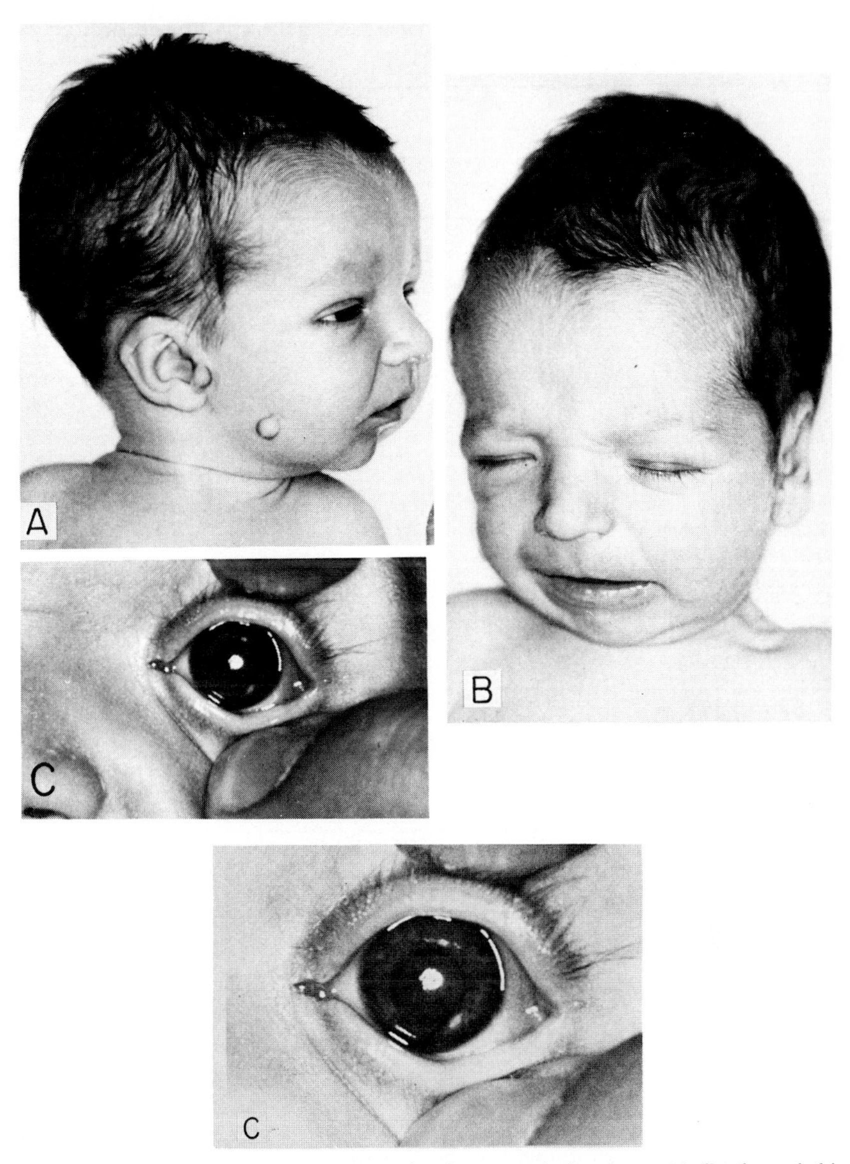


**Fig. 6.** Infant with cat eye syndrome showing preauricular sinus with fistula, and skin appendix in the middle of the right cheek, peculiar facies, and coloboma of iris (Fryns *et al.*, 1972).

**TABLE VII**
**The Frequencies of the Associated Phenotypic Manifestations in Patients with Cat Eye Syndrome[a]**

| Clinical feature | Complete cat eye syndrome ($n$=15) | Incomplete cat eye syndrome ($n$=15) | Cat eye syndrome with normal chromosomes ($n$=4) | Combined ($n$=34) | % |
|---|---|---|---|---|---|
| 1. Anal atresia | 15/15 | 10/15 | 4/4 | 29/34 | 85.3% |
| 2. Mental retardation | 6/10 | 7/7 | 3/3 | 16/20 | 80.0% |
| 3. Preauricular skin tag or sinus | 14/15 | 10/15 | 2/4 | 26/34 | 76.5% |
| 4. Coloboma | 15/15 | 4/15 | 3/4 | 22/34 | 64.7% |
| 5. Growth retardation | 6/10 | 6/10 | 2/4 | 14/24 | 58.3% |
| 6. Hypertelorism | 8/15 | 6/15 | 3/4 | 17/34 | 50.0% |
| 7. Abnormal and/or low-set ears | 6/15 | 7/15 | 4/4 | 17/34 | 50.0% |
| 8. Renal anomalies | 9/15 | 6/15 | 1/4 | 16/34 | 46.1% |
| 9. Skeletal anomalies | 3/15 | 8/15 | 3/4 | 14/34 | 41.2% |
| 10. Congenital heart disease | 6/15 | 4/15 | 4/4 | 14/34 | 41.2% |
| 11. Antimongoloid slant of eyes | 6/15 | 5/15 | 0/4 | 11/34 | 32.3% |
| 12. Microphthalmia | 3/15 | 2/15 | 2/4 | 7/34 | 20.6% |
| 13. Epicanthal folds | 1/15 | 2/15 | 3/4 | 6/34 | 17.6% |
| 14. Depressed nasal bridge | 2/15 | 2/15 | 2/4 | 6/34 | 17.6% |
| 15. Strabismus | 2/15 | 2/15 | 1/4 | 5/34 | 14.7% |
| 16. Cleft palate | 0 | 2/15 | 1/4 | 3/34 | 8.8% |

[a]The frequencies of the associated phenotypic manifestations in the typical cases (15) and incomplete cases (15) of cat eye syndrome and in 4 cases of cat eye syndrome with normal chromosomes are listed by frequency of occurrence.

were under 6 years of age. Six died during infancy. The average life-span of patients with cat eye syndrome is not known.

Thus far, a clinical diagnosis of cat eye syndrome has been made by the following minimal clinical criteria:

1. Combination of the two major features, namely coloboma and anal atresia, with or without other associated abnormalities.
2. Combination of one major feature, coloboma or anal atresia, plus at least one of the most frequent associated specific anomalies, i.e., preauricular skin tags or sinuses, renal anomalies.
3. Combination of one major feature plus two less frequent features such as

antimongoloid slant of eyes, skeletal anomalies, congenital heart disease, and other eye defects.

4. Combination of five or more minor specific features.

## C. Etiology

Of thirty-four known patients with clinical cat eye syndrome, an extra small chromosome with a morphology ranging from a very small acrocentric chromosome to a small submetacentric or metacentric chromosome (smaller than F group chromosomes) was found in thirty patients (Table VIII). Four patients had normal chromosome complements [Zellweger *et al.*, 1962; Neu *et al.*, 1970; Franklin and Parslow (two cases), 1972]; the two cases reported by Franklin and Parslow had normal banding karyotypes as well (with Q-banding). The extra chromosome was acrocentric in twenty-one cases and submetacentric or metacentric in nine cases (Table VIII). Satellite or satellite association of this abnormal chromosome was observed in twenty-four cases; five of these were submetacentric or metacentric chromosomes. Mosaicism of 46/47,+M (marker chromosome) was found in seven patients with the cat eye syndrome. The extra chromosome was found to be late replicating in nature in three cases (Pfeiffer *et al.*, 1970; Weber *et al.*, 1970; Kunze *et al.*, 1975) and relatively late replicating in one (Fryns *et al.*, 1972). Pfeiffer *et al.* (1970) suggested that the extra chromosome in their case might be part of a No. 14. Q-banding was used in ten cases [Gerald *et al.* (six cases), 1972; Bühler *et al.* (two cases), 1972; Kunze *et al.* (one case), 1975; Pierson *et al.* (one case), 1975]. G-banding was used in one case (Cory and Jamison, 1974). Both Q- and G-banding were used in two cases (De Chieri *et al.*, 1974; Bofinger and Soukup, 1977). Identification was only possible in three cases from two families. One family was comprised of an uncle and a niece (Bühler *et al.*, 1972) and one family had an affected daughter and a phenotypically normal mother who was a carrier of a balanced translocation (Bofinger and Soukup, 1977). In the family reported by Bühler *et al.*, identification was made possible by finding a partially deleted No. 22 chromosome (22q−) transmitted through three generations (Fig. 4); the proband's mother and maternal grandmother both carried a partially deleted No. 22 chromosome most likely from a balanced reciprocal translocation. The proband and her maternal uncle were affected with cat eye syndrome and carried an extra 22q− chromosome. Thus, the extra chromosome in this family was identified to be a partially deleted No. 22 chromosome, i.e., 22q−. The extra chromosome in the patient of Bofinger and Soukup (1977) was also identified to be a 22q−, but derived from the phenotypically normal mother who was a carrier of a balanced reciprocal translocation 46,XX,t(11q+;22q−). The extra chromosomes in all other cases were not identifiable.

**TABLE VIII**
**Morphology, Replicating and Banding Patterns of the Extra Chromosome Found in Patients with Cat Eye Syndrome**

| Authors Reference | Case No. | Extra chromosome: Metacentric or submetacentric | Extra chromosome: Acrocentric | Satellite or satellite association | Mosaicism | Late replicating | Banding | Parental chromosome: Mother | Parental chromosome: Father | Identification of the extra chromosome |
|---|---|---|---|---|---|---|---|---|---|---|
| Schachenmann *et al*. (1965) | 1 | | + | + | – | | – | N1 | Nl | |
| | 2 | | + | + | – | | – | Nl | Nl | |
| | 3 | | + | + | – | | – | 47,+M | Nl | |
| | 3a | | + (Mother of 3) | + | – | | – | 46/47,+M | Not studied | |
| Zellweger *et al*. (1962b) | | | + | + | – | | – | Nl | Nl | |
| Taft *et al*. (1965) | | + | | – | – | | – | Nl | Nl | |
| Ishmael and Laurence (1965) | | + | | | – | | – | Nl | Nl | |
| | | | | + | – | | – | Nl | Nl | |
| Curcio *et al*. (1967) | | + | | + | – | | – | Nl | Nl | |
| Ginsberg *et al*. (1968) | | + | | ? | + | | – | Nl | Nl | |
| Beyer *et al*. (1968) | | + | | ? | – | | – | Nl | Nl | |
| Gerald *et al*. (1968, 1972) | 1 | | + | + | + | | Q-banding | Nl | Nl | Identification not possible (4 other members of family 2 were mosaic) 46/47,+M |
| | 2 | | + | | + | | Q-banding | Nl | 46/47,+M | |
| | 2a | | + (Sister) | + | + | | Q-banding | Nl | 46/47,+M | |
| | 2b | | + (Father) | + | + | | Q-banding | Not studied | Not studied | |
| | | | | | | | Q-banding | 46/47,+M | Not studied | |
| | 2c | | + (Half niece) | + | + | | Q-banding | Nl | Nl | |
| | 3 | | + | + | – | | – | Nl | Nl | |

| | | | | | | | | | | |
|---|---|---|---|---|---|---|---|---|---|---|
| Thomas *et al*. (1969) | | | + | – | – | | – | Nl | Nl | |
| Pfeiffer *et al*. (1970) | | | + | + | – | | – | Nl | Nl | |
| Noel and Quack (1970) | | + | | ? | | | + | Nl | Nl | Suggestion for 14q– |
| Weber *et al*. (1970) | | | + | ? | – | + | – | Nl | Nl | |
| Darby and Hughes (1971) | | + | | + | – | | – | Nl | Nl | |
| Fryns *et al*. (1972) | | | + | + | – | Relatively late | – | Nl | Not studied | |
| Bühler *et al*. (1972) | 1 | | + | + | + | | Q-banding | 46,22q– | Nl | 22q– |
| | 1a | | + (Uncle) | + | – | | Q-banding | 46,22q– | Nl | |
| Petit (1973) | | + | | + | | | ? | Nl | Nl | |
| De Chieri *et al*. (1974) | | | + | + | – | | Q & G banding | Nl | Not studied | Not possible |
| Cory and Jamison (1974) | | | + | + | – | | G-banding | Nl | Nl | Not possible |
| Kunze *et al*. (1975) | | + | | + | – | + | Q-banding | Nl | Nl | Not possible |
| Pierson *et al*. (1975) | | | + | + | – | | Q-banding | Nl | Nl | Not possible |
| Bofinger and Soukup (1977) | | | + | + | – | | Q & G banding | 46,t(11q+; 22q–) | Nl | 22q– |

Eight patients with cat eye syndromes were derived from three families (family of Case 3 of Schachenmann *et al.*, 1965; family of Case 2 of Gerald *et al.*, 1972; family of Bühler *et al.*, 1972). The pedigrees of these three families are shown in Fig. 4. In all three families, the abnormal chromosome was clearly inherited through three generations. In the family of Schachenmann *et al.*, the two affected members showed an extra chromosome in every cell, whereas mosaicism of 46/47,+M was found in three phenotypically normal members, i.e., the maternal grandmother, maternal uncle and brother. In the family of Gerald, five members had cat eye syndrome; the extra chromosome in this family originated from the affected father who transmitted the abnormal chromosome to at least three of his nine children. One of the other six had cat eye syndrome but was not studied cytogenetically. Interestingly, one daughter (II-1 in the pedigree, Fig. 4) who, although mosaic for this extra chromosome (46/47,+M), was phenotypically normal but gave birth to a female child with 46/47,+M and cat eye syndrome. In the family described by Bühler, the phenotypically normal mother and the maternal grandmother of the proband may have been carriers for a balanced reciprocal translocation involving 22q− and each produced an offspring with an unbalanced chromosomal complement of 47,+22q−. The proband who was mosaic for 47,+22q− most likely initially had a 47,+22q− complement which underwent anaphase lag, resulting in mosaicism of 46/47,+22q−.

Chromosome 22 appears to have been involved in cat eye syndrome, because trisomy 22 and cat eye syndrome do share several clinical features in common, namely preauricular skin tag or sinus, antimongoloid slant of eyes, and skeletal abnormalities. Anal atresia is not a common feature for trisomy 22, yet one confirmed case of trisomy 22 (Punnett *et al.*, 1973) and two cases of 47,+22q− (Bühler *et al.*, 1972; Bofinger and Soukup, 1977) had anal atresia. The patient reported by Ishmael and Laurence was clinically compatible with both cat eye syndrome and trisomy 22 (Table II and Table V); this patient had anal atresia and many features for both cat eye syndrome and trisomy 22.

Coloboma and microphthalmia have not been observed in patients with trisomy 22, but it is well known that these eye features are associated with trisomy 13 (Smith, 1970). Mental and growth retardation and congenital heart disease are associated with all three syndromes, i.e., trisomy 13 and trisomy 22 and cat eye syndrome.

While incomplete cat eye syndrome without coloboma might be caused by a partial trisomy 22 as shown in the cases of Bühler *et al.* (1972) and Bofinger and Soukup (1977), it is possible that the complete cat eye syndrome may involve a part of chromosome 22 and a part of chromosome 13, possibly in a translocation containing a short arm, centromere and the proximal long arm of a No. 22 (22pter→22q11) and a small euchromatic region of the long arm of a chromosome 13 (13q32→13q34). As we have discussed earlier, the short arm, centromere, and proximal long arm of chromosome 22 are possibly responsible for trisomy 22 syndrome, and a distal euchromatic region of the long arm of chromosome 13 may be responsible for the cat eye trisomy 13 syndrome. The translocation chromosome is transmissible to the offspring of the carrier as shown in the families just discussed. Individuals with a balanced translocation or a low degree of mosaicism containing an abnormal cell line may be phenotypically normal, and individuals with an unbalanced translocation, i.e. with an extra translocation chromosome, would be affected with cat eye syndrome. In the majority of patients with cat eye syndrome and the extra chromosome, the parents were found to have normal chromosomes. Thus, in these patients, the abnormal chromosome was most likely produced *de novo* during gametogenesis in one parent.

If our hypothesis is correct, the explanation for those patients with typical cat eye syndrome and normal chromosomes would be

1. Finding only the normal cell line from an initially mosaic embryo with 46/47, since structurally abnormal chromosomes tend to undergo anaphase lag.
2. Difficulty in identification of a small region of duplicated chromosomal material of a euchromatic region of No. 22 and No. 13, even with the banding techniques.

Obviously, with further careful cytogenetic studies using various banding techniques in these patients with cat eye syndrome and their family members, the understanding of the chromosomal cause of cat eye syndrome can be anticipated. Another method could be the study of gene products known to be coded by these chromosomes, in order to search for dosage effects, as has already been demonstrated in trisomy 21 (Tan *et al.*, 1974).

## ACKNOWLEDGMENTS

We wish to thank Mrs. Minnie Woodson for her help in preparation of the manuscript and Drs. Pilar Reyes and Sara Kaffe for proofreading.

Research was supported by U.S.P.H.S. Grants (GM-19443 and HD-02552) and a Medical Service Grant from the National Foundation (C-155).

## REFERENCES

Alberman, E., Polani, P. E., Fraser-Roberts, J. A., Spicer, C. C., Elliott, M., and Armstrong, E. (1972). Parental exposure to X-irradiation and Down's syndrome. *Ann. Hum. Genet.* **36,** 195–208.

Alfi, O. S., Sanger, R. G., and Donnell, G. N. (1975). Trisomy 22: A clinically identifiable phenotype. *Birth Defects Conf.* Kansas City, Missouri, p. 93. Nat. Found. March of Dimes.

Bass, H. N., Crandall, B. F., and Sparkes, R. S. (1973). Probable trisomy 22 identified by fluorescent and trypsin-giemsa banding. *Ann. Genet.* **16,** 189–192.

Beyer, P., Ruch, J. V., Rumpler, Y., and Girard, J. (1968). Observation d'un enfant débile mental et polymalforme dont le caryotype montre la présence d'un petit extra-chromosome médiocentrique. *Pediatrie* **23,** 439–442.

Biesele, J. J., Schmid, W., and Lawlis, M. G. (1962). Mentally retarded schizoid twin girls with 47 chromosomes. *Lancet* **2,** 403–405.

Bofinger, M. K., and Soukup, S. W. (1977). Cat eye syndrome: Partial trisomy 22 due to translocation in the mother, *Am. J. Dis. Child.* (in press).

Bühler, E. M., Mehes, H., Muller, H., and Stalder, G. R. (1972). Cat-eye syndrome, a partial trisomy 22. *Humangenetik* **15,** 150–162.

Cory, C. C., and Jamison, D. L. (1974). The cat eye syndrome. *Arch. Opthalmol.* **92,** 259–262.

Crawfurd, M. d'A. (1961). Multiple congenital anomaly associated with an extra autosome. *Lancet* **1,** 22–24.

Curcio, S. (1967). Malformazione del retto e della vagina associata ad anomalia cromosomica (47,XX,?G+). *Clin. Ostet. Ginecol.* **72,** 533–539.

Darby, C. W., and Hughes, D. T. (1971). Dermatoglyphics and chromosomes in cat-eye syndrome. *Br. Med. J.* **3,** 47–48.

De Chieri, P. R., Malfatti, C., Stanchi, F., and Albores, J. M. (1974). Cat-eye syndrome: Evaluation of the extra chromosome with banding techniques. Case report. *J. Genet. Hum.* **22,** 101–107.

Dunn, H. G., Ford, D. K., Auersperg, N., and Miller, J. R. (1961). Benign congenital hypotonia with chromosomal anomaly. *Pediatrie* **28,** 578–590.

Ferguson, J., and Pitt, D. (1963). Another child with 47 chromosomes. *Med. J. Aust.* **1,** 546–547.

Fialkow, P. J. (1966). Autoimmunity and chromosomal aberrations. *Am. J. Hum. Genet.* **18,** 93–108.

Franklin, R. C., and Parslow, M. I. (1972). The cat-eye syndrome review and two further cases occurring in female siblings with normal chromosomes. *Acta Paediat. Scand.* **61,** 581–586.

Fryns, J. P., Eggermont, E., Verresen, H., and van den Berghe, H. (1972). A newborn with the cat eye syndrome. *Humangenetik* **15,** 242–248.

Gerald, P. S., Davis, C., Say, B. M., and Wilkins, J. L. (1968). A novel chromosomal basis for imperforate anus (the "cat's eye" syndrome). *Pediatr. Res.* **2,** 297 (abstr.).

Gerald, P. S., Davis, C., Say, S. M., and Wilkins, J. L. (1972). Syndromal associations of imperforate anus: The cat eye syndrome. *Birth Defects, Orig. Artic. Ser.* **8,** No. 2, pp. 79–84.

Ginsberg, J., Dignan, P., and Soukup, S. (1968). Ocular abnormality associated with extra small autosome. *Am. J. Ophthalmol.* **65,** 740–746.

Giorgi, P. L., Paci, A., and Ceccarelli, M. (1975). An extra chromosome in a case of Tay-Sachs disease with additional abnormalities. *Helv. Paediatr. Acta* **22,** 28–35.

Goodman, R. M., Katznelson, M., Spero, M., Shaki, R., and Padeh, B. (1971). The question of trisomy 22 syndrome. *J. Pediatr.* **79,** 174–175.

Gustavson, K.-H., and Hook, O. (1961). The chromosomal constitution of Sturge-Weber syndrome, *Lancet* **1,** 559.

Gustavson, K.-H., Hitrec, V., and Santesson, B. (1972). Three nonmongoloid patients of similar phenotype with an extra G-like chromosome. *Clin. Genet.* **3,** 135–146.

Haab, O. (1878). Beitrage zu den angeborenen Fehlern des auges. *Albrecht von Graefes Arch. Opthalmol.* **24,** 257–281.

Hall, B. (1963). Mongolism and other abnormalities in a family with trisomy 21-22 tendency. *Acta Paediatr. (Stockholm), Suppl.* **146**, 77–91.

Hayward, M. D., and Bower, B. D. (1960). Chromosomal trisomy associated with the Sturge-Weber syndrome. *Lancet* **1**, 844–846.

Hayward, M. D., Bower, B. D., Gustavson, K.-H., Hook, O., and Hall, B. (1961). The chromosomal constitution of Sturge-Weber syndrome. *Lancet* **1**, 558–559.

Hirschhorn, K., Lucas, M., and Wallace, I. (1973). Precise identification of various chromosomal abnormalities. *Ann. Hum. Genet.* **36**, 375–379.

Hsu, L. Y. F., Shapiro, L. R., Gertner, M., Lieber, E., and Hirschhorn, K. (1971). Trisomy 22: A clinical entity. *J. Pediatr.* **79**, 12–19.

Ishmael, J., and Laurence, K. M. (1965). A probable case of incomplete trisomy of a chromosome of the 13-15 group. *J. Med. Genet.* **2**, 136–141.

Jacobs, P. A., Melville, M., Ratcliffe, S., Keay, A. J., and Syme, J. (1974). A cytogenetic survey of 11,680 newborn infants. *Ann. Hum. Genet.* **37**, 359–376.

Kim, H. J., Hsu, L. Y. F., Goldsmith, L., and Hirschhorn, K. (1975). Familial translocation with double trisomy of 13 and 22. *Pediatr. Res.* **9**, 314 (abstr.).

Koulischer, L., and Perier, J. (1962). A-propos d'un cas de trisomie 22. *Bull. Acad. R. Med. Belg.* **2**, 329–344.

Kruger, E., Witkowski, R. and Piedre, U. (1968). Partielle trisomie $D_1$ -eine seltene chromosomenanomalie. *Humangenetik* **6**, 181–188.

Kucera, J. (1970). Down's syndrome and infectious hepatitis. *Lancet* **1**, 569–570.

Kunze, J., Tolksforf, M., and Wiedemann, H. R. (1975). Cat-eye syndrome. *Humangenetik* **26**, 271–289.

Lehmann, O., and Forssman, H. (1960). Chromosomes in the Sturge-Weber syndrome. *Lancet* **2**, 1450.

Magenis, R. E., Overton, K. M., Reiss, J. A., Macfarlane, J. P., and Hecht, F. (1972). Partial trisomy 15. *Lancet* **1**, 1365–1366.

Neu, R. L., Assemany, R. S., and Gardner, L. I. (1970). "Cat eye" syndrome with normal chromosomes. *Lancet* **1**, 949.

Noel, B., and Quack, B. (1970). Petit metacentrique surnuméraire chez un polymalforme. *J. Genet. Hum.* **18**, 45–46.

Paris Conference 1971 (1972). Standardization in human cytogenetics. *Birth Defects, Orig. Artic. Ser.* **8**, No. 7.

Penchaszadeh, V. B., and Coco, R. (1975). Trisomy 22. Two new cases and delineation of the phenotype. *J. Med. Genet.* **12**, 193–199.

Perez-Castillo, A., Abrisqueta, J. A., Martin-Lucas, M. A., Goday, C., Del Mazo, J., and Aller, V. (1975). A new contribution to the study of 22 trisomy. *Humangenetik* **30**, 265–271.

Petit, P. (1973). Identifying the extra chromosome in "cat eye" syndrome with Q, G and C technique. Symposium on "karyotype–phenotype," Pavia, 10–11. 1973. *Bull. Eur. Soc. Hum. Genet.* pp. 70–73.

Pfeiffer, R. A., Heimann, K., and Heiming, E. (1970). Extra chromosome in "Cat Eye" syndrome. *Lancet* **2**, 97.

Pierson, M., Gilgenkrantz, S., and Saborio, M. (1975). Syndrome dit de l'oeil de chat avec nanisme hypophysaire et développement mental normal. *Arch. Fr. Pediatr.* **32**, 835–848.

Punnett, H. H., Kistenmacher, M. I., Toro-Sola, M. A., and Kohn, G. (1973). Quinacrine fluorescence and giemsa banding in trisomy 22. *Theor. Appl. Genet.* **43**, 134–138.

Schachenmann, G., Schmid, W., Fraccaro, M., Mannini, A., Tiepolo, L., Perona, G. P., and Sartori, E. (1965). Chromosomes in coloboma and anal atresia. *Lancet* **2**, 290.

Schmid, W. (1967). Pericentric inversions. *J. Genet. Hum.* **16,** 89–96.
Smith, D. W. (1970). "Major Problems in Clinical Pediatrics," Vol. VII, pp. 42–45. Saunders, Philadelphia, Pennsylvania.
Sparkes, R. S., Veomett, I. C., and Wright, S. W. (1966). Trisomy G without Down's syndrome, *Lancet* **1,** 270–271.
Sparkes, R. S., Muller, H. M., and Veomett, I. C. (1970). Inherited pericentric inversion of a human Y chromosome in trisomic Down's syndrome. *J. Med. Genet.* **7,** 59–62.
Stoller, A., and Collmann, R. B. (1965). Incidence of infective hepatitis followed by Down's syndrome nine months later. *Lancet* **2,** 1221–1223.
Taft, P. D., Dodge, P. R., and Atkins, L. (1965). Mental retardation and multiple congenital anomalies. *Am. J. Dis. Child.* **109,** 554–557.
Tan, Y. H., Tischfield, J. A., and Ruddle, F. H. (1974). The genetics of the antiviral state in human cells. *Birth Defects, Orig. Artic. Ser.* **10,** No. 3, 158–159.
Thomas, C., Cordier, J., Gilgenkrantz, S., Reny, A., and Raspiller, A. (1969). Un syndrome rere: Atteinte colobamateuse du globe oculaire, atresie anals, anomalies congenitales multiples et presence d'un chromosome surnumeraire. *Ann. Ocul.* **202,** 1021–1031.
Turner, B., and Jennings, A. N. (1961). Trisomy for chromosome 22. *Lancet* **2,** 49–50.
Turpin, R., and Lejeune, J. (1961). Chromosome translocations in man. *Lancet* **1,** 616.
Uchida, I. A., Ray, M., McRae, K. N., and Besant, D. F. (1968a). Familial occurrence of trisomy 22. *Am. J. Hum. Genet.* **20,** 107–118.
Uchida, I. A., Molunga, R., and Lawler, C. (1968b). Maternal radiation and chromosome aberrations. *Lancet* **2,** 1045–1049.
Vianello, M. G., and Bonioli, E. (1975). Trisomy 22. *J. Genet. Hum.* **23,** 239–250.
Vital Statistics of the United States. (1962). "Natality Characteristics," Vol. 1, Sect. 2. U.S. Gov. Printing Office, Washington, D.C.
Walbaum, R., Samaille, G., Scharfman, W., and Maillard, E. (1970). Le problème de la trisomie 22. *Pediatrie* **25,** 133–143.
Weber, F. M., Dooley, R. R., and Sparkes, R. S. (1970). Anal atresia eye anomalies, and an additional small abnormal acrocentric chromosome (47,XX,mar+): Report of a case. *J. Pediatr.* **76,** 594–597.
Zackai, E., Aronson, M., Kohn, G., Moorhead, P., and Mellman, W. (1973). Familial trisomy 22. *Am. J. Hum. Genet.* **25,** 89a (abstr.).
Zellweger, H., Mikamo, K., Hokkaido, X., and Abbo, H. G. (1962a). Two cases of nonmongoloid trisomy G. *Ann. Paediatr.* **199,** 613–624.
Zellweger, H., Mikamo, K., and Abbo, G. (1962b). Two cases of multiple malformations with an autosomal chromosomal aberration–partial trisomy $D?_1$. *Helv. Paediatr. Acta* **17,** 290–300.

# 12

# Phenotypic Mapping in Man

RAYMOND C. LEWANDOWSKI, JR. and JORGE J. YUNIS



## I. INTRODUCTION

The constellation of congenital abnormalities in the classical chromosomal syndromes is usually distinctive, suggesting that a direct relationship exists between certain phenotypic anomalies and specific chromosome defects (Table I). Early attempts at analyzing this relationship (de Grouchy, 1965) attracted criticism because (1) chromosome segments could not be positively identified, (2) most of the patients had ring chromosomes which are unstable and do not necessarily reflect the original chromosome defect, and (3) the phenotypic effects described overlapped with those observed in other chromosomal syndromes (Hamerton, 1971). Since the advent of the banding techniques, a great deal of information has become available to make such an analysis feasible.

**TABLE I**
**Discriminating Phenotypes of Chromosomal Syndromes**

| Phenotype | Syndrome |
|---|---|
| Laryngomalacia, premature graying of hair | del 5p |
| Retinoblastoma | del 13q |
| Polydactyly | dup 13 |
| Agenesis of thumb and 1st metacarpal | del 13q |
| Bony syndactyly of 4th and 5th metacarpals | del 13q |
| Hypoplasia of 1st metacarpal | del 18q |
| Delayed ossification carpus, tarsus, pelvis | del 4p |
| Persistence of fetal hemoglobin and neutrophils | dup 13 |
| Thymic aplasia | dup 1q |
| Absent patella | dup 8 |
| Anal stenosis | dup 22q, del 13q |
| Holoprosencephaly | dup 13, del 13q, del 18p |
| Orbital hypotelorism | dup 21, del 13q, del 18p |

Some of the "classical" chromosomal syndromes have been divided into subtypes and many new syndromes resulting from partial duplication or deletion have been established (Lewandowski and Yunis, 1975; Chapters 3–11) (Fig. 1). These new entities have distinctive and reproducible phenotypes.

## II. DISCRIMINATING PHENOTYPES

Although none of the chromosomal syndromes have truly pathognomonic features, there are congenital anomalies which appear to be associated with only one or a few chromosome defects (Table I). In general, these discriminating abnormalities can be classified in three categories: (a) single phenotypic effects with limited chromosomal association, such as laryngomalacia and premature graying of hair which are correlated with deletion of the short arm of chromosome 5; (b) multiple anomalies which are most likely related to a single developmental error, such as holoprosencephaly occurring with defects in the long arm of chromosome 13 and the short arm of chromosome 18; and (c) abnormal features which are secondary to a more basic impairment in development, such as orbital hypotelorism resulting from midface hypoplasia in trisomies 13 and 21.

As will be seen below, several chromosome deletions with distinctive features have counterparts in single gene defects which are inherited in an autosomal dominant manner. Since dominant disorders express themselves in a hetero-

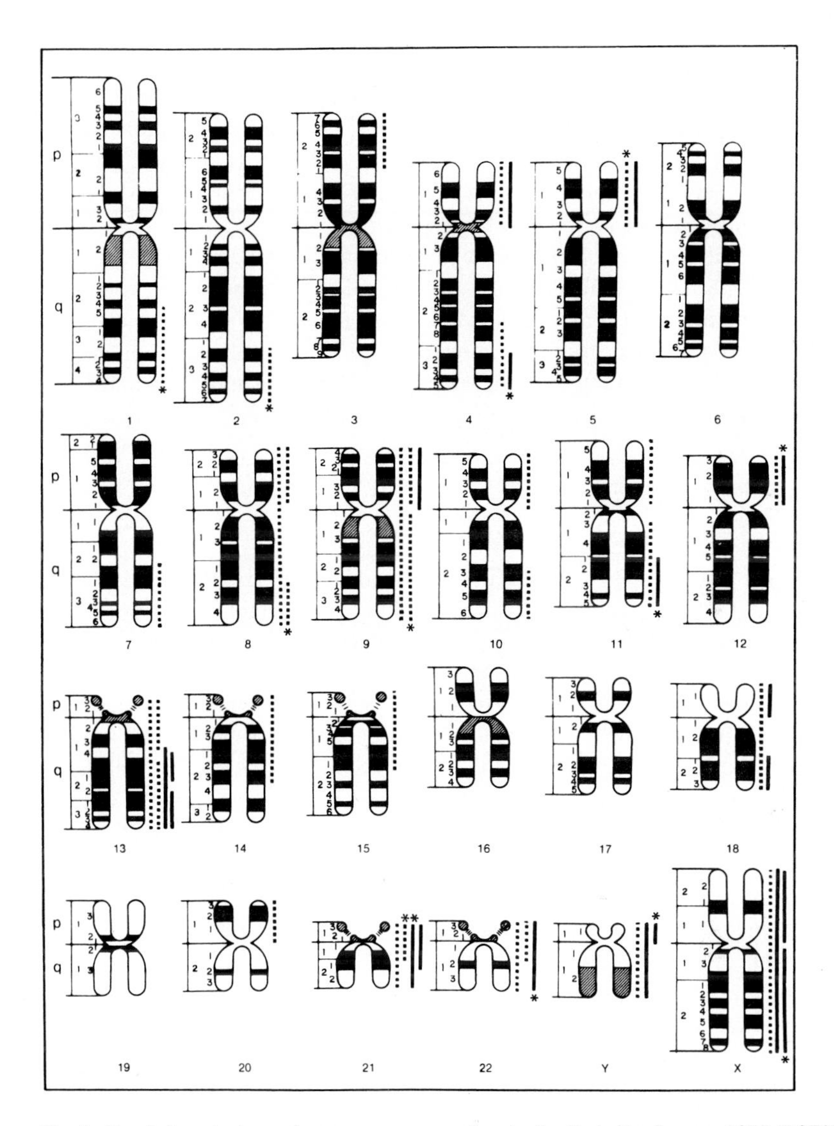


**Fig. 1.** Banded metaphase chromosomes according to the Paris Conference 1971 (1972). To the right of each chromosome are represented presently known syndromes: dotted vertical lines represent trisomic state; solid vertical bars represent monosomic state. An asterisk (*) above or below a line indicates that the full clinical expression is not yet firmly established. (See Table II of Chapter 1 for references.)

zygous state, it is plausible that at least some of the chromosomal deficiencies result in abnormalities subsequent to loss of one or more dominant loci. This appears to be the case with retinoblastoma and an interstitial deletion of the long arm of chromosome 13, as well as syndactyly of the fourth and fifth metacarpals and metatarsals in association with deletion of the telomeric end of the long arm of chromosome 13. A second possibility is that a chromosome deficiency may uncover or unmask a recessive mutation as postulated for holoprosencephaly and 18p− (Gorlin *et al.*, 1968).

A more difficult problem is the mechanism by which the trisomic state produces phenotypic effects. Since there is no known molecular counterpart, any explanation must remain conjectural. Nevertheless, it is conceivable that an extra gene dosage may result in increased production, utilization, or inhibition of products essential to specific developmental stages such as incomplete septation of midbrain and midface in trisomy 13 (see Section II,B).

## A. Single Phenotypic Effects

### *1. Laryngomalacia and Premature Gray Hair*

The cat cry or 5p− syndrome, first described in 1963 (Lejeune *et al.*, 1963), is characterized by a "catlike" cry with typical spectrographic findings (Vuorenkoski *et al.*, 1963; Legros and Van Michel, 1968). As the patients increase in age, 30% of them develop gray hair prematurely (Breg *et al.*, 1970; Niebuhr, 1971). These two features are found frequently only in association with the 5p− syndrome. In 1968, Ward *et al.* studied four patients with the cat cry syndrome. On direct laryngoscopy, three of the four patients in whom the "catlike" cry remained prominent had laryngomalacia. The epiglottis was long, curved, and flappy; the larynx was narrow, and the vocal cords were diamond-shaped during inspiration and left a large air space in the posterior commissure during phonation. The fourth patient had a high-pitched voice without the "catlike" quality and laryngoscopy revealed only a juvenile V-shaped epiglottis. Several other cases with the 5p− syndrome have been studied and noted to have laryngomalacia (Lejeune *et al.*, 1964; MacIntyre *et al.*, 1964a; Faed *et al.*, 1972). Since many of the patients with the 5p− syndrome lose the "catlike" cry with age (Gordon and Cooke, 1968; Breg *et al.*, 1970), it is felt that the larynx becomes normalized as the child grows older. A similar situation is observed in congenital laryngeal stridor where laryngomalacia produces stridor that usually disappears in time (Jackson and Jackson, 1937; Ferguson, 1970; Ward, 1973).

The foregoing suggests that the short arm of chromosome 5 contains a segment of the genome important for laryngeal development. In this regard, it is interesting to note that congenital partial atresia of the larynx and congenital laryngeal stridor do occur in an inheritable fashion, probably autosomal dominant (Seifert, 1889; Glas, 1908; Cleft, 1931; O'Kane, 1936; McHugh and Loch,

1942; Finlay, 1949; Crooks, 1954; Smith and Bain, 1965; Baker and Savetsky, 1970).

Recently, we have studied three patients in a family with most of the features of the 5p− syndrome, including the "catlike" cry and premature graying of hair, and with deletion of only the distal half of the light band 5p15 of the short arm (J. J. Yunis and R. C. Lewandowski, personal observation, 1976). It is proposed that the distal half of this light band of the short arm of chromosome 5 is the phenotype-determining chromosome segment for the cri-du-chat syndrome (Fig. 2).

### 2. *Retinoblastoma*

Retinoblastoma is an uncommon tumor in children with an incidence of 1:20,000 (Warkany, 1971). Most cases have been sporadic and unilateral. The familial type is usually bilateral and inherited as an autosomal dominant with 80% penetrance (Kaelin, 1955; Smith and Sorsby, 1958). Fourteen percent (10/72) of the patients with partial deletion of the long arm of chromosome 13 have had retinoblastoma (Lele *et al.*, 1963; Thompson and Lyons, 1965; Van Kempen, 1966; Wilson *et al.*, 1969, 1973; Gey, 1970; Taylor, 1970; Grace *et al.*, 1971; Howard *et al.*, 1974; Orye *et al.*, 1974; O'Grady *et al.*, 1974). In six patients, the tumor was bilateral as is seen in the familial form. The high frequency of retinoblastoma among patients with 13q− implies a relationship between these two defects.

Three patients with retinoblastoma and a small interstitial deletion allow for tentative sublocalization of the related segment. Wilson *et al.* (1969, 1973) reported a patient with deletion of either 13(q14→q22) or 13(q22→q32). Although not certain of which segment was involved, the authors felt the latter was most likely. Orye *et al.* (1974) described a patient with deletion of 13(q14→q22) and argued on the basis of Wilson's published partial karyotypes that his patient had a similar segment involved. Howard *et al.* (1974) described a third patient with deletion of 13(q12→q14) and possibly the uppermost part of 13q21. It would appear, therefore, that the segment associated with retinoblastoma may be tentatively localized to band 13q14 and the proximal portion of band 13q21 (Fig. 2).*

What role the deletion plays in the formation of the tumor is not clear. It is possible that loss of this segment results in faulty development of the retina which, in turn, predisposes to retinoblastoma. That the tumor has only been identified in 14% of patients with 13q− may be related to several factors. Since more than one-half of all patients with 13q− had ring chromosomes or were

*Most recently, Yunis studied a patient with deletion of approximately half of band q14 (involving sub-band q14.2 and a portion of sub-bands q14.1 and q14.3), bilateral retinoblastoma, and otherwise normal phenotype.

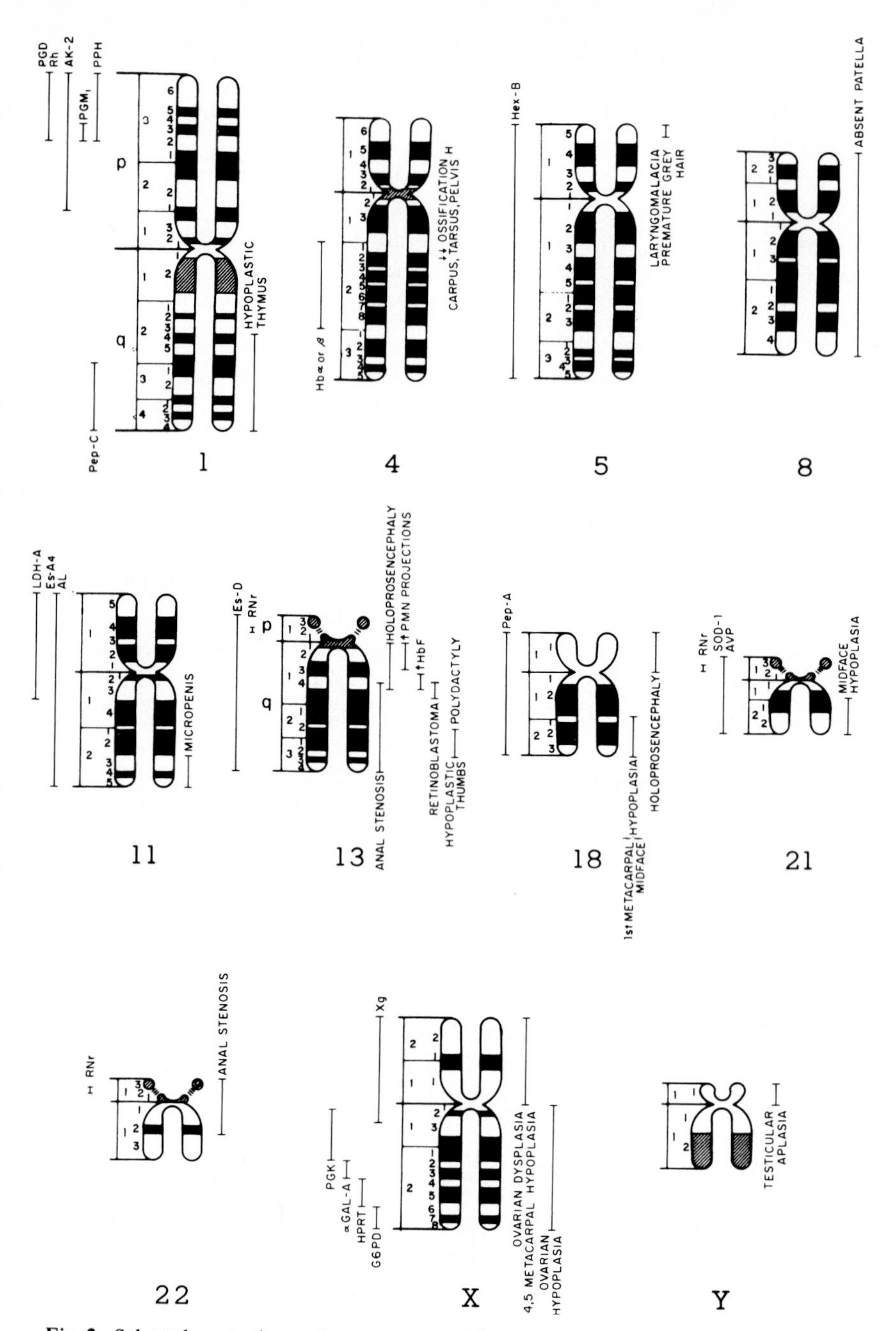


**Fig. 2.** Selected metaphase chromosomes with nomenclature according to the Paris Conference. To the left of each chromosome are found the confirmed gene mapping according to Rotterdam Conference 1974 (1975). To the right of each chromosome are found the provisional phenotypic mappings. For the phenotypic mapping, entries above the bars indicate trisomic state and entries below the bars indicate deleted state.

studied before the advent of the banding techniques, it is not clear if the pathogenic segment was involved. Furthermore, many cases with 13q− die during the first year of life, while the tumor usually manifests itself during the second year of life.

Retinoblastoma has also been found in five patients with trisomy 21 (Taktikos, 1964; Jackson *et al.*, 1968; Miller, 1970), two of which have had a double aneuploidy (Day *et al.*, 1963–XXX, +21; Rethoré *et al.*, 1972–XXY, +21). Since Down syndrome has a 20-fold increased risk for leukemia (Miller, 1970) and tumors of brain and testis also occur (Carter, 1958; Holland *et al.*, 1962; Turner, 1963; Matsaniotis *et al.*, 1967; Jackson *et al.*, 1968), it is not clear if these tumors are directly related to +21 or whether they represent a more generalized tendency for neoplasma in this condition.

### 3. *Polydactyly, Hypoplasia of Phalanges and Metacarpals*

Polydactyly is frequently seen only in association with trisomy 13 (75%). It has been reported occasionally in +18 and most recently in a large kindred with partial duplication for the long arm and partial deletion of the short arm of chromosome 3 (Allderdice *et al.*, 1975). In cases with partial duplication 13q in which careful banding studies have been performed (Hauksdóttir *et al.*, 1972; Niebuhr and Ottosen, 1973; Escobar *et al.*, 1974), the patients with polydactyly have been trisomic for the segment 13(q31→qter). Polydactyly has not been observed in patients trisomic for the proximal one-third to one-half of the long arm (13q11→13q22) (Yunis and Hook, 1966; Escobar and Yunis, 1974; Schinzel *et al.*, 1974). These findings suggest a tentative localization of the chromosome segment associated with this defect to bands 13q31→q34 (Fig. 2).

Patients with deletion of the long arm of chromosome 13 demonstrate oligodactyly (agenesis of the thumb and first metacarpal) (27%) and bony syndactyly of the fourth and fifth metacarpals and metatarsals (10%) (Orbeli *et al.*, 1971). These defects occur concurrently only in association with 13q−. Occasionally, these patients also have absence of the midphalanx of the fifth fingers and short first toes. Careful analysis has shown that the long arm deletion always involves bands q31 to q34 (Noel *et al.*, 1976) (Fig. 2). This localization corresponds to the chromosome segment involved in polydactyly in the trisomic state.

Proximal implantation of the thumb due to hypoplasia of the first metacarpal occurs in 92% of patients with 18q− (Schinzel *et al.*, 1975). Hand deformities have been frequently found in Turner syndrome including hypoplasia of the fourth metacarpal, abnormal carpal bones, slender metacarpal shafts, brachyphalangia, brachymetacarpia, and flattening of epiphyses (Kosowicz, 1965). Also, patients with the 4p− syndrome show marked delay in ossification of carpal and tarsal bones as well as the pelvis. Recently, two unrelated patients with the classical findings of 4p− syndrome were found to have a partial loss of subband 4p15.6. This allows for tentative sublocalization of the causative segment to this very small portion of the short arm of chromosome 4 (J. Yunis, unpublished data, 1976). These examples suggest that there are segments on chromosomes 4, 13, 18, and X which are important in the development of different parts of the hands and feet (Fig. 2).

It is of interest to note that some types of polydactyly (Frazier, 1960; Walker, 1961), brachymesophalangy V (absent midphalanx of the fifth finger) (Bauer, 1907; Dutta, 1965), and syndactyly of the fourth and fifth or third and fourth metacarpals and metatarsals (Kemp and Ravn, 1932) are transmitted as autosomal dominants, giving credence to the concept that some of the phenotypes of chromosomal deletions act like dominant gene defects.

### 4. *Persistent Fetal Hemoglobin and Neutrophils*

As early as 1966, Yunis and Hook studied three unrelated patients with partial trisomy 13 and constructed a tentative map for the persistence of fetal hemoglobin (HbF) and multiple nuclear projections of neutrophils observed in trisomy 13 (Fig. 3). Patient A was considered trisomic for the distal 4/5 of the long arm. She was missing the short arm, centromere, and proximal portion of the long arm. The patient had elevated HbF but did not have increased nuclear projections of PMN's. Patient B had 47 chromosomes and was considered partially trisomic for the proximal portion of the long arm of chromosome 13 and had increased nuclear projections and a normal level of HbF. Patient C (MacIntyre *et al.,* 1964a) was trisomic for the proximal one-fourth of the long

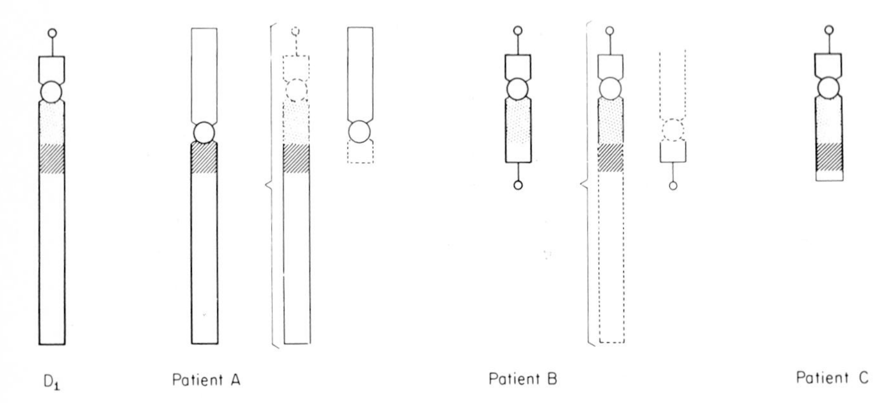


**Fig. 3.** Mapping of chromosome 13. The dotted area was felt to be associated with persistence of neutrophils with multiple nuclear projections and corresponds well with regions 13q11→13q13. The hatched area was felt to be associated with persistence of fetal hemoglobin and corresponds to region 13q13→13q14 (see Fig. 2). From Yunis and Hook (1966).

$D_1$ represents the normal chromosome 13. Patient A has a D/G translocation, Patient B has an abnormal chromosome originating from fusion of part of a $D_1$ and another D or G chromosome, and Patient C represents a partial $D_1$ trisomy.

arm of chromosome 13 and had both abnormal parameters. On the basis of these three patients and the patient of Craig and Luzzati (1965) with trisomy for the long arm of chromosome 13 and normal neutrophils, Yunis and Hook postulated that the segment of chromosome 13 responsible for increased polymorphonuclear projections was located close to the centromere on the proximal portion of the long arm, while the segment affecting HbF was also close but further apart from the centromere. These findings were later supported by Bloom and Gerald (1968).

After the banding techniques were discovered, Escobar and Yunis (1974) and Escobar *et al.* (1974) studied two additional unrelated patients with partial trisomy for the most proximal portion of the long arm (13pter→13q12 or q13?) of chromosome 13 and the distal one-third of the long arm (13q22→13qter), respectively. The first patient had increased nuclear projections and normal HbF and the second patient had normal values for both findings, lending further support to the original hypothesis. Additional cases which allow for more precise sublocalization are those of Taysi *et al.* (1973) with partial trisomy for 13q14→13qter and elevated HbF; Case 2 (Schinzel *et al.*, 1974) with partial trisomy for 13q14→13qter and normal values for both HbF and neutrophils; Crandall *et al.* (1974) with partial trisomy 13q13→13qter and normal neutrophils and elevated HbF; and Case 3 (Schinzel *et al.*, 1974) with partial trisomy 13pter→13q14 and Noel *et al.* (1976) with partial trisomy 13q12→13qter, both of which had elevated HbF and increased nuclear projections.

Based on these observations, it appears that the segment of the long arm of chromosome 13 associated with the persistence of increased nuclear projections of neutrophils is located in band 13q12 and the upper most part of 13q13, and the segment associated with elevated HbF is located in 13q14 (Fig. 2). These conclusions are essentially identical to those initially proposed by Yunis and Hook (Fig. 3).

Persistence of HbF and of neutrophils with increased projections are found in premature and newborn infants but disappear within a few months (Walzer *et al.*, 1966). In patients with trisomy 13, these values remain elevated for a few years (Marden and Yunis, 1967). It is conceivable that there are genes on the long arm of chromosome 13 involved in the formation of HbF and fetal neutrophils which when present in an extra dosage require a longer time to be "shut off."

### 5. *Thymic Hypoplasia, Absent Patella, and Anal Stenosis*

Aplasia or severe hypoplasia of the thymus has been noted in four unrelated patients with partial trisomy for the distal one-half of the long arm of chromosome 1 (Neu and Gardner, 1973; Van den Berghe *et al.*, 1973a; Norwood and Hoehn, 1974; Finley *et al.*, 1975). These findings suggest that this segment of chromosome 1 (1q21→1qter) is related to normal thymic development. Aplasia or severe hypoplasia of the patella has been noted in 8/35 patients with trisomy 8, again suggesting a relationship. Anal stenosis or atresia has been reported in 85% of patients with partial trisomy 22 (22pter→22q12) (Schachenmann *et al.*, 1965; Hsu and Hirschhorn, this volume, Chapter 11) and in 16% of patients with 13q– (Noel *et al.*, 1976). In all of these examples, more cases with careful analysis are needed. However, it does appear that these represent examples of relatively discriminating phenotypic effects associated with one or a few chromosomes or chromosome segments. It should be noted that aplasia of the thymus and parathyroids (DeGeorge, 1968), absent patella with coxa vara and tarsal synostosis (Geominne and Dujardin, 1970), and anal atresia with eye defects (Rieger, 1935; Pearce and Kerr, 1965; Crawford, 1967) may occur as dominant mutations. It is interesting that both chromosome defects and gene mutations may result in absence or severe hypoplasia of isolated organs, suggesting a poorly understood relationship between genes, chromosome defects, and organ development.

## B. Closely Related Multiple Effects

### *Holoprosencephaly*

Isolated anomalies, such as cleft palate and hypotelorism, are found in several chromosomal syndromes (Table II). Occasionally these anomalies occur together and are related to a more basic defect in embryogenesis, limited to only a few chromosome errors, such as holoprosencephaly. These defects occur in 17% of patients with the 18p– syndrome (Schinzel *et al.*, 1975), 10% of patients with 13q– (Niebuhr and Ottosen, 1973), and 60% of patients with trisomy 13 (Magenis and Hecht, 1974).

Holoprosencephaly with associated facial dysmorphia represents a single developmental field with cyclopia at one end of the spectrum and aplasia or hypoplasia of the corpus callosum at the other (DeMeyer and Zeman, 1963; Cohen *et al.*, 1971). There are examples of holoprosencephaly with autosomal recessive inheritance (François, 1961; Cohen *et al.*, 1972) and examples suggesting an autosomal dominant mode of inheritance with incomplete penetrance and variable expressivity (Grebe, 1944; Dallaire *et al.*, 1971; Patel *et al.*, 1972; Cohen, 1974; Lowry, 1974).

The greater than chance association between 18p– and holoprosencephaly has been postulated to represent the expression of a recessive mutant gene on the intact chromosome 18 (Gorlin *et al.*, 1968). This same argument could be applied to patients with 13q–. However, four of the seven patients with 13q– and holoprosencephaly had ring chromosomes of various sizes, and it was not possible to tell if the rings contained a duplication-deficiency (Bain and Gauld, 1963; Biles *et al.*, 1969; Juberg *et al.*, 1969; Rethoré *et al.*, 1970). The

**TABLE II**
**Selected Overlapping Phenotypes of Chromosomal Syndromes**

| | dup 1q | dup 2q | dup 3p | del 4p | dup 4p | dup 4q | del 5p | dup 5p | dup 7q | dup 8 | dup 9 | dup 9p | del 9p | dup 9q | dup 10p | dup 10q | dup 11p | dup 11q |
|---|---|---|---|---|---|---|---|---|---|---|---|---|---|---|---|---|---|---|
| Mental retardation | + | + | + | + | + | + | + | + | + | + | + | + | + | + | + | + | + | + |
| Growth retardation | + | | + | + | + | + | + | + | + | | + | + | + | + | + | + | + | + |
| Low birth weight | | | | + | | + | + | | + | | | | | | | | + | |
| Microcephaly | | + | + | + | + | + | + | | | | + | + | | | | + | | + |
| Downward slanting eyes | | + | | + | + | | + | | | | | | | | | + | + | |
| Upward slanting eyes | | | | | | | | + | | | | | + | | | | | |
| Hypertelorism | | + | + | + | + | | + | + | + | | | + | + | | | | + | |
| Microphthalmia | | | | + | | + | | | | | | | | | | + | | |
| Small palpebral fissures | | + | | | | + | | | + | | + | | | | | + | | |
| Strabismus | | + | | + | | | + | | + | + | | | | | | | + | |
| Broad nasal bridge | | + | | + | + | | + | | | | | | + | | + | | + | |
| Cleft upper lip | | | | + | | | | | | + | | | | | + | | + | |
| Cleft palate | | | | + | | | | | + | | | | | | + | + | + | + |
| Micrognathia | + | + | + | + | | + | + | | + | + | | | | + | | + | | + |
| Low-set ears | | | | + | + | + | + | | + | + | + | | | | + | + | | + |
| Short neck | | | + | | + | | | | | + | | + | + | | | + | | + |
| Clinodactyly | | | | | | | | | | | | | | | | | | |
| Camptodactyly | | | | | + | | | | | + | | | | | + | + | | |
| Congenital heart disease | + | + | + | + | | | + | + | | + | + | | + | + | + | + | | + |
| Cryptorchidism | | + | | + | | + | | | | | + | | | + | + | | + | |
| Elevated axial triradius | | | | | + | | + | | | | | | | | | | | |
| Transverse palmar crease | | | | + | | | + | | + | | | + | | | + | + | | |

*continued*

**TABLE II (*continued*)**

| | del 11q | dup 12p | del 12p | dup 13 | dup 13q$^{p}$* | dup 13q$^{d}$* | del 13q | dup 14q | dup 15q | dup 18 | del 18p | del 18q | dup 20p | dup 21 | del 21q | dup 22 | del 22 |
|---|---|---|---|---|---|---|---|---|---|---|---|---|---|---|---|---|---|
| Mental retardation | + | + | + | + | + | + | + | + | + | + | + | + | + | + | + | + | + |
| Growth retardation | + | + | + | + | + | + | + | + | + | + | + | + | | + | + | + | + |
| Low birth weight | | + | | + | | | | | | | + | + | | | + | | |
| Microcephaly | + | | | + | | + | + | + | | | | + | | | + | + | |
| Downward slanting eyes | | | | | | | | | + | | | | | | + | + | |
| Upward slanting eyes | | | | | | | | | | | | | + | + | | | |
| Hypertelorism | + | | + | | | | + | | | | + | | + | | | + | |
| Microphthalmia | | | | + | + | | + | + | | | | | | | | | |
| Small palpebral fissures | | | | | | | | + | + | | | | | | | | |
| Strabismus | | | | | | + | | | + | + | + | + | + | + | | | |
| Broad nasal bridge | | + | | | | | + | + | | | | | | | + | | |
| Cleft upper lip | | | | + | | | | | | + | | + | | | | | |
| Cleft palate | | | | + | + | | | + | | + | | + | | | + | + | |
| Micrognathia | + | | + | + | | | | + | + | + | + | | | | + | + | |
| Low-set ears | + | + | + | + | + | + | | + | + | + | + | + | | | + | + | + |
| Short neck | | | | + | | | + | + | + | + | | | | + | | | |
| Clinodactyly | | | + | | + | | + | + | + | | | | | + | | | + |
| Camptodactyly | | | | + | | | | + | | + | | | | | | | |
| Congenital heart disease | | | + | + | | | + | + | | + | | + | + | + | | + | |
| Cryptorchidism | | | | + | | | + | + | + | + | | + | | + | + | | |
| Elevated axial triradius | | + | | + | | | | | | + | | + | | + | + | | + |
| Transverse palmar crease | + | + | | + | + | | + | | | + | | + | | + | | | |

*"p" refers to partial trisomy for the proximal one-third, and "d" to partial trisomy for the distal two-thirds of the long arm of chromosome 13.

chromosomes of the three remaining patients were not banded but appeared to have deletions involving the distal one-third to two-thirds of the long arm (Opitz *et al.*, 1969; Orbeli *et al.*, 1971; Leisti, 1971).

Cases reported with partial trisomy 13q do not yet allow for the sublocalization of the segment related to holoprosencephaly, since none of the patients with partial trisomy for the distal portion of the long arm below 13q14 have had holoprosencephaly (Hauksdóttir *et al.*, 1972; Talvik *et al.*, 1973; Taysi *et al.*, 1973; Escobar *et al.*, 1974; Schinzel *et al.*, 1974; McDermott and Parrington, 1975) and patients with partial trisomy for the most proximal portion of the long arm have not been closely scrutinized for holoprosencephaly. Fryns *et al.* (1974) described a patient with holoprosencephaly and partial trisomy for "most of the long arm" of chromosome 13. Based on the above, it can only be suggested that the proximal portion of the long arm q11→q14 may contain the segment related to holoprosencephaly.

In patients with trisomy 13, the degree of holoprosencephaly (60%) varies from cyclopia to no defect in septation (40%). This is similar to the variable expression found in the dominant form of holoprosencephaly. In trisomy 13, expression of the defect may be related to an extra dose of a structural or regulatory gene(s) product that disrupts the critical metabolic pathways and/or cellular processes involved in midbrain, midface septation at the end of the first month of embryonic life.

## C. Secondary Phenotypes

### *Orbital Hypotelorism*

Orbital hypotelorism, an abnormal narrowing of the interorbital space, is only found frequently in trisomy 13, 18p−, and trisomy 21. The interorbital space is occupied mainly by the ethmoid bone, although it is not clear if primary hypoplasia of this bone or some more basic cause is the etiology of orbital hypotelorism. Currarino and Silberman (1960) identified orbital hypotelorism in patients with some degree of holoprosencephaly or with trigonocephaly, explaining the occurrence of hypotelorism in patients with the trisomy 13 and 18p− syndromes. In trisomy 21, there are no cases with associated holoprosencephaly and only a few with trigonocephaly. Rather, it has been proposed that the orbital hypotelorism is a result of primary hypoplasia of the central facial structures (Benda, 1946; Gerald and Silverman, 1965). Orbital hypotelorism, then, represents a phenotypic anomaly which results secondarily from a more basic defect in development.

There are other defects which may also be explained in a similar fashion. For example, microcephaly usually results from decreased brain growth and size (Warkany, 1971). Its frequent appearance among the chromosome syndromes

(Lewandowski and Yunis, 1975) (Table II) may relate to a more basic defect in brain development. Other examples may include (a) clinodactyly resulting from an abnormality of the middle phalanx as in +21, and (b) cleft lip/cleft palate in trisomy 13 as a result of holoprosencephaly.

## III. OVERLAPPING PHENOTYPES

It appears from the foregoing that a specific relationship exists between certain phenotypes and chromosome segments. Further studies are necessary to confirm these findings and to provide for more detailed sublocalization.

There are, on the other hand, many phenotypic effects which are present in most of the chromosomal syndromes (Table II). Several reasons may account for this phenomenon. Since many abnormalities, such as growth and mental development, can result as end products of one of many metabolic errors, duplication or deletion of a chromosome segment containing hundreds of genes could result in generalized effects (Hamerton, 1971). Furthermore, the defects may be at the regulatory level (Vogel, 1973), resulting in disharmonic growth of various tissues (Smith, this volume, Chapter 2). The Minute mutants in *Drosophila* offer still a further possible explanation (Bridges, 1919, cited in Bridges and Morgan, 1923; Schultz, 1929). There are at least 41 Minute loci localized to different regions on all four chromosomes. Mutations and chromosome deletions involving these loci produce a similar phenotype in the heterozygous state with small or missing bristles, slow development, small body size, large and somewhat rough eyes, missing aristae, thin textured wings, and low fertility. These loci likely represent the sites of synthesis of transfer RNA and the resultant decreased rate of protein synthesis yields slow development in the heterozygous state (Ritossa *et al.*, 1966; Steffensen and Wimber, 1971). In man, to explain generalized effects due to abnormal tRNA dosage, it would be important to learn about the chromosomal localization of the 62 known tRNA sequences and give a plausible explanation of how, in the trisomic state, extra sequences are able to disrupt basic cellular processes.

### A. Mental Retardation

The most ubiquitous of all phenotypes among the chromosomal syndromes is mental retardation. The etiologic relationships between chromosome defects and mental retardation still await elucidation. However, it is known that patients with trisomies 13, 18, and 21 have small brains and fewer than normal numbers of cells in the brain and other organs. Patients with trisomy 13 have the smallest and those with trisomy 21 the largest brain (Naeye, 1967). Furthermore, various gross and microscopic abnormalities have been observed. In trisomy 13, the most frequent defect is holoprosencephaly, although patients have also been described with cerebellar anomalies and hydrocephalus (Norman, 1966; Terplan *et al.*, 1966; Warkany, 1971). In trisomy 18, the brain is usually small and major malformations are less common and include hydrocephalus, abnormal cerebellum, abnormal and asymmetrical gyri and lumbar myelomeningocoele (Neühauser and Usener, 1966; Passarge *et al.*, 1966; Warkany *et al.*, 1966). The findings in trisomy 21 are similar to those in trisomy 18 (Apert, 1914; Brousseau, 1928; Davidoff, 1928; Hilliard and Kirman, 1965), although some authors feel that there is some specificity (Benda, 1960). In his studies, Benda outlines disorders of convolutionary and fissural patterns which appeared to be characteristic of the mongoloid brain in addition to degeneration of brain tissue, loss of nerve cells, and atrophy of the cortex with gliosis and pachymeningeal fibrosis. Information describing the anatomic and microscopic appearance of the brain in the partial duplications and deletions is sparse. This is related in part to many of the syndromes being new and the patients living longer.

### B. Growth Retardation

Growth retardation occurs in all chromosome syndromes with the possible exceptions of trisomy 8 and partial trisomy for the short arm of chromosome 20 (Francke, this volume, Chapter 8). It is not understood how chromosome defects cause growth retardation. Endocrinologic studies performed to date have not been illuminating (G. Giovannelli, personal communication, 1976). It could be related in part to structural brain defects and small brain size, since such patients often manifest growth retardation (Swaiman and Wright, 1975). It is also known that patients with trisomy 21 have a decreased rate of DNA synthesis in fibroblasts (Mittwoch, 1967) and lymphocytes (Mellman *et al.*, 1970) and that patients with trisomies 13, 18 and 21 have fewer than the normal numbers of cells in different organs (Naeye, 1967). Patients with trisomies 13, 18, and 21 also have prolongation of the cell cycle (primarily $G_1$) in tissue culture (Kuliev *et al.*, 1973; Porter and Paul, 1975). It has been postulated that patients with chromosome defects may suffer growth retardation on the basis of abnormal cell cycles, since an abnormal rate of cell division at critical times in embryogenesis may affect certain developmental stages when a given metabolic activity in specific cells is required (Klinger *et al.*, 1971; Porter and Paul, 1975).

Growth appears to be specifically related to loci on the short arm in the X chromosome. Deletion of the short arm of X(p11→pter) usually results in gonadal dysgenesis and short stature (Jacobs *et al.*, 1963; Court-Brown *et al.*, 1964; Bjøro, 1965; Atkins *et al.*, 1965; Tveteras, 1965; Lejeune *et al.*, 1966;

Steinberger *et al.,* 1966; Golob *et al.,* 1967; Aarskog, 1967; Goldberg *et al.,* 1968; Barakat and Jones, 1970; Backman *et al.,* 1971; Weed, 1972). Isochromosome for the long arm of chromosome X almost invariably shows gonadal dysgenesis and short stature (Simpson, 1975). Deletion of the long arm of the X chromosome has been described 17 times (de Grouchy *et al.,* 1961; Jacobs *et al.,* 1961; Court-Brown *et al.,* 1964; Aarskog, 1967; Bovicelli, 1968; Baughman *et al.,* 1968; Hecht *et al.,* 1970; Bocian *et al.,* 1971; Luthardt and Palmer, 1971; Newton *et al.,* 1972; Boczkowski and Mikkelson, 1973; Jenkins and O'Rourke, 1974; Sarto *et al.,* 1974). These patients usually have gonadal dysgenesis but normal height (Ferguson-Smith, 1965; Ferguson-Smith *et al.,* 1965; Simpson, 1975). Also, five patients with isochromosome for the short arm of the X chromosome have been described (Ferguson-Smith *et al.,* 1964; Ferguson-Smith, 1965; Caspersson *et al.,* 1970; de la Chapelle *et al.,* 1972; Van den Berghe *et al.,* 1973b; Grumbach and Van Wyk, 1974). All had primary amenorrhea and normal stature.

## C. Congenital Heart Disease

Congenital heart disease is a common defect and appears to be relatively nonspecific. Although it is possible to argue for a tendency toward certain types of defects in a given chromosomal error, such as atrioventricular commune in trisomy 21 (German *et al.,* 1972), a significant overlap remains. For example, among patients with a chromosome defect, ventricular septal defect (VSD) occurs in 47% of trisomy 13, 74% of trisomy 18, and 25% of trisomy 21. In addition, defects of almost every kind have been described in these trisomies (Warkany, 1971). A similar situation exists in the deletion syndromes. For example, patients with atrial septal defect (ASD), VSD, overriding aorta and tetralogy have been reported in association with deletion for the long arm of chromosome 13 (Orbeli *et al.,* 1971).

The apparent lack of specificity may be related to difficulty in understanding the genetics of congenital heart disease itself. For example, Noonan syndrome is inherited as an autosomal dominant (Bolton *et al.,* 1974) and right side heart lesions, especially pulmonary valve disease, clearly predominate (Celermajer *et al.,* 1968; Koretzky *et al.,* 1969). However, there are patients with Noonan syndrome with different kinds of heart abnormalities such as patent ductus arteriosus (PDA), myocardiopathy, and septal defects (Diekmann *et al.,* 1967; Noonan, 1968). Another example is the Holt-Oram syndrome (Holt and Oram, 1960) which also behaves as an autosomal dominant. Although atrial septal defect is most commonly found, several types of heart defects have been described (Holmes, 1965; Lewis *et al.,* 1965; Gall *et al.,* 1966). Furthermore, families with multifactorial inheritance for congenital heart defects do not "breed true" (Nora and Meyer, 1966).

It thus appears that the pathogenesis of congenital heart defects is obscure and the lack of specificity of congenital heart disease due to single gene defects, multifactorial inheritance, or chromosome defects is difficult to explain.

## D. Dermatoglyphics

Uchida and Soltan (1963) first noted that dermatoglyphic abnormalities of some type occur in most of the chromosomal syndromes. This is not surprising, since it is known that the ridge patterns, total ridge count, and creases are strongly influenced by many genes (Holt, 1952, 1968) and most chromosome defects involve large amounts of genetic material. Indeed, there are findings which are frequently found, such as elevated axial triradius or abnormalities of fingerprints, and there are few if any pathognomonic features. Despite these findings, many dermatoglyphic abnormalities are in fact not the same in the various chromosomal syndromes. For example, marked displacement of the axial triradius is seen in trisomy 13 ($t'''$), moderate displacement in trisomy 21 ($t''$) and mild displacement in trisomy 18 ($t'$) (Penrose, 1963, 1968; Penrose and Smith, 1966). In XXY and XXYY, the triradius is displaced in an ulnar fashion (Uchida *et al.,* 1964; Alter *et al.,* 1966). In regard to fingerprint patterns, increased ulnar loops are frequently seen in trisomy 21, increased radial loops and arch fibular S in trisomy 13, increased number of arches in trisomy 18, and increased whorls in deletion of the long arm of chromosome 18. Whether these are related to specific DNA segments remains to be elucidated. This may be possible to elucidate with the analysis of patients with small trisomies and deletions of chromosomes where dermatoglyphic abnormalities are well known in the full syndromes, such as the trisomy 18 and 18q– syndromes.

## E. Microphthalmia

Anophthalmia or microphthalmia is frequently found in several chromosomal syndromes. In patients with holoprosencephaly, microphthalmia is a commonly associated feature and this may account for its occurrence in trisomy 13. Since holoprosencephaly is not seen in most of the other syndromes in which microphthalmia occurs (Table II), the etiology is probably different. Indeed, this defect could be related to specific loci on different chromosomes. In *Drosophila,* for example, it is known that there are gene mutations of chromosomes 1, 2, 3, and 4 which produce small or absent eyes (Lindsley and Grell, 1968); in the mouse, genes on chromosomes 2, 6, and 10 do likewise (Green, 1966).

## IV. GENE MAPPING AND LINKAGE STUDIES

Gene mapping and linkage studies have played an important role in mapping the chromosomes of *Drosophila* and the mouse. In the mouse, for example, more than 300 mutants are found in 20 linkage groups, 18 of which have not been assigned to specific chromosomes (Miller and Miller, 1975; Green, 1975) (Fig. 4). In man, there are over 2000 known mutants with an established mode of inheritance (McKusick, 1975). However, linkage studies in man have been most concerned with the X chromosome with 95 loci presently assigned to it (McKusick, 1975; Race and Sanger, 1975). The recent use of somatic cell hybridization has allowed for the rapid localization of 50 genes to add to an equal number established by family studies (Grzeschik *et al.*, 1972, 1973; Ricciuti and Ruddle, 1973; Pearson *et al.*, 1974). This technique in combination with family linkage studies has also helped in the tentative localization of Becker's muscular dystrophy and hemophilia to a segment on the long arm of the X chromosome (Rotterdam Conference, 1974).

Somatic cell hybridization will continue to aid in the mapping of genes active after birth. However, the most important breakthroughs for the understanding of the molecular mechanisms involved in the pathophysiology of chromosome defects will come when gene products essential to critical stages of embryogenesis are mapped. By analysis of possible enzymes involved in specific developmental defects such as laryngomalacia in the 5p– syndrome, it may be possible to begin to understand the basic molecular mechanisms involved.

## V. FUTURE DIRECTIONS AND MODEL

Specific chromosome segment/phenotype relationships exist in man. In some instances, such as retinoblastoma and laryngomalacia, sublocalization of the phenotype determining chromosome segment appears possible. In other cases, such as in microphthalmia, there may be several loci which appear to be involved. There are also some developmentally complex phenotypic abnormalities, such as growth and mental retardation, which are present in most chromosomal aberrations. As more chromosomal defects are identified with detailed clinical and biochemical analysis, it may become possible to extensively map phenotype-chromosome segments. Nevertheless, since the known chromosome

**Fig. 4.** Gene map of the mouse (from M. Green, 1975). Arabic numbers at top refer to chromosomes; Roman numerals at bottom refer to linkage groups. Loci whose order is uncertain are not italicized. Brackets indicate that the order within the bracketed group has not been established. Knobs indicate the locations of the centromeres where they are known.

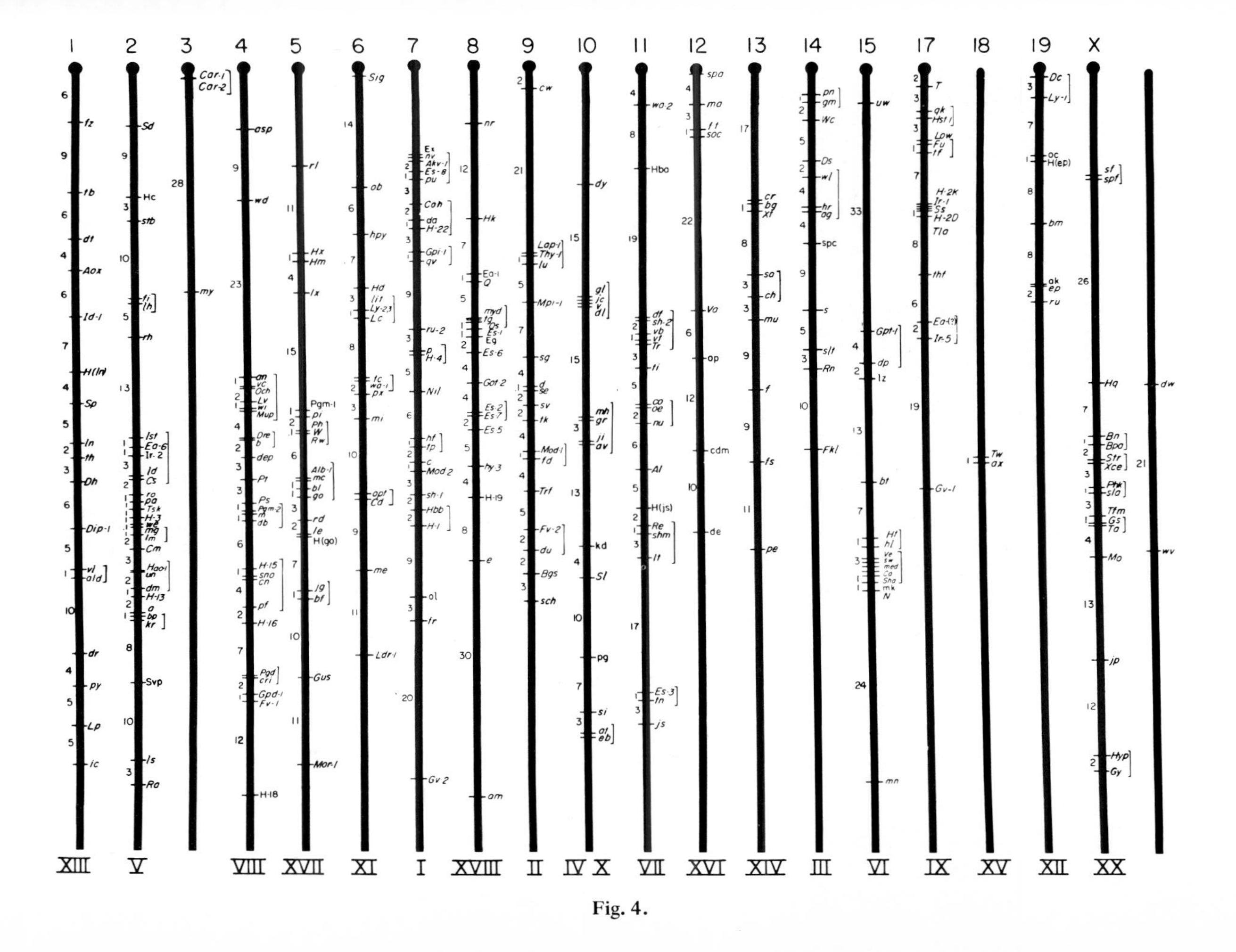

1
2
3
4
5
6
7
8
9
10
11
12
13
14
15
17
18
19
X
XIII
V
VIII
XVII
XI
I
XVIII
II
IV X
VII
XVI
XIV
III
VI
IX
XV
XII
XX


Fig. 4.

defects likely involve hundreds of genes, there is a need to develop high resolution chromosome techniques for more precise localization. Improvements have most recently been made over the metaphase techniques with 1256 bands now identifiable in late prophase (Yunis and Sanchez, 1975; Yunis, 1976). What is ultimately needed is a means of progressing from the chromosome-segment-phenotype to the chromosome-gene-phenotype relationship, as appears possible in the giant chromosomes of *Drosophila.* In addition, to better understand morphogenesis, there is a dire need to study genes important in development. For this, extensive studies in man and/or an animal models will be of crucial importance. Recent advances in the study of chromosome defects in mouse have been made and are worthwhile noting because of their potential relevance in this area (Tettenborn and Gropp, 1970; Gropp, 1973; Stolla and Gropp, 1974; Gropp and Kolbus, 1974; White *et al.,* 1974a, b; Gropp *et al.,* 1975). Trisomic embryos for chromosomes 1, 3, 4, 11, 12, 13, 16, 17, and 19 have been observed. These embryos usually show general developmental retardation, facial dysplasia, and early fetal death. Despite these overlapping features, the developmental profiles differ considerably, depending on which autosome is involved (Gropp *et al.,* 1975). To date, three reproducible syndromes have been established: trisomies 1, 12, and 19 (White *et al.,* 1974a, b; Gropp *et al.,* 1975). Trisomy 1 shows hypoplasia, facial dysmorphy, marked retardation, and absence of gross malformations; trisomy 12 shows exencephaly and microphthalmia without predominant general developmental retardation; trisomy 19 shows cleft palate. As mice with small duplications or deficiencies are studied, and enzymes important in development are hopefully elucidated, the understanding of the molecular basis of chromosome defects may finally unfold a promising avenue of research.

## ACKNOWLEDGMENT

This work was supported in part by NIH Grant Nos. HD01962 and GM05625.

## REFERENCES

Aarskog, D. (1967). *Acta Paediatr. Scand., Suppl.* **177**, 69.

Allderdice, P. W., Browne, N., and Murphy, D. P. (1975). *Am. J. Hum. Genet.* **27**, 699–718.

Alter, M., Gorlin, R., Yunis, J., Peagler, F., and Bruhl, H. (1966). *Am. J. Hum. Genet.* **18**, 507–513.

Apert, E. (1914). *Monde Med.* **24**, 201–211.

Atkins, L., Santesson, B., and Voss, M. (1965). *Ann. Hum. Genet.* **29**, 89–95.

Backman, R., De La Cruz, F., Al-Aish, M., and Santell, F. (1971). *Am. J. Ment. Defic.* **75**, 435–441.

Bain, A. D., and Gauld, I. K. (1963). *Lancet* **2**, 304–305.

Baker, D. C., and Savetsky, L. (1970). *Laryngoscope* **76**, 616–620.

Barakat, B. Y., and Jones, H. W., Jr. (1970). *Obstet. Gynecol.* **36**, 368–372.

Bauer, B. (1907). *Dtsch. Z. Chir.* **86**, 252–259.

Baughman, F. H., Vander Kolk, K. J., Mann, J. D., and Valdmanis, A. (1968). *Am. J. Obstet. Gynecol.* **102**, 1065–1069.

Benda, C. E. (1946). *In* "Mongolism and Cretinism," p. 310. Grune & Stratton, New York.

Benda, C. E. (1960). *In* "The Child with Mongolism," pp. 63–78. Grune & Stratton, New York.

Biles, A. R., Lüers, T., and Sperling, K. (1969). *Med. Welt* **20**, 1771–1775.

Bjøro, K. (1965). *Acta Obstet. Gynecol. Scand., Supp.* **4**, 44.

Bloom, G. E., and Gerald, P. S. (1968). *Am. J. Hum. Genet.* **20**, 495–511.

Bocian, M., Krmpotic, E., Szego, K., and Rosenthal, M. (1971). *J. Med. Genet.* **8**, 358–363.

Boczkowski, K., and Mikkelsen, M. (1973). *J. Med. Genet.* **10**, 350–355.

Bolton, M. R., Pugh, D. M., Mattioli, L. F., Dunn, M. I., and Schmike, N. (1974). *Ann. Intern. Med.* **80**, 626–629.

Bovicelli, L. (1968). *Riv. Ital. Ginecol.* **62**, 83.

Breg, W. R., Steele, M. W., Miller, O. J., Warburton, D., de Capra, A., and Allderdice, P. W. (1970). *J. Pediatr.* **77**, 782–791.

Bridges, C. B., and Morgan, T. H. (1923). *Carnegie Inst. Washington Publ.* **327**, 1–251.

Brousseau, K. (1928). "Mongolism," Williams & Wilkins, Baltimore, Maryland.

Carter, C. O. (1958). *J. Ment. Defic. Res.* **2**, 64–74.

Caspersson, T. A., Lindsten, J., and Zech, L. (1970). *Hereditas* **66**, 287–292.

Celermajer, J. M., Bowdler, J. D., and Cohen, D. H. (1968). *Am. J. Dis. Child.* **116**, 351–358.

Clerf, L. H. (1931). *Ann. Otol., Rhinol., & Laryngol.* **40**, 770–779.

Cohen, M. M., Jr. (1974). *Am. J. Dis. Child.* **127**, 597.

Cohen, M. M., Jr., Jirasek, J. E., Guzman, R. T., Gorlin, R. J., and Peterson, M. Q. (1971). *Birth Defects, Orig. Artic. Ser.* **7**, No. 7, 125–135.

Court-Brown, W. M., Harnden, D. G., Jacobs, P. A., Maclean, N., and Mantle, D. J. (1964). "Abnormalities of the Sex Chromosome Complement in Man," Med. Res. Council. No. 305. Stationery Office, London.

Craig, A. P., and Luzzati, L. (1965). *Am. Soc. Hum. Genet.* p. 32 (abstr.).

Crandall, B. F., Carrel, R. E., Howard, J., Schroeder, W. A., and Müller, H. (1974). *Am. J. Hum. Genet.* **26**, 385–392.

Crawford, R. A. (1967). *Br. J. Ophthalmol.* **51**, 438–440.

Crooks, J. (1954). *Arch. Dis. Child.* **29**, 12–17.

Currarino, G., and Silverman, F. (1960). *Radiology* **74**, 206–217.

Dallaire, L., Fraser, F. C., and Wiglesworth, F. W. (1971). *Birth Defects, Orig. Artic. Ser.* **7**, No. 7, 136–142.

Davidoff, L. M. (1928). *Arch. Neurol. Psychiatry* **20**, 1229–1257.

Day, R. W., Wright, S. W., Koons, A., and Quigley, M. (1963). *Lancet* **2**, 154–155.

DeGeorge, A. M. (1968). *In* "Immunologic Deficiency Diseases" (R. A. Good, ed.), pp. 116–123. Natl. Found.–March of Dimes, White Plains, New York.

de Grouchy, J. (1965). *J. Pediatr.* **66**, 414–431.

de Grouchy, J., Lamy, M., Yaneva, H., Salomon, Y., and Netter, A. (1961). *Lancet* **2**, 777–778.

de la Chapelle, A., Schroder, J., and Pernu, M. (1972). *Ann. Hum. Genet.* **36**, 79–88.

DeMeyer, W., and Zeman, W. (1963). *Confin. Neurol.* **23**, 1–36.

Diekmann, L., Pfeiffer, R. A., Hilgenberg, F., Bender, F., and Reploh, H. D. (1967). *Muench. Med. Wochenscher.* **109**, 2638–2645.

Dutta, P. (1965). *Acta Genet. Stat. Med.* **15,** 70–76.
Escobar, J. I., and Yunis, J. J. (1974). *Am. J. Dis. Child.* **128,** 221–222.
Escobar, J. I., Sanchez, O., and Yunis, J. J. (1974). *Am. J. Dis. Child.* **128,** 217–220.
Faed, M. J. W., Whyte, R., Paterson, C. R., McCathie, M., and Robertson, J. (1972). *J. Med. Genet.* **9,** 102–105.
Ferguson, C. R. (1970). *Otolaryngol. Clin. North Am.* **3,** 185–200.
Ferguson-Smith, M. A. (1965). *J. Med. Genet.* **2,** 142–155.
Ferguson-Smith, M. A., Alexander, D. S., Bowen, P., Goodman, R. M., Kaufmann, B. N., Jones, H. W., Jr., and Heller, R. H. (1964). *Cytogenetics* **3,** 355–383.
Finlay, H. V. L. (1949). *Arch. Dis. Child.* **24,** 219–223.
Finley, W. H., Garrett, J. H., and Finley, F. C. (1975). *27th Annu. Meet. Am. Soc. Hum. Genet.* p. 35A.
François, J. (1961). *In* "Heredity in Ophthalmology," p. 173. Mosby, St. Louis, Missouri.
Frazier, T. M. (1960). *Am. J. Obstet. Gynecol.* **80,** 184–185.
Fryns, J. P., Eggermont, E., Verrensen, H., and van den Berghe, H. (1974). *Humangenetik* **21,** 47–54.
Gall, J. C., Stern, A. M., Cohen, M. M., Adams, M. S., and Davidson, R. T. (1966). *Am. J. Hum. Genet.* **18,** 187–200.
Geominne, L., and Dujardin, L. (1970). *Acta Genet. Med. Germellol.* **19,** 534–545.
Gerald, B. E., and Silverman, F. N. (1965). *Am. J. Roentgenol., Radium Ther. Nucl. Med.* [N.S.] **95,** 154–161.
German, J., Ehlers, K. H., Crippa, L. P., and Engle, M. A. (1972). *Birth Defects, Orig. Artic. Ser.* **8,** No. 5, 96–103.
Gey, W. (1970). *Humangenetik* **10,** 362–365.
Glas, E. (1908). *Wien. Klin. Wochenschr.* **21,** 603–606.
Goldberg, M. B., Scully, A. L., Solomon, I. L., and Steinbach, H. L. (1968). *Am. J. Med.* **45,** 529–543.
Golob, E., Fischer, P., and Kunze-Mühl, E. (1967). *Dtsch. Med. Wochenschr.* **92,** 71.
Gordon, R. R., and Cooke, P. (1968). *Develop. Med. Child. Neurol.* **10,** 69–76.
Gorlin, R. J., Yunis, J. J., and Anderson, V. E. (1968). *Am. J. Dis. Child.* **115,** 453–476.
Grace, E., Drennan, J., Colver, D., and Gordon, R. R. (1971). *J. Med. Genet.* **8,** 351–357.
Grebe, H. (1944). *Erbarzt* **12,** 138.
Green, M. C. (1966). *In* "Biology of the Laboratory Mouse" (E. L. Green, ed.), 2nd ed., pp. 329–336. McGraw-Hill, New York.
Green, M. C. (1975). *46th Annu. Rep. Jackson Lab.,* Bar Harbor, Maine.
Gropp, A. (1973). *Excerpta Med. Found., Int. Congr. Ser.* **278,** 326–330.
Gropp, A., and Kolbus, U. (1974). *Nature (London)* **249,** 145–147.
Gropp, A., Koblus, U., and Giers, D. (1975). *Cytogenet. Cell Genet.* **14,** 42–62.
Grumbach, M. M., and Van Wyk, J. J. (1974). *In* "Textbook of Endocrinology" (R. H. Williams, ed.), 5th ed., Saunders, Philadelphia, Pennsylvania.
Grzeschik, K. H. (1973). *Humangenetik* **19,** 1–40.
Grzeschik, K. H., Allderdice, P. W., Grzeschik, A. M. Opitz, J. M., Miller, O. J., and Siniscalco, M. (1972). *Proc. Natl. Acad. Sci. U.S.A.* **69,** 69–73.
Hamerton, J. L. (1971). *In* "Human Cytogenetics" (J. L. Hamerton, ed.), Vol. 2, pp. 295–309. Academic Press, New York.
Hauksdóttir, H., Halldórsson, S., Jensson, O., Mikkelsen, M., and McDermott, A. (1972). *J. Med. Genet.* **9,** 413–421.
Hecht, F., Jones, D. L., Delay, M., and Klevit, H. (1970). *J. Med. Genet.* **7,** 1–4.
Hilliard, L. T., and Kirman, B. H. (1965). "Mental Deficiency," 2nd ed., Little, Brown, Boston, Massachusetts.

Holland, W. W., Doll, R., and Carter, O. (1962). *Br. J. Cancer* **16,** 178–186.
Holmes, L. B. (1965). *N. Engl. J. Med.* **272,** 437–444.
Holt, M., and Oram, S. (1960). *Br. Heart J.* **22,** 236–242.
Holt, S. B. (1952). *Ann. Eugen.* **17,** 140–161.
Holt, S. B. (1968). "The Genetics of Dermal Ridges," Thomas, Springfield, Illinois.
Howard, R. O., Breg, W. R., Albert, D. M., and Lesser, R. L. (1974). *Arch. Ophthalmol.* **92,** 490–493.
Jackson, C., and Jackson, C. L. (1937). "The Larynx and its Diseases," pp. 64–74. Saunders, Philadelphia, Pennsylvania.
Jackson, E. W., Turner, J. H., Klauber, M. R., and Norris, F. D. (1968). *J. Chronic Dis.* **21,** 247–253.
Jacobs, P. A., Harnden, D. G., Buckton, K. E., Court-Brown, W. M., King, M. J., McBride, J. A., MacGregor, T. M., and Maclean, N. (1961). *Lancet* **1,** 1183.
Jenkins, M. B., and O'Rourke, W. J. (1974). *Lancet* **1,** 210.
Juberg, R. C., Adams, M. S., Venema, W. J., and Hart, M. G. (1969). *J. Med. Genet.* **6,** 314–321.
Kaelin, A. (1955). *Arch. Julius Klaus-Stift. Verebungsforsch., Sozialanthropol. Rassenhyg.* **30,** 263–485.
Kemp, T., and Ravn, J. (1932). *Acta Psychiatr. Neurol.* **7,** 275–296.
Klinger, H. P., Kosseff, A. L., and Plotnick, F. (1971). *Adv. Biosci.* **6,** 207–222.
Koretzky, E. D., Moller, J. H., Koons, M. E., Schwartz, C. J., and Edwards, J. E. (1969). *Circulation* **40,** 43–53.
Kosowicz, J. (1965). *Am. J. Roentgenol., Radium Ther. Nucl. Med.* [N.S.] **93,** 354–361.
Kuliev, A. M., Kukharenko, V. I., Grinberg, K. N., Vasileysky, S. S., Terskikh, V. V., and Stepanova, L. G. (1973). *Humangenetik* **17,** 285–296.
Legros, J., and Van Michel, C. (1968). *Ann. Genet.* **11,** 59–61.
Leisti, J. (1971). *Ann. Acad. Sci. Fenni., Ser. A4* **179,** 1–69.
Lejeune, J., Lafourcade, J., Berger, R., Vialotte, J., Boesivillward, M., Serlinge, P., and Turpin, R. (1963). *C. R. Hebd. Seances Acad. Sci.* **257,** 3098–3102.
Lejeune, J., LaFourcade, J., de Grouchy, J., Berger, R., Gautier, M., Salmon, C., and Turpin, R. (1964). *Sem. Hop.* **40,** 1069–1079.
Lejeune, J., Doumic, J. M., Berger, R., and Rethoré, M.-O. (1966). *Ann. Genet.* **9,** 132–133.
Lele, K. P., Penrose, L. S., and Stallard, H. B. (1963). *Ann. Hum. Genet.* **27,** 171–174.
Lewandowski, R., C., and Yunis, J. J. (1975). *Am. J. Dis. Child.* **129,** 515–529.
Lewis, K. B., Bruce, R. A., Baum, D., and Motulsky, A. G. (1965). *J. Am. Med. Assoc.* **193,** 1080.
Lindsley, D. L., and Grell, E. H. (1968). *In* "Genetic Variations of *Drosophila melanogaster,*" p. 627, Carnegie, Washington.
Lowry, R. B. (1974). *Am. J. Dis. Child.* **128,** 887.
Luthardt, F. W., and Palmer, C. G. (1971). *J. Med. Genet.* **8,** 387–391.
Lutzner, M. A., and Hecht, F. (1966). *Lab. Invest.* **15,** 597–605.
McDermott, A., and Parrington, J. M. (1975). *Ann. Hum. Genet.* **38,** 305–307.
McHugh, H. E., and Loch, W. E. (1942). *Laryngoscope* **52,** 43–65.
MacIntyre, M. N., Staples, W. I., LaPolla, J., and Hemple, J. M. (1964a). *Am. J. Dis. Child.* **108,** 538–542.
MacIntyre, M. N., *et al.* (1964b). *Am. Soc. Hum. Genet.* p. 21 (abstr).
McKusick, V. A. (1975). "Mendelian Inheritance in Man," 4th ed. Johns Hopkins Univ. Press, Baltimore, Maryland.
Magenis, E., and Hecht, F. (1974). *In* "Birth Defects: Atlas and Compendium" (D. Bergsma, ed.), pp. 251–252.

Marden, P. M., and Yunis, J. J. (1967). *Am. J. Dis. Child.* **114**, 662–664.
Martin, J. P., and Bell, J. (1943). *J. Neurol. Psychiatry* **6**, 154–157.
Matsaniotis, N., Karpodzas, J., and Economou-Mavrou, C. (1967). *J. Pediatr.* **70**, 810–812.
Mellman, W. J., Younkin, L. H., and Baker, D. (1970). *Ann. N.Y. Acad. Sci.* **171**, 573–542.
Miller, O. J., and Miller, D. A. (1975). *Fed. Proc., Fed. Am. Soc. Exp. Biol.* **34**, 2218–2221.
Miller, R. W. (1970). *Ann. Acad. Sci. Fenn., Ser. A4* **171**, 637–744.
Mittwoch, U. (1967). *In* "Mongolism" (G. E. W. Wolstenholme and R. Porter, eds.), p. 51. Churchill, London.
Naeye, R. L. (1967). *Biol. Neonat.* **11**, 248–260.
Neu, R. L., and Gardner, L. I. (1973). *Clin. Genet.* **9**, 914–918.
Neühauser, G., and Usener, M. (1966). *Z. Kinderheilkd.* **95**, 244–262.
Newton, M. S., Jacobs, P. A., Price, W. H., Woodcock, G., and Fraser, I. A. (1972). *Clin. Genet.* **3**, 215–225.
Niebuhr, E. (1971). *J. Ment. Defic. Res.* **14**, 277–291.
Niebuhr, E., and Ottosen, J. (1973). *Ann. Genet.* **16**, 157–166.
Noel, B., Quack, B., and Rethoré, M.-O. (1976). *Clin. Genet.* **9**, 593–602.
Noonan, J. A. (1968). *Am. J. Dis. Child.* **116**, 373–380.
Nora, J. J., and Meyer, T. C. (1966). *Pediatrics* **37**, 329–334.
Nora, J. J., Tores, F., Sinha, A., and McNamara, D. (1970). *Am. J. Cardiol.* **25**, 639–641.
Norman, R. M. (1966). *Dev. Med. Child Neurol.* **8**, 170–177.
Norwood, T. H., and Hoehn, H. (1974). *Humangenetik* **25**, 79–82.
O'Grady, R. B., Rothstein, T. B., and Ramano, P. E. (1974). *Am. J. Opthalmol.* **77**, 40–45.
O'Kane, G. (1936). *Laryngoscope* **46**, 550–554.
Opitz, J. M., Slungaard, R., Edwards, R. H., Inhorn, S. L., Muller, J., and deVenecia, G. (1969). *Birth Defects, Orig. Artic. Ser.* **5**, 93–99.
Orbeli, D. J., Lurie, I. W., and Goroshenko, J. L. (1971). *Humangenetik* **13**, 296–308.
Orye, E., Delbeke, M. J., and Vandenabelle, B. (1974). *Clin. Genet.* **5**, 457–464.
Paris Conference 1971 (1972). Standardization in Human Cytogenetics. *Birth Defects, Orig. Artic. Ser.* (D. Bergsma, ed.), **3**, No. 7, 317–360.
Passarge, E., True, C. W., Sueoka, W. R., Baumgartner, N. R., and Keer, K. R. (1966). *J. Pediatr.* **69**, 771–778.
Patel, H., Dolman, C. L., and Byrne, M. A. (1972). *Am. J. Dis. Child.* **124**, 217–221.
Pearce, W. G., and Kerr, C. B. (1965). *Br. J. Opthalmol.* **49**, 530–537.
Pearson, P. L., Van der Linden, A. G., and Hagemeijer, J. M. (1974). *Birth Defects, Orig. Artic. Ser.* **10**, No. 3, 136–142.
Penrose, L. S. (1963). *Nature (London)* **197**, 933–936.
Penrose, L. S. (1968). *Brit. Med. J.* **2**, 321–325.
Penrose, L. S., and Smith, G. F. (1966). "Downs Anomaly." Little Brown, Boston, Massachusetts.
Porter, I. H., and Paul, B. (1975). *Birth Defects, Orig. Artic. Ser.* **11**, No. 5, 273–275.
Race, R., and Sanger, R. (1975). "Blood Groups in Man," 6th ed. Lippincott, Philadelphia, Pennsylvania.
Rethoré, M.-O., Praud, E., Le Loc'h, J., Joly, C., Soraux, H., Aussannaire, M., and Lejeune, J. (1970). *Presse Med.* **78**, 955–958.
Rethoré, M.-O., Saraux, H., Prieur, M., Dutrillaux, B., Meer, J.-J., and Lejeune, J. (1972). *Arch. Fr. Pediatr.* **29**, 533–538.
Riccuiti, F. C., and Ruddle, F. H. (1973). *Genetics* **74**, 661–678.
Rieger, H. (1935). *Albrecht von Graefes Arch. Opthalmol.* **133**, 602–635.
Ritossa, F. M., Atwood, K. C., and Spiegelman, S. (1966). *Genetics* **54**, 663–676.

Rotterdam Conference 1974 (1975). Intern. Workshop on Human Gene Mapping. *Birth Defects, Orig. Artic. Ser.* **11**, No. 3.
Sarto, G. E., Therman, E., and Patau, K. (1974). *Clin. Genet.* **6**, 289–293.
Schachenmann, G., Schmid, W., Fraccaro, M., Mannina, A., Tiepolo, L., Perona, G. P., and Sartori, E. (1965). *Lancet* **2**, 290.
Schinzel, A., Schmid, W., and Mürset, G. (1974). *Humangenetik* **22**, 287–298.
Schinzel, A., Hayaski, K., and Schmid, W. (1975). *Humangenetik* **26**, 123–132.
Schultz, J. (1929). *Genetics* **14**, 366–419.
Seifert, O. (1889). *Berl. Klin. Wochenschr.* **26**, 24–26.
Simpson, J. L. (1975). *Birth Defects, Orig. Artic. Ser.* **11**, No. 4, 23–59.
Smith, I. I., and Bain, A. D. (1965). *Ann. Otol., Rhinol., & Laryngol.* **74**, 338–347.
Smith, S. M., and Sorsby, A. (1958). *Ann. Hum. Genet.* **23**, 50–58.
Steffensen, D. M., and Wimber, D. E. (1971). *Genetics* **69**, 163–178.
Steinberger, E., Steinberger, A., Smith, K. D., and Perloff, W. H. (1966). *J. Med. Genet.* **3**, 226–229.
Stolla, R., and Gropp, A. (1974). *J. Repro. Fertil.* **38**, 335–346.
Swaiman, K. F., and Wright, F. S., eds. (1975). "The Practice of Pediatric Neurology" Vol. 1, C. V. Mosby Co., St. Louis, Missouri.
Taktikos, A. (1964). *Br. J. Opthalmol.* **48**, 495–498.
Talvik, T., Mikelsaar, A.-V., Mikelsaar, R., Käsosaar, M., and Tüür, S. (1973). *Humangenetik* **19**, 215–226.
Taylor, A. I. (1970). *Humangenetik* **10**, 209–217.
Taysi, K., Bobrow, M., Balci, S., Madan, K., Atasu, M., and Say, B. (1973). *J. Pediatr.* **82**, 263–268.
Terplan, K. L., Sandberg, A. A., and Aceto, T., Jr. (1966). *J. Am. Med. Assoc.* **197**, 557.
Tettenborn, U., and Gropp, A. (1970). *Cytogenetics* **9**, 272–283.
Thompson, H., and Lyons, R. B. (1965). *Hum. Chromosome Newslett.* **14**, 21.
Turner, J. H. (1963). Phd. Thesis, Univ. Microfilms No. 63-7539. University of Michigan, Ann Arbor.
Tveteras, E. (1965). *Acta Paediatr. Scand., Suppl.* **159**, 131.
Uchida, I. A., and Soltan, H. C. (1963). *Pediatr. Clin. North Am.* **10**, 409–422.
Uchida, I. A., Miller, J. R., and Soldan, H. C. (1964). *Am. J. Hum. Genet.* **16**, 284–291.
Van den Berghe, H., Van Eygen, M., Fryns, J. P., Tanghe, W., and Verresen, H. (1973a). *Humangenetik* **18**, 225–230.
Van den Berghe, H., Fryns, J. P., and Devos, F. (1973b). *Humangenetik* **20**, 163–166.
Van Kempen, C. (1966). *Maandschr. Kindergeneeskd.* **34**, 92–95.
Vogel, F. (1973). *Humangenetik* **19**, 41–56.
Vuorenkoski, V., Lind, J., Partnen, T. J., Lejeune, J., and Waez-Hockert, O. (1963). *Ann. Paediatr.* **12**, 174.
Walker, J. T. (1961). *Ann. Hum. Genet.* **25**, 65–68.
Walzer, S., Gerald, P. S., Breau, G., O'Neill, D., and Diamond, L. K. (1966). *Pediatrics* **38**, 419–429.
Ward, P. H. (1973). *In* "Otolaryngology" (M. M. Paparella and D. A. Shumrick, eds.), Vol. 3, pp. 603–608. Saunders, Philadelphia, Pennsylvania.
Ward, P. H., Engel, E., and Nance, W. E. (1968). *Trans. Am. Acad. Opthalmol. Otoloryngol.* **72**, 90–102.
Warkany, J. (1971). "Congenital Malformations." Yearbook Medical, Chicago, Illinois.
Warkany, J., Passarge, E., and Smith, L. B. (1966). *Am. J. Dis. Child.* **112**, 502.
Weed, J. C. (1972). *In* "Female Sex Anomalies" (C. M. Dougherty and R. S. Spencer, eds.), Harper, New York.

White, B. J., Tjio, J. H., Van de Water, L. C., and Crandall, C. (1974a). *Cytogenet. Cell Genet.* **13**, 217–231.

White, B. J., Tjio, J. H., Van de Water, L. C., and Crandall, C. (1974b). *Cytogenet. Cell Genet.* **13**, 232–245.

Wilson, M. G., Melnyk, J., and Townes, J. W. (1969). *J. Med. Genet.* **6**, 322–327.

Wilson, M. G., Townes, J. W., and Negus, L. D. (1973). *J. Med. Genet.* **7**, 164–170.

Yunis, J. J. (1976). *Science* **191**, 1268–1270.

Yunis, J. J., and Hook, E. B. (1966). *Am. J. Dis. Child.* **111**, 83–89.

Yunis, J. J., and Sanchez, O. (1975). *Humangenetik* **27**, 167–172.

# Index

## C

## D

## M



## N

O



P



Q



R



S

U



V



W



X



Y